Alarm Management for Process Control

Alarm Management for Process Control

A Best-Practice Guide for Design, Implementation, and Use of Industrial Alarm Systems

DOUGLAS H. ROTHENBERG

MOMENTUM PRESS, LLC, NEW JERSEY

Alarm Management for Process Control

First published in 2009 by
Momentum Press, LLC
222 East 46th Street, New York, N.Y. 10017
www.momentumpress.net

ISBN-13: 978-1-60650-003-3 (hard back, case bound)
ISBN-10: 1-60650-003-1 (hard back, case bound)

ISBN-13: 978-1-60650-005-7 (e-book)
ISBN-10: 1-60650-005-8 (e-book)

DOI forthcoming

Cover Design by Jonathan Pennell
Interior Design by Scribe, Inc.

First Edition March 2009

10 9 8 7 6 5 4 3 2 1

Printed in the United States of America.

To my dearest wife, constant companion, champion, and best friend, Katarzyna Gustaw, under whose sheltering roof and within whose protecting walls the first words to this book, nearly the final words, and much in between were written.

Książkę tą dedykuję mojemu najlepszemu przyjacielowi, drogiej żonie mojej Katarzynie Gustaw. Za jej to przyczyną i w jej przyjaznym domu pracowałem przez długi czas, rozpocząłem i zakończyłem pisanie.

About the Author

Douglas H. Rothenberg is the president and principal consultant of D-RoTH, Inc., a technology consulting company providing innovative technology and services for industry. His background includes 10 years as an independent consultant to Fortune 1000 companies; over 20 years with Standard Oil, BP Oil, and BP Amoco, where he was responsible for new, state-of-the-art technology to support advanced manufacturing solutions; and 10 years in academia at Case Western Reserve University. Current areas of specialty with D-RoTH include alarm management, fit-for-purpose product design, and innovation development for new products and services.

Rothenberg has a PhD in systems and control engineering from Case Western Reserve University, an MS in electrical engineering from Case Institute of Technology, and a bachelor of electrical engineering degree from Virginia Polytechnic Institute. He has several patents in instrumentation and controls. He is active in the Instrumentation, Systems, and Automation Society (ISA, formerly the Instrument Society of America) and a member of Sigma Xi, the Scientific Research Society. He is the recipient of 2005 Educator of the Year Award from Cleveland Technical Societies Council, Cleveland, Ohio.

Contents

Foreword

The control room of a process plant is by turns either a boring or terrifying place, much like the cockpit of a fighter jet or a large passenger aircraft: "hours of boredom punctuated by moments of sheer terror," as the old aphorism goes. The boredom comes from processes running properly. The terror comes when they do not. And, as the litany of process plant disasters shows, after the terror sometimes comes picking through the ruins, looking for bodies.

What we've learned in a generation of studying those horrifying days after plant disasters is that more often than not, the final straw has been the plant operators consistently making wrong decisions based on the information they think they have about what is going on in the process. Many of these disasters have been blamed on cascades of alarms that made it impossible for operators to figure out which alarms actually were important and what they meant.

Safety instrumented systems are designed to be emergency shutdown systems, and often they work properly. There are specific standards worldwide that define how a safety-instrumented system should work. What does not exist is the same kind of standard for alarms and alarm management.

Doug Rothenberg has been in the forefront of research and standards making for alarm management in the process industries for many years. He did the pioneering work in the development and deployment of distributed control system (DCS) alarm management technology from 1989 through the present. He also was a founding member of the Alarm Management Task Force and the ASM Consortium and is a voting member of ISA 18, the Alarm Management Standards Committee.

His book is a comprehensive treatment of the current best practices in industrial process control alarm management. Doug covers the entire alarm management process from how to recognize the level of performance of existing systems through the methodology and procedures for redesigning (or designing new) state-of-the-practice alarm systems. You do not need any special or detailed experience in the configuration or specification of process control equipment. The ability to appreciate technical issues is important, but no prerequirement exists for any specific technical, educational, or experiential background.

In this book, Doug elevates alarm management from a fragmented collection of procedures, metrics, and trial and error to the level of a technology discipline. This is the first book about alarm management to do so. Doug gives you the fundamental underpinnings that will provide a level of understanding that is independent of opinion and partial experiences. All critical tasks are explained, with examples and insight into what they mean. Alternatives are everywhere to enable industrial users to tailor-make their solutions for their particular sites.

Using this book, you will be able to understand how to rationalize alarms and how to work toward the same sort of human-factors engineering that has revolutionized cockpit design but is now applied to control rooms in industrial process plants.

If you work in a process plant, design process plants, or design, operate, or maintain control systems, this is your indispensable reference book.

Walt Boyes
Fellow, International Society of Automation
Editor in Chief, *Control* magazine

Acknowledgments

The author gratefully acknowledges the contributions of the following friends and colleagues:

Ari Bar-on—for his unflagging friendship, unrelenting constructive advice, and lending more ear than one ear might bear, all with compassion and class

Ian Nimmo—for believing in alarm management, steadfast friendship, and advice along the way

Steve Apple, Walt Boyes, Steve Elwart, Alan Phipps, and Chris Wilson—for their unrelenting push to write this book and make it a best of class and their guidance to help make it so

Greg Morris—for believing that alarm training is empowering

Jack Pankoff—for teaching me the business of running a business

Angela Stump—for reviewing drafts of this work with such heart and mind

David Gaertner, John Bogdan, and Diego Izarra—for believing in alarm management as a vital service

Joel Stein—for being as much a colleague as a publisher

Rachel Paul McGrath—for preparing the design and page layout with care and talent that brought words in files into visual life

There are a large number of professionals who took up the challenge of better production through better alarm systems. These pioneers are to be found in appendix 8 of this book, as their enumeration would take much more space than a simple acknowledgment might allow.

Douglas H. Rothenberg
Shaker Heights, Ohio, USA

Credits

The author wishes to thank the generosity of the following organizations for permission to use copyright materials that they own and may have previously published.

ControlGlobal.com, Itasca, IL, for Figure 3.3.2

Chevron NA, New Orleans, LA, for appendix 9

Ergon Refining, Vicksburg, MS, for Figure 7.6.2

Honeywell, Inc., Phoenix, AZ, for Figures 2.3.1, 2.4.1, 2.6.1, 2.9.1, 2.9.2, 4.10.2, 5.3.1, 11.3.1, 12.6.4, 12.6.5, 12.6.8, 12.6.9, 12.8.1, and 12.10.1 to 12.10.8

Henry Holt, Inc., New York, for Figure 12.4.2

Human Centered Solutions, for Figure 12.10.9

PAS, Inc., Houston, TX, for Figure 10.3.1

Power Engineering, Tulsa, OK, for Figure 1.3.2

TiPS, Inc., Austin, TX, for use of ACE and LogMate software to prepare numerous alarm configuration and operational graphs

The author and publisher gratefully acknowledge the license granted by Matrikon, Inc., Edmonton, Alberta, to reproduce in this book, Table 4.7.3 and Figures 4.7.1 to and including 4.7.3 taken from Matrikon. For further information about Matrikon, Inc., and its Alarm Manager product, please see http://www.matrikon.com.

Introduction

We warm ourselves by fires we did not build and drink from wells we did not dig.

—Ancient Semitic wisdom

This book embodies the current best practices of process control alarm systems for industrial manufacturing facilities. It is a comprehensive guide developed to help you understand, design, evaluate, and use alarm systems. The coverage is accurate and complete and, at the same time, easy to grasp. The book contains all the "what is" information about alarm management so that you will fully understand. There is an extensive "how to" that you can use to perform every aspect of alarm system redesign. The style is low key. The technology is down to earth, solid, and based on strong design fundamentals.

Some of you have experience with alarm projects. You've done work before others even knew alarm improvement was such an extraordinary opportunity to better operations. No doubt when you read this, you will find differences in what is suggested here in comparison to what you've done for your site. Please understand that this situation does not mean to imply that either of us might be wrong. It is just that now we understand the methodology better and have some very powerful and useful tools and procedures at our disposal. We do not have to do as much trial and error. New standards and practices are in place. We better understand how alarm management really works.

Right about now others might be asking, how in the world can a topic as obscure as alarm management possibly lead to an entire book? What can be so important and useful? It just so happens that alarm improvement is a powerful means toward a valuable end. It provides a useful way to better plant operations. Alarm management is one of those lucky finds that yields wonderful prizes. Let me tell you why. Think about a time, many years ago, when travel was done mostly on foot, without maps, on dusty roads with forks and no signposts or mile markers. There were a number of ways to try to get to a town or village. One could simply follow a road and watch to see if it were becoming more traveled as one went along. One might be fortunate enough to happen upon a fellow traveler and inquire. If there were no traveler or no road, one might follow a stream to see what could be found. Depending on the lay of the land, it might be possible to head down into an inviting valley, or in the direction of chimneys smoking, or follow the hubbub to a busy market.

Like the stream, road, or fellow traveler, alarm management leads to the village—but in our case the village is significantly improved plant operations. There are, of course, other roads to better operation. Examples would be a desire to reduce long-term costs, a need to improve product quality or delivery schedules, a requirement to manage environmental exposure, or the need to provide a significantly safer enterprise. Choose any one. Along the way, you will touch most of the same aspects you are going to touch by taking the alarm improvement path. Much of the benefits will be similar. But a most important difference will be that the alarm improvement road, this road, is traveled enough so that there are good maps, effective signposts, and lots of fellow travelers. This road has also been carefully planned with good restaurants and comfortable motels. Enjoy the trip.

NOT A HANDBOOK

This is not a handbook.

> Handbooks may deal with any topic, and are generally compendiums of information in a particular field or about a particular technique. They are designed to be easily consulted and provide quick answers in a certain area. (Wikipedia)

This book is not a handbook because alarm management is not about affirming business as usual. Nothing is wrong with business as usual, except that is how process control systems (PCS) alarm capabilities drifted away from their intended purpose and into the center stage of poor operations performance. Getting better is less about tinkering and polishing and more about rethinking and redesign. Alarm management now has a foundation and a body of implementation experiences. This text covers the entire alarm management process from recognition to action. The reader learns how to recognize the level of performance of existing systems as well as how to take action through the methodology and procedures for designing new state-of-the-practice alarm systems. To do this job well, you will need to know more than the highlights and have more than lots of lists and procedures.

AUDIENCE

This book was created for individuals with a general familiarity of modern process control systems and how operators use them to manage their plants. Readers need not have any special or detailed experience in the configuration or specification of process control equipment. The ability to appreciate technical issues is important, but no prerequirement exists for any specific technical, educational, or experiential background. The book is a comprehensive treatment of the current best practices. The text covers the entire alarm management process from how to recognize the level of performance of existing systems through the methodology and procedures for redesigning (or designing new) state-of-the-practice alarm systems.

You will find the style and content useful and understandable. The majority of the material has been presented around the world to a wide audience in a variety of formats as industrial and professional short courses and workshops. Audiences have included plant operators, operations supervisors and managers, process controls technicians, instrument and control technicians and engineers, health and safety personnel, process engineers and engineering supervisors, all manner of support staff, and notably senior plant management. The material of this book has met with enthusiastic reception. If you are interested in alarm management, this is your source!

USEFULNESS

This work elevates alarm management from a fragmented collection of procedures, metrics, and trial and error to the level of a technology discipline. Fundamental underpinnings provide a level of understanding that is independent of opinion and partial experiences. All critical tasks are explained, with examples and insight into what they mean. Alternatives are everywhere to enable industrial users to tailor-make their solutions for their particular sites.

Many of the leading power, chemical, mining, pharmaceuticals, and petroleum manufacturing companies contributed to this best practice. There are a growing number of alarm management applications and improvement programs. They serve to excite the industrial community with the importance of good alarm system design. This text is a one-stop shop for alarm management practices from start to finish. It includes important material for understanding and managing abnormal plant operations. It illustrates the serious importance of process control graphics in the management of plants.

Redesigning alarms will ensure that plant operators have an improved notification system to provide warning of abnormal operation. However, alarm improvement cannot stand alone. Warning alone fails to ensure operational success. This broad coverage exposes the practitioner to all the additional key aspects and practices that work together. The value of this treatment derives from the combination of the delivery of a clear, workable, comprehensive alarm system design and the coverage of intimately related "enablers" that empower the operator to fully leverage the improved alarm system.

CONTENTS

This book is organized in three parts. Part I covers the alarm management problem. Part II lays out the solution. Part III provides the pathway to make it all real. There are twelve chapters. There is a natural progression of the work, with each chapter covering a specific area. It is suggested that the reader cover the material in that order. However, each chapter has value if read separately. This can be especially useful for those with a working knowledge of the technology who are looking for more detail or greater depth in selected topics. The book provides a working guide for project planning and execution. Certain chapters work especially well as stand-alone treatments. Chapter 2, "Abnormal Situations," chapter 5, "Permission to Operate," and chapter 12, "Situation Awareness," are designed with this in mind. These topics bring out aspects of enterprise management that go beyond alarm systems in their overall importance. They elevate alarm improvement effectiveness to a level of value capable of delivering demonstrable benefit.

Each chapter begins with the mention of the key concepts that underlie the topic.[1] These key concepts are provided to assist the reader to clearly separate the concepts from the explanatory discussion. Taken as a whole, the set of key concepts could make up a shorthand bible of alarm management.

Part I: The Alarm Management Problem

Chapter 1, "Meet Alarm Management," explores the basic reasons for considering improving your alarm system, introduces the four fundamental concepts that guides the process, and bridges the implementation over the distributed control system (DCS) and programming logic controllers (PLC) controls platforms and the process types: continuous, discrete, and batch. Chapter 2, "Abnormal Situations," links process operation abnormalities to alarm performance requirements. Along the way we pick up the important concept of how to use time in setting alarm activation levels. Chapter 3, "Strategy for Alarm Improvement," brings us up to speed with the who, what, when, and how to effect alarm system redesign. The existing standards and best practices are also covered. Chapter 4, "Alarm Performance," covers the useful scales for measuring both alarm system design and operational performance.

Part II: The Alarm Management Solution

Chapter 5, "Permission to Operate," provides a framework to cover an important need in plant operational protocol to ensure that unmanageable situations are avoided. Chapter 6, "Alarm Philosophy," covers how each enterprise will specify their chosen alarm design. Chapter 7, "Rationalization," is the heart of building the new alarm designs. It covers the technical procedures for how alarms are chosen, configured, documented, and folded back into the rest of the plant infrastructure. Chapter 8, "Enhanced Alarm Methods," builds on the first-level alarm design to ensure that the alarm design accommodates to changes in plant situations.

Part III: Implementing Alarm Management

Chapter 9, "Implementation," gets into the realities of taking a new alarm design and producing a new working alarm system ready to fully assist the operator. Chapter 10, "Life Cycle Management," clears the way to understanding what is needed to keep a new alarm system working down the road so that it can deliver continuous operational benefit. Chapter 11, "Project Development," provides alternative ways to produce a program for comprehensive alarm improvement, from start to finish, that matches the enterprise's way of conducting work projects. Chapter 12, "Situation Awareness," rounds out the entire work by providing the understanding and technology for improving the operator's ability to manage a process without undue reliance on alarms.

BOOK DELIVERABLES

Upon completion, you will have a solid, clear foundational understanding of the purpose of alarms, the rationale behind a state-of-the-practice design, and sufficient how-to knowledge to competently perform in the technology. This book is designed to provide a basis for the following competencies:

- Understanding the proper use of process control alarm systems
- Knowing the underlying defining attributes and purpose of an alarm
- Appreciating the importance of effective alarm management
- Recognizing alarm system performance problems, including alarm flood
- Becoming knowledgeable in the best practices for alarm system design
- Learning the value of alarm data diagnostic tools
- Understanding the entire process for designing and executing an alarm improvement project
- Understanding the influence of good graphic interface design on plant operability and effective situation awareness
- Becoming a qualified participant in alarm improvement teams

IMPORTANT WORD

Please note that this book is not intended in any way to offer advice or recommendations as to the appropriateness or lack of appropriateness of including or excluding any specific alarm for any site. The choice of which aspects to alarm for the plant or process, the parameters of that alarm, the proper operator response to correct the alarm condition, and all other details of that alarm must be retained wholly by qualified, authorized members of the plant staff, who must act with full knowledge of their specific plant configuration, process conditions, equipment, and applicable statutory practices and requirements.

No single work is capable of conveying the entire collective experience and important nuances necessary for success. After you read this book, it is recommended that plants with intentions or plans for alarm improvement seek additional specific guidance and experience from knowledgeable experts.

NOTE

1. "The Use and Abuse of the Cept," *Time*, March 26, 1965.

PART 1

The Alarm Management Problem

CHAPTER 1

Meet Alarm Management

> If you need a new machine and don't buy it, you pay for it anyway but never get to use it.
>
> —Henry Ford

An alarm is an announcement to the operator initiated by a process variable (or measurement) passing a defined limit as it approaches an undesirable or unsafe value. The announcement includes audible sounds, visual indications (e.g., flashing lights and text, background or text color changes, and other graphic or pictorial changes), and messages. The announced problem requires operator action. An alarm is a construction by which an aspect of manufacturing operation is identified and configured in a binary way to be either "in alarm" or "cleared" (i.e., not in alarm). The condition of *in alarm* is passed to an operator via intrusive sounds and notices placed on video display units or other devices to gain attention. The operator can manage these sounds and notices only via specific "silence the alarm" or "acknowledge the alarm" actions using the existing, planned infrastructure of the alarm platform. Usually, this alarm platform is an integral part of the process control system (PCS) infrastructure.

The PCS alarm system is a vital and productive tool for managing industrial process control plants. Through several unique cooperative endeavors, industry has identified a best practice for alarm system design. This design utilizes configuration changes, alarm reprioritization and balancing, alarm reductions, graphics modifications, online filtering, and decision support aids. Alarms perform the vital function of operational integrity monitoring. Properly designed alarms will notify the operator of abnormal situations with enough time to successfully manage them.

Figure 1.0.1. Alarms are an intrusive notification to the operator

1.1 KEY CONCEPTS

Alarms are for the operator	The alarm system must be off-limits for all plant uses that do not directly require the operator to process actively the situation or information.
All of the objectives of alarm improvement are just good engineering	There is nothing that alarm improvement asks of plant operators that is over and above what constitutes effective plant design and operation. Somehow, bits and pieces are overlooked or shortcuts are taken. Poor or inadequate alarm performance is just the way we find out about these things. The technology of alarm improvement provides a focused, compact way of getting that job done.
Alarm redesign is based on important fundamental concepts	Alarm redesign is based on four powerful concepts: *only notify important conditions*, *notify in time*, *respond*, and *provide guidance*.
Initial alarm system performance matters little	Few, if any, unimproved alarm systems have been designed to meet the fundamental concepts; therefore, reducing alarm annoyance and activation rates only treats symptoms of a nonperforming design.
Improving alarms alone will not provide enough benefit	Good alarm systems only work when the entire plant infrastructure supports good operation.

1.2 ALARM PERFORMANCE PROBLEMS

Ask almost any operator if the alarm system is working for or against good process operation. The response you are likely to get is surprise—not because the question is unclear but because it took you so long to ask. Ask control engineers or process engineers that same question and they are more and more often going to say that the alarm system needs fixing. Ask experts in industrial accident investigation and they will tell you that the lack of adequate performance of the alarm system is contributory to a significant number of industrial accidents and major calamities. They will quickly suggest that you make plans to evaluate your PCS alarm system.

Symptoms

Right off the bat, there are clear indicators of alarm problems. If your site has *any* of these symptoms, there is cause for concern. If you can observe three or more of them, there is much serious work to be done.

- Alarm activations occur without need for operator action.
- There is no plantwide philosophy for the alarm system.
- There are no clear guidelines for when to add an alarm and how to do it.
- There are no controls for removing existing alarms.
- Operating procedures are not tied to alarm activations.
- When alarms activate, the operator is not always sure what to do about them.
- Seemingly routine operations produce a large number of alarm activations that serve no useful purpose.
- Minor operating upsets produce a significant number of alarm activations.
- Significant operating upsets produce an unmanageable number of alarm activations.
- Some alarms remain active for long periods of time.
- When nothing is wrong, there are active alarms.

Evidence

We look for evidence of alarm problems in four places. Two are quite objective; two are subtle.

- *Number of alarms configured in the PCS database.* How many tags are alarmed? How many alarms are configured for each alarmed tag? How close are the alarm limits to actual process limits? How closely is priority matched to actual operational risk?

- *Number of alarm activations.* What is the hour-by-hour average occurrence of alarms? How often do alarm floods occur? How many alarms occur during a flood? What is the distribution of alarm priorities during floods?
- *Operators' ability to gain insight and guidance from the alarm system prior, during, and after an upset.* If alarms are not active, how sure are the operators that the process is normal? When alarms activate, do they provide assistance for the operator to diagnose and remedy problems? Does the alarm system itself—by excessive activations, lack of proper activation, or existence of meaningless active alarms—interfere with or delay proper production management?
- *Operators' ability to determine how the process is actually performing from other tools and PCS capabilities.* What is in place that permits operators to view and understand how the process is actually performing? If there were no alarm system at all, how easily could the operator determine how the process was performing and how close it might be to an abnormal condition?

1.3 REASONS FOR ALARM IMPROVEMENT

No alarm system should be asked to overcome the intricacies and power of the uncontrolled effects of nature or the failed constructs of man. Pandora's box cannot be closed. Humpty Dumpty cannot be put back together. Yet the future is not grim. History is not written soley for the purpose of conveying the worst from our past. Alarm management is thus charged with aiding and abetting our best efforts for accommodating the worst and setting reasonable courses for recovery. Within these pages, you will find the concepts, ideas, and practical approaches to bring what is possible to you. You are encouraged to recognize that your best will neither prevent nor minimize all the effects of misdirection. You are empowered to believe that your best efforts will yield nourishing fruit. As Larry O'Brien and Dave Woll say in *Alarm Management Strategies*,

> Alarm management is one of the most undervalued and underutilized aspects of process automation today. In most cases, alarm systems do not receive the attention and resources that are warranted. This is understandable, because alarming appears to be a deceptively simple activity. Many plants still use the alarm management philosophy developed by the engineering firm when the plant was built.
>
> As alarm systems become less effective, they diminish the effectiveness of all automation.[1]

How Alarms Fit into Process Operating Situation

Safety shutdown systems are designed to close down affected plant operations in the unfortunate situation where that operation is too close to an unsafe situation, an environmentally challenging condition, or a mode of operation that threatens the financial integrity of the plant. Once a shutdown occurs, the plant operation is significantly curtailed. This usually results in a loss of production, some equipment damage, and a considerable degree of internal investigation. Prior to restart, there might be repairs, changes to equipment and procedures, and of course lots of administrative work.

Most plants would choose to avoid a shutdown if it could be done without risk. Abnormal situations rarely present themselves without some warning. But those warning messages and signs are not always picked up by the operator. They are often subtle. Sometimes they are downright abstract or confusing. Rarely are they front and center enough to be found in time and remedied. However, in a well-designed alarm system, one designed from the point of view of alarming abnormal situations rather than just abnormal variables, a significantly greater number of those early warning messages about abnormal plant operation can be seen. In this way, the alarm system might be considered to serve as a presafety shutdown system. Alarms are a way for the operator to see an important problem building and have an opportunity to keep it from leading to a more serious plant situation.

The process control system is set to the operating target (fig. 1.3.1). These controls are intended to keep production within the good operating region. Operation is still fine in the safe operating region, but the product is less valuable or more costly to produce. Few, if any, alarms should be designed to activate as the process approaches and appears likely to cross the boundary between good operation and safe operation. Alarming here is technically difficult and often results in many unnecessary alarm activations. The control system should be capable of responding to changes and disturbances that cause movement away from the operating target. Let it do its work. If tighter control is needed, either modify the controls or ask the operator to be more diligent and make more frequent adjustments to the controls (e.g., setpoint adjustments, supplemented with minor field adjustments).

Most alarms should be set to activate as soon as the process moves close to and appears likely to cross the boundary between safe operation and upset operation. At this point, there is a very high likelihood that the control system is unable to manage the situation properly. Manual operator intervention is required. The alarms are there

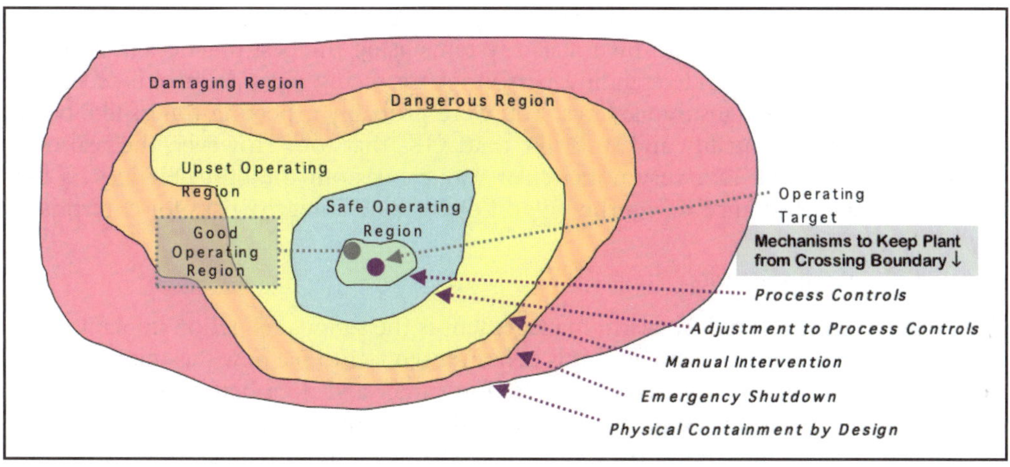

Figure 1.3.1. Operating region versus modality of remediation

to ensure that the operator does not miss these situations. If manual intervention by the operator is still not likely to ensure that the process remains within the upset region, alarms are be used to mark the likely movement through the boundary from upset operation to dangerous operation. Any plant movement from dangerous operation into damaging operation should be managed by the emergency shutdown system. Alarms are too late here, so none are needed.

Alarm Management

Alarm management is all about the understanding, design, implementation, and operation of an effective alerting capability for production plant operators. These alerts are intended to notify operators of situations and events that require operator attention in an explicit way within an acceptable time frame. This book is written for users and designers of industrial process control systems. These systems are traditionally comprised of equipment to measure plant conditions, other equipment to transfer those measurements to devices that are capable of interpreting the measurements, still other devices that employ these interpretations and other information to manipulate yet other process conditions in order to realize an appropriate plant operation, and finally equipment that permits operators and others to view all measurements and intervene as needed.

Being more pragmatic, alarm management is therefore the determination of (a) all plant conditions that will be alarmed, (b) the parametric setting for the activation of each alarm event, (c) the classification of the importance of each alarm event, and (d) the collation and presentation of information that documents the best understanding of how to successfully manage the event.

Benefits

Alarm improvement works. It can deliver clear, auditable benefits. Let's cut to the chase, as they say. Plants that redesign their alarm systems using the best practices in cooperation with energetic and understanding personnel are seeing payouts in reduced maintenance costs and lower insurance rates. You were probably expecting things like better operation and higher quality and stuff like that. OK, that too. However, the real message, and the one that in a very noncircular way proves out the best practices, is that maintenance and insurance reflect the biggest operational integrity risks for enterprises.

Reduced Maintenance Costs

One of the biggest surprises in alarm management is the serious reduction in plant maintenance costs after improvement. Yes, it is easy after the fact to see how this might be true. With the operator able to catch errant operation earlier and do a better job of bringing things back in line, it is not too great a leap to see that the equipment is less stressed. Therefore, operators encounter fewer overpressure events, fewer high-temperature excursions, and less pump cavitation. Less-stressed equipment breaks down less often. The interesting thing here is that in the past, operators did not expect to see these benefits. But now it has become obvious—alarm improvement works.

As you will discover later on in this book, improving plant maintenance is a vital requirement for gaining improved alarm system performance. It is a subversion of the alarm system to ask it to cover for all of the broken stuff, all of the missing but important stuff, and all of the other aspects that were designed into a plant but over time got lost. Maintenance is important. Going into alarm improvement, one might be concerned that maintenance costs would likely go up. Plants would need to fix everything that was broken. Repairs cost money; pay a little now. But what most do not realize is that the costs of fixing up return the favor in kind!

General results: within 6 months after alarm improvement work is commissioned and smoothly operating, plants see a 5% to 15% sustained reduction in unplanned maintenance costs.

Lower Insurance Costs

You might be new to alarm management. Your insurance company more than likely is not! Figures bouncing around in Europe reveal that the lack of an alarm management program increases premiums by substantial amounts. Figures for the United States are less, but U.S. experiences are limited. Insurance companies have followed the pacesetters in implementing alarm management. The 1999 publication of the Engineering Equipment Materials Users' Association (EEMUA) 191 guidelines[2] put out the word and showed everyone how to get started. As the numbers came in, they told the story. Again, this points to the fact that alarm improvement works.

General results: European manufacturers are seeing a 20% to 30% reduction in their risk operations insurance due to implementation of EEMUA-compliant alarm improvement.

Capture Workforce Knowledge

Manufacturing in many parts of the world is diminishing as industrial activity geographically redistributes in our mobile world. For a variety of reasons, the workforce in important segments of industry is aging. Fewer young workers are entering. At the same time, a disproportionate number of highly experienced workers will be leaving their positions through retirement, illness, or attractive cost-cutting management incentives. A great deal of unwritten knowledge will leave when they leave. According to *Power Engineering*[3] (fig. 1.3.2), nearly 20% of the entire workforce will have reached normal retirement age by 2015 and nearly 40% by 2020.

Alarm redesign, by design, provides a comprehensive, structured way to capture important operating knowledge. The entire activity of identifying abnormal situations through alarms and what the operator is supposed to do to manage them properly is a key part of the rationalization activity. So, not only will the present operating team have the benefit of this information, but as each new member moves to the board, they too will be able to understand and use this valuable asset. For the reader who would like to have an advanced view of what this knowledge looks like, you may refer to the alarm response sheet in chapter 7.

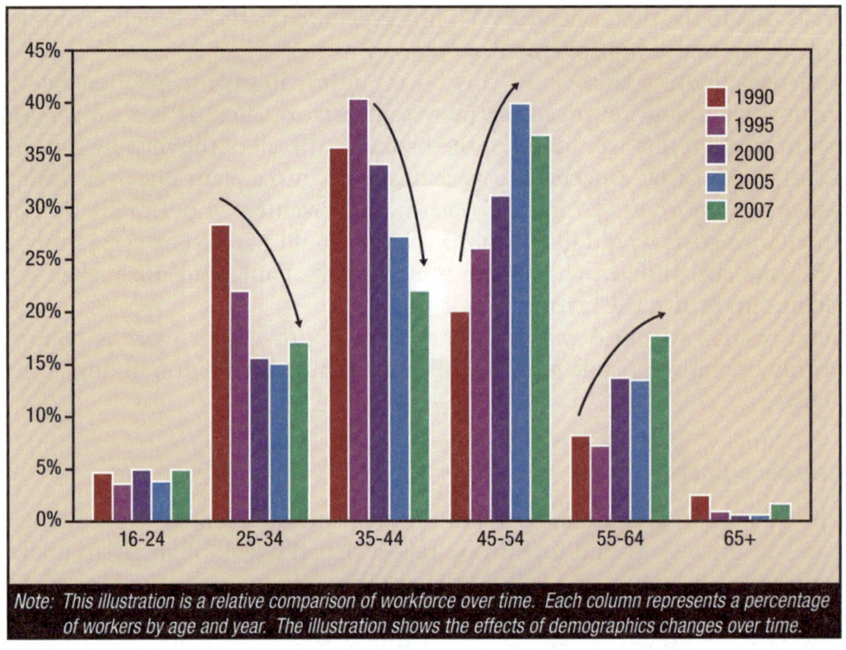

Figure 1.3.2. Aging workforce in U.S. power industry

Shift Handover

If you are an operator, you know how important it is to receive a full briefing from the operator you are replacing. Even if you are coming on after a relatively quiet shift, a lot might have happened earlier that you would benefit from knowing about later. As it happens, the actual process of shift handover can be a very haphazard event. For some plants, the two operators barely have time to wave as they pass. Other plants have found the benefit of doing this well and arrange for an informal overlap of operators' time. Interestingly enough, operators have found it so useful that they do it without additional pay!

Plants have discovered that a well-designed alarm system will provide an important orientation edge to the handover. Operators discuss alarms that have activated during the shift. They discuss any alarms that might have been disabled, inhibited, or otherwise outside of their normal operating regime. The alarm information provides a checklist of items not to be missed. Moreover, those items are likely at the heart of the operability of the enterprise.

1.4 A BRIEF HISTORY OF ALARM MANAGEMENT

The year 1988 is the generally accepted earliest date for the first recognition of alarm system problems. Yes, there were those few early pioneers who faced operational difficulties and traced some or most of them to alarm system deficiencies. And these pioneers certainly made improvements to their plants based on this understanding. However, their impact was limited to their parochial interests. Little alarm was raised to the body technical, as it were. This all changed in late 1988 and into 1989. A surprisingly large number of individuals from a broad range of industries who were working to better understand the way process upsets were being managed by operators came to realize that the alarm system itself might be a problem. By 1990, sites had worked out the basics of alarm improvement, and a few actually had begun to execute projects. But all this came to a surprising halt in early 1991. It was discovered that it was not possible to actually implement most alarm improvement plans within any currently used, conventional process control platform.

In 1991, they joined together in what is known as the Alarm Management Task Force (appendix 8). On a warm afternoon in mid-October that year in Phoenix, Arizona, the first of many meetings took place. It was represented by oil refining companies, chemical manufacturing companies, a controls equipment manufacturing company, and power companies from almost all segments of their respective industries. Over the next few years, they worked out the fundamentals of alarming, helped shape the requirements for alarm analysis tools and controls infrastructure, and plowed their way through the realization that prevention was the best alarm response possible. Along the way, the Abnormal Situation Management (ASM) Consortium was born and important paradigms were set for production management. By the mid-1990s, third-party software companies developed and refined important alarm capture and analysis tools.

In a parallel track, Brookhaven National Laboratory[4] became fully involved in developing a better understanding of alarm systems following the Three Mile Island accident.

Using the collective knowledge from the Alarm Management Task Force, the ASM joined with leading technologists in Europe to support an EEMUA initiative. In 1999, EEMUA collated and published the first and most-recognized guide to alarm systems. It was revised in 2007. To date, Publication 191 functions as the de facto best practice. Unfortunately, the availability of good guidance does not necessarily lead to a widespread adoption. Thus, many industrial nations are now seeking to codify requirements that they believe will lead to improved safe operation of industrial plants.

1.5 THE "MANAGEMENT" IN ALARM MANAGEMENT

The management part of the phrase *alarm management* is a bit of a misnomer. Alarm management is a design and implementation process for the *entire redesign* of the portion of the process control system capability that is used to alert operators to conditions where an alarm is needed. The minute-by-minute managing of an alarm is but one small part of this much more encompassing technology. The full process includes the following:

1. Benchmark analysis of present alarm system performance, including its impact on production, safety, and environmental
2. Development of a philosophy governing the operation of the enterprise sufficient to specify a design basis for the required alarm system and supporting plant infrastructure
3. Selection of which variables to alarm

4. Setting of alarm limits
5. Setting of alarm priorities
6. Determination of recommended operator actions
7. Design of advanced techniques to facilitate improved alarm performance
8. Addition of plant condition monitors and decision support tools
9. Incorporation of new alarm system design back into the plant infrastructure
10. Continual audits, assessments, and modifications for improvement

Alarm management is a process. When we successfully do alarm management, we end up with a fully functioning alarm system suitable to meet production requirements to better realize enterprise goals.

1.6 ALARM DESIGN ROADMAP

Here is the basic roadmap we will use to structure the alarm system improvement work. It captures the key steps and order found to be most useful and best at leading to success. See Figure 1.6.1. The entry ticket into the alarm redesign game is to know your alarm problem and understand how large it appears to be. Assemble some data, analyze it, and reach preliminary conclusions. If there are sufficient grounds (and that is part of what this book is all about), you will take your story and go to management.

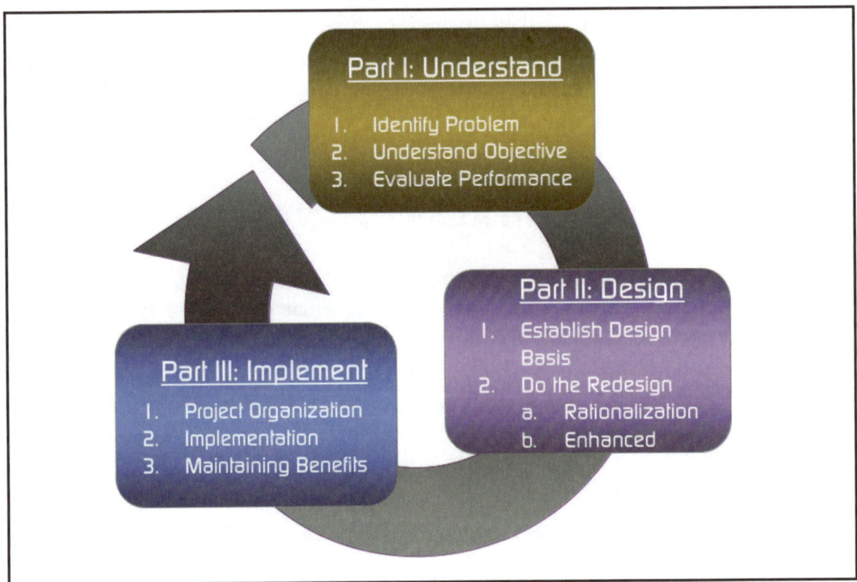

Figure 1.6.1. Alarm redesign roadmap

If your power of persuasion is strong enough and your case evident enough, management should offer up a commitment (let you propose a project and approve it) and help with the development of a site alarm improvement philosophy. Project in hand, you will continue (clockwise) and assemble a large amount of data sufficient to demonstrate most, if not all, of your site's alarm and related deficiencies. Next step is to analyze that data and reach conclusions that will assist you to focus your redesign project efforts where they are most needed.

The first actual changes to be made in your alarm management work appear in the next step—rationalization. This is a three-legged item. And the first leg is *housekeeping*, a vernacular word that means "fix everything that is broken." It is simple. We should not ask the alarm system to act as a substitute for correctable defects. The next two legs involve the redesign of alarms to fit the new requirements and the addition of any specialized alarm processing needed to better handle unnecessary alarms and alarm floods. Except for housekeeping, everything we have done so far is on paper. We are ready to do it, but not yet. Implementation is when we take all of our new design and make it real. The last step is actually the one that we will use continuously: maintain benefits by monitoring, improving, and correcting.

1.7 AUDIENCE FOR THIS BOOK

This book is written for individuals with a general familiarity with modern distributed process control systems and how operators employ them to manage their plants. It is a comprehensive treatment of the current best practices. The work covers the entire alarm management process from how to recognize the current level of performance of existing systems through the methodology and procedures for redesigning (or designing new) state-of-the-practice alarm systems. Readers need not have any special or detailed experience in the configuration or specification of PCS equipment. The ability to appreciate technical issues is important, but there is no prerequirement for any specific technical, educational, or experiential background. If you are interested in alarm management, this is your book!

Anyone with a desire to better understand and improve PCS alarm systems will find the style and content useful and understandable. The majority of the material has been used for several years and presented to a wide audience in a variety of formats as industrial and professional short courses and workshops. Audiences have included plant operators, operations supervisors and managers, PCS technicians, instrument and control technicians and engineers, health and safety personnel, process engineers and engineering supervisors, all manner of support staff, and, notably, senior plant management. This material has been presented to international audiences with enthusiastic success.

1.8 IMPORTANCE OF ALARM MANAGEMENT

At the end of the day, when the alarm system has been redesigned and implemented, the plant will have a tool that significantly improves production performance. This improvement will be seen both during normal operation (these periods will be longer and less eventful) and during upset periods (they should be less severe, less frequent, and better managed). We will see improved operator display screens that assist production personnel to ascertain plant operating conditions before most significant upsets occur. And, to return to the opening message, good alarm systems pay off in reduced maintenance and insurance costs.

When you think about it, a large percentage of currently installed PCSs provide only limited information regarding normal operation. It is up to the operator to search out those trends and critical variables and use intuition, experience, and insight to determine how normal things really are. As simple as it is to say, this is a difficult task to do. Not many of us can do it; even fewer can do it well enough to consistently manage even everyday problems. To make matters more difficult, plants are becoming more complicated, production requirements more demanding, and qualified personnel harder to justify and retain.

Still not convinced? Let us take a look at some of the statistics. It is not unusual for a medium-sized petrochemical company consisting of about six separate sites (refineries, chemical plants, etc.) to accumulate US$50,000,000 to $100,000,000 in yearly losses due to plant upsets and other production accidents. Alternatively, calculated another way, the generally recognized loss figure is 3% to 5% of throughput. On a U.S. national level, this number is pegged at around $20,000,000,000 annually, year in and year out. If this is not significant enough to have an impact on your decision to be concerned, consider that many experts believe that these figures dramatically underestimate the real situation. Moreover, not included in these numbers are the rare tragic incidents like Bhopal and Valdez. Increasingly, the performance of the alarm system itself has been identified as a significant contributor.

Alarm improvement is a big deal. It involves careful planning, steady commitment, adequate resources, full cooperation of site players, ability to modify weak infrastructure supporting components, and an understanding that this is a lifestyle change, not just a once-through project. So now you ask, "What if there are only one or two things that I can do? I cannot support a full alarm redesign; in fact, I'm a few years away from that. Can you recommend a plan in the interim?"

The answer might surprise you. Before we get to that, let us understand that the bottom line here is not to get another alarm system designed. The prime objective is to significantly improve production integrity. Improved integrity is the mantra of a good alarm system, but the alarm system itself it is not necessarily the first step. Here is my list; stop at any place where you run out of money. But stop at the risk of enterprise operational integrity.

- Fix all broken equipment and keep it running well.
- Implement a policy of only operating when the plant is known to be operable (see chapter 5).
- Improve the operator's ability be aware of the plant operational situation (see chapter 12).
- Redesign the alarm system (see this book).

1.9 FUNDAMENTALS OF ALARM MANAGEMENT

We design successful alarm systems by building our work from four fundamental precepts. I am sure that when you first became aware of alarm management, you might not have been certain that it had much to do with anything real, much less vitally useful, to you and the plant you support. Alarm improvement is a discipline with important technology that, when used, will significantly improve success.

As you do the job correctly, you will come to appreciate the importance of ensuring an effective operator support infrastructure. Operating procedures, training, safe operation, screen graphics and design, and maintenance will take on an entirely new level of importance. New concepts called "safe park" and "permission to operate" will become an integral part of your lexicon. You will come to believe that it not only will work, but its work will provide your plant with a level of operational integrity that is essential to a safe and profitable enterprise. You will wonder how it was possible that anyone would expect a plant to run well without this new approach.

Proper alarm design is a straightforward engineering process. Everything you know about engineering, about system design, and about project execution remains useful. Intuition is going to be useful. Moreover, while there are some who believe that alarm improvement is a complicated, delicate, and unforgiving process, it is certainly not. Yes, experience can be useful. If it is available, embrace it. At the end of the day, doing things well is easily within your personal reach.

Bottom Line of Alarm Management

This book is all about how to identify and configure vital alarms. Please, do not confuse "vital" with "emergency" or any other identifier that conveys extreme danger or loss. Vital simply announces to the world that this aspect of the plant, if not managed, will lead to operation that the enterprise has deemed to be unacceptable. Nothing more. If every alarm configured for a plant represented a vital aspect of plant operation and was engineered well, there would be no alarm management problems. No matter how many alarms there were, they all would be needed. If that many alarms would be too much for an operator to handle, then either the plant, the controls infrastructure, or the operations management would need modification.

Fundamentals

Right from the beginning, you are going to see every fundamental of importance for proper alarm management. This might be the first book you have ever read where the punch line—the essence of the story—is revealed in the very first chapter. There are four fundamentals—four precepts. Together they form the foundation of everything we need to know about the subject. They govern all successful alarm system designs. Knowing them should resolve almost every simple and usually every difficult decision you will face in understanding and designing new systems that work.

Figure 1.9.1. Alarm design foundation fundamentals

- *Precept 1: Require action.* Every alarm requires timely operator action, and that action must be a necessary one.
- *Precept 2: Provide enough time for success.* Every alarm activation must occur in time to permit the operator to successfully remedy the situation, if that remedy is at all a reasonable outcome (given the realities of the situation).
- *Precept 3: Provide information.* Adequate information must be provided to the operator to work the alarm.
- *Precept 4: Alarm only important things.* Only alarm important conditions/situations.

Precept 1: Require Action

Precept 1 means that only when operator action is required should the potential alarm be configured as an alarm. This does not work both ways, however. The point will quickly be made that not every (and certainly not even most) operator action need be preceded by an alarm. Rather, the operator is primarily responsible for understanding the operation of his unit and maintaining good order in that unit. Alarms are not to be considered as any indicator of good or bad operation. Even perfect alarm selection and design are not meant to be surrogates for effective operator vigilance in maintaining awareness of the true operational situation of the plant.

Precept 2: Provide Enough Time for Success

Now that we know what constitutes an alarm, the second precept tells us how we must configure and support the alarm so that it will provide the requisite benefits. For this requirement to be met, every alarm must be configured to activate in time for the operator to understand and implement proper corrective action and give the process enough time to respond. This, in turn, tells the alarm designer when to activate the alarm and what information and other operator support are needed to ensure that the abnormal situation could be successfully managed.

Precept 3: Provide Information

An alarm activates, the operator is alerted, and the system is designed so there is enough time to manage the situation. At the same time, we need to provide all of the background information and current "what to do" information so that the operator can do the job.

Precept 4: Alarm Only Important Things

Only important abnormalities should be alarmed. Why? Well, without this restriction, we can be true to the preceding three precepts yet be free to alarm any event for which the operator has an action and for which we can provide a timely warning. There are so many nice things to warn the operator about that no one would have the slightest trouble providing a large number of alarms, and we did just that in the past. We provided wake-up alarms everywhere we could and we realarmed them if we thought the operator might have forgotten about them. Now, with precept 4, we place a value on each alarm. By appropriately understanding value, the alarm system is self-constrained in the number of alarms.

What should define a candidate alarm? A candidate alarm must represent any unwanted situation of importance that rises above a minimum threshold of impact and that can be announced by an alarm. Moreover, without an alarm, the operator is unlikely to observe this situation on his own.

Operator Action

Precept 1 requires all valid alarms to have an operator action. But to what operator actions does it refer? Suppose we always require operators to look at their watches and write down the difference between their watch time and the alarm time. That is an action, isn't it? A wave is an action. A look is an action. OK, you get the idea. We will need to nail down the meaning of operator action to get this right.

Operator actions can be broken down into two broad categories: primary and secondary. Operator action in response to an alarm must be a primary action. That is, in most instances, when an alarm activates, if the operator must usually respond with a primary operator action (see below), then the alarm is a proper alarm. Any alarm that is not a proper alarm should not have been configured as an alarm in the first place. But you already knew that! Let us formalize what actions are all about.

Primary Action

A primary operator action is one that directly modifies or changes something physical in the plant. Primary actions include starting something, stopping something, modifying something, and closely examining something with the intent of doing any of the earlier things on this list. Typical primary actions include putting a controller in manual, starting a pump, shutting a valve, reducing a controller setpoint, breaking a cascade loop, waiting longer to initiate a task or initiating a task earlier, or any number of the other things operators do to manage plants with problems.

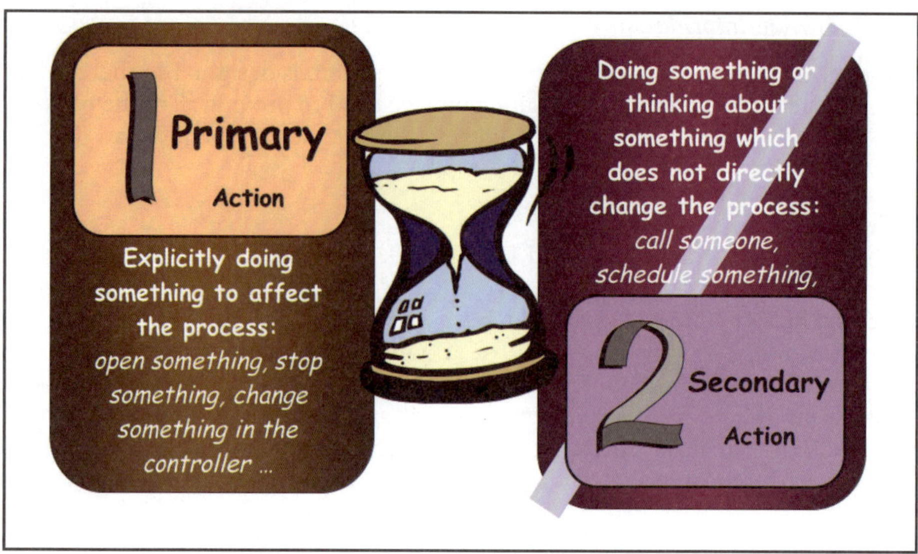

Figure 1.9.2. Operator action types

Just to be sure we get this right, an operator can closely watch an abnormal process variable with the intent of jumping in and making adjustments should it range too far out of bounds but never actually make an adjustment. This counts because it is a situation that probably needs adjustments when it becomes abnormal. The operator retains control over when, how much, and even if an action is needed.

Secondary Action

A secondary operator action is any act that is not a primary action. Secondary actions include communicating with others, scheduling something contemporaneously, thinking about something, taking note of an event or condition or situation, and the like. Secondary actions may involve calling maintenance to come and do something routine, reminding an outside operator to blow down a sump, or any number of other things that occupy operators during the normal course of their shifts. We can split hairs on this, but let's move on.

A secondary operator action is not considered secondary when it is combined with a primary operator action—that is, if an operator performs a primary action along with a secondary one. Together, they are to be considered a primary action.

Importance of the Fundamentals

OK, that is it. These concepts reveal it all. Everything else in this book, and in alarm management in general, is commentary. At any time, when there is uncertainty about what to do or how to do it, refer to the precepts. Any decision of consequence can be derived from them. This is not to say that the rest of the book is now unimportant. This book will show you how the fundamentals lead to the necessary understanding, tools, and insight to be successful. Contained within these chapters is a wealth of insight and practical methods. Using them will keep you honest to the precepts as well as save an enormous amount of time and effort in doing the important work of alarm design. This book brings to life what you will need to understand alarms and to design effective alarm systems. The precepts are about what to do. The rest of this book is about how to do it well.

1.10 DESIGN FOR HUMAN LIMITATIONS

A painter or writer can start with a blank sheet of paper, an empty screen, or a blank palate. An engineer or system designer can start with an approach that is limited only by what is believed to be in conformance with the rules of nature, economic viability, and good citizenship in a community. Of course, this situation is not as entirely unmanaged as these opening words would suggest. Good engineering practice is important. To follow what has been done successfully before is wise counsel, but that is not enough. The defining guidance will easily yield to discovery as soon as the designer insists that his creation be usable by the humans intended to use it. When all decisions have been made, the resulting design must fit the user's ability to successfully use it. Hence, you will find that all design goals and performance metrics for alarm management will be easily derived by considering whether or not a person can use them.

1.11 ALARM MANAGEMENT AND SIX SIGMA

One need not understand or apply any of the specific terminology and explicit practices of Six Sigma to do successful alarm management work. However, for sites with an active Six Sigma program, it is nice to know how the two relate. The Motorola Company developed Six Sigma as a business management strategy in 1986. They sought a more structured and updated methodology to focus the concepts and philosophy of the early quality giants: Deming, Crosby, Taguchi, Juran, and others.

With one important clarification, alarm management processes are an excellent fit with Six Sigma. That clarification is that alarm improvement (or design) is not driven by the actual performance data. Alarm improvement requires an explicit design methodology. The normal alarm events are usually considered to be data. However, any attempts to improve alarm performance cannot transcend a design that fails to comply with the fundamentals. Consider a civil engineering example. Improving highway bridges would be driven, not by bridge degradation and collapse data, but by the process of taking that information, uncovering root causes, and then modifying designs. Alarm improvement is this way as well. See Chris Wilson's article[5] for a broader discussion on this topic.

The term Six Sigma is derived from the use of σ (the Greek alphabetic character sigma) in statistics. Very simply stated, if something is within a tolerance of one σ, it is 31% efficient (31% close to what is desired), which is not very close. Two σ would be better than 69%. Six σ is 99.9997%—extremely close! A process that is Six Sigma is

DMAIC		DMADV	
D	**Define** goals consistent with enterprise and business	D	**Define** goals consistent with enterprise and business
M	**Measure** key aspects of process from collected data	M	**Measure** (figure out how to) the future characteristics of process
A	**Analyze** the data and make appropriate inferences for results	A	**Analyze** and develop design alternatives to meet needs
I	**Improve** the process based on the data analysis results	D	**Design** the new design to meet requirements (may require some form of simulation)
C	**Control** the continuing process to ensure that continuous improvements are made	V	**Verify** that the design meets requirements and is ready for use

Table 1.11.1. Terminology for Six Sigma

Six Sigma	Alarm Management
Define/ Design	Construct and use a comprehensive alarm design philosophy
Measure	Alarm performance studies; configuration studies; incident studies
Analyze	Compare against best practices
Improve	Use proven alarm improvement technology to improve existing design
Verify	Compare new design to requirements and ensure they are met
Control	Use continuous improvement methodologies to evaluate and improve ongoing alarm system

Table 1.11.2. Six Sigma approach compared to alarm management

understood to have been designed to deliver 6σ benefits. Six Sigma utilizes two variants of a work process with acronyms DMAIC and DMADV. DMADV is used for processes that have not yet been developed; DMAIC is for processes that exist but need improvement.

Table 1.11.2 shows how alarm management and Six Sigma line up. Both alarm management and Six Sigma share the importance of top-down leadership, the necessity of dedicated champions, and the requirement for proper training in and use of the technology. They are both designed for success.

1.12 CONTROLS PLATFORMS

Automatic process controls come in a wide variety of shapes and sizes. The variations start at the single loop contained in a stand-alone, separate box and progress up to tens of thousands of loops in a tightly integrated electronic and communications infrastructure. They include configurable built-in alarms and alerts. They might interface with an operator via a simple display of switches and lights. Or they might employ a sophisticated electronic display of panels, screens, sounds, and more. But they all have the same objective: keeping the plant properly operating. Since keeping something properly operating is not always easy, the equipment provides a way to inform the operator that things are not quite right. These we call alarms. Any way you look at it, it is not only possible but usual that the alarms part of things can become a problem.

Whenever alarms start to get in the way of effective operator activities, there is an alarm management problem. It matters little which controls platform or controls design is used. The concepts and technology advanced in this book are going to be useful and valid for the various controls designs and hardware/software platforms.

PLC versus DCS

The two workhorses of industrial controls are the programmable logic controller (PLC) and the distributed control system (DCS). At one time, PLCs were rarely used for heavy-duty continuous manufacturing operations and DCSs were seldom used for high-density sequencing and other intensive, stepwise production operations. The scale of PLCs was mostly on the small side. This meant that a lot of PLCs were required to manage an operation with a large number of items. The scale of DCSs was mostly on the very large side. This meant that they required too much cost and infrastructure for smaller operations. However, today, both structural platforms have been developed to be more scalable, and both have evolved to provide more of the other's features. Consequently, it should not be surprising to find that both alarm problems and alarm management solutions are not substantially different between them. This is not to say that there are not differences that bear discussion and special approaches. There are and they do.

The key differences derive from both inherent differences in the hardware platforms and associated equipment and the tendency to use PLCs for manufacturing facilities that have a large amount of equipment that is used for only parts of the manufacturing cycle or used for only specialized products that share the same controls infrastructure. The list includes the following:

- PLCs have either a dedicated hardware interface panel, no dedicated interface, or a more highly structured (and therefore less user configurable) electronic interface.
- PLCs have limited options for dealing with alarm matters.

- PLCs have limited ability to dynamically link information databases to operational displays.
- PLC-controlled older plants can have far fewer configured alarms.
- PLC-controlled plants have a significant number of pieces of equipment that are not used during portions of the manufacturing cycle.
- PLCs usually control operations that often produce different products at closely spaced but different times.

For a DCS, the first statistics that announce alarm performance problems are the large number of configured alarms, the high rate of alarm activations, and the dearth of useful information and guidance. It isn't uncommon to find several thousand alarms configured for each operator station.

For PLCs, the alarm performance problems are usually solved by ensuring that alarms are relevant to the current stage of manufacturing operations and that they provide useful operator support information and guidance. It is not at all unusual to find only a few hundred or less alarms configured for an operator area. Moreover, these processes very often involve portions of a plant that are used at different times (and so parts are shut down) or used in significantly different operating modes.

PLC Special Considerations

Most of the important benefits from enhanced alarm control will come from the ability to eliminate alarms from equipment that is not currently being used for production and spared equipment. Another benefit will come from the ability to tailor alarm configurations to the current production states. Thus, the key logic portions of interest will be *out-of-service plant* and *operating mode*.

In order to facilitate these logical judgments, it is very important that the alarm control/management engine be able to determine exactly where the plant is and exactly what equipment is part of the current campaign and which is not.

1.13 CONTINUOUS VERSUS DISCRETE AND BATCH

Alarm management came into the forefront because of the serious problems we saw in the continuous manufacturing industries. Many of the examples are taken from those applications. However, the methodology and technology is equally applicable to discrete manufacturing and batch production. The foundations are identical. The approaches are the same. Yet the practitioners in the discrete and batch industries are quick to say that their problems are different. Yes, the problems are different. But alarms are still used for the same reason. Operator overload is the same. The basic differences are illustrated in appendix 10. Please refer there for a detailed comparison between the manufacturing modalities.

Generally, the batch process industries will be spending more time defining the operating regimes and different situations under which the same or similar alarms are active and how the operator responses might differ for the same alarm due to the plant being in different operating situations. Also, there might be a different methodology for notifying operators of alarm activations, since they often are away from PCS operating screens during production runs.

1.14 APPLICATION EFFECT ON ALARM DESIGN

There is no form of manufacturing operation that cannot benefit from an effective alarm system. Chemical plants, petrochemical plants, breweries, bakeries, food processors, pulp and paper operations, power and distribution plants, pharmaceutical plants, metals and mining operations, upstream petroleum, and others too diverse to list here, all experience abnormal operation to the extent that the operator requires proper notification. Even to the inexperienced eye, most of these plants differ from others greatly. It is natural to ask how applicable all of this alarm management technology is to these broad classes of manufacturing. Do the different industrial segments have differences that affect how their alarm systems are designed? The answer is yes, but not strongly.

Current alarm design is sufficiently rich to be fully applicable to industry across the board. The fundamental concepts are valid. The process is viable. The end results are effective. But, the alarm systems are different. They are responsive to different needs. They utilize different hardware and software and they look different. Space and energy do not permit this book to approach a full discussion of the detailed differences between the many forms of application. Therefore, a decision was reached to target the general forms of petrochemical, power, and related manufacturing.

A brief discussion of how alarm system design would be responsive to differing manufacturing forms can be useful to illustrate how universal this technology can be. Consider how the alarm system might be designed to handle shutdown events. We examine a traditional chemical plant operation and contrast it with a remote offshore crude oil production platform.

Onshore plants are generally designed to take advantage of adequate space between dangerous parts of the plant and to provide multiple escape routes for personnel and multiple access routes for emergency preparedness and response activities to commence. Offshore platforms cannot be. Offshore, both the close proximity of equipment and personnel and the almost complete isolation from escape and rescue combine to require a more conservative design of conditions that prompt shutdown and a more stringent requirement for a proper shutdown. The operator has significantly different duty between onshore and offshore, and the alarm system design must be responsive to that need. Nonetheless, the alarm design principles and execution are quite similar for both.

	Chemical plant	Offshore platform
Normal operation	Alarms are designed and documented based on EEMUA and other best practices.	Alarms are designed and documented based on EEMUA and other best practices.
Before shutdown	Alarm system is designed to avoid shutdowns by enabling appropriate operator intervention when safe and possible. No shutdown confirmations are used as alarms.	Alarm system is designed to avoid shutdowns by enabling appropriate operator intervention when safe and possible. All shutdown confirmations are used as alarms.
After shutdown	Alarms are still active for all plant conditions that contain stored energy or dangerous materials or represent a risk to the integrity of the plant. Preshutdown alarms and (early-out) shutdown causative event(s) alarms are examined.	Alarms are still active for all plant conditions that contain stored energy or dangerous materials or represent a risk to the integrity of the plant. Postshutdown alarms are carefully examined to ensure that 1. the shutdown was complete and proper; 2. the shutdown contains no residual danger or risk; 3. the shutdown event raised no additional danger or risk. Preshutdown alarms and (first out?) shutdown causative event(s) alarms are examined.

Table 1.14.1. Example of differing alarm design requirements for chemical plant versus offshore oil platform

1.15 TIME AND DYNAMICS

By now you are becoming familiar with the idea that alarm management is about production events where either something has gone wrong, the operator isn't really sure whether something has gone wrong, or things are OK. It turns out that both process dynamics and time are going to be very important to the fabric of successful alarm design. Let's take a look at the general time-dynamics situation. We will do this using a fictional plant and a hypothetical form of view into the nature of that plant. The view is shown in Figure 1.15.1.

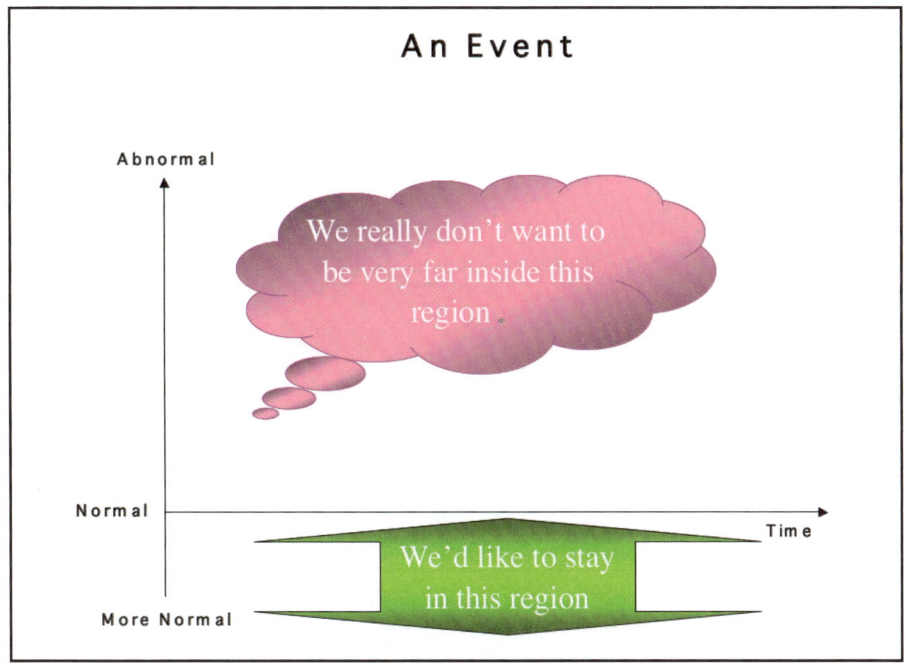

Figure 1.15.1. Magical viewer stage

In the figure we see a horizontal time line with the passage of time depicted as movement to the right. There is a label for "normal" at the left end of the time line. "More normal" which means that our magical viewer is telling us that the process is moving in a direction to be more normal, is below the time line. "Abnormal," meaning our process is moving in the direction of becoming more abnormal, is above the time line. Let's look into our "actual" fictional process timeline shown in Figure 1.15.2. There are eight segments in the response.

Normal We start out with the process just a bit in the normal region (segment label NORMAL). For a while, the process moves into the abnormal region and then back into the normal region.

Abnormal (no alarms) This segment (ABNORMAL—NO ALARMS) picks up from the NORMAL segment and moves deeper into the normal region before heading into the abnormal region. We are abnormal, but notice there are no alarms due to the process being in this state!

Abnormal (alarms) Time progresses. This segment (ABNORMAL—ALARMS) sees alarms activate. At this point, the operator has received notice of a problem. The process continues to become more abnormal.

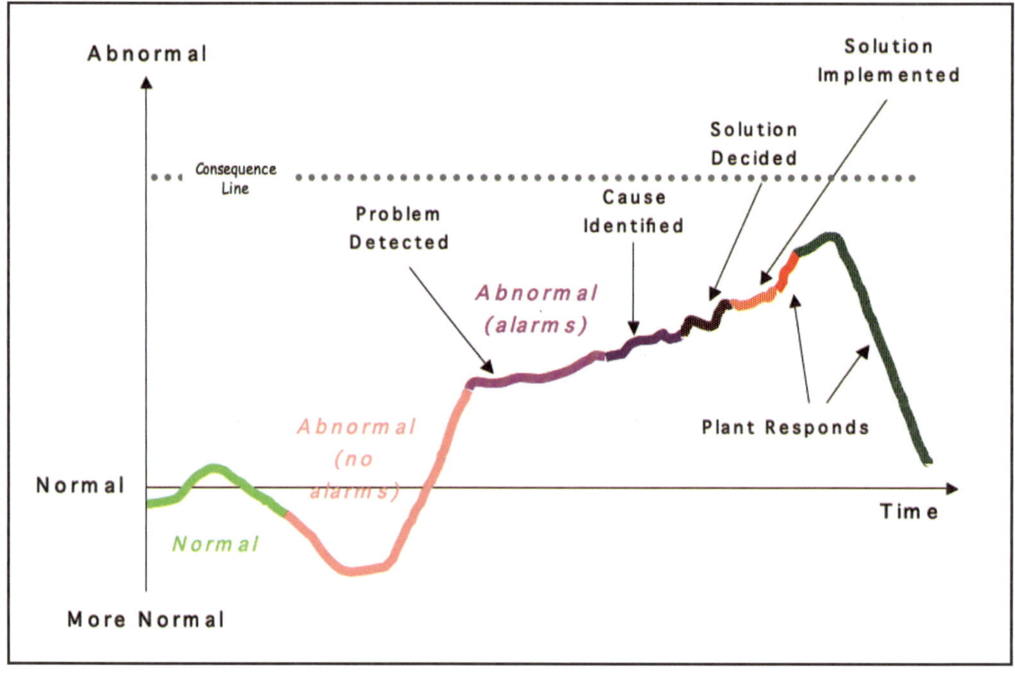

Figure 1.15.2. Looking into our process

Cause identified In this segment, the operator identifies the causes of the alarms. That is, the operator is now aware of what (is believed to be) the cause(s) of the abnormality. The process continues to move deeper into the abnormal region.

Solution decided Time continues to progress while the operator decides what to do to remedy the problem(s). He decides here. But the process continues deeper into the abnormal region. Why?
(A decision by the operator is not anything that can be "felt" by the process. The process must wait for the action.)

Solution implemented Here the operator implements the remedial plan. Changes have been made, yet the process continues to deteriorate. Why?
(An action is only a starting point. The process has dynamics or inertia. It will change in response to changes by the operator, but only in directions and with speeds that are inherent to its fundamental makeup and in some proportion to the magnitude of operator change.)

Plant responds Next, finally, the plant responds—yet that response moves it deeper into the abnormal region before eventually and rapidly returning to nearly normal.

What does this little fictional account tell us? First of all, it illustrates that it is possible for plants to become abnormal without producing alarm activations. Next, everything takes time. It takes time to discover the problem, understand it, decide what to do to correct it, and implement it, as well as for the plant itself to respond. Therefore, alarm redesign must take into account that plants can only respond in a limited way and that response will take time. Both the ways it can respond and the time it will take are basic and inherent to each particular process.

1.16 HISTORICAL INCIDENTS

The inadequate performance of PCS alarm systems has become a significant cause of industrial incidents and serious accidents. Often either process plant operators are kept unaware of abnormal conditions due to the failure of appropriate alarms to activate or they did not activate with sufficient time to permit the operator to react effectively. We also see the "blinding" of operators by the tremendous avalanche of alarms during upsets. This effectively prevents the operator from identifying which alarms are important and which are not. In both cases, the very tool that was supposed to warn and guide the operator during an abnormal situation not only failed to do either but created a distraction and additional stress. Eventually, this interference leads to escalation of seriousness by causing what started out as a minor manageable upset to produce major accidents, some progressing to serious disasters.

Three Mile Island

At about 4 a.m. on Wednesday, March 28, 1979, a series of failures and operational missteps occurred that resulted in the release of a small but measurable amount of radioactive material into the air. While this incident did not lead to serious physical loss, the enormous political reaction would curtail all building of nuclear power reactors in the United States for the next four decades.

Interestingly, the specific alarm management issues relating to this incident presaged most of the modern approach to alarm management today.

- Alarms are not applied properly due to a misunderstanding of the purpose of alarms and a failure to appreciate the scale of using them without careful consideration.
- The use of alarms is not sufficiently well understood. One must measure relevant data, infer performance (metrics), know what to do in the event of alarm activation, and know how to build alarms.
- Alarm design reaches deep into the existing infrastructure: alarms must be coordinated with plant design and culture.
- Alarm systems can really work—they were ineffectual here, but a good design could have made a meaningful difference.

- Alarm redesign is not simply an add-on—appropriate lead time is needed to arrive at a new working alarm system.

Milford Haven

In the early hours of the morning of 24 July 1994, at a Texaco refinery in Milford Haven, England, lightning struck the unit causing problems with the vacuum unit, the alkylation unit, the Butamer, and the FCC. Immediately after the strike, a number of units shutdown due to lack of utility power. A production control valve closed (we think moved to its failure position) and a unit started to fill with liquid hydrocarbon. That control valve showed an erroneous state of being "open" on a process display graphic.

In response to the hydrocarbon build up, and as designed for safety protection of the vessel, the overpressure relief valve "popped" three times. The escaping liquid entered the relief system eventually ending up in the knockout drum. The knockout drum eventually overfilled (in part due to an earlier modification that had not been properly assessed at the design time, nor appreciated during the upset). The overfilled knockout drum then spilled liquid into a relief line that was not designed to contain such a flow. The resultant failure released about 20 metric tons of hydrocarbon into the operating plant. The vapor cloud produced by this release eventually ignited about 110 meters away from the rupture, causing a major explosion equivalent to 4 metric tons of high explosive. The explosion caused $80 million damage and injured 26 people (thankfully, none seriously).[6]

There was an investigation. The Health and Safety Executive (HSE) identified a number of contributing factors to the refinery's inability to recognize and contain the abnormal operation. They are summarized in Table 1.16.1.

Specifically, paragraph 69 of the HSE Report states, "Warnings of the developing problem were lost in the plethora of instrument alarms triggered in the control room, many of which were unnecessary and registering with increasing frequency, so operators were unable to appreciate what was actually happening."

HSE's specific recommendations directly point to the necessity to provide tools and technology so that the production operator is able to detect abnormal process conditions in a way that clearly extends beyond monitoring individual values of variables and their alarms. Moreover, we see here, perhaps many of us for the first time, the addition of another important element to the operating equation: the expectation that operators be able to manage the sometimes charged issue of continuing to operate versus initiating controlled shutdowns. The implied expectation of this particular item suggests that plants clearly visit, decide, plan for, and train for situations where operators will be expected to manage the explicit issue of continuing to operate or not. Additional HSE targeted items appear as numbered items below. The reference numbers refer to specific numbered items in the formal report.

Cause factor	Cause category
The "valve open" was a command signal, not an actual feedback indication.	Faulty indication to the operator of true status of important process equipment
There was no good overview display for the operator. Existing displays focused on unit details, not situations and imbalances.	Lack of effective situation indication to the operator of true status of the process
They tried to keep the process running, hoping to find and fix the problem.	Failure of management to set high-level operating rules and procedures, thus paving the way to unsafe operating situations
The original plant design would have coped with the knockout drum overfilling; the modification was intended to reduce "slops."	Inadequate management of change (design and operation)
There were far too many alarms; some critical alarms were "lost" in the flood.	Inadequate alarm system design

Table 1.16.1. Contributing factor analysis of Milford Haven accident

#3 Display systems should be configured to provide an overview of the condition of the process including, where appropriate, mass and volumetric balance summaries.

#4 Operators should know how to carry out simple volumetric and mass balance checks whenever level or flow problems are experienced within a unit.

#5 [Provide] clear guidance [to the operator] on when to initiate controlled or emergency shutdowns.

#6 The use and configuration of alarms should be such that safety critical alarms, including those for flare systems, are distinguishable from other operational alarms; alarms are limited to a number that an operator can effectively monitor; and the ultimate plant safety should not rely on operator response to a control system alarm.

Texas City

At five minutes after three on the morning of 23 March, 2005 [at the BP refinery in Texas City, Texas], a high-level sensor alarm went off indicating rising levels of flammable hydrocarbons in a distillation tower, but a redundant alarm

> never went off on the day of the accident, the investigator said. The [investigating] board estimated that liquid inside the tower ultimately reached a point of 120 feet or more. The tower normally operated with less than 10 feet of liquid at the bottom.[7]

The incident involved the raffinate splitter (a distillation column that separates gasoline-blending components) and the blowdown drum and stack (F-20), designed to handle pressure relief and vent streams. The investigation concluded that the explosions were most likely the result of ignition of hydrocarbon vapors released from the F-20. These hydrocarbons were discharged when the pressure in the splitter column increased rapidly and exceed the set pressure of the overhead line relief valves. The F-20 was unable to handle all of the fluids, vapors, and liquid discharged from the top of the stack. An unknown ignition source from the numerous potential ones present in the uncontrolled area (vehicles, trailers, etc.) ignited the resulting vapor cloud.[8]

> The [incident] resulted in 15 deaths, about 170 injuries, and significant economic losses, and was one of the most serious U.S. workplace disasters of the past two decades. Key alarms and a level transmitter failed to operate properly and to warn operators of unsafe and abnormal conditions within the tower and the blowdown drum.[9]
>
> Federal investigators say managers authorized the start-up of a unit in March despite knowing key alarms weren't working. That start-up killed 15 people.[10]

According to an Associated Press article, "The Occupational Safety and Health Administration agency also is considering whether to refer some violations to the Justice Department for possible criminal prosecution, said John Miles, Jr., regional administrator for OSHA."[11]

Of a total of $20,720,000 in fines levied against BP, the instrumentation and alarm system portion alone was $2,170,000. We are led undeniably to value the importance of the instrumentation and control systems for safe process operations and the critical role played by the alarm system.

Why Now?

It must be clear to the reader at this point that the alarm system has two important functional requirements. First, it must detect and warn the operator of abnormal operating conditions that require attention. Second, it must not mislead, overload, or distract the operator while meeting the first requirement. As the above incidents tell and, unfortunately, similar ones reinforce, this situation seems to be getting worse. Why are we getting more alarms now than ever before? The answer points to an interesting illustration taken from nature: an iceberg (Figure 1.16.1). An iceberg floats with a bit less than 10% showing above water and the remaining 90-plus percentage hidden below. We see alarm problems "above the waterline." We do not see most of what really causes them "below the waterline."

There are a number of suggestive industry trends that bear on the causes below the waterline. The affected industries have been mostly mature ones. Their products tend to be commodities. Over the past two decades, there have been serious economic pressures in these segments. The response of the enterprise managers was to cut costs, try to force operating responsiveness to more closely follow fast-changing market pressures, and adopt the expectation that new investments would rarely be forthcoming. This in turn led to the reduction of engineering and other support staff and the outsourcing of an increasing portion of what little technology applications remained. The causal chain leads to plants having individuals with much less experience in both design and maintenance. The ultimate effects of all of this, eventually and proximately, wind up in decreased process reliability. Decreases in reliability show up as increases in abnormal operations. Abnormal operations are usually announced first by alarms. Hence, we see more and more challenges to the alarm system, just at a time that we have come to learn that the "tried and true" design is itself inadequate! We see (fig. 1.16.1) larger numbers of alarms (ice above the waterline). But those alarms are very often the result of underlying deficiencies (ice hidden below).

Figure 1.16.1. The iceberg of alarm management

1.17 THE NEW DESIGN

Not by Subtraction Alone

It will be very helpful for us to recognize that alarm management problems will not be, nor are most problems generally, solved by subtraction alone. That is, we should not expect to improve the alarm system simply by deciding which of the already configured alarms must be removed. Many of those old alarms had been configured to assist the operator in some way. Unfortunately, those alarms did not deliver the expected aid. There are new alarms that need adding. Therefore, other, more effective means must be sought to fill in that void. Redesign is a package deal. It will change the look and feel of many of the operator's tools. It first requires the establishment of a philosophy with a set of clear alarm system design objectives and operating parameters. This philosophy includes specific guidelines and practices for the new alarm design. It also will visit what, if any, important tools will be added to provide operator assistance to manage situations that were, in error, assigned to the alarm system alone. Permission to operate and the situation awareness tools are examples.

Starting Alarm Improvement

Data gathering and analysis are used to provide a clearer understanding of existing deficiencies. Baseline performance metrics are calculated and used to specifically understand exactly what the current alarm system is doing. Later on, we will use these same measures to evaluate the new design behavior. Alarm conventions are established to determine the minimal required alarm configuration for the plant. Rules for priority and alarm activation points are established. Each alarm is fully documented, which, incidentally, results in a very good training and diagnostic tool for operators. Other advanced techniques are developed to help control alarm flooding as well as to ensure that extraneous alarms are reduced. Graphics are evaluated and improved to provide operators with the ability to understand the degree of normal of their plant operation.

Alarm Philosophy

All enterprises have goals for their operation and recognized limitations as to what they can accomplish. The alarm philosophy will recognize both and incorporate them into the alarm improvement process. In the alarm philosophy, alarms are defined, appropriate operator responses are identified, success criteria are established, and the roles that all other parts of the enterprise should provide to support the alarm system redesign and subsequent operation are explained. This is where the complete design is framed. Think of the philosophy as really meaning "fully expanded design specification" sufficient to provide the alarm improvement teams with the clear guidance necessary to produce the new alarm design.

Data Gathering and Analysis

No problem is understood until its true magnitude is known. Much of our information up until now is anecdotal and subjective. Hard data will be needed for the plant to understand the problems and generate the redesign activities to make the extensive modifications and improvements necessary to provide the proper operator support. Yet it is the very nature of the problem that makes the capture and analysis of the data hard to do. The amount of data is extensive. For example, it is not unusual in any given 24-hour operating period for a typical plant (one board operator's area) to produce 1000 to 2000 entries in activity logs. These logs normally include alarm activations, alarm acknowledgments, alarm clearings, operator actions, and system status events. Three months of data are usually used. This normally means from 125,000 to 250,000 entries.

Let's look at some typical data. Figure 1.17.1 shows the frequency of occurrence of the top eleven alarm events over a period of 30 days. It has been sorted from highest to lowest. For example, the highest event, an off-normal alarm, occurred nearly 20,000 times. The eleventh highest event, also an off-normal alarm, occurred over 2000 times. Just considering the top eleven events, the operators of this area had to face an alarm about every 30 seconds, hour in and hour out. We will see in a later metrics portion of this book that this rate far exceeds any operator's ability to manage. And just as clearly, we conclude that either such frequent interruption will severely stress the operator or, what usually happens, alarm events will be relegated to background noise and largely ignored.

During upset conditions, alarms can activate so quickly that they actually appear to flood the system, hence the term *alarm flood.* Figure 1.17.2 depicts how a typical alarm flood might look. In our example, during a 5-minute period at the onset of flood, as many as 900 alarm events occur. Interestingly, alarm floods actually are defined to begin

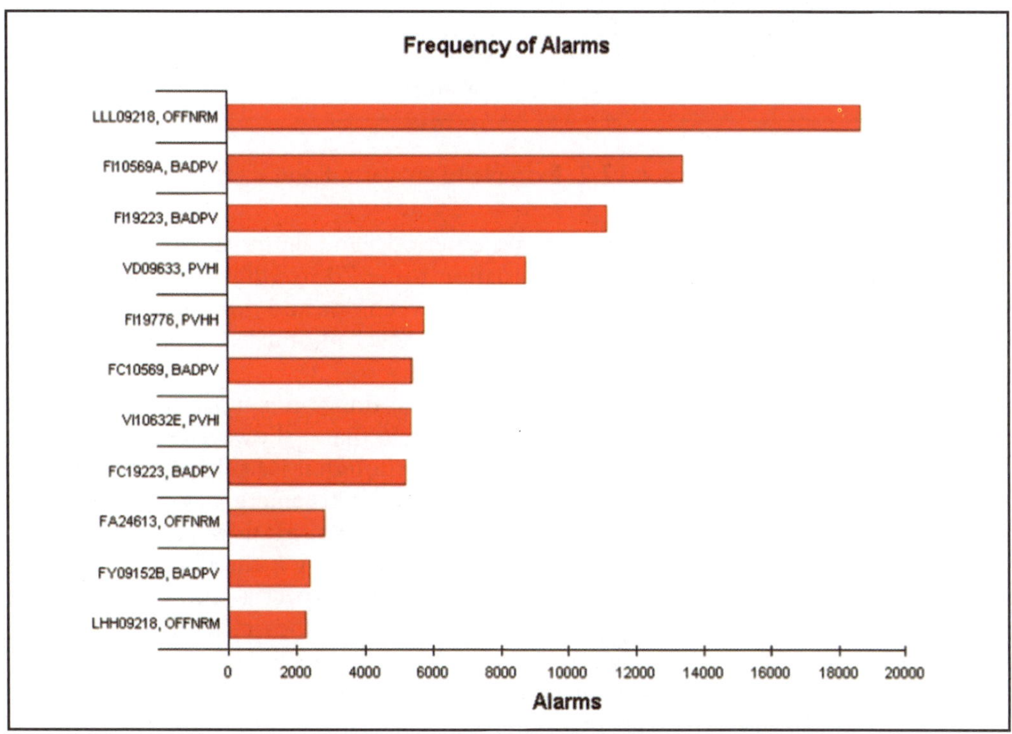

Figure 1.17.1. Number of alarm occurrences within last 30 days (by frequency)

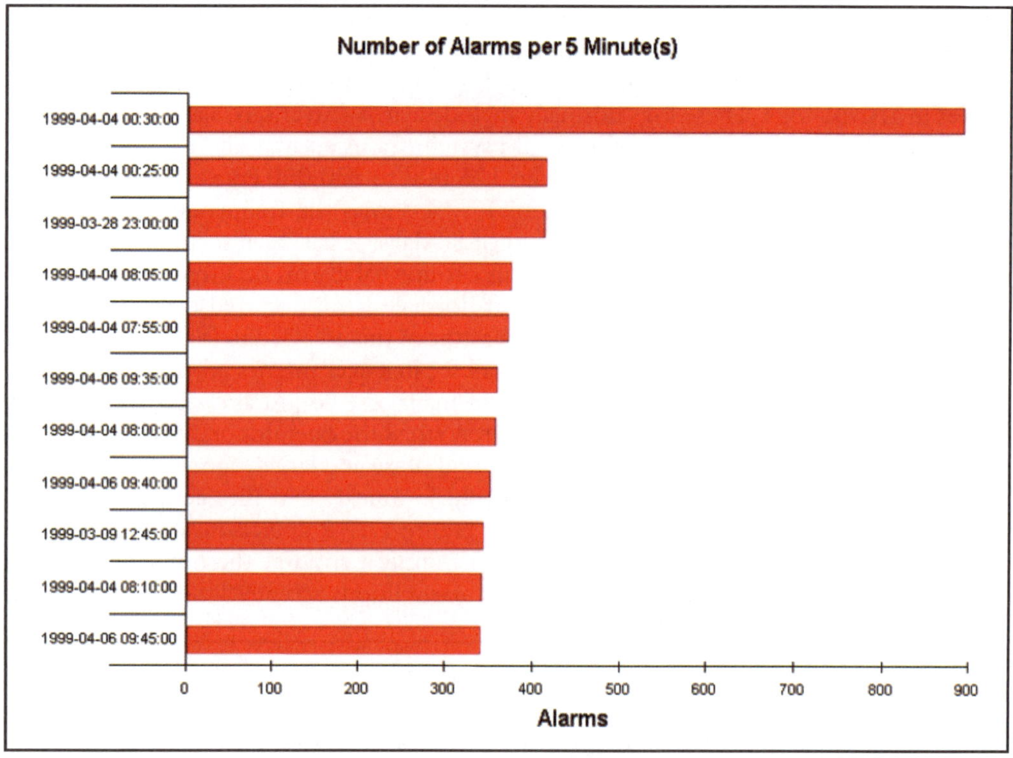

Figure 1.17.2. Example of an alarm flood

with ten or more alarms in a 10-minute time period. The figure shows typical alarm activation rates during an "alarm flood" situation to be 70 to 180 per minute (350 to 900 per 5-minute period). This calculates to be between one and three alarm activations per second. This is far beyond anyone's ability to even read, much less understand and respond to.

Figure 1.17.3 illustrates the effects of the lack of uniform standards for setting of alarm priority. This figure differs from the earlier two in that it shows data from the PCS configuration. The bars do not represent activations, though it is likely that there will be some relationship. Note also that the ratio of low-priority to emergency-priority alarms is somewhat out of balance (it should be closer to ten or more to one, as opposed to two to one, as shown). However, the most startling discovery is the very large number of high-priority alarms in proportion to low-priority. We expect a ratio of four to five low-priority alarms to each high-priority alarm. In the case shown, the high priority is actually over two-and-a-half times more than the low priority.

Figure 1.17.4 demonstrates a clear lack of correlation between alarm activations and operator actions. Recall our fundamental definition of an alarm: ALARM ACTIVATION = OPERATOR ACTION. In this example, the alarm activations are a factor of 25% to 400% greater than operator actions. A good design has them about equal. It can be estimated

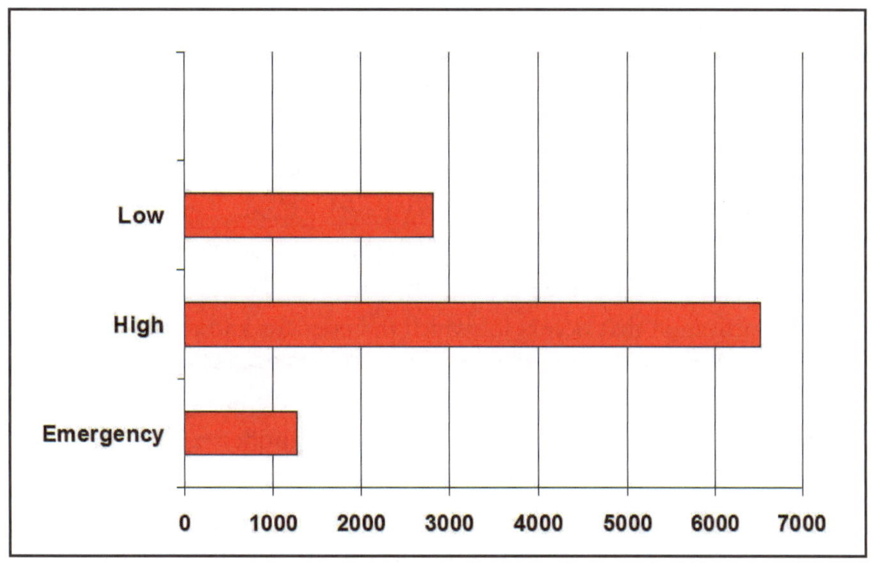

Figure 1.17.3. Typical alarm priority spread

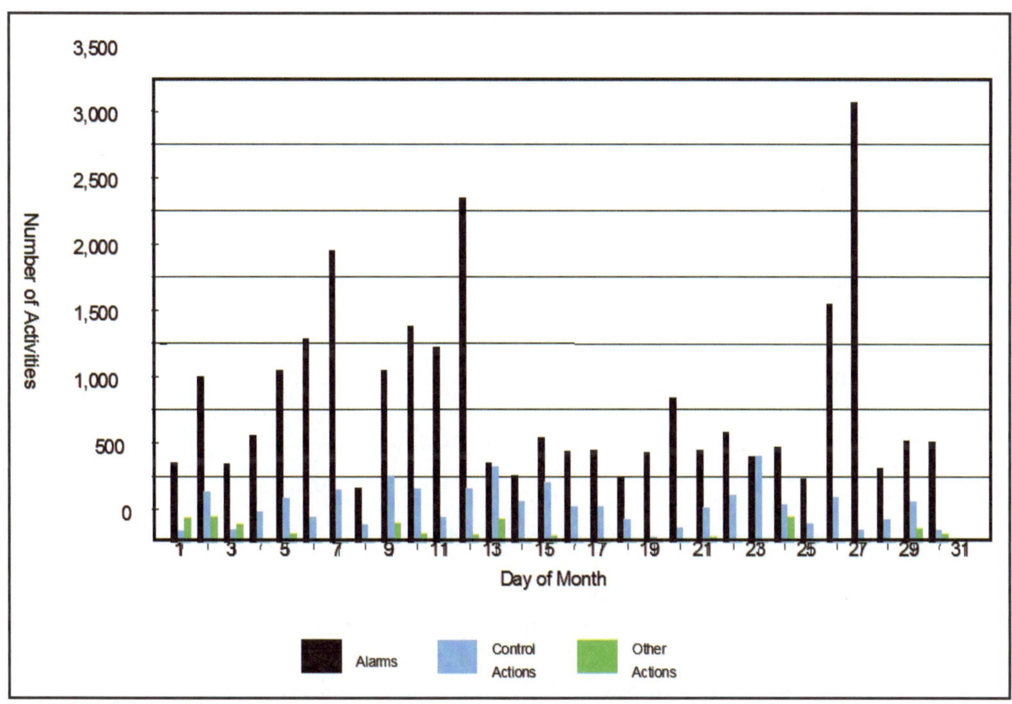

Figure 1.17.4. Link between alarms and actions (3 months)

that actions occurred for only about 30% of the alarms. A conservative estimate would conclude that the alarms could be pared down to less than one third. The alarm system should not substitute for operator vigilance. Alarms are not "wake-up" calls. Useless alarms serve no purpose.

- Any alarm activation that occurs for which the operator has no understanding of what action is called for is a useless alarm.
- Any alarm activation that occurs for which the operator knows the action but cannot implement it is again a useless alarm.
- Any alarm activation that occurs for which the operator knows the action and implements it but cannot determine whether the action has been effective is a useless alarm.
- Alarm activations that occur too quickly for all to be either acted on or understood are useless. (The alarm system itself has been identified as a prime cause of an operational upset for many of these cases.)

Now that we have had a chance to look at typical data, we are in a better position to understand the importance of the basics of effective alarm management. A key benefit derived from the alarm performance data is the identification of "bad actors." Quite often, a relatively small percentage of configured alarms will be responsible for a large percentage of the actual alarm activations. Refer again to Figure 1.17.1. Looking at the first tag (the one at the top in the rank-ordered list), activations occur about ten times the activation rates of the last two tags in the list. To place these first eleven in perspective, the median number of all activations for the month is approximately 1/10,000th of the highest one. This exponential form of falloff is typical. To eliminate bad actors, focus on the underlying issues that led to the alarms: wrong point to alarm, incorrect alarm limit, unresolved process or instrumentation problem, inconsistent alarm activation parameters (lockup, etc.), and more. But, as troublesome as they are, bad actors are only part of our problem. Simply fixing them all will not resolve your basic alarm management limitations. Redesign is in order.

Your data can be instrumental in assisting the improvement teams to develop a proper and useful alarm philosophy. A good philosophy is the blueprint for a good design. Once the nature of the problem is better understood, it aids the process of focusing on the major contributing aspects to the problem. Understanding the data requires tools. Chapter 4 discusses the major providers and their products.

Alarm Conventions and Redesign Guidelines

A useful measure you will use to evaluate the design of your alarm system will be to compare your site's data against established norms for the industry. There are two categories of data: static and dynamic. *Static data* are the data contained in the PCS configuration that specifies what is alarmed and how. This type of data is also referred to as *configured alarm data*. *Dynamic data* are the data contained in alarm activation logs. Each time an alarm occurs, it is called an alarm activation. A single configured alarm can activate many times (it will generally do so every time the process variable crosses over the alarm limit). Many configured alarms never activate. Clearly, there is a link between configured alarms and alarm activations. However, the link is rather indirect, though the more that are configured, the more will likely activate. As we have seen, this is especially true during upset situations.

The list of static alarm metrics includes the following:

- Number of alarms configured; the priority spread of configured alarms
- Number of alarms configured per control loop
- Number of alarms configured per analog measurement (not part of a control loop)
- Number of alarms configured per digital measurement

The list of dynamic alarm metrics includes the following:

- Short-term rates of activations
- Long-term rate of activations
- Priority spread of activations
- Time for the operator to acknowledge alarms
- Time for the alarm to return to normal
- Rate of activations during alarm floods; number of stale (or standing) alarms
- Ratio of alarm activations to operator actions

The alarm system is redesigned to bring the plant's data closer to industry standards for the related applications. Alarm system redesign involves a complete review of all alarms. It uses the alarm philosophy to enable system designers to set standards and implementation criteria to design a properly functioning system. Most major aspects of the enterprise will be examined. They include operating procedures and practices, control graphics, training, and maintenance. The process of redesign (alarm rationalization) relies on a hierarchy of specialized procedures. In the final analysis, it is really quite straightforward:

1. Eliminate all redundant alarm points or start from no configured alarms and configure only the minimum needed alarms.
2. Determine the alarm activation point(s).
3. Determine the correct priority for each alarm.
4. Develop a plan to handle each alarm—what it means, how to recognize its symptoms, and what to do to remedy the problem announced by the alarm activation.
5. Design the strategy to handle alarm flood.
6. Design an enhanced ability of the operator to detect early malfunctions of the plant.

7. Design the implementation plan.
8. Document all changes (manuals, graphics, management of change, training, etc.).
9. Work out the process for keeping the design effective down the road.

Using the above process, you will find either the number of points in the new alarm system is modestly less than it was originally or they are substantially less. It all depends on how carefully the alarm philosophy was developed and how dedicated its application was to the alarm improvement process. If the results achieved do not appear to be sufficient, the process is repeated, with more attention to truly understanding the essential need for each alarm.

1.18 EXAMPLE ALARM REDESIGN (RATIONALIZATION) RESULTS

One of the more surprising realizations to relative newcomers to alarm management is how many alarms are eliminated or modified by a proper rationalization. In the beginning, they are prepared to find from a quarter to a half of the original alarms eliminated. In fact, the expected reduction is more like 75% or 80%. At this point, the reader must be coming face-to-face with the realization that there is more to alarm improvement than a bit of handy engineering.

As an example, and to illustrate the process and results to themselves, a chemical plant developed a test site case for alarm improvement. This particular test was a small part of a single-operator area. They start out with a total of 154 configured alarms. What follows is a line-by-line explanation of where the alarm system was modified. Please take care to note that while each line item is correct, many of the categories overlap. Therefore, simple mathematically sums will be misleading.

Start with 154 configured alarms.

- **62** (of the 154) were deleted outright. They were unnecessary.
- **59** (of the remaining 92) had documentation errors that were corrected.
- **52** (45 from the 92 remaining; 7 from the 62 deleted) changed to alerts (an alert is a message to the operator that does not use the alarm system to deliver; see chapter 12).
- **50** (of the 47 remaining, some more than one each) had configuration corrections (some aspect of the point needed to be changed to conform to the original engineering design requirements; changes included priority and alarm setting below).
- **26** (of the 47 remaining) had their priority changed.
- **19** (of the 47 remaining) were reduced to 7 alarms.
- 7 (of the 35 remaining) were reduced to a single alert.
- **3** (of the 28 remaining) had the alarm settings changed.
- **2** new alarms were added (making a total of 30 alarms).

- **1** new alert was added.
- **14** (of the 30 alarms) were "toggled" on or off based on plant state.

RESULTS:

- **30** configured alarms
- **53** alerts (52 initially, which counts the ones reduced to an alert, plus the added alert)

This is an 80% reduction in alarms. They were able to gain this impressive improvement only by going back to basics. Tinkering just prolongs the process. Success is attained by understanding the real purpose of alarms and seeking fundamentals by which to approach a redesign. Applying the new knowledge and using good engineering practices can attain truly impressive results.

1.19 COMPLETING THE DESIGN

It has often been recognized that good alarming practices can be achieved only after the production plant process itself is better understood. As a result, not only is the plant on the way to an improved operational tool (the alarm system), but the operators better understand what they are doing and how best to do it. It is a nice synergy.

There is more to be gained if you are to realize the full potential of your redesigned alarm system. The remaining topics provide the ingredients for task completion.

Advanced Techniques

Up until this point, all of our focus and discussion about alarms has been about alarm points individually. We decide whether or not to alarm the point. If we alarm it, we assign activation points and priority. Then we decide what to do about the alarm. But the alarm points individually are only part of the story. Just as the plant is interrelated and interconnected, so are the alarm points for the plant. We now take into understanding these interrelationships. Using advanced techniques (discussed later), you will eliminate alarms for equipment that is not operating at the time. You will also manage alarms that can occur individually but often occur together. There are various means to do this. Most employ some logic to detect operating states of equipment or alarm status of other alarm points.

Situation Awareness

About now careful readers are about to reach a conclusion:

> Now that I've carefully done all that was needed to redesign my alarm system, I feel that I have ended up with many fewer alarms than I thought I needed to run

my plant. How am I supposed to find out what is really happening deep inside it? How will I know that something might be wrong? What do I need to know to tell the operator when anything does go wrong?

If alarm system design were as far as it goes, the answer would have to be, "You'd just have to take your chances with fate." Fortunately, this isn't as far as things go. Our next step will be to improve the operators' ability to watch the process and understand how it is actually performing, which is, after all, one of the key reasons why the alarm system is there and why it somehow grew into that awful monster that caused most of the alarm management problems we needed to fix. This ability to understand the operating condition of the plant is called *situation awareness*. To acquire situation awareness we provide the operator with tools for early detection of operational problems. Such tools include sensor validation, valve monitoring, control loop integrity monitoring, and process condition monitoring (tower flood, reaction runaway, compressor surge, etc.). These tools provide a significant benefit over the old way of hoping that either the operator can spot the problem or, if he does not, the alarm system will detect it for him in enough time for disaster to be avoided. There are a growing number of techniques and commercial tools now in the marketplace to assist situation awareness and thus improve production.

Operator Screen Design

Key to better plant performance is the operator's ability to see what's going on in the process. Graphic control screens are the primary means to do that seeing. When the screens do not guide the operator, when the information is too hard to find, when the information is confusing, or when the information is not there, the operator will not be able to manage as well as required. The current best practices in operator graphics involve a hierarchical structure of displays, clear navigation between displays, and a straightforward causal organization of information on each display (limits on use of color, control of flashing items, and exploitation of icons over text and figures, to name a few). Chapter 12 contains complete coverage of this material.

Operational Integrity Improvement

Anyone who spends time looking over the shoulder of a production plant operator will quickly realize what an enormous task it is to watch the many hundreds of variables, keep track of how they interact, and detect developing problems. Most try. Very, very few are able to do it well. The new approach relies on some very solid, traditional engineering:

1. Design and build the physical plant for operability
2. Develop and implement comprehensive procedures for operation and maintenance
3. Train extensively
4. Keep things maintained

5. Monitor performance
6. Feed results back and make improvements

Condition Monitoring

As important as they are, the items above are not sufficient to do the job. Consequently, there are a number of specialized tools to assist the operator. They range all the way from sophisticated mathematical techniques to pragmatic algorithms. An abbreviated list includes the following:

1. Controller tuning monitors
2. Controller mode tracking and utilization monitors
3. Extensive mass and energy balances
4. Sensor validation
5. Unit operation integrity monitors
6. Multivariate statistical analysis (multivariable state estimation)
7. Advanced controls
8. Adaptive controls

1.20 ALARM IMPROVEMENT PROJECTS

We have recognized the need and have the desire to improve our alarm system. Here are the key steps in the process.

Phase I: Problem Awareness and Solution Framework

- *Assess the current alarm situation.* Assemble the data-gathering tool kit, gather alarm performance statistics, identify upset histories, and assess equipment maintenance relationships with operational integrity and alarm performance.
- *Develop an alarm management philosophy.* Make the effort plantwide, include relevant enterprise goals, and provide guidance and specifications for alarm redesign.

Phase II: Alarm Redesign

- *Do housekeeping.* Fix all of the other problems in the facility that have been engineered and installed and should be working but are not.
- *Perform alarm rationalization.* Conduct actual redesign of alarm system including configuration and graphics.
- *Incorporate enhanced alarming techniques.* Design any enhanced alarm capabilities needed for managing alarm flood and so on.

Phase III: Implementation

- *Reconfigure PCS for new alarm system design.* Install new configuration changes (including parameters and priority modifications) and graphics.
- *Modify operating procedures to align with new alarm design.*
- *Modify training to align with new alarm design.*
- *Complete a management of change.*

Phase IV: Continuous Benefit

- *Perform periodic follow-up alarm performance studies.*
- *Investigate alarm floods and process upsets.*

1.21 LESSONS FOR SUCCESSFUL ALARM MANAGEMENT

As you have come to expect, this book is going to reach straight for the problem of what's wrong and then march directly onto the path of a successful resolution to get it right. To do this well, you will need the eyes of a military lookout to see the real problems, the wisdom of a seer to get to the heart of the matter, the strategy and planning of a general to get it done, and the heart of a healer to keep the heady success vision from overcoming the needed reality to get it done with style and respect.

Lesson 1—THEORY

- *Beautiful theories are often destroyed by ugly facts.* Improving the alarm system alone will not improve the control room.

Lesson 2—PROGRESS

- *Real progress often requires a change in direction.* It is time to end the "blame the operator" approach to control room problems.

Lesson 3—HISTORY

- *Do not forget about history.* Incidents and accidents are powerful teachers but not the only ones.

Lesson 4—HUMILITY

- It is *always wise to maintain some humility.* Almost everything we think we know about alarms will change.

Lesson 5—TRUTH

- *Technologists must always have a single agenda—the truth.* Get good data, perform adequate analyses, and believe the results.

Lesson 6—EVIDENCE

- *Incredible results require incredible evidence.* Plants without a lot of alarms actually work better.

Lesson 7—FUNDING

- *A good idea does not always attract funding.* EEMUA, OSHA, ISA 18, and NAMUR all need a business case as well.

Lesson 8—PREPARE

- *Be prepared to be unpopular and uncomfortable.* Every activity in the plant will be touched by alarm improvement work.

1.22 IMPORTANT DESIGN AND SAFETY NOTICE

In addition to the broad and comprehensive approach to developing a working solution to the alarm management problem, an important strength of this book is the wealth of examples, alternatives, and suggestions for your consideration. All control schemes, design suggestions, displays, diagrams, tables, figures, trend charts, and the like described and illustrated in this book refer to materials that have been designed to amplify and explain concepts and practices used for alarm management understanding. They are provided for training and understanding purposes only and are not intended for implementation. The choice of which aspects to alarm for the plant or process, the parameters of that alarm, the proper operator response to correct the alarm condition, and all other details of that alarm must be retained wholly by qualified, authorized members of the plant staff, who must act with full knowledge of their specific plant configuration, process conditions, equipment, and applicable statutory practices and requirements.

No single work is capable of conveying the entire collective experience and important nuances necessary for success. After reading this book, it is recommended that plants with intentions or plans for alarm improvement seek additional specific guidance and experience from knowledgeable experts.

1.23 CONCLUSION

Alarm management problems announce themselves. There are always alarms on the screen. Alarms occur too often or too quickly when things go wrong. When lots of alarms do activate, it is rarely clear what to do about them. Sometimes, the very alarm system itself contributes to poor upset response.

Successful alarm management works because it points the way to significant plant improvements. Not only are the alarm activations less frequent but they are more meaningful when they do occur. And quite apart from individual alarms, the redesign process points out the need for operator support tools to identify early process problems and to better convey the plant information through the graphics. Alarm management is truly a synergistic activity!

1.24 NOTES AND ADDITIONAL READING

Notes

1. Larry O'Brien and Dave Woll, *Alarm Management Strategies*, ARC Strategies (Boston: ARC Advisory Group, 2004).
2. Engineering Equipment Materials Users' Association, *Alarm Systems—A Guide to Design, Management and Procurement*, EEMUA Publication No. 191 (London: EEMUA, 2007), 91. http://www.eemua.co.uk.
3. Kevin McCarthy, "Facing a Long-Term Memory Loss," *Power Engineering*, October 2008, 112, 100.
4. John M. O'Hara and William S. Brown, "Human Factors Engineering Guidelines for the Review of Advanced Alarm Systems," Office of Nuclear Regulatory Research, U.S. Nuclear Regulatory Commission, NUREG/CR-6105 BNL-NUREG-52391 (Washington, DC: U.S. Nuclear Regulatory Commission, 1991).
5. Chris Wilson, *Applying Six Sigma to Alarm Management*, TiPS TechDoc White Paper (Georgetown, TX: TiPS, Inc., 2008).
6. Health and Safety Executive, *The Explosion and Fires at the Texaco Refinery, Milford Haven, 24 July 1994* (Sudbury, Suffolk, UK: 1994)
7. *Fatal Accident Investigation Report—Isomerization Unit Explosion Interim Report, Texas City, Texas USA* (London: BP, 2005).
8. Ibid.
9. U.S. Chemical Safety and Hazard Investigation Board, "Urgent Recommendation [BP Texas City Explosion and Fire, March 2004]," news release, August 17, 2005.
10. InTech, Instrumentation, Systems, and Automation Society (now the International Society of Automation) (Research Triangle Park, NC: 2005).
11. U.S. Chemical Safety and Hazard Investigation Board, Investigation Report, Washington, DC, Refinery Explosion and Fire, BP Texas City, Texas, Report No. 2005-04-I-TX, March 2007.

Recommended Additional Reading

Brabazon, Philip, and Helen Conlin. *Assessing the Safety of Staffing Arrangements for Process Operations in the Chemical and Allied Industries.* HSE Contract Research Report 348. London: HSE, 2001.

Bransby, M., and J. Jenkinson. *The Management of Alarm Systems.* HSE Contract Research Report 166. London: HSE, 1998.

Nimmo, Ian. *The Safety Issues of Batch (and Other) Controls.* Phoenix: User Centered Design Services, 2005. http://www.mycontrolroom.com.

Shook, D. *Alarm Management—What, Why, Who and How?* Matrikon White Paper. Edmonton, Alberta: Matrikon, Inc., 2007.

Smith, W. H., C. R. Howard, and A. G. Foord. *Alarms Management—Priority, Floods, Tears or Gain?* London: 4-Sight Consulting, 2003.

Thomas, Brent J. "Six Sigma Alarm Management." *ControlGlobal.* 2008. http://www.control-global.com.

Wilson, Chris. *The Operations Excellence Puzzle—The Alarm Management Piece.* TiPS TechDoc White Paper. Georgetown, TX: TiPS, Inc., 2005.

CHAPTER 2

Abnormal Situations

> After plant startup, the cost challenges placed upon the ethical pursuit of lawful chemical manufacture can dwarf the chemical product line capitalization cost.
>
> —William K. Lutz, Union Carbide

If production plants always behaved as expected, alarm systems would not be needed. If they mostly behaved as expected, alarm systems could be useful, but their design would be unimportant. Nature being unpredictable, people being less than perfect, equipment being vulnerable, and the market being fickle combine together to produce actual plant behaviors that exhibit all manner of usual and unusual effects. Therefore, well-designed alarm systems are an important first-line tool for operators. Even the most carefully designed alarm system alone is not enough.

A plant is abnormal when it is in a state, or status, different from what is expected. A plant is expected to produce the required product in the required manner at the required rate and quality. When it is not, it is abnormal. Even when a plant is shut down, it may be in an abnormal state if the plant has a mismatch between its actual state and its expected state. The ideas and concepts are discussed here in order to open up awareness to the breadth of operator support and the important needs it places on system designers to do the job right. This chapter is not intended to be a complete discussion of all of the aspects of abnormal situation management. Rather, it presents specific areas and concerns that directly impact our understanding of how best to approach alarm management.

Operator support. At present, a dearth of tools support the decisions made by operators and adequate, timely process condition information supplied to plant operators. Modern plants can be highly interactive and significantly complex. There is some degree of regularity in their operation, but there are often times when operation moves into unfamiliar and challenging states. Past history and operational experience provide little guidance for real understanding and management of proper production. Yet, because our production plants are physical, they possess fundamental operational relationships and causal effects. The ability to provide a deeper understanding and guidance to operators by utilizing advanced state information and decision support tools is a key goal of abnormal situation management.

A different way of thinking. The abnormal situation management approach calls into question the current generally accepted thinking about the simplicity of root causes. It suggests a new protocol for examining and certifying safe practices. Several methodologies for evaluating safe operation illustrate the benefits as well as reinforce the utility of the approach. A typical "upset" event is examined to illustrate the existence and vital importance of time delays affecting the ability of the plant's control equipment to detect problems, the operator's need to understand and react to problems, and the plant's ability to respond to corrective actions—and do all this quickly and accurately enough to prevent substantial (and potentially catastrophic) losses. Most importantly, out of the shadow of possible disaster comes a pivotal, essential benefit: a fundamental way to determine appropriate settings for each alarm activation level.

This chapter lays the foundation for understanding what an abnormal situation is. It moves on to review the broad impacts of abnormal situations on plant operations. This discussion provides a fresh awareness of how difficult it is to operate plants safely and productively. Along the way, we pause to make a vital observation that the operator is subject to an inescapable aspect of human nature that, if not understood and properly managed, will almost certainly place him or her in the fault path for many otherwise preventable events. Key to the understanding of the importance of this topic is the recognition of the substantial magnitude of operational integrity risks and resulting losses experienced routinely by industry. For many enterprises, the ability to better manage production operations will determine the difference between subsistence and profitability, between danger and safety, and between a society menace and public acceptance.

2.1 KEY CONCEPTS

Abnormal situations	Understanding what things can go wrong and how they do and the potential damage they might do is the key to being able to accept new ways of preventing and managing them.
Root cause vs. symptoms	Incidents rarely repeat themselves in the same way. By understanding and managing root causes, operators can improve underlying plant infrastructure.
Everything takes time	Accommodating process and equipment dynamics will be essential for abnormal situation management.
Design must fit people	Recognizing the limitations on human understanding and decision making and designing systems to accommodate them can significantly improve the operator's ability to manage.

2.2 INTRODUCING ABNORMAL SITUATIONS

A typical day at a production facility.

Two Scenarios

The "Stage"

If you are not already an operator when reading this, think of yourself as one. Imagine that you, at this very moment, are on the job as the lead inside operator at Ajax Specialty Chemicals. Ajax is one of several industrial manufacturing plants located in a thriving industrial district around Lake Charlene. The company has been in business for just under 12 years. It is an owner-operator plant. You have been with the company for just under 4 years. This was your first employment after earning your associate's degree in manufacturing and industrial technology. You worked your way up to the position of inside operator. After months of on-the-job training by the retiring lead operator, you have been lead board operator for 4 months.

Typical Day 1

The day is shaping up to be really fine. There is a bit of a wind, but other than that, the weather is clear and the outside temperature is climbing up to a very comfortable 74°F. For the past several shifts, the plant has been operating as smoothly as ever. The process feels like it is "running like a top." You bring up the latest lab values from the last shift's analytics. They are all in spec. Throughout your shift, you continue to keep the usual eye on things. It is near the end of the shift, so you've turned your attention to preparing the end-of-shift reports and are getting ready to leave. Two hours into the next shift, the laboratory technician suddenly calls the control room to report very troubling analytical values.

Typical Day 2

Today is shaping up to be a bit on the windy side, but other than that, the weather is clear and the outside temperature is climbing up to a very comfortable 74°F. For the past several shifts, the plant has been operating as smoothly as ever. However, for some reason, something does not feel exactly right to you. You check, for the fifth time, all of your temperature profiles and key flows. They all appear to be exactly where you want them. You bring up the latest lab values from the last shift's analytics. They are all in spec, though to be honest, one value is a little different than it was on your last rotation. Nevertheless, the value is well within the acceptable range. In fact, it is almost exactly in the middle—which is a bit unusual. Throughout your shift, you continue to keep a close eye on things. Nothing materializes. By the end of the shift, you've stopped looking and direct your efforts to preparing the end-of-shift reports. The handover goes well and you leave. The next shift passes uneventfully.

So, which of these two days were abnormal? Was it typical day 1? Or was it typical day 2? Was it neither? Was it both? Yes, both.

The Two Sides of Abnormal Situations

Side 1: What Is the Plant Doing?

Typical day 1 ended with our first operator leaving shift with the feeling that the plant was doing fine. It was into the next shift that it was discovered that the plant must have been operating abnormally during the previous shift, but the operator was not aware of it. The first side of an abnormal situation is determined by what the plant is actually doing.

> An abnormal situation exists whenever the operation of the *plant is upset* or disturbed, regardless of whether the *operator is or is not aware* of the situation. The existence of the process being not normal is sufficient to classify it as abnormal.

This side of abnormal is what most of us think about when we think about operation that is not normal. It is about as straightforward as it can get.

Side 2: What Is the Operator Thinking?

Typical day 2 ended with the plant operating without incident. During the shift, our operator felt that something was amiss. Despite careful attention to the plant, nothing important or notable was identified. The plant continued to produce a proper product. However, during the entire shift, our operator remained uncomfortable. It is this existence of a tangible level of discomfort regarding the true nature of the plant that forms the second side of an abnormal situation.

> An abnormal situation exists whenever the operator thinks, suspects, or feels that the operation of the plant is *questionable*, regardless of whether the *actual true nature* of the plant is abnormal or not. The existence of uncertainty by the operator regarding the true nature is the sufficient condition to classify it as an abnormal situation.

This second part is important for our understanding of abnormal situations. It will shape our appreciation of the concepts because it provides valuable insight that leads to the provision of productive technology for operator support. OK, that is a mouthful. What does it all mean?

Abnormal situation management has two facets. The first is the ability to recognize the true nature of the plant condition. Let us call this *identification*. This is sometimes called, inappropriately, *state estimation*. This facet asks if the plant is operating normally. Is it normal but likely to slip off normal and head for difficulty? Is it far from normal but the operator is unaware of this? If it is not normal, how not normal is it? What are the current dangers of continued operation?

The second facet is the provision of information and guidance. This provision is necessary to aid the operator to restore the process to normal. Or, if that is not possible, the operator can guide the movement of the process to a safer place to provide adequate time to manage the situation later. Information and guidance, provided by the infrastructure, is then used by the operator to obtain a better understanding of the problem. This is the place where the operator is assisted to determine what is likely to happen if the situation is not successfully managed. This is the place where the operator can gain information and knowledge regarding what he might do to actually manage the abnormal situation and thereby return to normal operation.

It should come as no surprise that a well-designed and fully functioning alarm system is a key asset in the operator's tool kit for production integrity—a process we have now called abnormal situation management. However, it is by far not the whole story. A good alarm system, as important as it is, will not be enough.

2.3 OBSERVING ABNORMAL SITUATIONS

Abnormal situations are the unfortunate part of plant operation. The goal of a manufacturing enterprise is to recognize an abnormal situation and manage it so that either incidents can be avoided altogether or their impact can be reduced to acceptable levels. Moreover, we study these historical events to derive important lessons learned from them. Learning a lesson, however, does not mean that we catalog the past events and write procedures and protocols that address each event. Documenting and rewriting manuals and training procedures to better manage past problems are not really good ideas. Major events, as a rule, do not repeat themselves. They are one of a kind. Consequently, we concentrate on understanding the underlying situations and develop approaches that address root causes. These approaches then form the core of abnormal situation management.

Figure 2.3.1 illustrates a typical situation against the backdrop of plant operations. The basic timeline is shown in the lower portion extending from the extreme left (the beginning of our observation period) to the extreme right (the end of our observation period). Also shown is an inset, greatly enlarged, of the center incident portion. First, let's understand the conventions of the figure and its meanings. We examine the inset.

The black graph line (better observed in the zoomed data) depicts a single composite production rate. The higher that rate, the more profitable the enterprise. The scale for the composite production rate is divided into three zones. From the bottom, we find a pink zone, which represents zero profit and extra costs due to damage from incidents. The central yellow region represents production that yields various profit levels from loss, to break even, to profit. The upper region represents a more profitable region than the yellow zone, but it will be difficult to operate there. The current operating target divides the yellow zone from the more profitable green zone. This place for the operating target is not selected arbitrarily. Far from it. Here's why.

At the very top is the theoretical production limit. Engineering analysis has determined that if all equipment and materials are as expected, then the plant should be able to operate at this value. But, of course, there are things here and there in the plant that are not up to the task of delivering their design performance. Nor might the raw

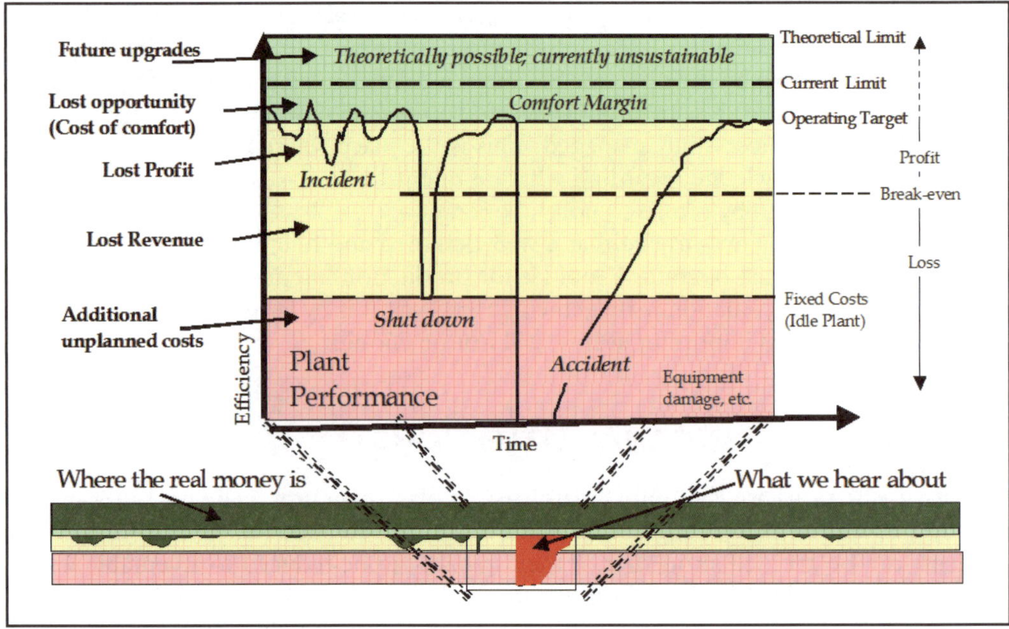

Figure 2.3.1. What an abnormal situation looks like

materials be close enough to the design assumptions to permit operation at these limits. Taken together these departures suggest that a lower current operating limit is more realistic. There is a reality here that must capture our attention—as it does for actual plant operators. The closer a plant is operated to an operating limit, the more profit is realized, yet the plant is more exposed to equipment damage, and the product to quality losses. Production variability is a fact of life. Therefore most plants select their operating targets so that most of the expected normal variability will not place it too close to an operating limit. In this way, equipment damage and product losses are minimized for normal operations. This missed opportunity region is referred to as a comfort margin. Comfort margins are costly. The better a plant is at reducing variability, the closer to the current limits it may reliably operate. Reduction of variability is both a plant design goal and an operating success result.

We return to our timeline. For most of the time, the plant is operating around the operating target level. However, even at this scale, there is considerable visible variability. Along come several incidents. Refer to the inset. The first incident results in an unplanned shutdown but no accidents or other damage—thankfully. This shutdown lasted 3 days (note that no time scale is actually present on the figure). Then production resumes only to experience a precipitous incident resulting in an accident. The damage was moderate (in the vicinity of several hundred thousand dollars), and there were only slight injuries (one operator with minor chemical burns, and another with scrapes and bruises after a fall while rapidly exiting the affected process area). Unfortunately, the plant was completely shut down for over a week while waiting for replacement equipment and cleanup. During the slow recovery, taking almost two weeks, the plant was carefully recommissioned, but little production was realized.

Fortunately, the previous example is hypothetical. No actual losses or injuries were experienced. On the other hand, this example is unfortunately all too typical. Events like this happen almost every week at one plant or another in the United States. By paying close attention to the causes of situations like this and by handling these situations correctly, the industry can become better at avoiding them whenever possible and managing them when they do happen.

2.4 UNDERSTANDING ABNORMAL SITUATIONS

As the reader knows, perhaps more than anyone else, modern industrial manufacturing plants are not inherently safe. That is not to say that they aren't safe at all. By and large, they are. The overall track record is quite good. But, it is not good enough. By recognizing that plants are not inherently safe, we acknowledge that safe operation cannot be assumed. It will be gained only through special care and attention. Without that care and attention, the processes themselves would be uncooperative with any goal of safety, efficiency, and economics. Here are a few key points along that line:

- Process media (i.e., raw materials, intermediate products, or discarded by-products) are often extremely hazardous.
- Processing equipment is under constant stress as physical limits are challenged.
- Process conditions are continually changing to meet available raw materials, schedule, and quality requirements.
- Costs are under significant "margin" pressures, which can lead to delayed maintenance and reductions of important infrastructure expenditure.

The first step in managing abnormal situations is understanding the magnitude of the problem. Here, real plant data are vital. Most plants are not set up to track these situations very accurately. Yes, of course, they all track the major events. Most of them are fully documented and extensively analyzed to ensure that they do not repeat. However, for the most part, that is about it. Therefore the first line of defense for better alarm management is information. This discussion will start with some background on the kinds of incident that set the stage. Then we get to the punch lines.

Incidents are often filed into one of five categories. They are listed in order of most impactful to least impactful. Figure 2.4.1 depicts a typical incident "triangle" showing relative ratios of types of events.

A *critical incident* has a broad impact on the plant (i.e., one or more major portions of plant shut down; significant area affected outside plant limits; usually some loss of life; serious injuries; usually one or more significant environmental releases; very serious financial impact—placing the very existence of the business at risk).

A *major incident* (or very close call to a major incident) has a major impact on a limited area of the plant (i.e., one or more units affected; limited affect outside plant

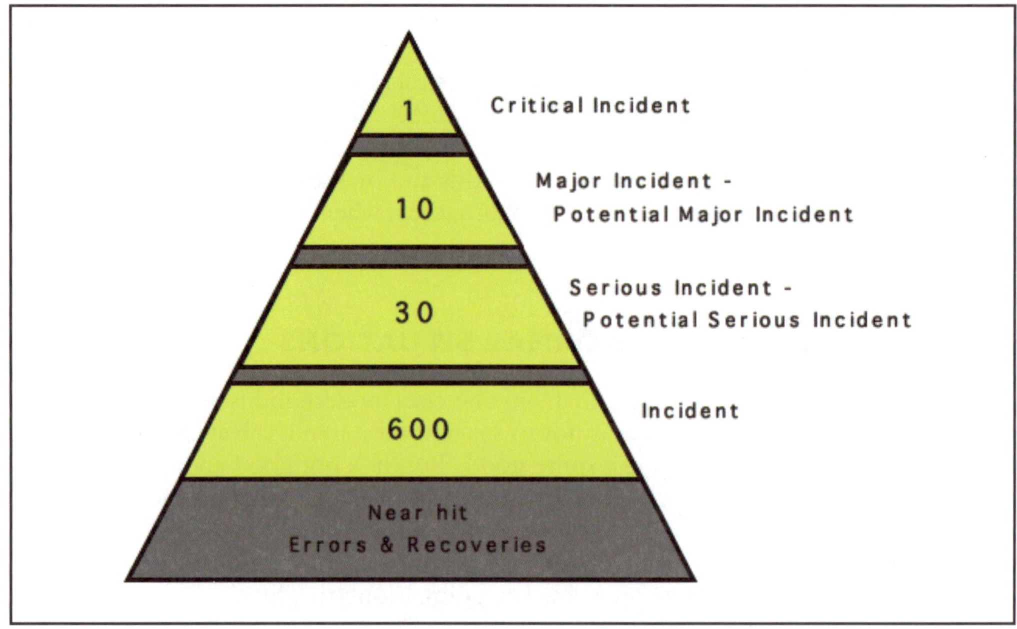

Figure 2.4.1. Incident severity triangle

limits; possibly serious injury, but no immediate loss of life; reportable environmental release potentially requiring targeted cleanup actions or other immediate responses; major financial or delivery impact).

A *serious incident* (or very close call to a serious incident) has a clear, defined impact on a limited area of the plant (i.e., if not successfully contained, this incident could have easily escalated to a major incident; one unit partially affected; potential for serious injury, but no loss of life; possible environmental release; noticeable financial losses include direct and consequential damages).

An *incident* is a clear event with defined impact on one or more plant production or support components (i.e., disruption of production is minor; potential injuries possible, none serious; potential reportable environmental release; most financial impact due to direct losses).

All incidents that were narrowly averted, regardless of whether the miss was due to efforts or actions (even significant) by plant operating personnel or due to chance luck, we call a *near hit* (or a near miss). The relevant message is that this was a situation that very nearly ended in an incident. Therefore the "hit" part is the real message, not the "miss."

You are very likely familiar with these classifications, unfortunately. The important message is the numbers and what they imply. Going back to Figure 2.4.1, we see one critical incident. That is our basis. For every critical incident, we usually see ten major incidents. That is a lot. For every critical incident, we usually see thirty serious incidents. Troubling. For every critical incident, we might expect to see about 600 incidents. For every critical incident, we will see a large number of near hits. Whether these near hits are recorded is another matter. And if they are recorded, whether they are investigated and put into action is yet another matter.

This discussion is not meant to scare anyone. Nor is it meant to suggest that industry is doing a terrible job managing production. What is intended here is to announce that all plants have at their disposal a very important way to make things better. This source of operating history, if "mined" and appropriately actioned, can lead to important, often fundamental, ways of improving plants.

2.5 UNDERSTANDING INCIDENTS

General Concepts Learned

There are three key concepts we learned from examining the history of incidents:

1. Plant data will point the way toward important problems, issues, and requirements that will be of significant value during alarm redesign.
2. Analysis of the process leads to an important missing link in conventional risk assessment practices.
3. Understanding the unfolding of events from the operator's perspective provides the single most important element of the entire alarm management practice.

Let's take a closer look at each of these concepts.

Your Plant Data

The data we are talking about is the information that may or may not have been saved. If it was saved, it may or may not have been examined. If saved and examined, it may or may not have been evaluated. And if saved, examined, and evaluated, it may or may not have been actioned to the point of being rolled into plant practices sufficient to improve awareness and response. Moreover, data will provide an important postimplementation benchmark for judging the value of any alarm improvement efforts that are made. Yes, as you will see in chapter 4, there is a well-structured program for evaluating the alarm system performance itself. That program has been designed to provide information to evaluate how well the new design works when compared both to standard industrial benchmarks and to the local sites' key indicators. However, that performance is secondary to the real issues here. What is of paramount concern is that plants operate better. No matter how well the new alarm system performs, if the impact on actual plant operations isn't demonstrably improved, we haven't gained much.

Now is a good time to begin your inventory of information. Try to identify all the sources that you can use to gather operating effectiveness information as reflected in all the unfortunate things that have gone wrong. Identify what data the plant has, locate the owners of the data and where they repose, and explore how that data and the information derived from the data are used by the plant. Where possible, review typical items and try to develop some ideas about what your situation might be. Please resist any temptations to conclude anything at this point. Premature judgments can be very misleading.

The next source of information is the way the plant identifies risks and specifies work procedures. Locate your procedures for doing hazardous operations (HazOps). Find out how well that process works. Identify plant individuals who are experienced and might be useful later to assist with the formation and work of alarm improvement teams. Look at some recent HazOps in your plant to get a feel of what they are all about. If possible, find a case where the HazOp had to be revisited due to a later problem. See how the process addressed new knowledge and amendments to procedures, if any.

All this information will be used later as part of the process of designing your alarm system. Details are found in chapter 6. However, even at this point, you will find it very useful to gather some of the preliminary data to use as you work up the justification case for alarm improvement work. We now move on to looking at incidents from a more global perspective.

2.6 GENERAL LESSONS FROM INCIDENTS

We table the examination of specific site data in favor of looking at a broader range of incidents that have affected manufacturing in general. Start by examining classes of incidents from a historical perspective. The first thing that jumps out is the sheer magnitude of damages. It has been conservatively estimated[1] that abnormal operations (OK, let's call them what they are: accidents, breakage, incorrect operation, lack of maintenance, and all the other contributors to production inefficiencies) cost the United States in excess of $19 billion annually. Annually! Incident investigations strongly suggest that a significant amount of this loss is unnecessary and therefore savable if future operations and policies were to change. The best path to change is through understanding.

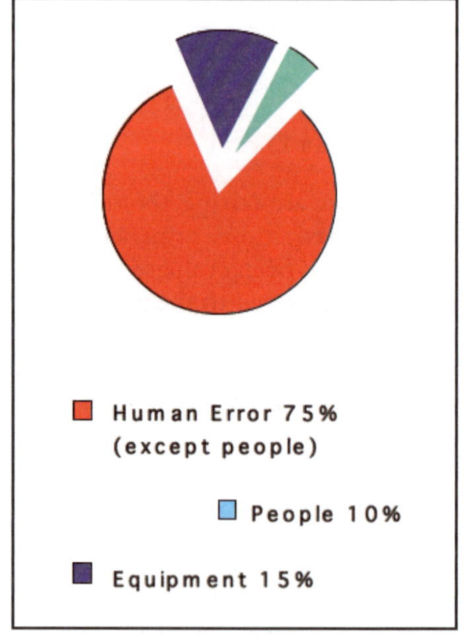

Figure 2.6.1. Causes of incidents

One of the more important and useful results to come out of the careful examination of abnormal situations has been to better understand their structural causes. Let us first examine a common belief that operator errors lead to most of the problems. An extensive review of incidents yielded a breakdown into their primary cause categories. Figure 2.6.1 illustrates the results.

The first thing to notice is that only 15% of the causes are equipment related. That apparently leaves the bulk of the cause on the shoulders of people. Seventy-five percent of the incidents were caused by general human error. Yet only 10% can be attributed solely to actual individuals performing badly. Performing badly means deliberate or negligent acts: working under the influence of performance-impairing chemicals (legal or not), working when overtired, working when significantly bothered by personal concerns, working without paying reasonable attention (e.g., daydreaming, etc.), working when too ill (physically or psychologically), working without disclosing important physical limitations that would otherwise disqualify from work, and other similar situations. This is all we can assign directly to human error as the proximate cause of the incident. The remainder of the "human error"–type causes lie deeper within the infrastructure. Let's take a look.

To what causes can the remaining 75% human error be attributed? When we take a closer look, new things come into focus. Look at the list below. Keep in mind that these are indirect "people" items. They are listed here because this list is often attributed to people acting poorly, yet these items are, in actuality, systemic in origin.

- People mistakes
- Inadequate or improper research
- Inadequate or erroneous design
- Faulty construction
- Improper installation
- Poor operation
- Inadequate or improper maintenance
- Improper manufacturing
- Missing or inadequate inspection
- Inadequate management

Examination for Cause

Taking a closer look at each of these categories, a surprising thing comes out. When you look at them, they all sound reasonable. Such a list would likely be what each of you would have prepared after thinking about it a bit. And perhaps that is so. However, the record seems to speak very differently. If these items were ingrained into the infrastructure and most plants paid appropriate attention to them, then they would not be present at incident investigations as proximate causes for such a large proportion of accidents! If everyone were familiar and used appropriate methods and procedures to ensure that they were adequately covered, then it would not be possible to attribute 75% of all incident causes to them. Yet we now know that these are the neglected true contributors.

Let's take a closer look at each of these components. We do this with the expectation that knowing the problem will be of valuable assistance to those responsible for the design and operation of industrial manufacturing enterprises. Knowing leads to improvement.

People Mistakes

People mistakes are mistakes either made by individuals making reasonable errors in judgments or caused by individuals overstretching their personal, physical, cognitive, or emotional limitations. They are also errors caused by equipment that was not adequately suited to accommodate the realities of a human individual; errors caused by lack of fitness of the person to perform due to inadequate training or missing, inadequate, or improper procedures; or finally errors due to the overcomplexity of the equipment or situation that placed proper understanding outside of reasonable human bounds. These are mistakes made by people. Yet the mistakes are not caused by people. Fixing the person will not eliminate the mistakes. Reducing the errors will involve a remedy that reaches deep into the enterprise infrastructure. Therefore caution is advised where attempts to assign cause to people *transforms* into an avoidance of examination of the infrastructure to uncover true cause.

Inadequate or Improper Research

Research is a process whereby members of a society attempt to examine and thereby discover the true nature of things with the goal of exploiting new understandings for gain. If the fruits of research are to be relied on, it is necessary that the research be conducted so that reliable results free from deceit are produced and that appropriate prudence is exercised to ensure subsequent reliance is well placed. Research that fails to uncover reasonable implications of truth runs the risk of empowering gain without sufficient discovery of risk.

Inadequate or Erroneous Design

The path to bringing an idea for a manufacturing enterprise into reality goes through a process we call design. Design produces the plan that can be used to understand what it takes to build the enterprise, estimate the construction costs, identify the appropriate site for the plant, and finally do the detailed build to produce the plant. If that design is not carefully done or if some of the assumptions for the work are incorrect or inadequate, the ensuing design is likely to be flawed.

Almost all engineering disasters—that is, disasters that involve a failure of an engineering entity that stands alone (like a building, a bridge, a dam, etc.)—can be traced directly to one or more design faults or significant assumption errors. We should not infer that because something is built it will be safe and adequate to use. When we turn our attention to examining manufacturing enterprise disasters, we find the added complexity of operating errors contribute to failures.

Faulty Construction

Even the best of designs do not build themselves. Even when all the construction materials are within specification, there is no guarantee that they will be put together properly. Even when the construction is thought to be appropriate, special circumstances might have been missed that would have dictated a different construction.

Improper Installation

Installation is an important aspect. It differs from construction. Generally, construction involves mostly fixed portions of the plant. These include foundations, major vessels, piping, and other major infrastructure. Installation would involve specialized equipment like pumps, valves, electrical equipment, wiring, and the like. Installation is a sensitive endeavor that requires careful attention to the specifics of the plant under construction. Seemingly innocuous items like proper shielding for sensor wiring, the careful alignment of rotating equipment, or the addition of mechanical shielding necessitated by the proximity of other equipment or hazard exposure can assume an importance well beyond their intrinsic cost and value.

Poor Operation

Readers of this book most likely have a very good idea of incident causes. In addition to the usual suspects here, we are reminded that even honest attempts at being careful may not be enough. Operating too close to key limits can overstress equipment and overstep operator skills. Operating in very bad weather conditions can negate normal safeguards and "cushions" and thus expose the safe operation of the process to danger. Using raw materials that differ in important ways from the materials that the plant was designed for can cause the plant to operate in ways that may not be adequately compensated for by the equipment and present operating situations, and so operators may not be able to manage due to inexperience or even lack of competence.

Inadequate or Improper Maintenance

Faulty maintenance is most often the largest contributor to incidents and accidents. Even if everything else is perfect, if the equipment is not ready or able to perform as designed, it will fail or cause the operation to move to places for which it is not appropriate. This is also the easiest to manage and for which the plants are always set up to do.

Improper Manufacturing

One of the last things most industrial plant operators even think about is how suited their plant is to make what they are trying to make. Nonetheless, more often than we'd like to think, either the plant is less than what is needed to do the job properly, safely, and economically; the raw materials aren't a good match; or there are other constraints that reflect directly back to the way the manufacturing facility has been designed to operate versus what it can actually do well.

Missing or Inadequate Inspection

To quote a famous man, "Trust but verify" should be a cornerstone of safety and site integrity. It is the second line of defense against error, mischief, or larceny. While it applies here primarily for mechanical equipment and material, it holds equally for raw materials, electronic equipment and software, and business systems. There is a story here that bears telling.

In a plant perhaps similar to yours, the site manager happened to read an article in a trade journal that reported the story of a very serious accident at a similar plant. It turned out that a section of pipe had ruptured with tragic results. Serious equipment damage and injuries were reported. As he read, he understood that the pipe in question was original but made out of a material that was not up to design specifications.

The manager said to himself, "If it can happen in that plant, it can happen in mine." Therefore he immediately set about on a program of positive material identification. Using special equipment, the material of construction for every piece of equipment, every valve, and every piece of piping was examined. Even insulated piping was opened at every other joint to be tested. The entire project cost a bit over a million dollars.

The ensuing inspection located a single length of pipe that was not to specification. This pipe section was almost about to rupture. That disaster, prevented by expenditure of one million dollars, returned over tenfold in averted costs. More important, the non-monetary costs of production disruption, environmental damage, personnel injuries, and loss of community reputation were all avoided.

Most all modern plants now have a very strict policy for ensuring that any new materials entering their facility are tested. Some have initiated a program for testing in-place materials.

Inadequate Management

This is last on our list of incident-contributing factors. It is the most important. It is most often the primary cause of most infrastructure failures. And make no mistake; all the above causes have roots in infrastructure failures. The quality and consistency of management set the entire playing field for the enterprise. Unfortunately, these management deficiencies do not show up until much later down the road. And just as often, we are not used to looking in that direction for faults.

Hazards Defined by the FAA

The U.S. Federal Aviation Agency (FAA) was one of the earliest examiners of complex accidents. A hazard is defined by FAA Order 8040.4 as a "condition, event, or circumstance that could lead to or contribute to an unplanned or undesirable event."[2] As the result of the broad experiences gained from the difficult task of determining root cause, the FAA learned a very difficult lesson. Yet it is this lesson that can profit us well down the road, should we choose to heed it. The vital lesson is this: seldom does a single hazard cause an accident. More often, an accident occurs as the result of a combination or sequence of causes.

Two Events

HazOp is an important risk management tool. It is well-structured and designed to expose most weaknesses in operations of plants and equipment. It is little wonder that its use is so well established. A pivotal assumption is that only proximate causes will be considered. Any additional causes or situations that are extremely rare and therefore most unlikely to occur are not considered. Unfortunately, we are now finding that this is a serious limitation in its effectiveness.

We are now ready to express the fundamental takeaway message. Most accidents involve more than one cause. The first cause usually comes from the list that can easily be made during examinations of risk. They are examined during the normal HazOp studies. The second event is usually rare and therefore not anticipated. Nonetheless, whenever both the primary, or proximate, cause and the secondary cause co-occur, the incident or accident is almost a guaranteed result. Said another way, if both things occur at the same time, the accident will certainly occur. Therefore, since no matter how unlikely both are to occur at the same time, when they do, you *will* have the accident. We must consider ways to keep them from occurring. Consider taking steps to examine your site's risk assessment and incident review programs to ensure that they adequately expose all risks.

2.7 CRITICAL CONTRIBUTORS TO INCIDENTS

If there is a silver lining from bad incidents, it is the lessons that can be learned from them. Unfortunately, these lessons were expensive. Fortunately, others who are humble can learn them without paying the high price. Best of all, the lessons are simple and easily learned. And if that weren't enough, they are easy to state succinctly. We take these items from Milford Haven, Texas City, and others.

Subtle Abnormalities

Alarms are well suited to warn of abnormalities that present themselves rather quickly, but not if they come instantaneously, like a shutdown. Generally, problems that occur and can reach their undesirable consequences in a matter of many minutes are what we are thinking about here. However, there are a number of abnormal situations that unfold very, very slowly. Consider leaks from contained systems or unnoticed biases to control objectives. The normal time scale is hours or even more. Because they unfold so slowly, operators are unaware of the drama unfolding. Because there is no obviously clear-cut threshold of trouble, the placement of an alarm activation point can be nearly impossible to do well. Place the alarm limit too tightly, and it always "cries wolf." Place it later on down the path to certainty, and one will need the expert services of wizards and crystal balls to come up with a workable value.

On the other hand, when this situation is looked at from a completely different perspective, nice alternatives emerge. In general, most subtle irregularities involve energy or materials "leaking" out of a place we expect it not to. If we are able to measure (either directly or by computation) the value of the leak, we simply add up (totalize) the values of the leak as time goes on. Once the total gets above a threshold, it is clear that something is wrong and it can be alarmed. Mass and energy "leaks" are the traditional irregularities to track using mass and energy balance calculations. The Health and Safety Executive in the United Kingdom has effectively mandated the appropriate use of energy and material balances to assist operators in these subtle situations.

The Human Nature of Operators

When accidents and disasters are investigated in the clear light of the afterward, a modest percentage of them are found to be caused by the operator. Or if they were not initially caused by the operator, they were nonetheless made worse by the operator's inappropriate action or lack of appropriate action. Further investigation has determined that most of the operators in that scenario are competent and unimpaired. Is there an inconsistency? No. Ultimately, we find the root of this situation lies in the inherent human nature of us all. Let's see how this works.

Industry places operators in the position and role of managing the production process. They are charged with keeping things running well. When they do, we encourage them to improve it. When upsets occur, they are expected to manage them and get things back on track. It is the very nature of upsets and other abnormal situations that they create a sense of time urgency. We inherently assume that the faster the thing is worked on, the better chance we have to find an acceptable outcome. So, the de facto rule is get the job done as fast as possible. Speed becomes the guiding parameter. But, excessive speed is the mother of disaster.

Now we find ourselves in the middle of what seems to be a no-win situation. If the operator waits, the process might reach disaster before he has a chance to act. If he acts too quickly, mistakes abound. The answer to this conundrum is to look at why the operator fails to appropriately manage. The answer is as simple as it is elegant.

Once an operator believes that he has found the cause of the upset, he rarely, if ever, pauses to rethink that belief. Even if new information comes to his attention that contradicts his earlier assessment, it is discounted. It is downplayed because it is not consistent with his earlier evaluation. Being inconsistent, it must not be true. If it is not true, the earlier conclusion must still be right.

And down comes the process. Self-recovery is rare. Psychologists have done extensive research into this phenomenon.[3] They found that not until after five major contradictions show themselves does the individual finally realize that his earlier belief was in error. For most industrial operations, that realization comes too late.

Putting this knowledge into practice, we then should do the following:

1. Train operators that understanding the situation must come before action.
2. Train operators in the basic competencies of managing abnormal operations, instead of managing events—most events do not repeat.
3. Ensure that the controls infrastructure provides the operator with sufficient information to make a proper determination of the cause and appropriate remediation action.

This new understanding now opens a door to managing and resolving this situation. The next section begins that discussion.

Stop in Time

Operators have long been observed mismanaging plant upsets by their failure to appropriately determine that they are facing unmanageable situations. So, they continue to keep trying to find a way to manage the abnormal situation rather than abandon the approach and shut it down. A disturbing number of serious industrial disasters have this situation as the proximate cause. Chapter 5 explores this concept and provides options to manage the issue.

2.8 THE IMPORTANCE OF TIME

If somehow, magically, we could stop time during an upset, all upsets could be successfully managed! If every disaster in the making could be paused, it would be possible to spend as much time as needed to understand what is going on, to assess the damage that would likely happen if it continues, and to fix it if we wanted to. Even if we could not reverse time, just the mere ability to stop time would be the wonder tool. Alas, such daydreams cannot come true. Yet exploring this fantasy has helped us to focus on something vital and important. During an upset, the most valuable commodity an operator has is time. How much time is available and how it is used to manage the problem and reduce its unwanted impact can be key to determining success or failure.

An Example

Consider a hypothetical upset progression. Our illustration is an operating unit that springs a leak in a combustible gas line. Figure 2.8.1 shows the timeline. We follow the two scenarios. Note that the first 5 minutes of time have been expanded to better illustrate the situation. The first scenario is the solid (red) one. This shows the actual gas concentration in the unit in percentage of lower explosive limit (LEL) as a function of time. The second scenario is the dashed (green) one. This shows not the actual conditions but the conditions that the instrumentation and therefore the operator will see. For

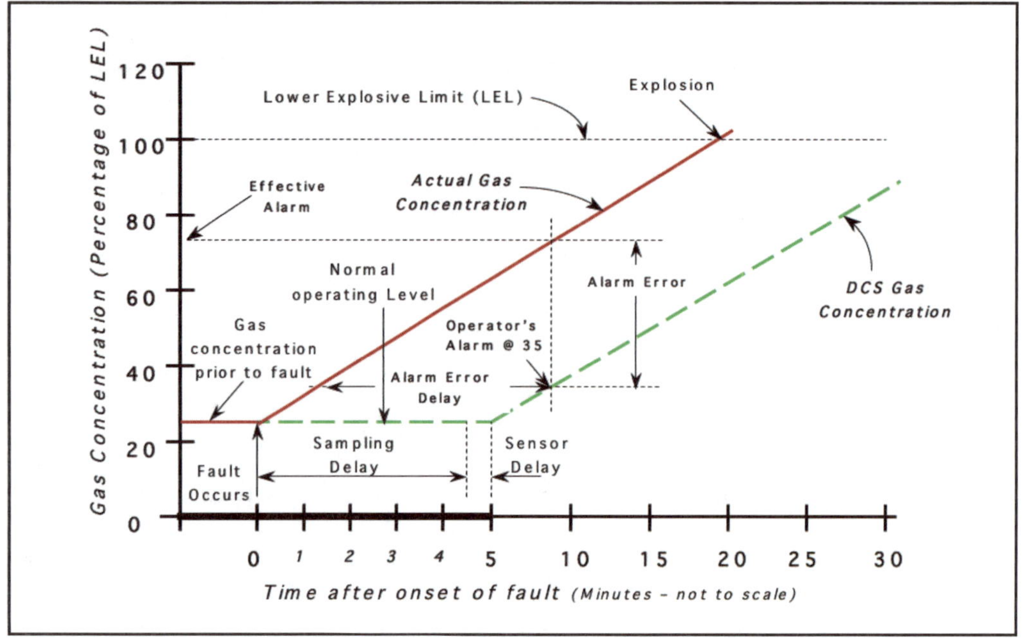

Figure 2.8.1. Timeline of an explosive gas leak

increasing gas leaks, the actual gas concentration is always significantly higher than what the operator will see. The actual concentrations solely govern all process events including fire and explosion. Operator-observed concentrations only govern most operator thinking and actions. To the extent that what the operator sees differs from the actual plant situations, the plant is at risk.

Solid timeline. Before the fault occurs, there is an ambient level of our gas at around 25% of the LEL. Once the LEL is reached in the vicinity of any ignition source, the gas will explode. At time zero, the gas line rupture occurs and our leak starts. We know from the control system configuration that the alarm activation point for this gas has been set to 35%. A single gas sensor is located in the unit. The particular location was selected as the most representative but, of course, cannot ensure that it will provide an adequate measurement under all conditions. By design, the alarm activation point is assumed to provide enough of a safety margin to ensure that any local concentrations would not be too close to the LEL. We should note that if nothing is done to correct the problem, at some point, an explosion is an almost certainty.

Dashed timeline. Even though the actual line rupture occurred at incident time zero, due to a lengthy sampling delay and the delay in the sensor's ability to perform the analysis, the control system, and therefore the operator, will not see the event until almost 5 minutes later. At time 5 minutes plus, the operator would see the gas level start to rise, if it were being watched at all. However, at this time, the actual gas concentration at our sensor in the unit is above 60%. Since the control system manages the alarm, there will be no alarm until about 7 minutes after the ideal alarm should have activated to inform the operator that the gas concentration is 35% (green line). However, by that time, the actual concentration is nearly 75% (red line). If the operator is not able to manage this problem, there will be an explosion in the unit before the observed value exceeds 60%. All of the operator's actions and assumptions regarding how much time is available to manage this situation are in error. It is unlikely that our operator will be successful.

OK, you've had a look at this and are probably thinking that the example might be too contrived to be useful. It is unlikely that any critical sensor would have a sampling delay so long. Most do not, but some do. The important thing here is to appreciate that this situation is not that far removed from situations that are all too real.

Process Safety Time

What the operator sees is the only important reality in the chain of production management. Therefore systems designers must take into account what is required to ensure that the operator's view best represents actual situations and events. System enhancements, tools, and modifications are required to provide adequate and accurate understanding of those actual conditions. One of the most important tools is the proper setting of alarm activation limits. We now take a closer look at the effects of time on operator actions. Figure 2.8.2 illustrates an operator-centric view of the timeline of an event.

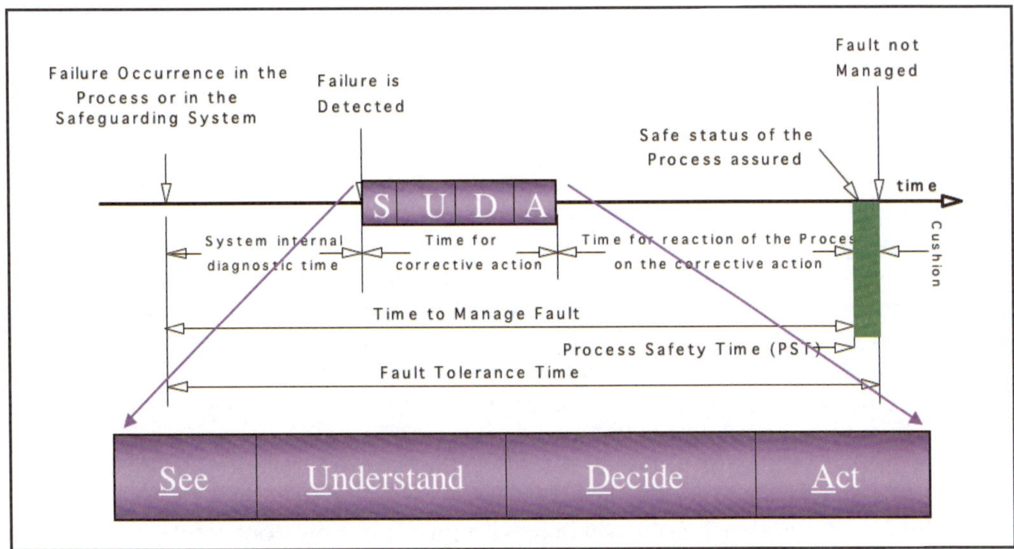

Figure 2.8.2. Process safety time and SUDA

SUDA

The time for corrective action is governed by these four key activities:

See	How easy is it for the operator to *see* that an alarm has activated and to distinguish it from all other alarm activations?
Understand	How easy is it for the operator to *understand* what the alarm means? "What the alarm means" includes being able to verify that the process state is such that it is likely that the alarm should have activated and being able to determine the consequences if this situation is not properly managed in time—that is, the operator has ensured that a software, hardware, or instrument problem did cause the alarm to activate but the plant is actually OK.
Decide	How successful is the operator at diagnosing the underlying cause for the alarm? How efficient is the operator at *deciding* the appropriate course of action to remedy or repair the underlying cause for the alarm?
Act	How effective is the operator at *acting* to implement the needed remedial actions to manage the upset?

We examine the event. From the left of Figure 2.8.2 we see a failure occurrence. This starts both the time-to-manage-faults clock and the fault-tolerance time clock. If the time-to-manage clock ever meets or exceeds the fault-tolerance clock, the damage will be done. We see a delay during which the raw field sensor detects the event, the detection is transmitted to the control system, the control system picks it up, and it is relayed to the calculation and display utilities. At that point, we assume that the fault has crossed the alarm activation point and an alarm is initiated. This is the operator's first formal notice that something is amiss. This also starts the hypothetical fault tolerance time clock. The operator might have already been concerned about the plant conditions or other aspects of operation; however, when the alarm activates, concern is confirmed. Our operator must now find out what actually happened (*S*ee).

Our operator must find out what the process will do if a remedy is not found or not found in time (*U*nderstand). Our operator must then determine what to do to remedy the situation and decide how to do it (*D*ecide) and finally actually make whatever changes and other actions are needed to implement the selected remedy (*A*ct). Once these actions are complete, and not before, the process will have a chance to respond. If the actions are correct, the process is on its way to recovery. The race is started. If the process can respond to the operator's corrective actions before it enters the region of improper or dangerous operation, all is well. The objective is to always ensure that the time to manage fault is always less than the fault tolerance time!

Sure, we can see what is going on here with time, but how does seeing it control it? We would like to be able to ensure that the operator will be able to remedy the abnormal situation before the process escalates into an incident. Let's take a closer look at the time line at the top of the figure. The system internal diagnostic time (or processing time for the signal) is controlled solely by the equipment in place. This is mostly fixed. The time for the process to appropriately react to the operator's corrective action is predetermined by the fundamental nature of each part of the process. The only thing left is to ensure that the time for corrective action be as short as possible. This is the procedure that alarm management will be using to assign proper alarm activation points. The clock time for the operator to complete each of the four tasks above is directly impacted by good alarm system design. Yes, training and procedures have their place. The underlying skill set that operators possess will certainly affect effectiveness at process management. But this is expected as part of good enterprise management. They are important parts of the infrastructure. Adding to that, a good alarm system design can reduce the time required to manage alarms, thereby increasing the likelihood that an abnormal situation does not escalate into an incident. Everything must work effectively and efficiently together.

Alarm Activation Point and Time

The previous discussion is important to the understanding of time. However, there is one more aspect to consider to drive this home. *The bottom line of alarm management is to provide the operator with enough time to properly respond to alarms.* The way we do this is to set the alarm activation point at a value sufficiently before the abnormal situation becomes an incident. The value is determined in order to provide the operator with enough time to do what is needed to manage the problem. Therefore, instead of blithely assuming that once an alarm activates, the operator will be successful in short order, we must activate the alarm early enough so the operator has the time to plan and act and the process has the needed time to respond.

As true as this discussion is, there is a problem with implementing it. Nowhere in the PCS alarm configuration is the ability to set an alarm activation point based on time. All alarms must be activated by a process measurement value or computed value that can be compared to a process measurement. We need a way to transform time into process conditions. Chapter 7 provides the technology and tools for doing just that.

2.9 WHY ABNORMAL SITUATIONS ARE IMPORTANT

The word *abnormal* sounds fairly innocent, does it not? After all, we could have used *accident* or *disaster*, and those certainly evoke an image of great loss. But let's not kid ourselves: abnormal situations often lead to those other terrible events. And they do cost plenty. Extensive investigation has determined that cumulative incident costs range from about $100,000 to over $1,000,000 per site per year. This cost is incurred year in and year out. One plant incurred 240 unplanned shutdowns in 1 year. For others, a cost of $8,000,000 is not unusual for a "bad" year. Annualized figures from the insurance claims show that industry claims exceed $2,200,000,000 per year in the Americas due to equipment damage alone. The estimate for lost production ranges around $10,000,000,000. This is really adding up.

The message is these costs are enormous and completely out of line with the modest costs of improving plant operability. Every plant that does not engage in such planning is betting against nature and the inherent limitations of man's constructs. You probably do not need this book to tell you how much abnormal operation costs your company. You certainly do not need this book to suggest that much of these costs are avoidable. By now you might already suspect that alarm system effectiveness is one of the key ways to improve operation.

Let's have a look at one of the ways that plants might recognize their current state of healthy operation. Figure 2.9.1 depicts a graph of daily production rate over a year's time. It is almost trivial to construct. Yet it is a powerful communicator. The only data needed is the actual total production, as a percentage of plant functional capacity, for each day of the production year. It is graphed by calculating the number of days that the plant produced at given production for each percentage capacity. For example, let us say that the plant illustrated in the figure had produced for 40 days (cumulative) at a rate of 95% of capacity. At that 95% point on the horizontal axis, the vertical height of the curve will be 40 (for the 40 cumulative days at 95% production). In this example, there were no days during which the plant produced at a rate of less than 60%. The sum of all of the number of days (ordinate values) will add up to 365. The area under the curve multiplied by the plant capacity is the total yearly production.

Observe the several markers on the horizontal axis (abscissa). The one at 100% is the plant operating limit. Moving down in capacity, we see markers at operational constraints, planning constraints, and plant operating target. The region between the operational constraints and the plant capacity limit is where the plant seeks to move into as a result of optimization efforts. Our figure indicates the plant has selected the 95% rate to be the nominal target. This target governs the setpoints for all the controllers. Much of the inability of this plant to achieve a higher rate of production comes from an inability to manage production variability and from an inability to manage abnormal situations. The associated losses come to between 3% and 8% of capacity. For most plants, this amount determines the plant's ability to be profitable or not. Alarm management is not only about safer and better production. It can contribute to the very viability of the enterprise. The decision to spend even 10% of those losses on alarm improvement will likely return significantly more in results.

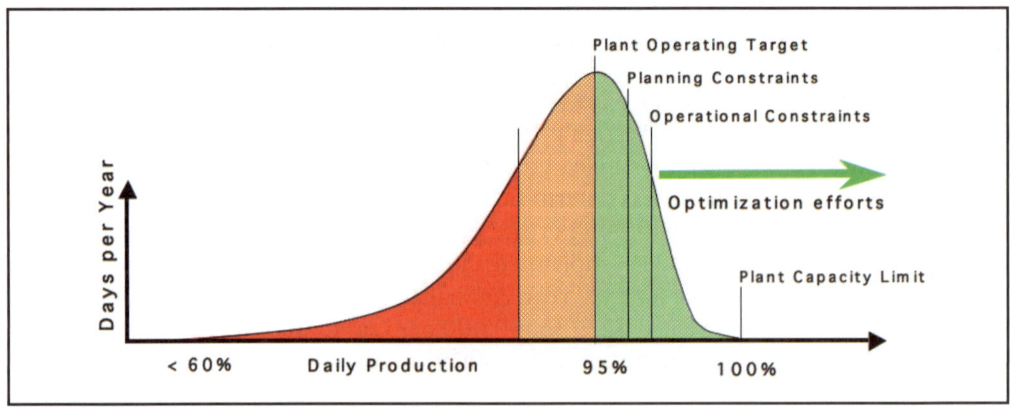

Figure 2.9.1. Mapping current production

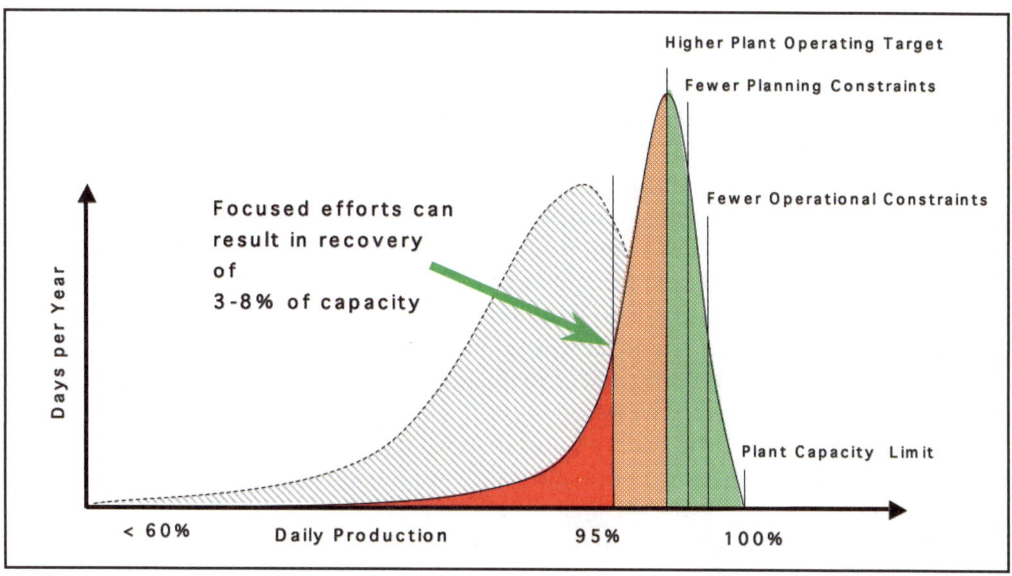

Figure 2.9.2. Improved production through better process management

In closing this section, we turn to what production might look like after successful efforts have been made to improve operations. Figure 2.9.2 shows what the production profile might look like for this case. What we see is the plant's success at moving the plant operating target and consolidating more production around that target, from an earlier value of 95% to a new value of about 97%. The better control also contributes to a "sharpening" of the peak, which provides improvement, well beyond the apparent 2% peak movement. This improvement is gained by more careful attention to better

- management decision making and enterprise controls;
- maintenance for the instrumentation and controls equipment;
- maintenance of the heavy production equipment used in the manufacturing process;
- alarm system design and performance;
- operator procedures and training.

2.10 MESSAGE OF ABNORMAL SITUATIONS

The day-to-day familiarity with the plethora of problems and ready appearances of limitations in all aspects of the enterprise usually lull most of us to focus on what we believe we can manage. Often it will move us away from thinking and doing of those things that can actually make a difference. It is easier to think about concrete things like alarms. However, the real issue is that the operator really needs to know what is abnormal in his plant. We use alarms to tell him this. But that is not the message. The message is that before we look for alarms that might be needed, we first must identify the abnormal situation and then, and only then, find the best alarm(s) to use to notify the operator. For most of us, this is a new way of doing business.

Most of us avoid changing the way we do things. But avoiding change can prevent progress and deter resolution. There is an illustration about a frog that comes to mind and bears telling. As you read this piece, please remember that it is an illustration, not a prescription for injury to small animals.

Take a frog and place it in a cool pan of water. Then gradually heat the pan. If it is done so very slowly, by the time the frog actually senses the danger of the situation, it is already too warm for it to be able to jump out.

On the other hand, take that same frog and place it into that very same pan, now with very hot water, and the frog will immediately jump out and thereby save itself.

So, drop the image; keep the lesson.

We are more like the frog in cool water than we'd actually like to admit. For most, it takes the shock of a dramatic event to awaken them. Unfortunately, that shock comes at a very high cost. The study of abnormal situations and the technology of abnormal situation management is a way to envision our exposure to danger and take steps to manage the unacceptable risks there. This awakening does not have to come through dramatic revelation. Quite the contrary, it comes from the powerful yet simple discipline of not accepting "business as usual." All of this might sound rather aloof and academic. Let's get concrete.

State of Control Loops

When we examine the operational state of the actual process controls for a typical plant, we often find that a full 20%–25% of the controls are not in their normal designed-to-function mode. This means that some of the simple regulatory loops are in "manual" mode instead of "automatic." It means that some of the cascade (or ratio or other coordinated) loops are in "local" instead of cascade, ratio, or track mode. It means that some of the advanced control loops are in other states instead of their designed mode of "calculate." All of these out-of-design situations are easy to identify. Your plant team can easily come up with similar numbers. All of these out-of-normal controls contribute directly to process operational difficulties. This leads to increased abnormal operating situations. These lead to alarm activations.

The causes for these out-of-normal control loops are not hard to find. Examination shows problems with controller tuning, valves and sensors, inadequate equipment design, and process operations being outside of engineered parameters but not necessarily out of equipment design limits. The difference between being outside of engineering limits and outside of design limits has been confusing for operators. The natural tendency is to think that as long as the design limits are respected, everything else should really work properly. Yet it is the engineering limits that are used to shape controls and procedures.

Understanding the concepts of abnormal situations has become a very useful methodology for garnering important lessons from the ordinary. It has been a way to reacquire the promise of better things to come. It will be an important tool for understanding and managing operations in the future.

The Magic in a Control Loop

How is it that a single feedback control loop, properly applied, can produce such a taming of the wild nature of Nature? What is it about such a loop that employs such a simple construction to produce effective operations over a very wide range of operating conditions? And all of this amazing efficiency really has little to do with the actual details of the algorithms used or the clever methodology of implementation. The amazement is fundamental. And once you see it, it should forever change how you understand and use the control loop.

A control loop works its magic by *moving* unwanted disturbances from where they cause harm or other unwanted effects to another place where they are not much in the way! The loop moves the trouble from where we do not want it to another place where it does not matter what it does.

OK, your imagination and interest is up here. But this probably still does not make enough sense for you to say, "Ah, I understand." So let's take a look at some simple examples.

Temperature Control Example

Consider a temperature control loop. Say we have a large drum partially filled with a liquid that we need to keep at an elevated temperature. Liquid is constantly flowing into and out of the drum. The drum is heated by steam that flows through a submerged coil. The steam flow is managed by a control valve. The control valve is managed by a controller that senses the temperature of the liquid in the drum. Let's look at what goes wrong. The liquid temperature is "disturbed" by changes in inflowing liquid, changes in outflowing liquid, changes in the pressure of the drum, changes in the weather outside the drum, or changes in the heating value of the steam itself. The place where we do not want the disturbances or variations is the temperature of the liquid within the drum. So to where does the temperature control loop move them?

The control loop manipulates the steam flow in order to respond to temperature variations. Regardless of what might be the cause of the temperature variations, the steam flow is manipulated to provide additional heat if the liquid in our drum is too cool or less heat if the liquid is too warm. The act of control is *moving* the temperature variations in the drum liquid into flow variations in the steam system. The steam system is designed to handle those variations.

Flow Control Example

The steady management of materials flow is important to the uniformity of many manufacturing processes. The way this is done is to measure the flow and then use a flow modulating device (usually a valve for liquids) to increase or decrease the flow until it is in agreement with the desired flow value. On the surface, this seems like the practical way for this to work. Below the surface, where the actual activity takes place, we have an entirely different story.

All efforts to modulate a flow in one part of a series process will upset the flow both upstream (at another processing station where they have the same type of flow controller that we have) and downstream (again at another processing station where the steady management of flow is important and might be a different value than our upstream one). We have a paradox. Engineers have long recognized this. To avoid it, the design includes places for the flowing material to accumulate between the flow control loops. They call those places surge. Surge usually takes the form of a separate surge drum, an accumulator, or simply a space at the lower end of processing or separation vessels for the liquid to ebb and flow.

In this manner, plant designers move the flow-affecting disturbances, which get in the way of good production, into level variations, where it is presumed they will not bother the process very much. This is a good plan. Unfortunately, sometimes the designers of the level control systems aren't privy to the master plan. Most control designers value level management in the same way the other control loops are valued. As a direct consequence, levels are controlled to be as close to a desired value as they can. But a desired value almost always has no process importance. In the end, we have tight level controls interacting with other tight flow controls. These loops fight each other by trading disturbances back and forth. Each one works very hard. None can work as needed.

Let us end this example with the suggestion that plants examine all of their level control loops for proper construction. Where strict level control is needed, the flow must be available to be modulated at will. Where the level is used as a disturbance catcher, then its control structure must be changed to enable managed swings that permit good flow control without resulting in level movements outside the physical limitations of the surge volume. Standard nonlinear level controls work well here.

Message of Moving Disturbances

More often than not, disturbances are being traded around instead of being removed.

The fundamental thing is to examine control loops everywhere to ensure that they are moving the disturbances to where they should be. They are usually passed around to other places where they are still not OK instead of away from the plant production process. There is a rich reward to understanding this concept, locating all the places where the disturbances are ineffectively or improperly moved, and moving them to other proper and better places.

Abnormal Situations in Perspective

As this chapter is brought to a close, let us be reminded that abnormal situations are part and parcel of enterprises. To better understand them—and by understanding, to better manage them—is the benefit.

2.11 NOTES AND ADDITIONAL READING

Notes

1. Abnormal Situation Management Consortium, "ASM Consortium Contacts," http://www.asmconsortium.com/asm/dashboard.nsf/ContactsPublic.
2. Federal Aviation Administration, *System Safety Handbook* (Washington, DC: Government Printing Office, 2008), appendix G.
3. James Reason, *Human Error* (Cambridge: Cambridge University Press, 1990), 89.

Recommended Additional Reading

Kletz, Trevor. *What Went Wrong.* Houston, TX: Gulf, 1994.

Perrow, Charles. *Normal Accidents—Living With High-Risk Technologies.* New York: Basic Books, 1984.

Petroski, Henry. *Design Paradigms—Case Histories of Error and Judgment in Engineering.* Cambridge: Cambridge University Press, 1994.

CHAPTER 3

Strategy for Alarm Improvement

I believe in taking the time to be right—it's a lot faster than being wrong.

—Unknown

The stage is now set for understanding and appreciating the alarm management problem in the context of abnormal situations. The current best practices for alarm system design and implementation will be discussed. This chapter develops a formal structure for the alarm improvement process—the big picture, if you will. It reviews each of the four phases of the work. The phases are further broken down into a total of fourteen tasks. The importance of identifying the owners and stakeholders is stressed. The single, formal, overriding, all-inclusive, and defining attribute of an alarm is set out:

> An alarm is the operator notification of any abnormal event that the operator might miss but should not and requires operator action.

This is the defining test. If it fits the test, you have an alarm candidate. If not, you do not.

Let's take a look at the fundamental purposes of alarms and alarm systems. We will examine the differences between the symptoms of poor alarm performance and the underlying root causes of poorly performing alarm systems. We move on to a formal, staged structure of the alarm improvement process. The chapter concludes with a discussion of the various regulations and standards for alarm management and the importance of humility.

3.1 KEY CONCEPTS

Alarm improvement is intuitive to understand but not to design	Alarm design is based on key fundamental foundations supplemented with solid engineering judgment.
We got into trouble honestly	Before alarm management was understood (early part of the 2000s), we did not understand what the true needs were.
There are excellent guidelines and standards	Alarm improvement advice is in good agreement, and following it properly will lead to valuable benefits.

3.2 HOW WE GOT OURSELVES INTO TROUBLE

In 1960 rarely did the number of installed alarms top 200 per operator. Most operator positions were in the neighborhood of 50 to 100 alarms. Alarm growth since then has been exponential (fig. 3.2.1).[1] Economists and naturalists will quickly suggest that this is what is normally found in an environment where one entity has found a way to dominate all others and is therefore free to grow at a rate proportional to its size. No wonder most industrial manufacturing enterprises find themselves not only overwhelmed by alarms but also lacking in the very guidance and notification that alarms were meant to provide.

It really is no one's fault! Nowhere in this book will you find a finger being pointed. Besides, blame cures nothing except goodwill. We came face-to-face with the deficiencies and limitations of alarming honestly. Alarm systems lost most of their effectiveness during the major controls revolution from single-case instrumentation to distributed, programmable control platforms.

Overheard in the Control Room after a $500,000 Mistake

"You mean that this shutdown could have been prevented just by pressing this button?"

"Yes, but . . . "

"Why didn't it get pressed?"

"There was a lot going on and they just couldn't tell what to do."

"Well, give them an alarm when they need to press the button!"

"But there is already an alarm on it."

"Make it a higher priority! Make it blink! Just don't miss it again!"[2]

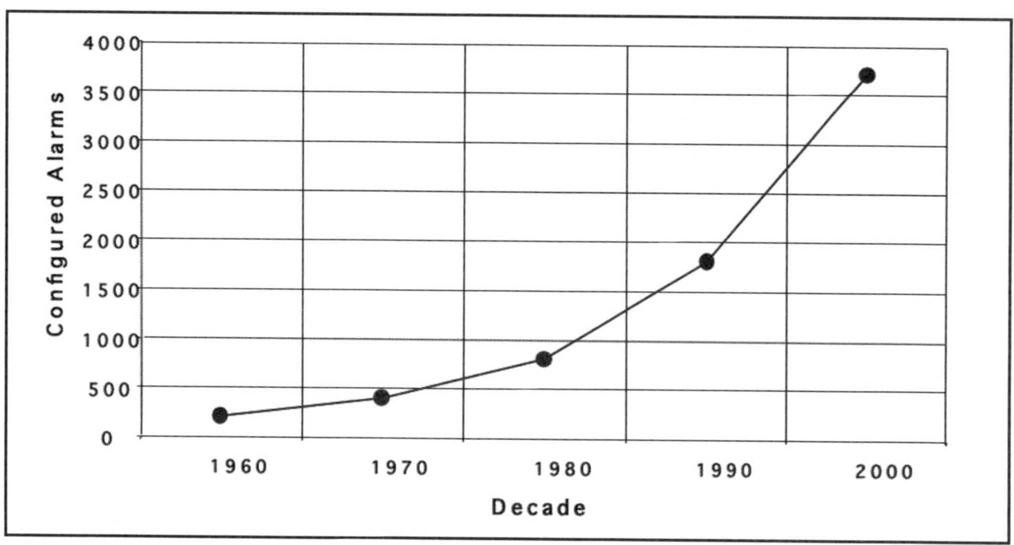

Figure 3.2.1. Exponential growth in configured alarms

Control engineers, electrical engineers, mechanical engineers, chemical engineers, and all manner of fully devoted technicians thought they were doing the right thing when they confronted the enormous list of control block configurations and went through them. There were places for high alarms, low alarms, high-high alarms, low-low alarms, and more and more and more alarms. They made mental calculations about what would be good values and used them. Never for a moment were they aware of the seeds they were planting for enormous trouble down the road. And designers weren't alone. Operators found that this wonderful "notifier" or "reminder" tool was really terrific. They added to them, not realizing what was truly going on. Only recently, as modern alarm design awareness grew, did it become painfully clear that our happenstance, evolutionary approach to alarms was leading into failure.

Controls Technology Evolution

In the mid-to-late 1970s, the central engine for monitoring and controlling industrial processes went through a startling revolutionary change. Overnight (and it was one of the best-kept industrial secrets ever) the Honeywell Company introduced the first fully commercial digital distributed controls system (DCS). Certainly, this paradigm shift was in the works. The Foxboro Company had already split the architecture between the conversion and conditioning of field signals, the algorithmic controls calculations, and the display of information to the operator. It first appeared in discrete pneumatic controllers, then in electronic controllers. The world community was preparing for this to eventually appear in a programmable controller form (PCS) a decade later than it actually did. Not only was the industrial controls community taken by surprise, end users were as well.

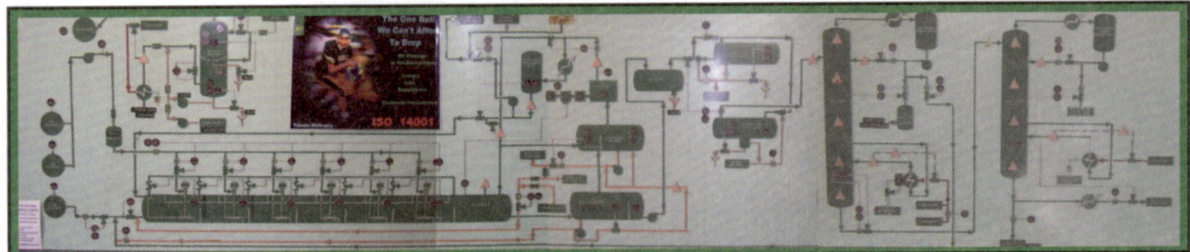

Figure 3.2.2. Panel board semigraphic

So instead of taking 10 years to get ready, the industrial manufacturing community was catapulted into the new way of doing things in as many months. DCSs offered an easy way to do what before had been difficult, hardware intensive, and inconvenient to modify. We were ready for the idea of flexibility but mostly unprepared for how to actually use it. When the DCS came, we did what we almost always do—we tried to use the new technology to do mostly what we did using the old technology. Alarms were one of the direct carryovers. Except now, instead of each new alarm costing nearly $2000, its cost was nearly nothing. Now there was a way to let the operator know about all the irregularities and deviations that might cause a problem, not just the critical few. And we did. The average number of alarms per operator exploded from about thirty or so to hundreds. Over the intervening years, they grew to thousands.

The industrial controls community fell into the trap of "scale." A principle exists in the history of science and technology that deals with how to identify the right path when the right path is unknown. This principle is called the *principle of scale*. Very simply said, it states that if we are on to the correct solution to remedy a difficulty, having more of that solution should improve things. That is, the more we have, the better off we become (before the point of absurdity, of course). For alarms, the more we had, the more we needed, and the worse off the operator's use of alarms became. Clearly, alarms needed to be rationed. Our unbridled use of alarms failed the principle of scale.

How We Think

How are we able to identify what is the truth from what is not? How can we know whether a particular solution is the way to go? These are easy questions to ask. Their

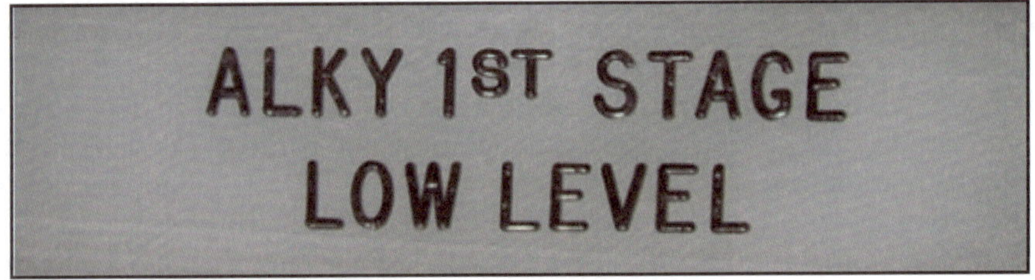

Figure 3.2.3. Engraved window for dedicated panel-mounted alarm

answers have vexed thinkers and decision makers for as long as there has been thought and important things to be decided. As you might imagine, definitive tests do not exist. Fortunately, there have emerged powerful principles to guide us.

Principle 1: Simplicity. If there are two competing theories or explanations for something important, we believe that the simpler one (that does not require suspension of intellect, of course) is the more correct one.

For those who might be familiar with the competing theories for the orbits of planets before Copernicus, there was an attempt to force circular orbits to work. To do so required modifications of a circular orbit with the addition of an epicycle. When one epicycle did not do the job satisfactorily, more were added. Eventually, even with a very large number of them, the orbits could not be adequately described. Enter Newton and his simple yet elegant elliptical orbits, and the problem was solved.

Principle 2: Not by subtraction alone. Most problems show themselves as too much of this or too much of that. It is tempting to approach the resolution by working hard to reduce the "too much" into "just right." That is OK, except, more often than not, the "too much" came about because there was a real problem. The "too much" was arrived at by trying to solve that problem. Cutting back will reduce the "too much" but often leaves the real problem unresolved. Therefore, as you approach a problem with too much of something (e.g., too many alarms), try to find the real need and add it to the cutting-back solution.

Reduction of configured alarms must be accompanied by profound increases in maintenance, better ways of recognizing abnormal plant operations and conditions, and more effective options for managing plant upsets and disaster avoidance. (Subtraction would be reducing alarms; addition would be more maintenance, better recognition of abnormal situations, and more ways to handle problems.)

Principle 3: Scale. Scale simply suggests that if one is on the right track to solve a problem, then more of the solution should solve the problem even better.

Here is where the traditional use of alarms got us into most of the trouble we are now in.

When there were few alarms, alarms were added to accommodate (or rather prevent) specific problems. Each new alarm was thought to prevent a specific problem. And for a while—a short while, unfortunately—it worked just that way. Then, as operations became more complicated and PCSs had the ability to add more and more alarms easily and without much cost, plants quickly passed the demarcation between useful alarms into the territory of not useful. Folk wisdom provided the nudges that told us to add more alarms because our operators were missing important things and the plants were getting into trouble because of it.

The Way Forward

Now, at a time when the use and configuration of an alarm costs next to nothing and can be done with laser-like precision, we find ourselves with the knowledge that less is better. Just when it is so easy to produce alarms, we recognize that what to alarm and how to do it are the most important considerations. To get value from an alarm system, it must be carefully managed. Mastering this management is the purpose and content of this book.

3.3 THE ALARM MANAGEMENT PROBLEM

The problem with most plant alarm systems is that they do not work. Let's examine both the symptoms and underlying causes for this situation.

Symptoms

The following list, sad to say, is only a limited amount of evidence that current alarm systems do not perform as needed for good enterprise management:

- Plantwide philosophy or guidelines for alarms do not exist.
- Operating procedures are not associated to alarms.
- When alarms activate, the operator is not sure what to do about them.
- Minor operating upsets produce a significant number of alarms.
- Some alarms remain active for a long period of time.
- Significant operating upsets produce an unmanageable number of alarms.
- Seemingly routine operations produce alarms that serve no useful purpose.
- Alarms occur without the need for operator actions.
- When nothing is wrong, there are active alarms.
- Alarms often occur at a rate that operators cannot keep up with.
- Too many alarms have a high priority.
- Lots of alarms switch in and out quickly and provide no value to the operator.
- Important alarms are missed during incidents.
- Records or accounts of why alarms were designed the way they were are poor or missing.
- Alarm testing procedures or records are poor or missing.
- There are no plant controls on alarm activation points or methods and practices for defeating alarms.

This list looks much like a "laundry list." That is the message. Any plant with even one of the above deficiencies is already experiencing operational difficulties. Any plant with three or more of the deficiencies is ripe for serious operational problems—and sooner rather than later, unfortunately. Seriously, if your plant is one of them, you have already had some warning incidents and lots of close calls. The "big one" is probably just around the corner. Please take heed.

Root Causes

Symptoms are useful for identifying problems. Root causes help us to understand how to fix them.

- No plantwide philosophy or design exists.
- Operator training is inadequate.
- Operating displays are poorly designed.
- Inadequate attention is paid to plant practices and procedures.
- Alarm limits and priorities specified at design are rarely visited for validity during actual plant operating conditions.
- Controls platforms (e.g., PCSs) contribute to overuse of alarms and overly complex alarm system designs.
- Alarms are constantly added (e.g., HazOps, etc.) but rarely deleted.

A Good Alarm

I often personify the task of figuring out what a good alarm might be by asking, if you were walking down the street and knew that there were alarms along the way, how would you identify a good one? Well, easy. They have all the following attributes:

Relevant. They are not spurious or of low operational value.

Unique. They do not duplicate another alarm.

Timely. They do not actuate too early before a response is needed; nor do they actuate too late for the operator to properly respond.

Prioritized. They indicate the importance to the operator for successfully managing the underlying problem they announce.

Understandable. They have a message that is clear and easily recognized.

Guiding. They facilitate the solution.

So Many Alarms, So Little Time

OK, so you've seen this type of phrase over and over again, mostly to illustrate a humorous aspect of one form or another. Here it really fits. Not enough time is where the operator and alarm system fail to match up. So forget whether spending a lot of time with an alarm system is a good use of operator resources; if the operator had enough time to spend, then any alarm system would do. Alarm improvement could be relegated to the "nice to do" column on a long to-do list. Sadly, this is not so. A poorly designed and poorly performing alarm system interferes with good process operation by demanding too much operator time, thus reducing efforts spent at other operations improvement tasks. It also interferes by failing to provide sufficient guidance during times of need. It is these two hits, so to speak, that are the heart of alarm problems, and this heart is itself at the heart of plant operability problems.

Benefits of Rationalization

In the article "How to Perform Rationalizations,"[3] the author lists the top ten reasons to rationalize alarm systems. The summary is provided in Figure 3.3.1. The list is right on the mark. We take a moment to observe that there are several items in the list that go well beyond a limited alarm-repair approach. This book shares that synergy. Let's review a few points.

Black Screen

Item 6 proposes that a "black screen" is an ideal. Black is, of course, a placeholder for ensuring that no alarms are active that should not be. Easy to say, but what it means is profound. It suggests that alarms should represent only abnormal situations or conditions that are relevant and that can be fixed and are to be fixed by the operator in the

1. To make sure operators get the information they need when they need it; formatted so they can recognize what's happening, its importance, and the appropriate action.
2. To reduce the total number of alarms to the minimum necessary to operate the plant safely and efficiently.
3. To prioritize alarms by importance or significance in terms of risk (safety, environmental, operational, and cost) and in relationship to other alarms.
4. To improve presentation, organization, and availability of alarms for safe and efficient operation of the facility and for effective troubleshooting.
5. To reduce the number of alarms occurring during abnormal conditions to the minimum required to diagnose or identify the indicated condition.
6. To approach the "black screen" concept: Bring the number of alarms during normal operation near zero by reducing standing, chattering, nuisance, and transient alarms; optimizing alerts; managing pre-alarms; improving field instruments; identifying process issues; etc.
7. To validate all alarm parameters including action, setpoint, deadband, test frequency, etc.
8. To verify alarm performance parameters including detection time, required action, appropriateness of action, required time to perform action, training, procedures, help screens or manuals, etc.
9. To identify process design issues and faulty field instrumentation.
10. To document the alarm system for internal use and regulatory compliance.

Figure 3.3.1. Top ten reasons to rationalize your alarm system

time frame of a shift or so. Obviously, spare and shutdown equipment equipment should be "blacked out," as well as broken things that await lengthy repair cycles, or relegated to the never-to-be-repaired situation.

Process Issues

Item 9 is prophetic. It advances the very real idea that alarms can be useful to expose fundamental flaws or gaps in the design and operation of a plant. You will discover that after a successful alarm improvement execution, new alarm activations can take on a dramatically different role. Yes, they warn the operator that something is abnormal. But now that warning is meaningful against the very fabric of process operation. We get to ask a real question after each activation: Did this alarm activate because of a normal abnormality, or was its cause deeper? Did it reflect defects or limitations in process design, raw materials, operational practices, training, or another aspect that goes right over the top of simple natural variations? It is like an MRI for the plant.

3.4 ALARM ACTIVATION PATH

We know what an alarm is. We understand that they are vital and necessary for good plant operation. But what equipment is allowed to produce alarms? How important is it to make sure that only qualified alarm producers are allowed? What do you think? Should we keep on the safe side of things and simply decide that only mainline controls platforms be permitted to produce them? Or is there another way to approach this?

Figure 3.4.1. Alarm activation path

There is a very simple yet elegant answer. It is called an alarm activation path. If we have one, then the mechanism that produced and delivered the alarm to the operator is qualified. If not, then forget about using this mechanism to notify the operator about conditions that must be treated as alarms.

In order to have a (qualified) alarm activation path, the following three things must exist:

1. A timely, secure, and reliable way to monitor the chosen aspect in the plant (physical variable, virtual construct, or other deterministic aspect of need)
2. A timely, secure, and reliable agent that compares the chosen aspect against a valid criterion to produce the alarm condition of (a) IN ALARM or (b) NOT IN ALARM
3. A timely, secure, and reliable way to transfer the alarm condition to the operator

Any equipment that has all three of these characteristics can be considered a qualified producer of alarms. Be the equipment a mainline brand of DCS, PLC, or a special purpose-built apparatus, it will be alarm qualified. The rest is up to convenience and preference.

3.5 THE GEOGRAPHY OF ALARM MANAGEMENT

All alarms have a home. Their place is in direct association with a single, tangible element in the plant. If it is a process-variable alarm, then it is associated with a flow, a temperature, a pressure, an analyzer value, a level, or other physically measured aspect of the plant. If it is a calculated or inferred value, then it is linked mathematically, but nonetheless linked, to a fabricated variable in the PCS. No matter what the alarm link is, that link refers to a physical location within the plant. It is this "geography" that we now examine.

Plant Area Model

Get ready for an important subject. It is so important that if you get it wrong, you will not even know anything is amiss until the final stages of alarm redesign implementation. It is truly one of those "pay me now or pay me later" situations. Only the later "payments" might be so costly that they won't be affordable. And then, just as your careful alarm design is poised to reap real benefit to the operation, you'll not be able to deliver. The reason for that failure has a lot to do with how your plant's component labels are constructed in the PCS. If the labels are constructed to incorporate clear information as to exactly where in the plant it is located and with which equipment it is associated, then your ability to modify alarm parameters, as the plant operates, can be easily and clearly managed. Otherwise, all of your management must be done from large lists of explicit equipment and situations. We call that process "management by enumeration." It is not a good way to do things.

A plant area model is all about geography and labels. Refer to Figure 3.5.1 for an illustration of the concept. The figure depicts a single plant. In this typical plant, there

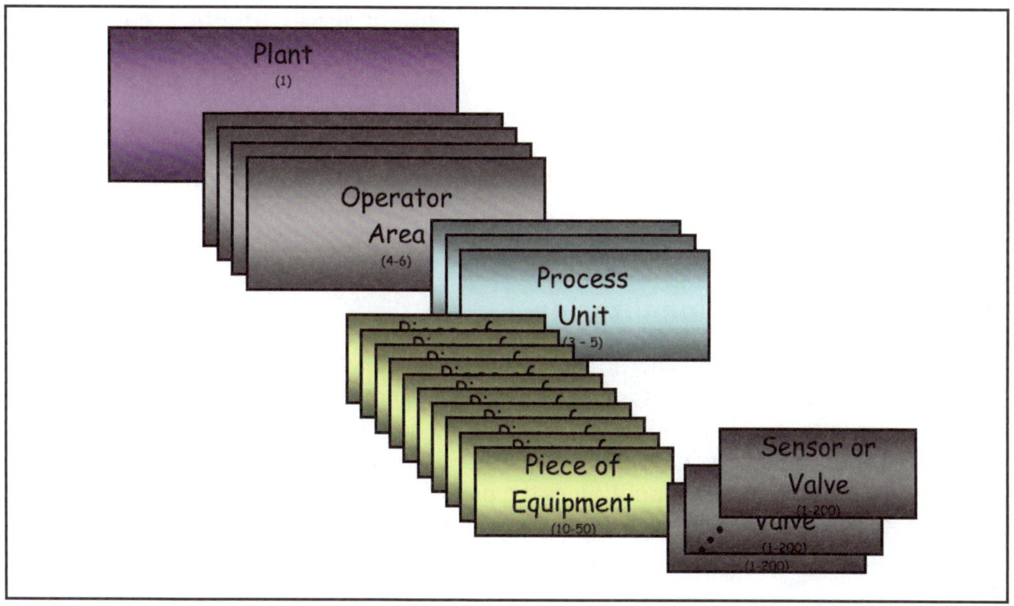

Figure 3.5.1. Conceptual plant area model

are four to six operator areas. Within each typical operator area, there may be three to five or so individual, significant process units. Each process unit might contain ten to even fifty individual pieces of equipment (e.g., pumps, heat exchangers, towers, etc.). Within each piece of equipment, there might be between one and several hundred individual sensors, valves, and so on.

But the plant area model is really not about numbers. It is about location and relationship to other locations and pieces of equipment. (Note: A tag is a label given for each unique entity appearing in the PCS that is labeled and later used for control or computation or view.) For example, if the tag contains the area model, it will be easy to manage alarms. Consider the following example flow controller tag.

Table 3.5.1 shows how the tag was constructed. There are likely many tags in the reactor mixer, all with a unique label and number. Plants need not use letters to identify geography. A tag can also be constructed of only numbers, after a few letters for a

FC-1-3-REACT-MIX-03452				
Plant ID	Operator ID	Process Unit ID	Piece of equipment ID	Sensor, valve, or element ID
1	3	REACT	MIX	03452
1	3	124	15	03452

Table 3.5.1. Construction of a tag using a plant area model

prefix. For situations with a fixed format, the dashes or spaces are not needed. Hence our example tag for the flow control loop shown in Table 3.5.1 might just as well read FC131241503452.

We are now ready to make use of this capability. Suppose that the mixer in the reactor was shut down and no alarms were needed there. It is a simple matter to disable all alarms with tag FC1312415 x x x x x where the five Xs refer to all numbers for all such tags in the database.

Smallest Area of Rationalization

You have certainly heard the rhetorical question, how does one eat an elephant? Please, no harm to elephants implied—it is just a saying. Well, how does one eat a huge proverbial elephant? One bite at a time. And that is exactly *not* how rationalization is done! Rationalization is all about redesigning an alarm system so that it works for the operator. Hence the operator, or rather, an operator position, is the unit of implementation. Please read carefully, because without care, it might appear that we are splitting hairs (to reference another proverbial metaphor). We must not *implement* any results unless it is an entire operator area.

> No significant alarm redesign results should be implemented for any plant area smaller than an entire operator area.

In practice, an operator area means one operator and his PCS with all related screens, input devices, alarm display devices, and other ancillary equipment within easy reach and in close sight. However, in some unusual circumstances, a single operator might be using a PCS or other similar system to manage one part of the responsibility and another clearly distinct interface for the rest. That other interface might range from an old legacy panel board to a completely different vendor's PCS equipment. In this case, the smallest unit of rationalization would be either one or the other, so long as they are completely separate in the operator's mind.

Design is entirely distinct and separate from implementation. Certainly, each alarm can be selected, designed, and configured as an individual entity. While you would always want to keep in mind that this alarm is for an abnormal situation that usually involves more than just a single, very narrow aspect, nonetheless, each alarm can be examined on its own. However, once the design is done, the implementation (i.e., the changing of the PCS configuration, the operational controls, the procedures, the graphics, and the training) must be done at the same time for all alarms in the operator area.

The reason why is very simple. It all boils down to trust. If some alarms were rationalized and some not, how would your operator know which were which? How would the operator know which alarms had the proper activation point, the proper priority, the proper remediation information, the proper relevance? There would be no way. Hence there could be no trust that any particular alarm was well designed. And, unfortunately, the alarm system would be as detrimental then as it was before—probably worse.

3.6 ALARM IMPROVEMENT TEAMS

Like your favorite sports team, alarm improvement teams are built. Like sports teams, each player has a role to fill and experience and expertise to bring to the game. Unlike sports teams, alarm improvement teams depend less on physical prowess than experience and expertise. Like successful sports teams, the players will need to function as a team.

Alarm improvement teams (or alarm project working teams) are composed of a unique cross section of the plant. The key members come from operations—the owners of the alarm system. Members also come from the frontline support area, including technology and maintenance. Finally, members come from plant infrastructure areas of health, safety, environmental, and training. The construction of the teams is not at all about political correctness. Each of these areas is intimately involved in the safe and effective operation of the plant. Each member has an essential role. The members will have important and productive input into the alarm improvement process. Figure 3.6.1 depicts the composition of the overall team.

Representation

At first glance, you might wonder how ten or more individuals would be properly utilized over the course of a project. Like team sports, the playing team is a lot smaller than the entire bench. The bench team has about ten or so members. They are asked to participate in the periodic reviews, but for most of the work, they are involved on an as-needed basis. The "A" team is operations and maybe a controls participant. As work progresses and as issues or problems come up, those who are needed are asked to join the work for the brief time it takes to resolve the particular issue or concern.

Process Operations

The primary representatives come from the specific process area that is being worked on. Experienced operators will review the specific operational nature of each abnormal

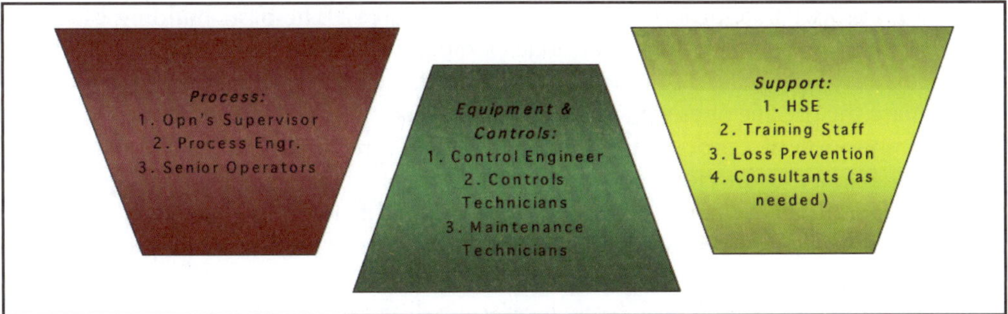

Figure 3.6.1. Plant representation on alarm improvement teams

situation and make recommendations to the team. Enhancing that experience will be the expertise and experiences from process engineers and designers as well as operator supervisors. As a collective group, they are asked to represent the alarm owners in the particular plant process area.

Their experience is augmented with the following two groups.

Equipment and Controls

It is the job of the equipment and controls people to keep the mechanical and controls infrastructure in good working order. The controls people will be the key implementers of any new alarm designs that relate to the process control and information systems. The mechanical and repair people monitor the equipment, respond to appropriate maintenance requirements and operational concerns, and keep everything in proper working order. According to plant custom, they may be charged with proposing equipment improvements as well.

Support

A health, safety, and environmental (HSE) plant group, training, and loss prevention might seem to be "odd couples" in the alarm improvement game. Please remember that alarm improvement is broader than alarms alone. In order for operational improvement to be delivered, it will be vital to reach into all related aspects of the plant infrastructure that support the operator. This will include the areas represented by this part of the team.

Let us not forget that each enterprise has aspects that benefit from the expert experiences to be had from consultants. If your plant has used them, do not forget that they may have a limited but important role in alarm improvement.

Local Teams

Local teams will be doing the hands-on work of redesigning the alarm system. There will be one for each separate process area (e.g., operator area). The basic building block for teams is a process area work team. The areas of representation are the same as the general team shown in Figure 3.6.1. However, since all of the teams will look similar, we will differentiate the specific ones with their relevance label. Figure 3.6.2 illustrates the combined composition of a process area work team. The three components will come from personnel assigned to or closely involved with the particular process area (e.g., operator area) under improvement. Therefore, while some of the individuals may be shared between various operator areas, most of the process representation and equipment and controls representation will be very local to the unit in question.

Each of the process areas will likely have some differences. However, the differences are mostly ones of assignment—that is, who is assigned to what plant area. The nice

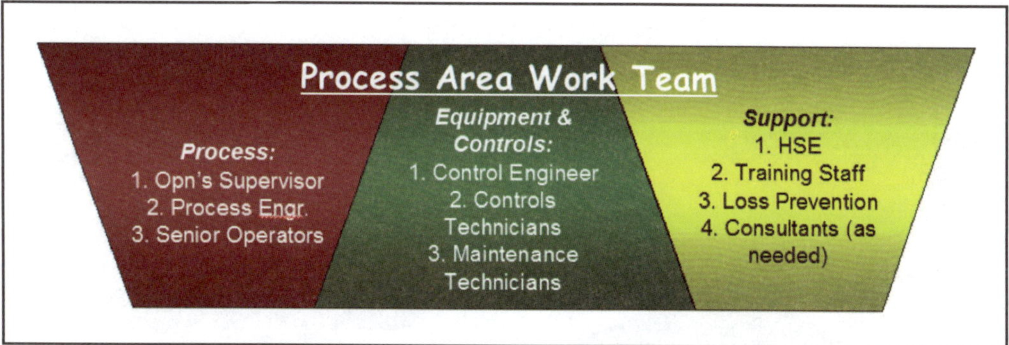

Figure 3.6.2. Process area alarm improvement team

thing about this is that if one or more individuals from one plant area are unavailable and it is critical for work to progress, individuals from related or adjacent areas should be well qualified to stand in their place.

Site Team

The site team is the main steering group for the alarm improvement work. The composition would involve similar skills and breadth as the process area team. Rather than starting again from scratch, we generally draw on representatives from each existing process area team to also serve on the plant site team (fig. 3.6.3). This is for two

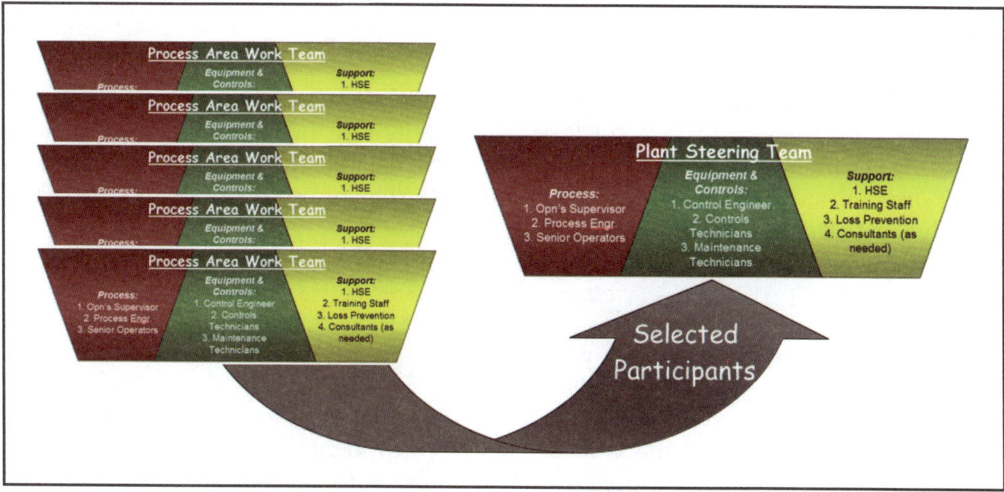

Figure 3.6.3. Formation of a site alarm team

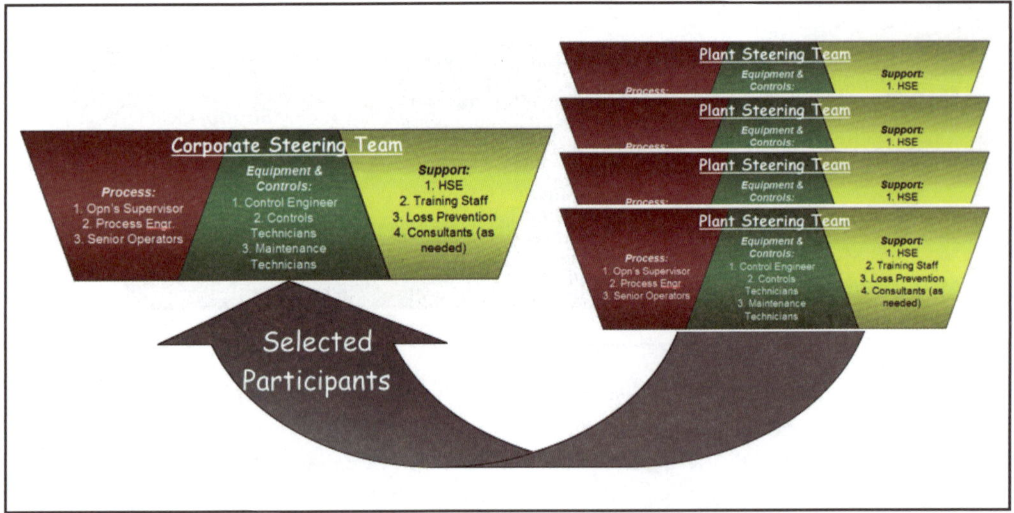

Figure 3.6.4. Formation of a corporate alarm team

reasons. First, there are only so many individuals with the skills and interest at any given plant. Too many projects can stretch them too thin. Second, and probably more important, having individuals participate on the site team fosters sharing of experiences between the process area projects and encourages a degree of uniformity and efficiency to the work.

Large Corporate Team

For larger enterprises that have a number of individual sites, and for much the same reasons as having a sitewide team, a corporate-wide team is formed. Members of this team are mostly key members of the individual site teams. The degree of control of this team over individual site work is governed, of course, by the culture of the corporation and individual enterprises.

Figure 3.6.4 illustrates how the individual plant teams work into a corporate team.

3.7 ALARM IMPROVEMENT PROJECTS

Improving your alarm system is going to look like a lot of other engineering projects at your plant. It will involve the familiar basic steps of identifying the need, obtaining ownership and buy-in, creating specifications and developing a scope and costs, funding, execution, and post audits. Other than requiring a full understanding of what makes a successful alarm system design, the rest is pretty conventional. This is not to say that it is matter-of-fact. Alarm improvement is a serious job. You'll need the right technology and the right talent to make it happen.

A piece-by-piece breakdown of the production process for alarm improvement is shown next. Each part is fully discussed in the appropriate chapters later in this book. The phases and tasks in the alarm improvement process are outlined below:

Phase I—Problem awareness and solution framework

Task 1—Assess the current situation.

Task 2—Develop an alarm philosophy.

Phase II—Alarm redesign

Task 3—Operationalize the philosophy.

Task 4—Develop the new alarm system design.

Task 5—Perform "housekeeping" repairs and modifications.

Task 6—Perform the alarm rationalization.

Task 7—Design the enhanced techniques.

Phase III—Implementation

Task 8—Reconfigure the PCS.

Task 9—Incorporate the alarm design into operating procedures.

Task 10—Incorporate the alarm design into plant training.

Task 11—Incorporate the alarm design into the remainder of the plant's infrastructure via established MOC practices.

Phase IV—Continuous benefit

Task 12—Perform routine follow-up alarm system performance studies.

Task 13—Investigate alarm flood events, process upsets, and near-hit incidents.

Task 14—Make modifications and improvements as indicated by the follow-up studies and investigations.

Guidance for how to successfully do the work comes from the experiences of others and the expectations of appropriate standards and regulatory sources. This chapter continues with the current standards and regulations. It provides good news.

3.8 STANDARDS AND REGULATIONS OVERVIEW

The sole purpose of this section is to be informative. No reliance should be placed on it as representing any practice, regulatory requirement, safety requirement, or any other safe or effective practice. No representations are offered that should be construed to be legal advice, safe operation, or any other advice. Moreover, we will not discuss any regulation or practice set by the various states within the United States or any other geographical or political location for the rest of the world with the exception of those explicitly mentioned by name.

Best Practices Summary

What follows is a very brief summary of the expectations and recommendations put forth by the known and generally accepted guides and recommended practices. Note that some are still in a development stage. However, enough is known to include them here. There is comfortably large agreement between all. The following is a summary of the key points:

- All alarm design and improvement work should be preceded by the formulation of a design basis (called the alarm philosophy).
- All alarm activations will announce to the appropriate operator(s) with appropriate sounds, messages, and other visual indicators placed on operator interface equipment.
- All alarm activations will be announced to the operator in time for effective management of the abnormal situation.
- All configured alarms will include a measure of its importance (priority) derived from its impact on predetermined plant aspects or the time urgency of properly managing the alarm condition.
- All alarm activations require explicit, primary operator actions.
- Alarm performance measures have been developed and should be used for preimprovement assessments, as a check for effectiveness of redesign work and as advice for periodic improvement.
- Effort should be expended to ensure that alarm documentation is available to the operator during plant operation.
- Alarm parameter modifications should be used to ensure alarms reflect plant state changes.
- Alarm management is a life-cycle process: an ongoing plan for monitoring and evaluating alarm performance is expected with the result that the alarm system will be modified to ensure adequacy.

Key Messages

After a decade of serious focus, the "messages" are now quite clear:

- The ability of operators to effectively monitor and respond to PCS alarms is key to effective production management.
- The existing and emerging "standards" are functionally close in interpretation and guidance; there are no inconsistencies or philosophical differences of any consequence.
- Manufacturing enterprises are being held accountable for the effectiveness of all of their alarm systems; fines and lawsuits are now directly linked to alarm performance.
- Controls equipment and software providers are expected to facilitate the best practices for alarm management; if the customer or user cannot implement proper alarm design as a result of equipment or software limitations, technology providers are now aware that they may be open to charges of joint liability.

Expectations of Alarm Users

Adequate monitoring of abnormal situations by manufacturers (i.e., users of alarm systems) is an explicit responsibility. It is expected that their alarm system will function as a key safety element assisting the operator to mitigate the escalation of plant abnormal situations. In some countries, adherence to accepted industry best practices can be used as an affirmative defense against negligence (i.e., for alarm design and use).

Expectations of Platform Providers

Controls (e.g., PCS) equipment and software providers are expected to provide a reliable capability for alarm management. This means that manufacturers who use the equipment and software "as intended" should be able to use that equipment straightforwardly, without paradoxes or other difficult-to-understand or hidden aspects. Moreover, all user choices normally made during setup and configuration need to be well informed and the resulting effects readily understood.

Guides, Standards, and Regulations

There are very few explicit regulations for alarm system design and operation. It is too new. There are studies and practices currently under construction. And this work is extremely encouraging. First of all, serious and productive work is occurring on both sides of the Atlantic and in other important locales worldwide. That work is driven by very similar understandings, philosophical underpinnings, and end effects. Consequently, the results are expected to converge to align together with generally understood best practices. There will be differences, but those differences are expected to be ones of how far to go, not what to do. This is important.

Regulations

With the exception of brief mention and coverage of OSHA 1910.119, there are no actual regulations in place as of early 2009.

3.9 PROPOSED REGULATIONS

Department of Transportation (United States)

The Pipeline and Hazardous Materials Safety Administration[4] has proposed specific alarm management-directed modifications for pipeline controllers (i.e., hands-on equipment operators). Changes to 49 CFR Parts 192, 193, and 195 include clear and specific requirements around situation awareness for the controller, alarm activation, and response. Moreover, this work includes attention points to controller fatigue and more effective shift turnover.

Areas of Interest: Situation Awareness and Alarm Management

Effective controller selection, training, and procedures should be in place to provide "appropriate timely data, a control system designed to aid in the prompt identification of abnormal conditions, and an understanding of the controller's authority to take appropriate actions."[5] (Author's note: This is a clear reference to the concepts in chapter 5.) Simulators or noncomputerized simulations for training in the recognition of abnormal situations will be required.

Operators must (a) routinely review alarm and event displays, (b) periodically audit alarm configurations and handling procedures to provide confidence in alarm signals and foster controller effectiveness, and (c) develop performance metrics with particular attention to response to abnormal operating conditions. Areas of needed improvement include display graphics and alarm management. Review of ISA 18, EEMUA 191, and API RP1167 is specifically encouraged to ensure operation according to best practices.

Key Takeaway

This proposed regulation represents the first public recognition of the importance of ISA 18 and EEMUA 191 and the expectation that they will set the standard of care. Alarm management is now a specific requirement for responsible operation of industrial enterprises that have the potential to affect public safety and environmental quality.

3.10 STANDARDS AND GUIDES

There are no standards published as of early 2009. However, EEMUA 191, a design guide, serves as a de facto standard. It was first published in 1999 (revised in 2007) and includes the consensus best practices then in use in the United States and Britain. No work or experiences since then have cast any shadow on this design guide. In October 2003, the ISA 18 Committee (renamed from ISA SP-18) was formed by the Instrumentation, Systems and Automation Society (ISA). ISA 18 is expected to issue during 2009. Also, in October 2003, NAMUR first published its guide NA 102 in Germany.

EEMUA 191

EEMUA 191, *Alarm Systems: A Guide to Design, Management, and Procurement*,[6] is produced by the Engineering Equipment and Materials Users' Association (EEMUA), based in the United Kingdom. The guide is neither a standard nor a regulation. That said, this document represents a watershed. The technical content and intent of purpose of this publication contains the backbone of almost every standard and regulation that we expect to see in the near term. Therefore this document is a very important resource for understanding what is needed in appropriate alarm system design and operation. Moreover, those who use this document to construct their alarm assessments and guide their planning for improvement can have confidence that most of the important aspects have been considered. This is not to say that simply incorporating all the practices and recommendations of Publication 191 will ensure a good alarm system design. This will not happen for several basic reasons. First, Publication 191 raises a number of alternatives from which practitioners must select the most appropriate for their plants. Second, there is much left to be read "between the lines" that needs to be discovered and understood. Third, it places all the responsibility on the equipment user, none on the equipment provider. However, using it as a guide will ensure that most of the appropriate user issues and concerns have been visited. The most important aspects—and the most impactful—are summarized next.

The Operator as the Unit of Interest

Once a process upset goes beyond the automation equipment's ability to resolve, it is up to the operator-in-charge to either step in, if not already involved, or otherwise escalate efforts at resolution or management. The alarm system must fit with the operator: message delivered, message understood, adequate focus on problem, assistance in remediation, and verification of results.

The Basis for Metrics

All performance metrics for alarm systems are based on the current best understanding of the capabilities of the human plant operator. The rate of activations, the diminution of annoying and unnecessary activations, and the requirement to identify urgency of activations must respect how fast and how effectively a human operator can act. All design metrics for alarm configuration are designed with the objective to control and manage alarm activations during operating conditions.

Metrics

Alarm metrics are divided into two components: performance and design.

The performance metrics (sometimes referred to as activation or dynamic metrics) lay out a guide for how often alarms can activate and what is the ratio of importance to rate of activation. For example, alarms are expected to activate no more than five or six times per hour, on average. However, the rate for high-priority alarm activation should be less than five per 8-hour shift. Take special notice that at these rates, the operator will be totally engaged in responding to alarms—no time is left over for any other duties.

The design metrics (sometimes referred to as configuration or static metrics) represent an extraordinary effort to lay out a design guide that should produce a working alarm system capable of meeting the *performance* metrics just discussed. Most of this guidance has been derived experientially. To date, field validation is consistent. As examples, the suggested total number of configured alarms, of all types, for a single-operator station is limited to 1000 or so. Of those, twenty total should be of the highest priority, 5% or less for the next, 15% for the middle level, and 80% for the remainder. Following these guidelines usually results in alarm activations that tend to meet the necessary performance guidelines.

Due Diligence

In the absence of published standards and valid regulation, special attention to comply with EEMUA 191 should represent good-faith efforts to comply with state-of-the-technology alarm system design.

Key Takeaway

EEMUA 191 provides a framework that should result in appropriate, fit-for-purpose PCS alarm systems. It should facilitate the adoption of an industry-wide effort for successful deployment and use of PCS alarm systems. A word of caution: EEMUA 191 may not do enough to realistically relieve the operator of a potentially heavy burden to locate and understand alarm management documentation, including procedures and linked concerns relating to plant condition, in time to be of value. Critically, you the reader should note that this guide must be supplemented with lessons and wisdom from experienced alarm management practitioners.

NAMUR (Germany)

NAMUR Publication NA 102, *Alarm Management*[7] (in German and English), is produced by the User Association for Automation in Process Industries, located in Germany. Much along the same lines as EEMUA 191, it is meant to be a guide for German controls equipment manufacturers and users. In contrast to EEMUA 191, it is a concise document without examples or extensive appendices, yet it covers most if not all the same bases. On the other hand, this recommended practice goes notably beyond EEMUA with requirements for advanced alarm filtering and processing. Whereas EEMUA mentions the possibility, NAMUR strongly suggests it.

The important additions, over and above EEMUA 191 will now be reviewed.

Time Synchrony

The importance of the controls equipment to properly time-stamp alarm events is stressed to the point of requiring cross-equipment time synchrony to ensure the ability to adequately understand the true time order of events and actions.

Plant States

Whereas EEMUA 191 suggests the opportunity to link plant states to alarm and process conditions, NA 102 lays out expectation to affirmatively do so. Moreover, the linking to plant states should include the ability to modify alarm conditions as well as provide plant-state-relevant guidance to the operator. Explicit suppression of cascade alarm activations is expected.

Operator Interface

The operator interface is a vital part of effective operator communication and hence good alarm management. This quotation frames the intent: "The operator must be given the best possible support for the fast recognition of alarms."[8] It requires the operator interface to be designed and configured to take explicit account of human requirements and limitations. Moreover, it includes discussion of use of color and flash and ability to readily distinguish relevant, important information. Effective linking of displays and operator navigation is covered; proper delineation of operating areas and their relationship to alarms is vital.

Manufacturers' Responsibility

Specific mention is made of the responsibility of controls equipment and PCS software providers to provide the architecture and capability to support the recommendations for effective alarm management. Examples include the following:

- modifiable alarm prioritization
- linking of alarms to alarm management information
- viewing and filtering of alarm events (historical) as well as alarm states (current condition)
- enhanced cause-effect visualization support
- multimedia capabilities of operator stations (to include a wide range of operator support information)
- linked horn control per operator and per control room
- distinct separation of process alarms from system alarms
- complete screen display or refresh time of less than a second
- enhanced configuration of alarms, including advanced functionality, via agents rather than individually or in small groups via enumeration

- alarm database integrity and modification management
- explicit control of alarms via suppression groups and other area model–related architectures

Due Diligence

In the absence of published standards and valid regulation, special attention to comply with NA 102 would represent good-faith efforts to comply with state-of-the-technology alarm system design.

Key Takeaway

NA 102 dictates a framework that explicitly includes significant advanced process management knowledge and suggests the full use of that knowledge for contemporaneous operator support. The expectation is to advance an industry-wide responsibility for aggressive design, comprehensive implementation, and diligent use of alarm systems.

ISA 18

In 2003, the ISA[9] convened a working committee to draft recommendations for a practice or standard for PCS alarm systems. Their work is built on the recommendations of EEMUA 191 with the intent to bring those practices up to date with current understandings and experiences obtained by the process industries in the years following 1999. The scope of this work to date closely follows EEMUA 191 without any substantive modification or amendment. The major contribution is in the areas of broadening the explanations for many of the practices and adding clarifications to areas that were sketchy before. Particular attention is paid to the contents and scope of the alarm philosophy (the design basis for each plant's alarm system), the incorporation of the methodology used by the major alarm analysis tool providers for assignment of priority, the expanded list of nomenclature, and the inclusion of aspects specific to the United States.

Due Diligence

ISA 18 has been approved by the working committee and has passed on to the formal standards board for final approval and publication. Until published and then accepted by the American National Standards Institute (ANSI), this work has no standing in the United States. Until then, alarm practices and systems relying on it possess no special benefit at law.

Key Takeaway

When ISA 18 issues as a recommended practice, anticipated in 2009, it passes to ANSI for recognition. If recognized, it is likely to be recognized also by the Occupational Safety and Health Administration (OSHA) as the recommendation of duly constituted standards bodies. OSHA's policy is to then declare the practice to be the nominative requirements for PCS alarm management. See the next section.

OSHA (United States)

The OSHA[10] is a formal aspect of the U.S. federal government and has no current specific standards, guidelines, or recommendations that govern alarm management for the general process industries. However, there are two general areas currently used by the agency (and enforced) and one area that is suggestive.

The Code of Federal Regulations (CFR)[11] includes the following:

- CFR 1910.119 (e)(3)(iii): "1910.119: Process Safety Management of Highly Hazardous Chemicals; (e): *Process Hazard Analysis*; (3): The process hazard analysis shall address (iii): Engineering and administrative controls applicable to the hazards and their interrelationships such as appropriate application of detection methodologies to provide early warning of releases. (Acceptable detection methods might include *process monitoring alarms* [emphasis mine], and detection hardware such as hydrocarbon sensors.)"
- CFR 1910.119 (j)(v): "1910.119: Process Safety Management of Highly Hazardous Chemicals; (j): *Mechanical Integrity* . . . shall apply to the following process equipment . . . : (v): controls (including . . . alarms, and interlocks)."
- CFR 1910.119 Appendix C 5 (nonmandatory): "1910.119: Process Safety Management of Highly Hazardous Chemicals; (appendix C): Compliance Guidelines and Recommendations for Process Safety Management (nonmandatory): 5: *Operating Procedures and Practices*: . . . Operating procedures will include specific instructions or details on . . . what alarms and instruments are pertinent if an upset condition occurs."

Basically, alarms and alarm systems are addressed only by their possible use; where they might be used, if chosen to be used; and for what, again, if chosen for use. There are other sporadic mentions of alarms, largely related to emergency plant alarms, mostly operator-initiated, that broadcast an ongoing serious situation like a fire or runaway equipment.

With all of this discussion, there appears to be no actual requirement to use alarms at all. There are lots of situations where a process alarm may be used, but none that requires one. Moreover, nowhere is the design of any alarm specified or indicated.

Due Diligence

OSHA-established requirements, including guidelines and practices, have the force of law. At present, the OSHA coverage of alarm management is very light and generally of little directive value. No reliance on them should convey much toward an affirmative suggestion of proper plant alarm management, either in fact or at law. Take note, however, that significant findings of noncompliance (i.e., negligence) and substantial fines have resulted from investigations of industrial incidents that find OSHA violations.

Key Takeaway

When ISA 18 issues as a recommended practice (as it is now anticipated), it is likely to be recognized by OSHA as the recommendation of a duly constituted standards body. OSHA's policy is often to declare the practice to be the nominative requirements for alarm management. That usually results in an "incorporation by reference." It will then take its place with the other OSHA regulations as lawful requirements.

HSE (UK)

The Health and Safety Executive (HSE)[12] is basically an OSHA parallel in the United Kingdom. HSE's operating methodology is a bit different. The standard for compliance is whether an enterprise is using what is known as "generally accepted best practice" for the concerns at hand. The existence of a generally accepted best practice then has the force of a requirement. EEMUA 191 is such a practice. This is not to say that the HSE cannot amend, modify, clarify, or extend the practice; it can. If a given plant's design, implementation, maintenance, and use of process control alarm systems generally meets the industry-accepted best practice, that plant is deemed to be in good-faith compliance.

Due Diligence

Plants and enterprises are encouraged to understand that some geopolitical governing entities (countries, states, and the like) rely on the role of *generally accepted industry best practice*. In the absence of statute, reliance on it should be sufficient to suggest responsible operation with respect to proper alarm management.

Key Takeaway

To date, the HSE has remained reluctant to set specifics regarding alarm practices. Rather, they appear to be more interested in documenting and sharing incidents and lessons learned. This is entirely consistent with the role of supporting "enforcement" by keeping the public aware of industrial practices.

EPRI (United States)

The Electric Power Research Institute (EPRI)[13] develops standards and practices for the power industry. An alarm management standard has been prepared but is not available at this time to the general public.

Remarks

There are a few other guidelines and such that are under consideration. The American Petroleum Institute (API) has announced work on standard API 1167, yet work is very formative. The Department of Transportation has announced the intention of incorporating the work of EEMUA and ISA into public law for certain pipelines.

Alarm management is in the very fortunate position of having a worldwide consensus on what should be done to deliver effective technology to manufacturing enterprises. So far, efforts have been able to build on each other's success with the result of a very strong understanding of what should be done and how to alarm it. It is expected that any regional variants that eventually emerge will be the result of the need for eventual harmony with the broad range of existing statutes and requirements for safe and effective operation. The differences are expected to be small.

The goal of this section is to reinforce the reader's confidence that this technology is well understood and agreed to by knowledgeable practitioners around the globe. Whether you choose to use EEMUA, NAMUR, or ISA as a reference, the major thrust and guidance will be quite similar. You are on solid ground!

3.11 CONCLUSION

This chapter has covered a lot of ground. Plants got into alarm trouble naturally enough. We know how to recognize the problems and who works to improve things. The alarm improvement path is a well-defined process. Most of it is good engineering practice. While there are few regulations, the existing standards and guides are in considerable agreement. This reinforces that the technology is in place and is expected to work.

In the next chapter, we will find out how to assess your plant's alarm performance. As you know from this chapter, alarm improvement is much more than fixing those alarms that are not performing.

3.12 NOTES AND ADDITIONAL READING

Notes

1. B. Hollifield, D. Oliver, I. Nimmo, and E. Habibi, *The High Performance HMI Handbook* (Houston, TX: PAS, 2008), appendix 4. http://www.pas.com.
2. Chris Wilson, *The Operations Puzzle: The Alarm Management Piece*, TiPS, Inc., White Paper (Austin, TX: TiPS, 2005).
3. William L. Mostia Jr., "How to Perform Rationalizations," *ControlGlobal*, 2005, http://www.controlglobal.com/articles/2005/320.html.
4. U.S. Department of Transportation, Part II Pipeline and Hazardous Materials Safety Administration, *Pipeline Safety: Control Room Management/Human Factors; Proposed Rule* (Washington, DC: Government Printing Office, 2008).
5. U.S. Occupational Safety and Health Administration, Washington, DC, http://www.osha.gov.

6. Engineering Equipment Materials Users' Association, *Alarm Systems—A Guide to Design, Management and Procurement*, EEMUA Publication No. 191 (London: EEMUA, 2007), 91. http://www.eemua.co.uk.
7. NAMUR, *Alarm Management*, NAMUR Publication NA 102 (Potsdam, Germany: NAMUR, 2005), http://www.namur.de.
8. Ibid.
9. Instrumentation, Systems and Automation Society, *Management of Alarm Systems for the Process Industries*, ISA 18.02 (Research Triangle Park, NC: ISA, 2008), http://www.isa.org.
10. U.S. Occupational Safety and Health Administration, Washington, DC, http://www.osha.gov.
11. U.S. Government Printing Office, "Code of Federal Regulations (CFR): Main Page," http://www.gpoaccess.gov/cfr/index.html.
12. Health and Safety Executive, Sudbury, Suffolk, UK, http://www.hse.gov.uk.
13. Electric Power Research Institute, *Alarm Management and Annunciator Applications Guidelines* (Palo Alto, CA: EPRI), http://www.epri.com.

Recommended Additional Reading

American Petroleum Institute. RP 1167 Alarm Management (as of spring 2009, proposed draft standard). http://www.api.org.

Campbell-Brown, D. "Alarm Management: A Problem Worth Taking Seriously." *Control*, July 1999, 52–56; August 1999, 62–66.

Gould, Jeff. *Institutionalizing Alarm Management*. Matrikon White Paper. Edmonton, Alberta: Matrikon, n.d. [ca. 2007].

CHAPTER 4

Alarm Performance

Data without generalization is just gossip.

—Robert Pirsig (writer)

You've suspected or even known for a while that something might be very wrong in the control room. Operators and others have been making noises about things for a while. You sense that whatever might be wrong is costing your company a lot. By this time you're pretty well on your way to being convinced that the alarm system is part of the problem or, more likely, most of the problem.

The goals of an alarm performance assessment (APA) is to get your "arms around" the performance of your present alarm system. After you have done the APA, the impact of an alarm philosophy (if one exists) should become clearer. It should be clear that operators face alarm problems. You will know which alarms cause an unusually large number of activations. You will know when the rates of alarm activations exceed any reasonable use to your operators. When the APA is complete, you will have a very useful benchmark of the existing performance of the alarm system. You can use this information to justify alarm improvement programs and provide a basis for "before" and "after" studies to quantify improvement programs.

Production facilities accumulate considerable alarm and operational data that can be useful for best-practice benchmarking. This information is used to place plant data into perspective with what's expected for good operation. The usual metrics include the number of tags, number of tags with alarms, and priority of alarms. Metrics for the activation data include priority spreads, alarm activation rates, patterns of activations, number of standing alarms, time to acknowledge, time to clear, number of disabled alarms, chattering alarms, repeating alarms, and the class of correlated or otherwise related alarms.

This chapter covers alarm system measurements and how they are compared against the best practices for alarm system design. The APA results represent typical base-level assessments for both the configuration of the alarms and the performance of alarms during plant operation. Alarm performance evaluations are also used extensively for forensic-type analysis. The data from these analyses are used not only for evaluative purposes but also to support the detailed process of examining individual alarms for purpose and to identify plant areas where alarm design or performance suggests problems. These will be covered in section 4.8, "Advanced Activation Analysis" (and extensively in chapter 10). Sample results are derived using some of the very effective and easy-to-use commercial "tools."

4.1 KEY CONCEPTS

Role of alarm performance studies	Unless the alarm system has been improved using the fundamental concepts and best-practice methodology, *alarm system performance studies are only useful to demonstrate that there is a problem that needs solving.*
Caution	Do not even think about improving alarm design by responding primarily to any alarm performance results from your plant.
Benchmarks exist	Alarm configuration and performance benchmarks are well accepted and, when used in the design process, can be extremely useful predictors for a productive alarm system after improvement.
Tools are essential	Few PCSs have enough onboard tools to either effectively analyze alarm performance or provide a good resource to support alarm rationalization.

4.2 ALARM PROBLEMS

Chapter 1 listed the symptoms of alarm performance problems. If your site had even one of the symptoms, it could be improved by alarm redesign efforts. If your site had more than one symptom, alarm improvement should be a priority for your to-do list.

In their article "Why Invest in Alarm Management?"[1] UReason provides a few more symptoms of poor or inadequate alarm systems to add to the list. Here are a few additional ones to consider:

- high alarm loading rates (alarms per time unit, alarms per operator, alarms per event)
- incidents or near-incidents where operators missed key data provided (or not provided but should have been) by the alarm system
- too many high-priority alarms

- track lost of alarm (i.e., activation) points or why they were set there in the first place
- not knowing when alarms were last tested
- large number of defeated alarms

In this chapter we measure the extent of these problems or symptoms your plant might actually have. This will be part of the assessment of the alarm system. We are not going to directly compute numerical values for many of these problems. Instead, we prepare queries to the data that, based on the way they are constructed and the answers we get back, should enable us to infer whether we have a problem and, if so, its severity and importance.

4.3 ALARM PERFORMANCE ASSESSMENT

Alarm performance assessment is the activity of finding out how well the alarm system is doing its job of assisting the operator. Two important databases form the core: configuration and performance. The performance data are the minute-by-minute captures of alarm activations, acknowledgments, operator actions, and the like. The configuration data are all aspects of alarms that appear in the PCS control configuration. This chapter reviews obtaining the data, analyzing the data, and assigning conclusions to the results via an APA, also referred to as an alarm performance review (APR) or, in some cases, alarm objective analysis (AOA). This is a powerful process. It will open the door to the ways that operators, your operators, are assisted in maintaining good operation or are prevented from understanding and responding to operational difficulties. It will also point the way forward to alarm improvement.

But you need to get beyond the numbers. The data contained in operator event logs, production interruption reports, incident reports, maintenance logs, and operator experiences can provide a wealth of information that will support valuable conclusions regarding current operating effectiveness. For example, if the operator logs reflect a shift without unusual problems or events, but the alarm data contain records of significant activity, there is the suggestion of unnecessary nuisance alarms. These data provide the situational backdrop to compare the performance data results against actual plant operational situations. Within this context, the data analyses can possess the relevance so important for engineering evaluation.

4.4 ALARM METRICS AND BENCHMARKS

In this section, we introduce the full set of alarm metrics or benchmarks. These metrics, or measurements and targets, are grouped into two categories: *configuration* (i.e., instructions to the PCS [DCS and PLC, etc.] on what alarms are present and how and when to activate them) and *performance* (the time history of actual alarm activations that occur during plant operations). The discussion about metrics is equally at home regardless of the manufacturing modality (continuous, batch, discrete). However, due to the nature of batch operations, there can be an abnormally high number of alarms.[2] Remember that it is the operator who must be able to manage. So irrespective of what the plant is designed to do, if the alarm system overloads or fails to lead the operator, we end up with trouble.

Metrics for alarm assessment provide a great deal of information about an alarm system's underlying strengths and weaknesses. But aren't they obvious? And the answer would be yes, they can be, but they are not unique. The metrics that one decides are important might not be the same that others choose. But even if they were mostly the same, chances are that they will not be measured in the same way. When they are not the same or measured the same way, comparisons can be tedious.

Why Have Metrics?

Metrics might sound just like another fancy word. In this case, it represents a powerful concept. Metrics together with benchmarks permit us not only to measure what we have but also to compare those measurements with accepted practices in ways that provide important benefit. The metrics are yardsticks that are laid down across the data. The benchmarks become the targets that can be used to identify where things are doing well and where performance might need improvement.

There are those who practice alarm management who suggest that no configuration metrics are useful. They say that at the end of the day, the only thing that is important is to ensure that the alarm system provides the operator with the needed operational support. They ask that the alarm rate be matched to operators' ability to use the alarms. That statement is absolutely correct. The whole discipline of alarm management is directed

The term *benchmark* originates from the chiseled horizontal marks that surveyors made in stone structures, into which an angle-iron could be placed to form a "bench" for a leveling rod, thus ensuring that a leveling rod could be accurately repositioned in the same place in the future. These marks were usually highlighted with a chiseled arrow below the horizontal line.

Wikipedia

Figure 4.4.1. Benchmark[3]

to that end. However, and this is a fundamental "however," goals do not design anything. Goals are a test for the design after it is done and evaluated. On the other hand, metrics can be and are used as important tools to guide the work. For without useful guidance and strong methods, alarm design is reduced to more trial and error than technically driven design.

With this foundation for the value of metrics, it is important to keep a perspective. There are poorly performing, unimproved alarm systems that meet or exceed the metrics. There are very successful alarm systems that do not come close to the metrics. They are the unusual.

Plant Area of Focus—A Single-Operator Area

We start by clarifying how large a part of a plant these measurements and norms apply to. The basic plant unit we will use from now on is a single-operator area. An individual single-operator area is the part of a plant that has been designed to be overseen and managed by a single (often, inside) operator. In general we find a single-operator area involves somewhere around 250 to 350 control loops, from 1000 to 1500 analog-only measurements, and in the neighborhood of 200 to 2000 digital measurements. Our rationale for considering one operator area to be the basic rationalization unit can be traced directly back to the two fundamental design precepts for proper alarm system design: (1) provide alarms that must be acted on and (2) provide sufficient time for successful action. Both the action and the available time (including thought resources, which might be affected by overload, confusion, etc.) must be allocated by an individual. Most often, that individual is the inside operator. The target metrics are designed to ensure that the operator can manage the regular tasks as well as those that arise during abnormal operation. In short, do the job effectively.

Basic Configuration Metrics

Table 4.4.1 shows the results from an early survey of various industries done by EEMUA. The values are the number of configured alarms for control loops (and most loops contain a control valve, hence that number is often used as representing the loop count), number of configured alarms of analog measurements only (ones that are not used *directly* for control), and number of configured alarms of digital measurements (on-off switches used for indication and perhaps start-stop control). The results are shown for the lowest plants, the highest plants, and the averages of all plants surveyed. Remember, this is a survey! The number of alarms per control loop varied from a high of 6 to a low of 1 with an average of 4. Similarly, the alarms per analog measurements varied between 2 and 0.5 for an average of 1; the number of digital measurements alarmed varied from 0.6 to 0.2 with an average of 0.4.

Experience has found that only the lowest survey result will satisfy requirements. This is depicted by the gray shaded column in the table. The number of configured

Metric	Low	Average	High
Alarms per control value	1	4	6
Alarms per analog measurement	0.5	1	2
Alarms per digital measurement	0.2	0.4	0.6

Table 4.4.1. Configuration metrics survey results (from EEMUA 191)

alarms should be limited to about one per control loop, to about 50% of the analog measurements, and about 20% of the digitals. At these values, the likely number of alarm activations should occur at a rate that the operator should be able to manage and respond to most of the time. The recommended target should come as no surprise. After understanding the need to manage alarm activations, one of the best ways is to control the number of configured alarms.

Now that we have an idea of the total numbers of alarms recommended for the configuration, we turn our attention to how alarms are divided up among the alarm priorities. Remember, alarm priority is only used by the operator to obtain a fast understanding of the relative importance of alarms as they activate. Where competing activations occur, the operator uses priority to decide the order that they will be worked. Table 4.4.2 provides the EEMUA recommended breakdown. You might notice that the illustration shows four distinct priorities. Some plants use more, some stop at three. Refer to chapter 7 for more discussion.

We now have a better idea of the recommended numbers of alarms from these metrics and the priority suggestions (Tables 4.4.1 and 4.4.2) for alarm design targets. Table 4.4.3 provides the current best-practices understanding of the numbers of configured alarms versus the required complexity needed to properly manage them. The average number of configured alarms obtained by following the guidelines from Table 4.4.1 usually ends up around 1000. This value is marked by the bold arrow along the left-hand border of the table. The configuration metrics study (Tables 4.4.1) and the operator responsibilities versus configured alarms (Table 4.4.3) reinforce each other and

Priority of alarm	Number configured
Critical (emergency)	About 20 (total)
High	5% of total configured
Medium	15% of total configured
Low	80% of total configured

Table 4.4.2. Recommended priority allocation of alarms (from EEMUA 191)

Number of installed alarms per operator	Number configured
Under 100	Simple technology OK. Places a large SAFETY load on operator.
100–300	Mixed annunciator and computer alarms can be effective. Need other tools for basic problems.
➡ 300–1000	Need sophisticated tools with powerful logical tools. Philosophy is very important.
Over 1000	Major system. Requires significant investment and industry best practices throughout.

Table 4.4.3. Number of configured alarms versus alarm system complexity (from EEMUA 191)

are consistent. Pay particular attention to the mention that even for this seemingly low number of alarms, additional logical tools to manage alarms are required to do the job well. Refer to chapter 8 for more discussion.

Prior to modern control platforms, most operator areas contained fewer than 100 alarms. The "simple technology" referred to in Table 4.4.3 was, of course, alarm light boxes or lights scattered along the semigraphic of a control board. That "large SAFETY load" refers to a very strong reliance on training, intuition, and unrelenting diligent attention to all process situations and conditions.

Summarizing this discussion, for a well-designed alarm system, we should see somewhere in the neighborhood of 1000 or so configured alarms per operator area. We usually obtain this result by providing one alarm per control loop, alarming at most one-half of the other analog measurements and one in five of the digital signals. Section 4.6 has more details and examples.

Basic Activation Metrics

Activation means alarms becoming active. Actually, this is the key aspect about alarm management. As long as operators are not faced with more alarms than they can properly manage, the plant should be OK with respect to alarm system design. Table 4.4.4 shows the various observed rates and identifies the recommended level by an arrow. Notice that an activation rate of less than one alarm per 10 minutes is judged acceptable. This is not to suggest, even for a moment, that this would be a good value to see day in and day out. Operators facing such rates would have little time left for other duties. Although they could manage the plant at this rate, that management would be primarily reactive. There would be little time or energy left for improvements and optimization. Any operator in this situation would be under constant stress and fairly devoid of confidence and job satisfaction.

Long-term average rate	How acceptable is it?
> 1 per minute	Not acceptable
1 every 2 minutes	Over demanding
1 every 5 minutes	Manageable
< 1 every 10 minutes	Acceptable

Table 4.4.4. Alarm activation rates benchmark (from EEMUA 191)

Rates of alarm activation are important. Their priorities are important as well. Recall that priority provides the single measure of how important the condition reflected by the corresponding alarm actually is. Too many highly important alarms produce extremely tense operation. Table 4.4.5 provides the basic accepted priority ratios for alarm activations. Pay particular attention to the very low expected rate of critical alarms. Moreover, just like the alarm activation rate metric, those shown here are maximums. Coming even close to them on a frequent basis is cause for concern. It suggests improper priority assignment or poor alarm system design. Also note that a primary rationale for the alarm priority ratios in the configuration (Table 4.2.2) is that this spread tends to produce alarm activations more in line with the recommendations (Table 4.4.4).

Now that we have these numbers, how do we get a plant in line with them? Well, that is the story for the rest of this book. Stay tuned. If you can't wait, turn to chapter 7. We now turn our attention to the tools that are used to efficiently produce the measurements for assessments.

Priority	Target maximum rate
Critical (emergency)	VERY infrequently
High	< 5 per shift period
Medium	< 2 per hour
Low	< 10 per hour

Table 4.4.5. Maximum alarm activation rates by priority (from EEMUA 191)

4.5 ALARM ASSESSMENT TOOLS

A tool is a purpose-built aid that can be used to examine alarm data and provide useful answers to questions we want to ask. The tools are software packages furnished either by the controls platform provider or, more likely, a third-party provider. They link into the extensive alarm and operator databases in an automated way. They provide a powerful and relatively simple way to query the data. They also provide the warehouse to store the results of the examinations.

Why Use a Tool?

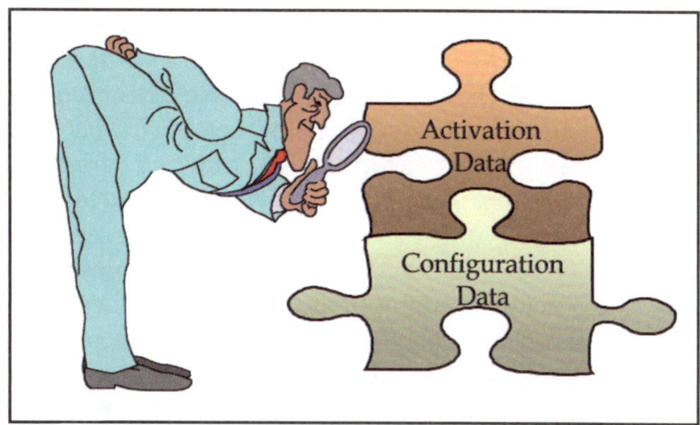

Figure 4.5.1. Alarm tools examine data

A careful analysis of about 3 months of running alarm data is required to benchmark a plant's performance. The first task is to actually archive the data. Depending on how the data are archived, we might be looking at a quarter-million messages. No one wants to do this even once by hand or even using Excel. There are several purpose-designed tools on the market to facilitate this process. All tools archive the configuration (static) data and the performance (dynamic) data to provide some level of analysis and diagnosis and assist in the documentation and rationalization process. Using tools, it will be possible to efficiently evaluate alarm performance both for an initial APA as well as for after-improvement tests and design.

Characteristics of Good Tools

Good tools must be competent—that is, when a query is formulated and driven against the data, the results must be correct. They must be easy and intuitive to use. It makes little sense to be constantly reviewing manuals or using a lot of trial-and-error testing to figure out how some specific task is done. Tools must be capable of tweaking to facilitate repetitive data testing—that is, they should not require the tester to completely reenter all of the search criteria when only a small portion is changed.

The tools must also interface extremely well to the PCS platform(s) used in the plant. This interface will be used to gather the configuration information and format it within its database for later use. The interface will permit all continuously accumulating real-time data including alarm activations, operator actions, and the like into the tool's database for analysis, reporting, and operations management activities. To provide additional utility, we expect the tool to format and archive information in such a way that it can be presented for use in convenient and useful ways. These ways include reports, online advice, and e-mail or other rapid notifications of problems that go beyond the immediate activities in the control room.

Tool Providers

There are three frontline providers and a fourth one of limited value. They are briefly reviewed below in order of their general market share for alarm management use. In addition, several of the PCS vendors provide their own versions of the software that are suited to their own specific platform. The PCS vendors' offerings are not discussed here.

Additionally, some PCS vendors have partnered with one or more of the tool providers discussed below for inclusion of their particular tools with the product. The degree of integration with the platform varies; however, those included tools should perform well.

LogMate from TiPS (Georgetown, TX). LogMate[4] includes an alarm analysis engine, a rationalization utility, a configuration enforcement tool, and an alarm key performance indicator (KPI) engine. The analysis engine supports standard EEMUA-based statistics that can generate contemporaneously or e-mail on a scheduled basis. The rationalization utility allows for calculation of alarm activation point and alarm priority. It also includes capabilities for documentation of alarm settings, rationale, and related operator instructions. Configuration enforcement includes the ability to compare the accepted design configuration to the active configuration to identify discrepancies. Alarm KPIs can be used to monitor trends in alarm behavior. KPI trends may also be integrated with data historian trends for better data interpretation. LogMate collects alarm and event data from a PCS via most current open communication standards including object linking and embedding for process control, open database connectivity, as well as other methods including TCP/IP or serial connections. Data are stored in a Microsoft Structured Query Language (SQL) server database. Users access LogMate through a standard Web browser; hence there is no need for any specialized user software.

AMO-Rt (PAS, Inc., Houston, TX). AMO-Rt[5] is a comprehensive alarm management solution that covers all the key aspects of alarm management, including documenting and defining customer alarm strategies relative to best practices (such as EEMUA guidelines), as well as the necessary tools for implementing, enforcing, and auditing these strategies in real time. AMO-Rt works on all industrial control systems, allows data to be viewed from multiple locations in the plant via PCS, and includes a variety of tools to view and analyze alarm data. These tools include the real-time alarm viewer, event historian and journal viewer, and performance metrics manager to provide timely and accurate access to information regarding KPIs. All of PAS's alarm software solutions are applications of PlantState Suite, their infrastructure to host performance monitoring and improvement products.

Process Guard and Alarm MOCCA from Matrikon (Edmonton, Alberta, Canada). ProcessGuard[6] addresses real-time alarm assessment. It connects to any PCS or SOE platform and consolidates the data within an SQL or Oracle database. The diagnostic package provides a thin-client Web interface that automates qualitative alarm load assessment. The product is firewall friendly and can be deployed in a multilayered corporate network infrastructure. Alarm MOCCA is the configuration, documentation, and rationalization tool. It stores PCS configuration data, engineered values, and documentation. Its primary focus points are to automate MOC and assess control system configuration.

An additional toolmaker bears mention. It provides a decent amount of analyses but is very limited on alarm knowledge warehousing and rationalization support. In the view of this author, this deficiency makes it of limited value for most alarm improvement applications.

iMAC from Industrial Control Software (Matlock, Derbyshire, United Kingdom). iMAC is a product that performs detailed alarm analyses and transfers seamlessly to Excel for more user control. It also provides its own interface to manage and display alarm information including the ability to capture and archive PCS screen shots. It provides limited alarm suppression. It does not support rationalization.

Getting the Data In

We now discuss how the data are provided to the tools—that is, how the data are transferred from the PCS into the tools.

Configuration Data

All the configuration and much of the alarm activation data reside within the PCS. The configuration data are exported from the PCS "build" files in one of many file export formats. So long as each export file is structured consistently and is represented by entries that are separated by appropriate separators, it will be importable to the tool. The actual physical transfer can take place either via a processor-to-processor data link or through the use of portable media (i.e., CD, DVD, etc.).

The key to the process is the ability to construct and then use import filters for the data. These filters are custom matched to the specific PCS and the particular version of software it uses. They are usually provided by the tool provider as part of the product delivery. As each line of the configuration is scanned, the filter parses the individual entries into labeled categories in a highly structured database. Figure 4.5.2 illustrates the import process schematically.

Once the configuration files reside within the tool, powerful search and analysis utilities are used to provide the desired analyses. Refer to sections 4.6, 4.7, and 4.8 for more details. A sample of the imported configuration data is shown in Figure 4.5.3. Notice

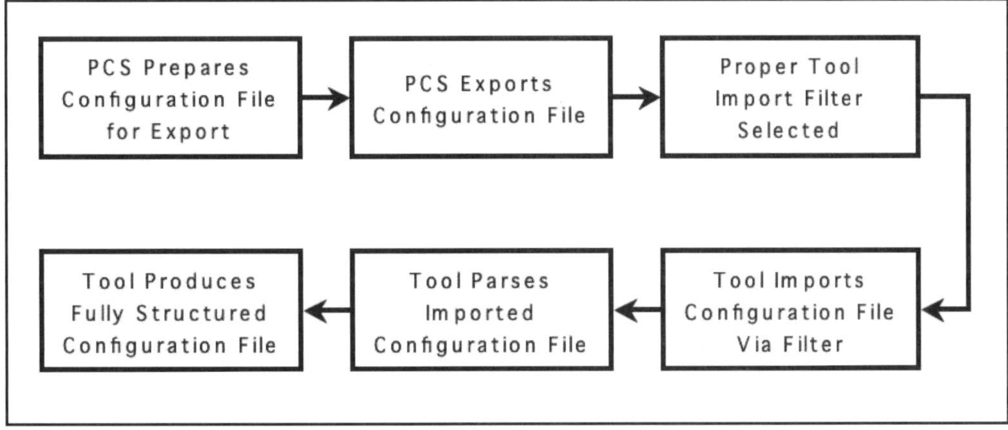

Figure 4.5.2. PCS configuration import process for alarm tools

Results | Details

Details for 'Low::64FC0030':

(1 to 10 of 17)

ID	Tag ▲	Type	Limit	Priority	Inhibited	Severity: Saf	Env	Fin	Time Available	Pri. Method
24	64FC0030	ADVDEV	0.000	Low	False	Low	Low	Low	30	Sum Severity
25	64FC0030	BADCTL	0.000	Low	False	Low	Low	Low	30	Sum Severity
26	64FC0030	BADPV	0.000	Low	False	Low	Low	Low	30	Sum Severity
27	64FC0030	BCLEAL	0.000	Low	False	Low	Low	Low	30	Sum Severity
28	64FC0030	BCLFAL	0.000	Low	False	Low	Low	Low	30	Sum Severity
29	64FC0030	CLEALM	0.000	Low	False	Low	Low	Low	30	Sum Severity
30	64FC0030	CLFALM	0.000	Low	False	Low	Low	Low	30	Sum Severity
31	64FC0030	CNFERR	0.000	Low	False	Low	Low	Low	30	Sum Severity
32	64FC0030	DEVHI	0.000	Low	False	Low	Low	Low	30	Sum Severity
33	64FC0030	DEVLO	0.000	Low	False	Low	Low	Low	30	Sum Severity

Results | Details

Details for 'Low::64FC0030':

(11 to 17 of 17)

ID	Tag ▲	Type	Limit	Priority	Inhibited	Severity: Saf	Env	Fin	Time Available	Pri. Method
34	64FC0030	PVHH	0.000	Low	False	Low	Low	Low	30	Sum Severity
35	64FC0030	PVHI	0.000	Low	False	Low	Low	Low	30	Sum Severity
36	64FC0030	PVLL	0.000	Low	False	Low	Low	Low	30	Sum Severity
37	64FC0030	PVLO	0.000	Low	False	Low	Low	Low	30	Sum Severity
38	64FC0030	PVROCN	0.000	Low	False	Low	Low	Low	30	Sum Severity
39	64FC0030	PVROCP	0.000	Low	False	Low	Low	Low	30	Sum Severity
40	64FC0030	PVSGCH	0.000	Low	False	Low	Low	Low	30	Sum Severity

Figure 4.5.3. Example of a structured configuration database entry (TiPS)

that lines 24 through 40 in Figure 4.5.3 contain the configuration information for only one tag, 64FC0030. Each horizontal line represents a different configured alarm for this single tag. For example (line 32), there is a DEVHI (high value of deviation) alarm with a priority of "Low" that is not inhibited but as yet has a zero value for its activation point. It would seem that this alarm is in construction. Clearly, this is a good example of a bad example of alarm configuration. With seventeen separate alarms available, no wonder more than a few get used.

Activation Data

While the configuration data might be transferred between the PCS and our tools in a casual manner, no such arrangement is appropriate for the performance data. The messages are produced second by second in real time and so must be captured contemporaneously. Before PCS manufacturers opened up their data structures, most of the transfer took place using a printer port on the PCS by connecting a computer to receive the print commands and translate them into files that were then structured and archived. That capability exists today but is rarely used. Current practice is for the tools and PCS to communicate directly through file and database transfer protocols over a network or other bridge connection. The input translation occurs via a similar process to the configuration, except now, since there are a number of differently formatted real-time events coming in, each event type has its own special filter. Figure 4.5.4 depicts this process.

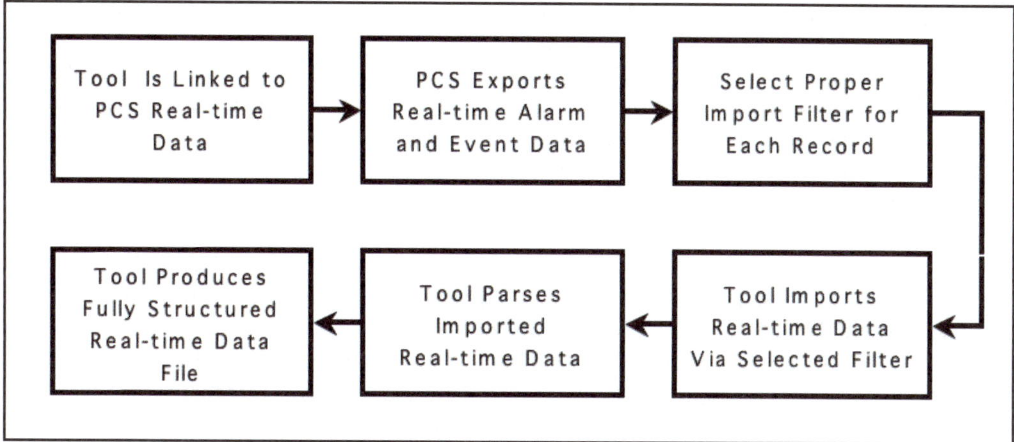

Figure 4.5.4. Real-time data import structure for tools

A sample structured database is shown in Figure 4.5.5. Each line represents a single time-stamped message containing a single event (e.g., alarm activation, alarm clearing, operator acknowledgment, operator manual action, etc.). The generally accepted amount of data for an APA of 3 months actually produces somewhere in the neighborhood of between 100,000 and 250,000 individual entries (lines). Quite a lot!

Now, with the data all imported and structured, we are ready to analyze.

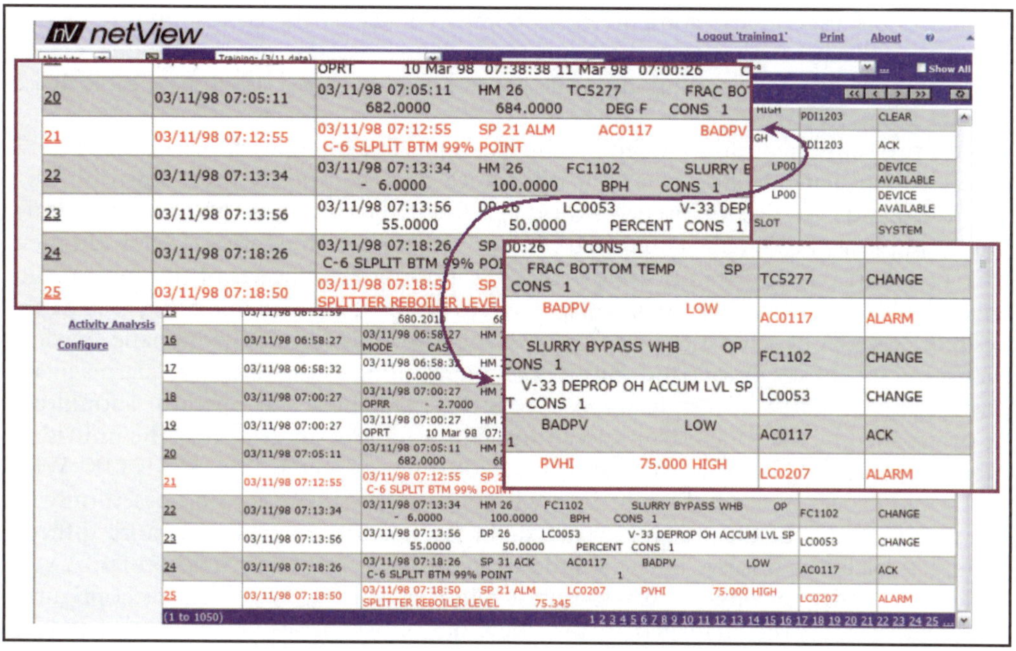

Figure 4.5.5. Example of structured real-time data (TiPS)

4.6 CONFIGURATION ANALYSIS

If most alarm activations were really useful to the operator, there would be no need to examine the configuration nor any reason to be concerned about the alarm system itself in any way. Such is not the case. So the first thing one might ask is, "How can any statistical-like examination of a configuration provide useful information about how the alarm system will perform?" We already know that the design of proper alarm systems is based on the limitations of the individuals sitting in the role of operators. Therefore we need to know how our alarm system is performing in order to learn whether its demands on the operator are compatible with the operator's ability. Of course, we are not taking about a specific operator—it's operators in general. No operator support system can be useful if it overloads the operator. The alarm system is the front-and-center operator support tool. No operator support system can be used if it does not provide consistent, accurate, and effective support. Make no mistake, as goes the configuration, so performs the alarm system.

The numbers and method of configuring alarms contribute directly to the numbers and usefulness of the eventual alarms that activate. The fewer alarms are configured, the fewer can activate. The priorities of the configured alarms must be apportioned to ensure that there is a balance between the most vital and the ordinary. When larger numbers of alarms activate, the operator will use priority to choose the ones to work first. Table 4.6.1 shows the more common metrics and their corresponding benchmarks. Remember that all metrics refer to a single operator. Most trace back directly to EEMUA 191. Some of these introduced earlier are repeated for completeness.

In order to make use of this evaluation process, the entire operator area will be examined and each metric computed and compared against the corresponding recommendation. Keep in mind that the benchmark figures are just guides. If your analysis comes up with priority distributions in the general neighborhood of those in Table 4.4.2, you are going to be fine.

We turn our attention to the following example. The example shown in Table 4.6.2 shows a particular operator area. You will notice that the priority breakdown differs a bit from EEMUA. This difference is not significant. Here comes the interesting part. Our example area is actually comprised of two smaller operations areas. Let us call them area 1 and area 2. It does not follow that the priority breakdown of the two smaller areas needs to match either the priority breakdown for the entire operating area or the desired EEMUA benchmark score. The reasoning for this seemingly dramatic freedom isn't some sort of sleight of hand. But it is simple. So long as the sum of all alarms in any given operator's area of responsibility meets the guidelines, our operator should not be overloaded by the total alarm activations. This properly suggests that the individual subareas are free to be what they actually need to be so long as the total is OK. What we should expect is alarm distributions within operator subareas to be based only on the specific needs of the subarea: more or higher priority ones in the more difficult parts, less or lower priority in the simpler parts. But do not carry this too far. A very difficult area that has only twenty tags in it might have many fewer alarms configured than a very large, easily managed area. Table 4.6.3 shows the results for our example of

Metric	Benchmark	Notes
Alarms	Alarms per loop: 1 or less Alarms per measurement: less than 50% Alarms per digital: less than 20%	
Alarms by priority	Emergency: less than 20 High: 5% of total Medium: 15% of total Low: 80% of total	For three priority levels, ignore the number for emergency; alarm systems with more than four priorities are not recommended
Disabled alarms	Less than 1%	Alarms that are disabled by explicit operator action; total number should not exceed 10
Shelved alarms	Less than 30	Specific operator action related to accommodating a short-term problem
Suppressed alarms	No limit	Includes only alarms that are disabled by logic or other non-operator-controlled procedure
Duplicate alarms	None	No configured alarm should duplicate another

Table 4.6.1. Configuration metrics and benchmarks

Metric	Example scores
Alarms by priority	Emergency: 15 High: 7% Medium: 18% Low: 75%

Table 4.6.2. Example configuration metrics; variability example

Metric	Example scores	Notes
Alarms by priority COMBINED OPERATOR AREA (of 1000 alarms)	Emergency: 15 High: 7% = 70 Medium: 18% = 180 Low: 75% = 750	For each category, both the percentages (of *1000* alarms) and the actual number of alarms are given.
Alarms by priority AREA 1 (400 alarms)	Emergency: 2 High: 10% = 40 Medium: 10% = 40 Low: 80% = 320	For each category, both the percentages (of *400* alarms) and the actual number of alarms are given.
Alarms by priority AREA 2 (600 alarms)	Emergency: 13 High: 5% = 30 Medium: 23% = 140 Low: 72% = 430	For each category, both the percentages (of *600* alarms) and the actual number of alarms are given.

Table 4.6.3. Configuration metrics; subunits example

two subareas. Note that area 2 has thirteen of the total fifteen emergency alarms; area 1 has 200 fewer total alarms but contains 25% more high-priority alarms.

In general, the specific evaluations described in this section should produce a picture good enough to judge the configuration effectiveness so that the designer can gauge the amount of work necessary to provide an efficient alarm system configuration.

The next section discusses the real-time analysis.

4.7 ACTIVATION ANALYSIS

The rate of activation as well as the difficulty of understanding and managing alarms is the real concern for operators. Just as we saw in the section on configuration metrics, we know that the design limits of all alarm systems are set around the human factors limits of people who sit as operators. No operator support system can be useful if it overloads the operator. Operator support systems should be designed to provide consistent, accurate, and effective support.

Table 4.7.1 shows the traditional alarm activation metrics and benchmarks. As before, the benchmarks are traceable to EEMUA 191.

In addition to the traditional alarm metrics in Table 4.7.1, there are a number of metrics that assist the understanding of the other aspects of alarm performance. Many of these go beyond the traditional EEMUA ones. Table 4.7.2 provides information that defines both the metric and the target benchmark. They are listed in alphabetical order.

Experience has shown that during alarm flood conditions, if the alarm activation rate climbs toward 100 in 10 minutes, the operator will likely abandon use of the alarm system entirely.

Metric	EEMUA benchmark	Notes
Normal operation	No alarms	
General activation rate	6 or less per hour	Without regard to priority; all for well-designed alarm system
Emergency priority	Very rare	For a 3-level priority system, ignore this category
High priority	1 or less per hour	Most hours should have none
Medium priority	2 or less per hour	Most hours should have none
Low priority	12 or less per hour	This should be rare; the general activation rate of 6 or less should be the norm

Table 4.7.1. Major alarm activation metrics and benchmarks

Activation Analysis across Industrial Segments

It is interesting to compare the basic EEMUA benchmarks against actual activation analyses from several industrial production segments. Matrikon[7] published a study of thirty-nine plants from oil and gas, petrochemical, power, and other industries. The results from the study are summarized in Table 4.7.3. EEMUA benchmarks are provided for reference. We notice immediately the enormous discrepancy between the recommended number of alarm activations per day (144 per EEMUA) and the actual activation rates found in the various industries (ranging from 900 to 2000). The actual activation rate exceeds the recommended by an order of magnitude or more. This isn't a curiosity. These results have very important implications for any operator's ability to properly understand and service plant upsets. On the surface, we can say these alarm systems will most likely fail to provide useful information for good operation. The large numbers of alarm activations have a deleterious effect on plant operation. Also observe that the study shows a much larger number of higher-priority alarm activations than EEMUA recommends. Such a skew in priority will prevent the operator from properly assessing the true importance of alarms, thus causing important ones to be missed amid the sea of unimportant ones. These survey results are typical of those found throughout a broad swath in industries, well beyond the thirty-nine sampled.

Deriving Implications from Activation Analyses

Let's go back a bit to consider what we might learn by taking a look at the more interesting measures in Table 4.7.2. As we discuss specifics, please refer back to the table for the appropriate metric definitions for terms. It is important to note that the results of

Metric	Benchmark	Definition
Acknowledgment ratio	99.5%	The percent of time an alarm activates AND the operator acknowledges it before it clears
Alarm flood (serious)	No standard; think 0	10 consecutive time periods of 10 or more alarms within 10 minutes; 100 or more alarms within 10 minutes regardless of how often it occurs
Alarm flood (short term)	No standard; think 0	10 or more alarm activations occurring within any 10-minute time period until the rate drops below 5 activations in a 10-minute time period
Chattering	No standard; think 0	10 or more within 1 minute (used for digital nuisance alarms)
Consequential alarms	No standard; think 0	1 or more alarms that follow the first one within 5 minutes 90% or more of the time
Related	No standard; think 0	Two or more alarms that occur simultaneously, without regard to activation order, within 5 minutes 90% or more of the time
Repeating	No standard; think 0	10 or more within 15 minutes (used for analog bad actors)
Stale	5 or less	Acknowledged but uncleared for between 8 and 12 hours—one shift
Standing	No standard; think 0	Acknowledged but not cleared for 24 or more hours, but none after a month
Time to acknowledge (nonflood situation)	Emergency: < 30 seconds High: < 1 minute Medium: < 3 minutes Low: < 10 minutes	Should meet this standard over 95% of the time; horn may be silenced at any time without effect
Time to clear	No standard	Useful to track general operability difficulty

Table 4.7.2. Additional alarm activation metrics and benchmarks

	EEMUA	Oil & Gas	PetroChem	Power	Other
Average Alarms per Day	144	1200	1500	2000	900
Average Standing Alarms	9	50	100	65	35
Peak Alarms per 10 Minutes	10	220	180	350	180
Average Alarms/ 10 Minute Interval	1	6	9	8	5
Distribution % (Low/Med/High)	80/15/5	25/40/35	25/40/35	25/40/35	25/40/35

Table 4.7.3. Cross-industry activation study

these analyses usually appear in the form of questions—that is, the score values for certain analyses "suggest" certain faults in the enterprise infrastructure rather than provide direct, useful, actionable information. It is left to the reader to probe more deeply and uncover root causes from these directional hints. The additional plant problems in section 4.2 should be very useful in that examination.

Acknowledgment Ratio

The acknowledgment ratio is defined as the average number of all alarm acknowledgments for a given time period divided by the average number of all alarm activations during that same time period, expressed as a percentage. The averaging time period normally chosen is a shift, a day, or a month. Other time periods are used as well. The acknowledgment ratio can point to alarms that are cycling due to excessive process noise, have poorly specified deadbands, or appear unimportant to the operator. A low acknowledgment ratio, when it is coupled with a long average time to acknowledge, can point to a large number of alarms that the operator knows are unimportant and therefore will not do anything for them. Of course, the horn will be either bypassed or silenced almost constantly.

Time to Acknowledge

The *time to acknowledge* is the average time lapse between alarm activation and its corresponding alarm acknowledgment by the operator. This average is computed using the same time periods discussed earlier. If relatively consistent during an operating campaign, time to acknowledge can be used as a probable indicator of the importance of individual alarms. Alarms that are acknowledged promptly will be ones considered by the operator as either totally unimportant, so they can be acknowledged reflexively, or very important. Probe to find out which each is. Highly variable time-to-acknowledge values for specific alarms can indicate the alarm is interpreted either differently by different operators or differently in different circumstances. It also might reflect simply a lack of consistency on the part of operators or, even more simply, a lack of respect for the alarm system itself.

Time to Clear

Time to clear is the average of time between the alarm activation and the corresponding alarm returning to normal. This average is computed using the same time periods discussed earlier. Time to clear is useful as a measure of the operator's ability to properly manage upsets, the general difficulty of managing upsets due to weaknesses in plant design, the inherent nature of the plant, the presence of equipment inadequacies, poor maintenance, or poor operator support such as training and diagnostic aids. Long values for time to clear must be investigated and their root causes exposed. These causes can point, with astute clarity, to infrastructure problems that require attention. No alarm system will be able to surmount these problems. All must be attended to, understood, and resolved.

Alarm Flood

Just like its watery namesake, an *alarm flood* is a situation where the alarm activations occur so rapidly that the operator is "flooded" by them. And like a water flood, the operator can quickly be drawn under water or swept away. In any event, the operator is not in any position to derive useful information from the alarms and therefore will either select inappropriate ones to work on in an attempt to struggle back to the surface or simply give up and pay no attention at all. Any alarm system in flood is one that ceases to be useful. Any plant experiencing alarm flood is unlikely to be operating properly. Measuring flood conditions is one way of objectively demonstrating to plant personnel that there are significant operational problems. Reliable plants will take steps to understand this and improve the situation.

Chattering and Repeating

A *chattering alarm* is an alarm that is activated and cleared many, many times within a very short time period by (almost exclusively) digital signals.

Repeating alarms are very similar but normally are caused by analog signals. Therefore, while they do not activate and clear nearly as fast as chattering alarms, nonetheless, they occur often enough to be a nuisance. All are usually completely unnecessary.

Both chattering alarms and repeating alarms suggest that the alarms either have been improperly configured or are announcing abnormal equipment conditions or plant operations that are extremely variable and need prompt attention. Often, the alarm activation loading for chattering and repeating alarms will account for 10%–60% of all activations. Clearly, fixing these can produce quick and effective results. Their presence and number indicate an important aspect of the general care exercised in the alarm system design. See "Nuisance Alarms" later on in this section.

Related and Consequential

Related and consequential alarms are alarms that (almost always) occur within a short time period of each other. These are often referred to as *correlated alarms.* This nomenclature is a bit confusing. However, the use is conventional by now, so we continue it here. See section 4.9 for additional breadth and coverage. Consequential alarms are ones that occur one before the other, with the same one activating first. Related alarms occur one after the other, but without a specific order.

Plants are usually complex, with a large number of interacting components and systems. It is not unusual for alarms to be placed on several variables, which can end up being either redundant or highly overlapping. Either way, there is the distinct possibility that the same abnormal condition will make its presence known by more than one alarm. Being able to tally the number of alarms that appear to duplicate each other can be useful. If the alarms are redundant, a single best representative is selected. If the alarms are sometimes redundant and other times not, then advanced alarm techniques can be used so they can be suppressed when they need to be.

Standing and Stale

Stale and standing alarms are alarms that are active (with or without being acknowledged) for long periods of time. Standing alarms are of shorter term; stale ones are around almost "forever." Since all alarms must be acted on, any alarm that remains active for a long period of time is either an ignored alarm, an alarm that cannot be acted on sufficiently to return the situation to normal, or one that cannot be returned to normal because the normal state is never reached (e.g., spared equipment). Either way, most standing and stale alarms probably need removal or additional logic so they are enabled only when useful.

Nuisance Alarms (Bad Actors)

One of the usual first objectives of an alarm activation assessment is a search for *nuisance alarms*, often referred to as "bad actors." Basically, when alarm activity is first examined at most plants that have not undergone alarm improvements, it can be easily shown that a relatively few number of alarms will have caused a significant number of alarm activations. Over and over again, lots of alarms activate. When they are examined, it is often found that over half of all alarm activations are caused by fewer than twenty individual configured alarms.

This topic was delayed until the last part of this section on purpose. As interesting and as useful as studying nuisance alarms might be, such a study is not the primary purpose or benefit of an alarm assessment. Yes, it is quite easy to eliminate most of them quickly and with gratifying results to the operator. And it can be shown that afterward, the alarm system will appear to work much better during normal operations. Unfortunately, that is about all that is improved. The remaining thousands of alarms are still not examined for need, for proper priority, for appropriate activation points, or for appropriate documentation and advice. Moreover, the act of just fixing the nuisances can confuse others into falsely concluding that enough alarm improvement has been made to obviate any need to spend real money going further! What started out as an honest effort to make things better might backfire to the point of getting in the way of serious work at fundamental improvements. Yes, fixing them is important. But strategically, it is recommended that it be done as the first step of an approved program rather than as a "gift" before starting. Doing it as the first step provides an immediate benefit to the plant and, at the same time, increases the project's creditability and good faith.

Let's have a look at some of the typical effects of nuisance alarms. Figure 4.7.1 shows a 16-week graph of the tags that caused the twenty most frequent alarm activations by week and all alarm activations for the week.[7] For most of the 16 weeks, the top twenty alarming tags comprised over half of all alarm activations. Quite a load on the operator.

For all of the bad things we've been saying about nuisance alarms, you might be ready to conclude that there might be little use for examining them. Not so. Examination will be very revealing. Let's see why. Figure 4.7.2 shows the relative plant area locations of them. This is the same plant as depicted in Figure 4.7.1. Figure 4.7.3 shows the results of a diagnostic examination of the top twenty bad performers for cause.

Our look into nuisance alarms was a good illustration of the examination power of alarm assessments and the tools that are utilized. The results are always useful. We close this topic with the gentle reminder that at the end of the day, no amount of alarm activation analysis will be useful for anything other than improving the existing alarm system. And that will improve basic operator support only slightly. The real improvement will come from assessing the situation as a whole and implementing a better design.

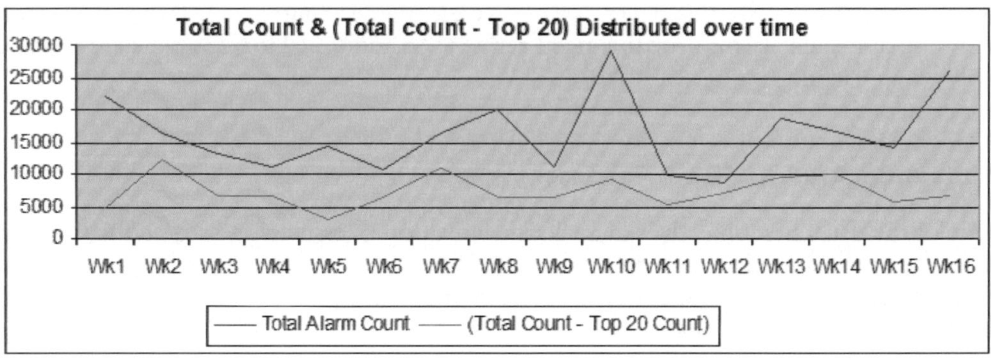

Figure 4.7.1. Top twenty nuisance alarms versus total alarms (Matrikon)

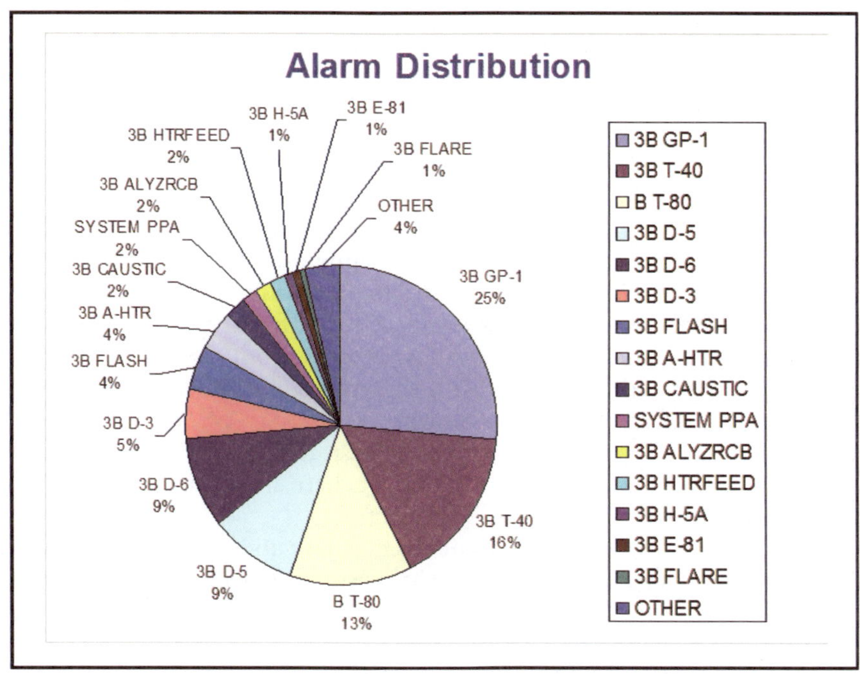

Figure 4.7.2. Plant locations for top twenty nuisance alarms (Matrikon)

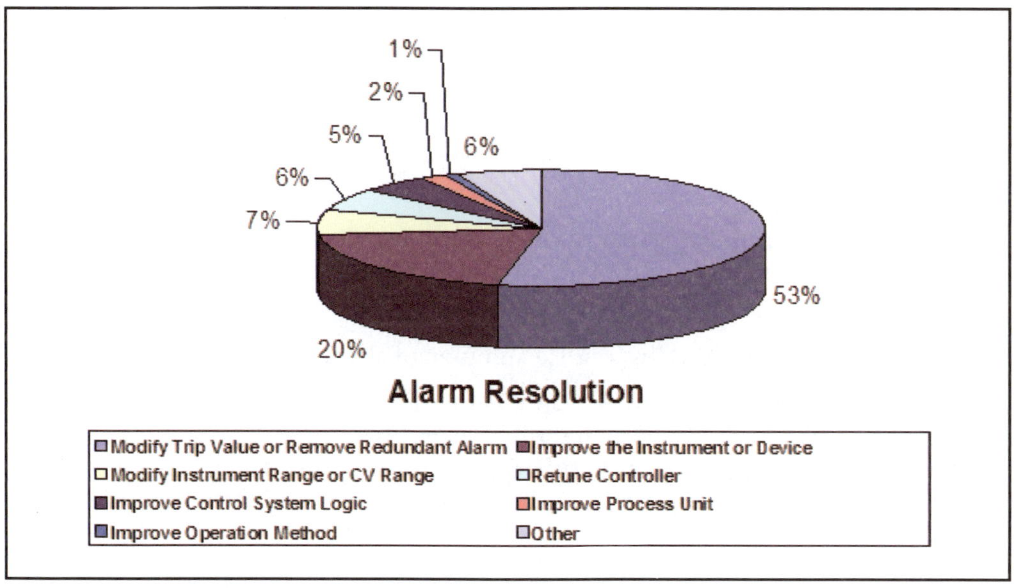

Figure 4.7.3. Problems remedied to remove top twenty nuisance alarms (Matrikon)

4.8 ADVANCED ACTIVATION ANALYSIS

Alarm performance is also used for forensic-type analysis. In this case, data from these analyses are used to support the detailed process of examining individual alarms for purpose and to identify plant areas where alarm design or performance suggests problems. This section will cover these tests. The list is rather extensive.

The analyses are constructed around two central themes: (1) identify whether the performance of the alarm system differs in any significant or suggestive way in different areas of the plant, in different plants at a site, or at different sites for an enterprise; and (2) identify differences in alarm performance for different types of alarms, different operators, or different shifts. For each of these themes, both overall averages and individual results are computed. The basic list is shown in Table 4.8.1. Many users develop additional items that reflect specific needs or problems.

For some plants, a subset of these forms the basis for their KPIs and is used to track alarm performance over time. Note that the targets here are specified by the plant alarm philosophy. There is no particular order to the items.

Chapter 10 extends this discussion for plants that have undergone alarm improvement. In that situation, the analyses are able to provide very valuable infrastructure evaluations.

4.9. ALARM CORRELATION ANALYSES

EEMUA 191 suggests that we consider "auto- and cross-correlation of alarm records." This can be a confusing subject. First of all, the notion of correlated alarms differs from any logical mathematical definition of correlation. And that is OK! The conventionally accepted understanding of alarm analysis usage of consequential and related alarms was covered in section 4.7. Here we expand on that notion to cover analyses that can be quite useful but might be beyond the scope of many conventional alarm assessments.

Situations

The discussion is perhaps best understood when broken down into three situations.

Situation 1

Description: Alarm activations, by Tag ID and time of alarm.

Benefits: Expose alarm patterns that help identify related alarms with the idea in mind of eliminating them (via deconfiguring), explicitly controlling them (via enhanced logic), or providing specialized situational information to the operator (again, via enhanced logic). *This is perhaps the strongest and most direct benefit from correlation.*

Situation 2

Description: Alarm activations, by Tag ID and time of alarm *and* referenced to a specific plant event like an equipment trip or dangerous or toxic release like liquid out the flare line.

#	Description (per 30-day time period)	Target	% of time target is met
1	Number of alarms per hour; 30-day average		
2	Peak number of alarms per hour		
3	Percent of time the process is in alarm flood		
4	Number of emergency-priority alarms		
5	Average time in alarm (hours) for emergency alarms		
6	Average time to acknowledge for all emergency alarms		
7	Average time to clear for all emergency alarms		
8	Number of high-priority alarms per hour		
9	Average time in alarm (hours) for all high-priority alarms		
10	Average time to acknowledge for all high-priority alarms		
11	Average time to clear for all high-priority alarms		
12	Number of medium-priority alarms per hour		
13	Average time in alarm (hours) for all medium-priority alarms		
14	Average time to acknowledge for all medium-priority alarms		
15	Average time to clear for all medium-priority alarms		
16	Number of low-priority alarms per hour		
17	Average time in alarm (hours) for all low-priority alarms		
18	Average time to acknowledge for all low-priority alarms		
19	Average time to clear for all low-priority alarms		
20	Number of alarms that are used for "information only" per hour		
21	Average number of stale alarms		
22	Average amount of time one or more alarms were chattering or repeating		

Table 4.8.1. Advanced alarm activation analysis list

Benefits: Some benefits of situation 1, with the added information about alarm patterns that might provide additional warning (more time) or additional information (better understanding of how to manage the abnormal event).

Situation 3

Description: Conventional statistical correlation calculations of alarm activations (no tag ID information is used) by number and time of activation (a conventional time series) with and without reference to any operator log "events," which are used as markers.

Benefits: Identification or "suggestion" of patterns that help to expose plant abnormal operations that are present but

1. might be observable by the operator but thought to be random and therefore mistakenly considered to be of little or no value; or
2. might not be observable by the operator but, if exposed and understood, might yield important information that leads to improved operations (something broken being fixed, procedures revised, or simply a better job done of monitoring indicator variables).

General Comments

These three situations are valid exercises, not only for initial alarm performance assessment, but also later on as a routine activity post rationalization. They might be especially useful after a significant plant event. They should always be useful for postevent analysis of alarm floods, when these were not associated with any accident or release. In any event, we are talking about surgical tools, not broad-brush techniques that can be applied without careful consideration and understanding of the benefits and limitations of the results.

To date, correlation analyses for batch processes are in their infancy. No good experiential information is available.

4.10 ONE DAY IN THE LIFE OF AN ALARM SYSTEM—CONFIGURATION

It is customary to gather about 3 month's worth of plant data for a proper assessment. And those 3 months should be representative, not the best and not the worst. For the moment, in this chapter, we examine only a single day. With a single day, we can take the luxury of a more detailed examination of minute-by-minute events. We can see what a typical day might look like when placed under our alarm analysis microscope. A day personalizes the data. We can understand a day. And by understanding a day, we can better extrapolate weeks, months, and more.

Before we take that look at the specific performance data, let us examine how the particular alarm system itself was designed. We do this by evaluating the PCS configuration database for alarm elements. Our example operator area is composed of a gas plant and a reactor/regenerator.

We examine the following data:

1. Number of tags
2. Number of tags with alarms
3. Number of alarms by alarm type
4. Number of alarms by priority
5. Number of duplicate alarms

Number of Tags and Tags with Alarms

Of a total of 2455 tags that might have alarms configured, 1836 of them have one or more. This amounts to 75%. So at a minimum, our plant has over 1800 alarms for this operator area. In actuality, there are 4339 individual configured alarms (not counting the journal alarms). This works out to about 2.5 alarms per alarmed tag. Let's take a look at how the configuration breaks down by plant area. The gas plant has 1691 configured alarms. The reactor/regenerator has 2302 configured alarms.

Please observe that the ratio of configured alarms is not the same for both plants. The gas plant has 91% of the tags alarmed. The reactor/regenerator has 64%.

You might also have noticed that the totals for alarms and tags that could have alarms for the two plants do not add up to enough for the entire operator area. This is, of course, due to the presence of alarms that are in our operator's area but not associated with either of the two major plants—a typical situation.

Number of Alarms by Alarm Type

The number of alarm types configured is shown in Table 4.10.1. Process-related alarms are shown shaded in the table. The first thing that "jumps out" from this data is the large number of system-type alarms. Actual process-related alarms amount to only 24%. This is not unusual. One of the early tasks in alarm improvement is to ensure that system-type alarms are not routinely routed to the operator.

Priority of Configured Alarms

We see that this plant is using three active priorities (low, high, and emergency) and one priority for logging only (journal). Table 4.10.3 shows the numbers.

Alarm	% of total	Meaning
BADPV	21	Bad process variable (measurement)
CNFERR	14	Configuration error
OFFNRM	13	Block off normal (not normal)
CHOFST	9	Change of state (for digital variable)
PVHI	8	Process variable high
PVLO	8	Process variable low
DEVHI	4	Deviation high
DEVLO	4	Deviation low
UNREAS	3	Unreasonable
CLEALM	2	
CLFALM	2	

Table 4.10.1. Configured alarm types in order of prevalence

Priority	EEMUA benchmark (%)
Emergency	5
High	15
Low	80

Table 4.10.2. EEMUA 191 "suggested" number of configured alarms for three priorities

Examining Table 4.10.3, we see that the entire operator area conforms fairly well to the EEMUA guidelines. The individual plants are quite another story. The gas plant is a bit low in the high-priority category and correspondingly high in the low-priority category. The reactor/regenerator numbers differ quite a bit. That is OK, since all will be managed by one operator.

Duplicate Alarms

Duplicate alarms are those that appear, by how they exist in the configuration, to be duplicating one or more other alarms. For example, in a controller loop, it is possible for the process measurement to be alarmed at both the input block and the controller block. Figure 4.10.1 depicts the duplicate alarms finding for our example.

We see that the first tag (PI 0202) appears to be duplicated four times. The others (only the first ten are listed) are duplicated twice. If we take a closer look at TI 5149 (third from the top in fig. 4.10.1), for example, we see via the configuration map (fig. 4.10.2)

Metric	Score	Notes
Alarms by priority COMBINED OPERATOR AREA (of 4339 alarms)	Emergency: 5% = 199 High: 17% = 733 Low: 78% = 30407	For each category, both the percentages (of *4339* alarms) and the actual number of alarms are given.
Alarms by priority GAS PLANT (1691 alarms)	Emergency: 4% = 62 High: 7% = 116 Low: 89% = 1513	For each category, both the percentages (of *1691* alarms) and the actual number of alarms are given.
Alarms by priority REACTOR/ REGENERATOR (2302 alarms)	Emergency: 5% = 110 High: 25% = 567 Low: 70% = 1625	For each category, both the percentages (of 2302 alarms) and the actual number of alarms are given.

Table 4.10.3. Configuration metrics; subunits example

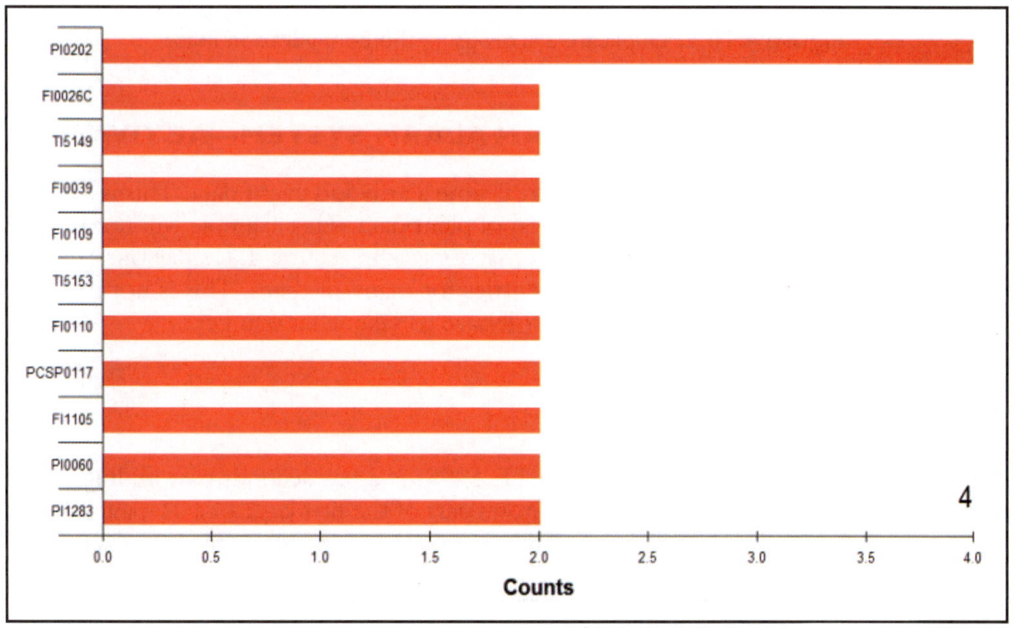

Figure 4.10.1. Duplicate configured alarms

that an alarm is configured for both the input block (TI 5149, in white) and the controller block (TC 5149, in black). Figure 4.10.2 was drawn by one of our tools directly from the imported configuration file.

We have completed the configuration alarm analysis and are now ready to look at the performance alarm statistics.

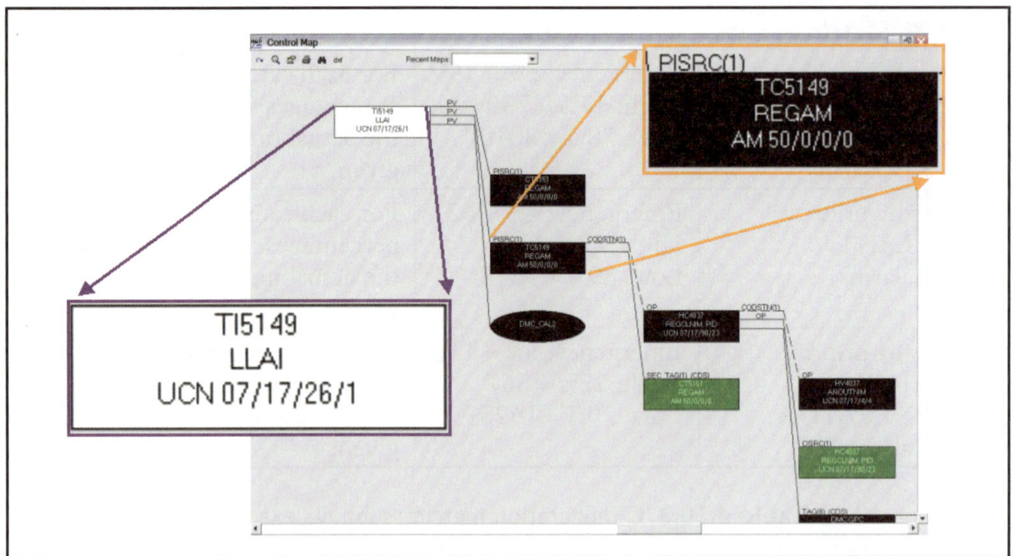

Figure 4.10.2. Duplicate alarms using the configuration map

4.11 ONE DAY IN THE LIFE OF AN ALARM SYSTEM—ACTIVATION

We now turn our attention to the plant's real-time alarm and event data. This is done to show what is expected from an examination of plant data. This example will not be able to provide the full list of activation metrics that appears in section 4.7, since our example database does not contain data category entries for many of them. This is not unusual. Differing PCSs log different things. You will have to take what you get.

The Raw Data

Our one day starts on Wednesday, March 11, 1998, at a few seconds past midnight and runs for the next 24 hours. Table 4.11.1 shows part of the first page of a 21-page activity list. Appendix 2 contains the full list of all relevant operator and alarm messages generated

```
03/11/98 00:10:38     SC 31 ACK        AI0114         PVHI
03/11/98 01:43:28     SY 19       NODE 46
03/11/98 01:43:28     SY 19       NODE 46
03/11/98 01:48:38     HM 26        TC0091        HGO RETURN TO BED 3
03/11/98 02:19:29     SC 21 ALM        AI0114         PVHI            50.000
03/11/98 02:33:37     SC 31 ACK        AI0114         PVHI
```

Table 4.11.1. March 11, 1988, real-time journal

for this day in a typical refinery operator area. The data are tabulated by time of event. This illustrates what the operators (more than one shift, obviously) saw and did during their time at the job. The data for this section is taken from that tabulation. You may wish to scan through this data at some point to see how difficult it is to glean importance from this amount of data. This is why alarm assessment tools are useful.

Each line of the journal represents a single message written by the PCS to one of the several real-time event journals. Each message occurred and was written only on March 11. The overall table is the union of all such journals. Table 4.11.1 enlarges, for ease of viewing, the first six entries of the combined log. Remember that an entry can be an alarm event (i.e., alarm activation, operator acknowledge of an alarm, or a return-to-normal), system alarm or message, operator action, or other message.

We take a closer look at Table 4.11.1 to better understand what each line actually tell us. The entry at time 00:10:38 (the first line) is read, from left to right as follows:

1. SC 31—refers to the plant area "SC 31"
2. ACK—indicates that the log entry is an operator alarm acknowledgment
3. AI0114—is the tag
4. PVHI—is the type of alarm acknowledged
5. HIGH—indicates that the alarm is a "high" priority
6. D-10 KO POT H2S—is the text description for the tag AI 0114
7. 1—is a redundant indicator for the alarm acknowledge event

The entry at time 01:43:28 (the second line) is read as follows:

1. SY 19—refers to the plant area "SY 19"
2. NODE 46—refers to data highway "node 46"
3. IF00—refers to a device identifier
4. DEVICE FAILED—indicates that the particular entry is a system alarm for a failure of a PCS infrastructure component
5. 00—is an address identifier

HIGH	D-10 KO POT H2S					1	
	IF00 DEVICE FAILED						00
	IF00 DEVICE FAILED						00
MODE	MAN	AUTO		DEG F	CONS	1	
HIGH	D-10 KO POT H2S		50.860				
HIGH	D-10 KO POT H2S					1	

Moving down the page a bit, the last entry we'll review is at time 01:48:38 (the fourth line) and is read as follows:

1. HM 26—identifies the plant area as "HM 26"
2. TC0091—is the tag being acted on
3. HGO RETURN TO BED 3—is the description for the tag
4. MODE—indicates that the entry describes an operator change of mode for the controller
5. MAN—indicates that the new mode is "manual"
6. AUTO—indicates that the previous mode was "automatic"
7. DEG F—shows that the units of measure for the loop are "degrees Fahrenheit"
8. CONS 1—shows that the change was done by an operator working on "Console 1"

That is probably enough to give you the idea of what an alarm and event log looks like. Where did this log come from? It was generated as a normal course of operation of the PCS. The PCS will usually provide logging of all alarm events. It also generally provides a log of some, if not all, operator actions. It usually provides a log of all system alarms and other results of internal consistency monitoring. Many times these logs are separated and archived in independent databases. However, so that they are useful for alarm activation analysis, we import them all and merge them into a single, time-of-event-ordered list. All the current alarm management support tools permit this functionality.

Amount of Data Produced in One Day

March 11 is a typical day. So how much is happening on our typical day? Let's see, we have about fifty entries on the first page. With 21 pages, that adds up to something over 1000 entries. Quite a busy day for our operator—but remember, this is an ordinary day! One thousand is pretty much the expected number of entries for events in most typical, uneventful days.

Now you see why your alarm analysis tool is going to be so useful.

Alarm Activations

One of the first things we'd like to know is what happened and when did it happen. Figure 4.11.1 shows all alarms for the day shown on a timeline from midnight to midnight.

There is a quiet period for the first few hours in the early morning. After that, we see alarms activating all throughout the rest of the day. Alarms average twelve per hour for an approximate total of around 300. Remember, the guideline metric suggests that the operator can only manage about six per hour. We're experiencing over twice that level. This typical day is turning out to be very taxing for the operator. There are numerous

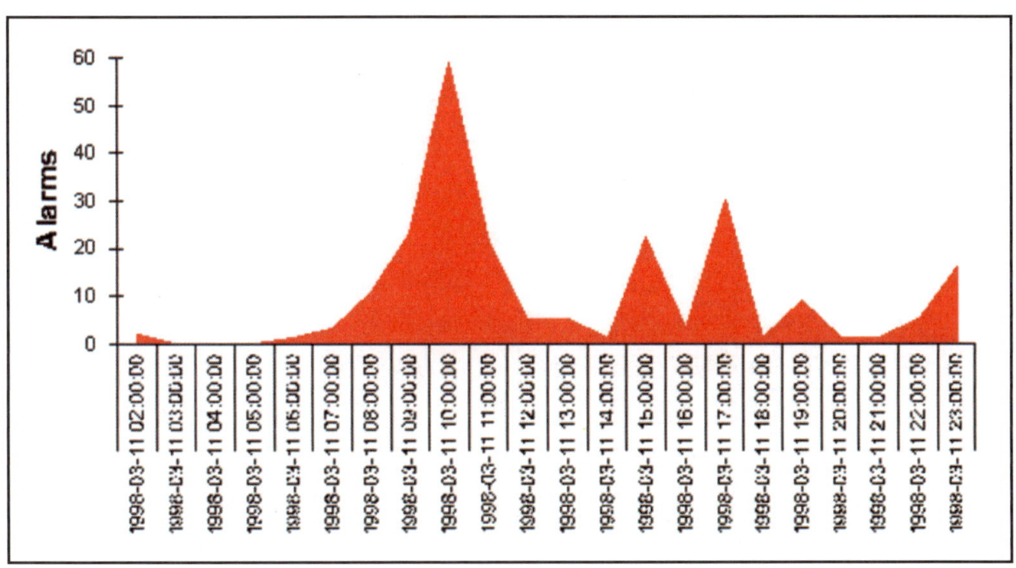

Figure 4.11.1. March 11 alarms per hour

very busy periods. The first is a rising crescendo starting around 7 a.m., peaking at a rate of nearly sixty per hour by 11 a.m. and then trailing off at noon only to plateau for the next hour. This is followed by four smaller but significant peaks at 3 p.m., 5 p.m., 7 p.m., and midnight. All in all, it was busy.

Let's take a more diagnostic look at the alarm activations. Figure 4.11.2 shows the alarm activations in the order from most to least activations for the first eleven most frequent.

It is easy to see which alarms are most active. LC 0207 PVHI activated (and cleared) twenty-five times during the day. FC 0022 PVLO did so twenty-three times. The eleventh most active alarm, AI 0114 activated three times. We could plot similar statistics for alarms by priority, by variable type (e.g., flow, temperature, etc.), or any other logged criteria. This is well beyond the scope of this chapter.

Time in Alarm

Time in alarm measures the cumulative time duration between an alarm activation and the corresponding alarm clearing (returning to normal), alarm by alarm. For March 11 we see (fig. 4.11.3) one alarm AI 0114 PVHI was active for a total of 50,000 seconds, or basically about 14 hours. The tag is an analyzer. Knowing that analyzers are prone to maintenance problems, this is not so unusual. In fact, the three longest duration alarms are all related to analyzers. Maybe the plant has problems with effective analyzer maintenance. Or maybe March 11 was the day that the technicians were devoting to analyzer calibration and other maintenance.

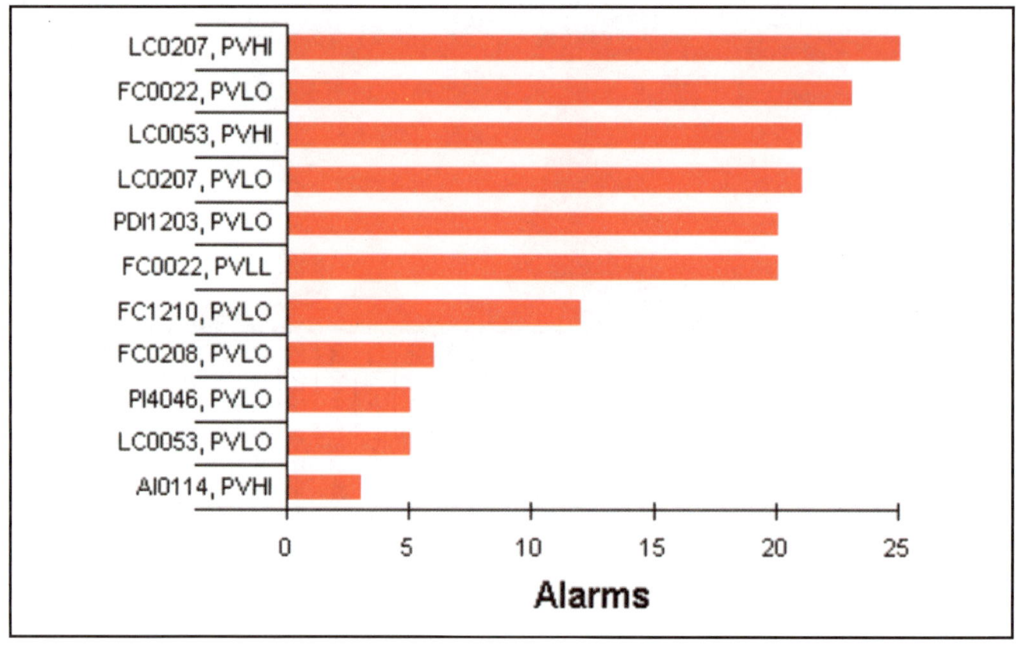

Figure 4.11.2. March 11 alarm activation by frequency of activation per tag

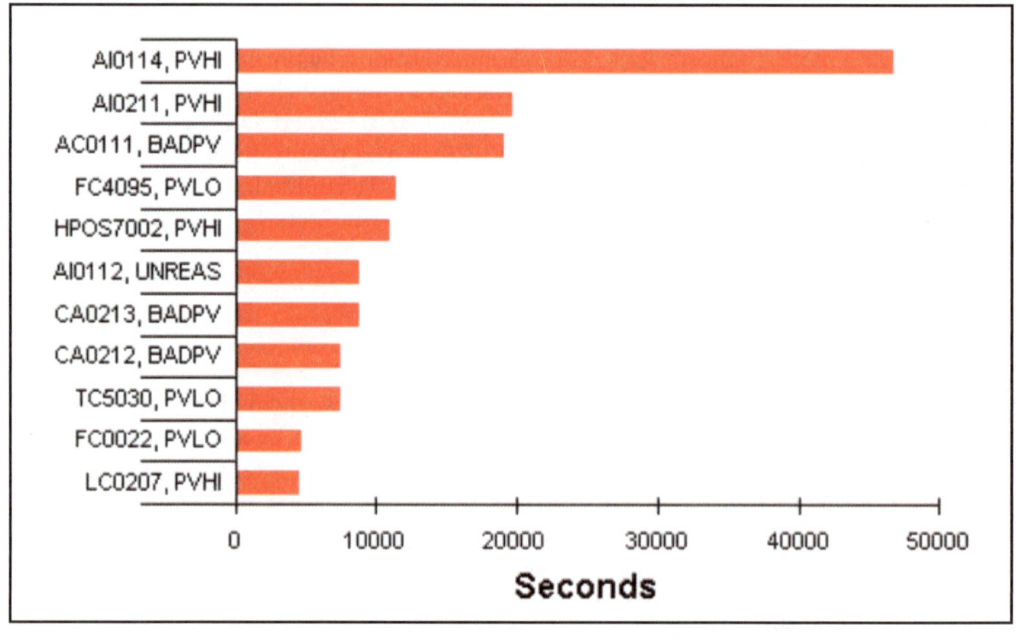

Figure 4.11.3. March 11 time in alarm

Time to Acknowledge

The amount of time it takes our operator to acknowledge any alarm in particular, or alarms in general, often is related to either the overall general workload of the operator or, just as plausibly, the operator's general respect for an alarm or the alarm system as a whole.

The longest alarm (fig. 4.11.4) is PD I1203; the operator took well over an hour to get to it. For the next ten longest time-to-acknowledge alarms, the average time is about 9 minutes. I leave it up to your experience to draw whatever meaningful conclusions you may from this.

Operator Actions

Now look at how often the operator took action during the day. Recall that the primary reason for an alarm is to elicit operator action. Figure 4.11.5 shows the number of operator actions per hour. For the time period between 1:00 and 1:59, there were only two operator actions. Between 7:00 and 7:59 there were thirty actions. There were eighty-two actions between 11:00 and 11:59. For twelve 1-hour periods, there were no operator actions at all.

Compare operator actions to alarm activations. The results for the entire day are shown in Figure 4.11.6. For each bar pair, the left bar depicts alarm activations, the right bar depicts operator actions.

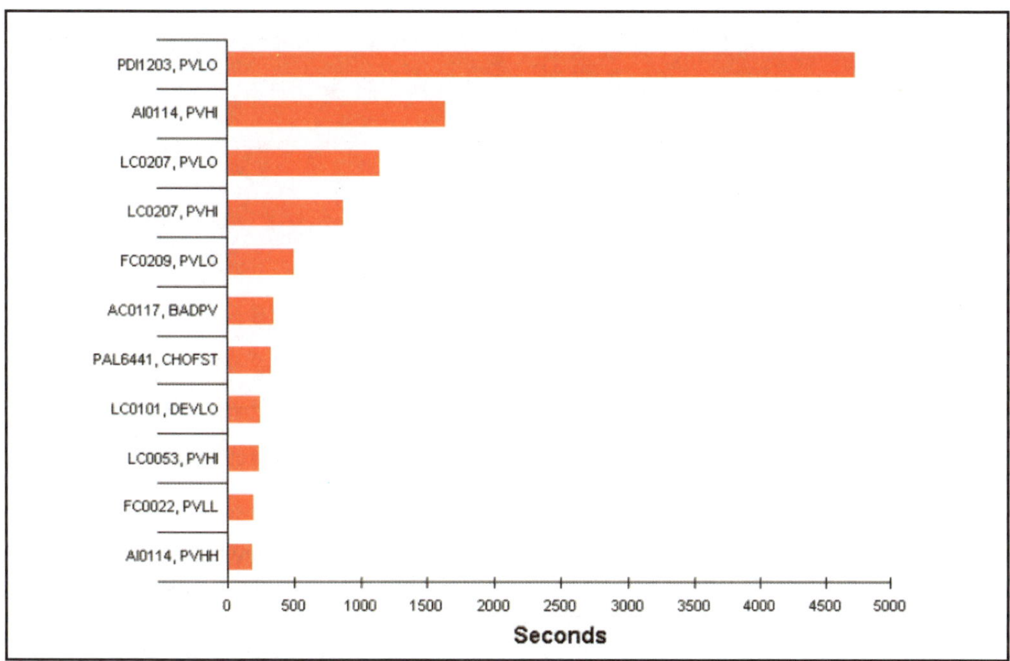

Figure 4.11.4. March 11 time to acknowledge

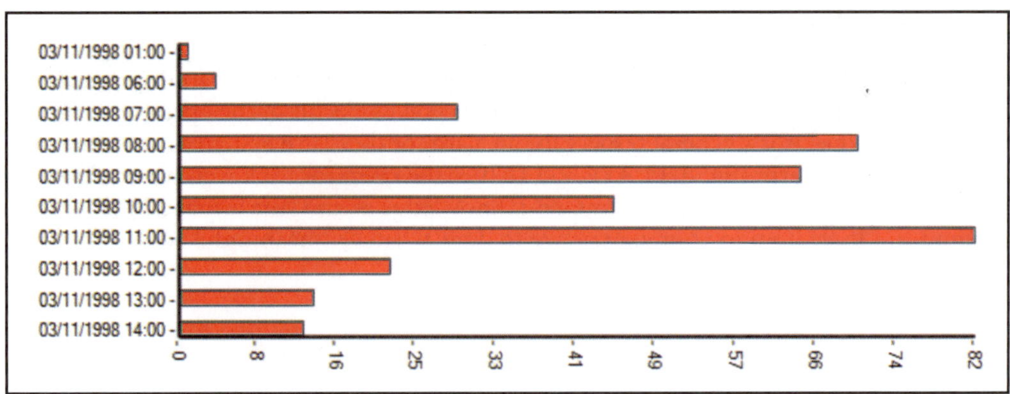

Figure 4.11.5. March 11 operator actions

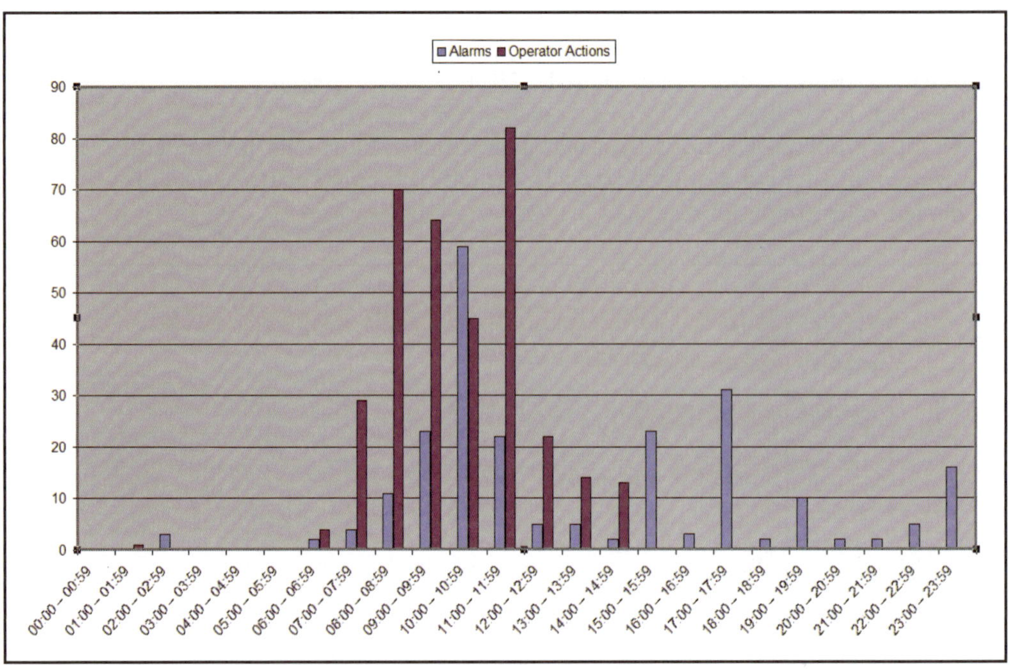

Figure 4.11.6. March 11 operator actions versus alarm activations

Earlier in the day, alarms and operator actions appear to be somewhat related. Around midday, actions surpass alarms by a large margin. This is OK, since it suggests that the operator was on the job. Then, after about 2 p.m., we have a steady cycle of lots of alarms one hour and a few the next, but with no operator actions at all. Something is amiss. Finding out what is actually going wrong is beyond the ability of any APA tool. Those answers must be found in operational logs, operator interviews, and the like.

4.12 ALARM SYSTEM PERFORMANCE LEVELS

We have seen and worked with a number of interesting and useful alarm performance and evaluation metrics. We have seen them applied to the configuration database. We have seen them used for the real-time alarm and events data. We have a good idea of what the good-to-meet targets are and why they have been set at their respective levels. We use these metrics and targets to monitor and evaluate alarm systems. Moreover, the results of monitoring can directly be translated into straightforward alarm improvement tasks. It is now time to move our view, and hence our evaluation, up a level.

Alarm performance rating, besides having a nice ring to its sound, provides a very useful and easily understood measure of the total condition of a plant alarm system. This measure communicates an important message to everyone on the enterprise team. The rating assigns a single word to the performance. The rating scale (from least effective to most effective alarm system) consists of these levels: overload, reactive, stable, robust, and predictive.[8] Table 4.12.1 summarizes the five different levels.

Performance level	Description
Overload	Operator is faced with a continuously high rate of alarms, with rapid performance deterioration during process upsets. The alarm system is difficult to use during normal operation and is essentially ignored during plant upsets as it becomes unusable.
Reactive	Some improvements compared to overload, but the peak alarm rate during upset is still unmanageable. The alarm system remains an unhelpful distraction to the operator for much of the time.
Stable	A system well defined for normal operations but less useful during plant upsets. Compared to reactive, there are improvements in the average peak alarm rate during both normal operation and upset. Nuisance alarms are resolved and under systematic control. Problems remain with the burst alarm rate.
Robust	Average and peak alarm rates are under control for the foreseeable plant operating scenarios. Dynamic alarming techniques are used to improve the real-time performance. Operators have a high degree of confidence in the alarm system, and have time to read, understand, and respond to (almost) all the alarms.
Predictive	The alarm system fully encapsulates the aspirations of the EEMUA guidelines. The alarm system is stable at all times and provides the operator with the right information at the right time. (Important) alarms are predictive and anticipate problems before they actually occur in order to avoid process upsets or minimize their impact on production.

Table 4.12.1. Alarm performance rating categories (Campbell-Brown)

Now that we have the ability to identify the level of alarm, this ability can facilitate an unambiguous assessment of how well the alarm system can be expected to do its job. How well the alarm system does its job can translate almost directly into how effective operators can be at keeping the enterprise safe and productive. Let's leave the profitable part to the business guys—that is their job, not ours! Go ahead, do your assessments on that current alarm system and assign the appropriate rating. More than likely your rating will be overload, reactive, or somewhere in the middle. The important thing is that you're not stable. Stable represents the minimal effective status for alarm systems. When you tell management that the plant is not stable, they should take you seriously when you tell them they need to improve the alarm system.

4.13 CONCLUSION

Alarm activation data suggests a rich source of interesting and illuminating information. By and large, it delivers on its promises and then some. Examination of the data exposes a broad picture of how the alarm system acts as the plant moves from normal operation to abnormal operation to upset conditions and beyond. The quantity of available data is large. Competent commercial purpose-built tools are available and should be used for these examinations. These same tools have important capabilities to warehouse alarm development and response information and assist with the documentation and rationalization process. The work is facilitated by the ability to recognize PCS configuration data and use this information to provide alarm structural analyses.

Nonetheless, data analysis alone cannot provide adequate knowledge for you to learn which alarms are important and which are not. Nor can it suggest proper alarm configuration or appropriate alarm understanding and response information. Data analysis alone cannot uncover underlying process weaknesses. The practitioner is encouraged to dig into the available data. However, the really hard work of significant alarm improvement lies ahead. Rationalization is the process by which the alarm system, as an entity, is reshaped into a significant and competent operator support tool. That comes later in this book.

4.14 NOTES AND ADDITIONAL READING

Notes

1. UReason Holdings, "Why Invest in Alarm Management?" *Automation in Petrochemicals.*
2. Joseph S. Alford, "Mining for Gold in Batch Alarm Records" (paper presented at the WBF Conference, Baltimore, MD, spring 2007).
3. *Wikipedia*, s.v. "Benchmarks," http://www.wikipedia.org.
4. TiPS, http://www.tipsweb.com.
5. PAS, http://www.pas.com.
6. Matrikon, http://www.matrikon.com.

7. The author and publisher gratefully acknowledge the license granted by Matrikon, Inc., to reproduce in this book Figures 4.7.3, 4.7.4, and 4.7.5, which are taken from Matrikon. For further information about Matrikon, Inc., and its Alarm Manager product, please refer to http://www.matrikon.com.
8. Donald Campbell-Brown, "Horses for Courses—A Vision for Alarm Management," *Proceedings of the Third Series of Seminars and Workshops on Alarm Systems, IS1172* (London: IBC Global Conferences, 2002).

Recommended Additional Reading

Larsson, Jan Eric, Bengt Öhman, Antonio Calzada, and Joseph DeBor. "New Solutions for Alarm Problems." *Proceedings of the Danish Automation Society Meeting*. Teleca Benima, Malmö, Sweden, September 11, 2002.

PART II

The Alarm Management Solution

CHAPTER 5

Permission to Operate

If you find yourself in a hole, stop digging.

—Will Rogers

An operator's assigned role is to operate. The enterprise only makes money when it produces a saleable product. To produce, it must operate. Therefore it is only natural that there is a strong inclination toward operating. For the most part, this works. However, when it does not, the results are often extremely costly. Careful studies of abnormal situations have determined that a very large percentage of incidents are directly attributable to operators not realizing in time that their unit upset was not going to be contained short of a catastrophic event.[1] By the time the operator does fully realize the shocking reality, it is almost always too late. It is too late either because he does not have enough time to do what it would take to return to a safe operating mode or because he still cannot determine what is wrong or what to do. In either case, the result is the same: equipment damage, production losses, personal injury, environmental degradation, or worse.

BP Texas City operators[2] were vainly attempting to manage their abnormal situation for over 10 hours before the tragic explosion. Milford Haven was in serious upset without the operators understanding what was wrong and what to do about it for over 5 hours before disaster struck. After the investigation of Milford Haven,[3] the HSE specifically identified the failure to understand when to continue operating or not as one of the proximate causes of the incident:

> #5—[Provide] clear guidance [to the operator] on when to initiate controlled or emergency shutdowns.

One of the more important and critical decisions an operator must face is whether and when to shut down rather than continuing to operate in the face of uncertainty. New understandings regarding abnormal plant operation have concluded that when the decision to operate in the face of uncertainty is made properly and if sufficient technology is in place to implement that decision, a significant reduction in operating losses can be realized. This chapter reviews the process by which a decision to operate in the face of uncertainty can be made effectively. The chapter includes an extension of these principles to the remaining key players in the manufacturing enterprise operation.

Please keep in the forefront that each individual plant or enterprise will be shaping guidelines for operators. This chapter is designed to provide grist for the mill of decision making and policy. The various examples here are intended to illustrate the range of alternatives and amplify the power of the approach, nothing more.

5.1 KEY CONCEPTS

Good operator intentions can cause serious problems	The recurring root cause of most serious plant disasters is maintaining operation too long during abnormal conditions.
Effective "rules" are powerful	It is possible and recommended that plants manage the control room operator's charter to operate.

5.2 MANAGEMENT'S ROLE

This chapter lays out the options for management to consider as they establish a policy granting plant operators permission to operate within their area of responsibility. The purpose of this chapter is twofold: first, to make the reader aware of an important part of the infrastructure for operating a production enterprise that heretofore was never thought of explicitly; second, to put forth a number of options and recommendations to assist the development of an explicit policy. This is not a decision that the alarm management teams make. This is one of the primary prerogatives of senior management—determine the rules of the game.

Up until now, we really didn't pay much attention to a decision of this type. No part of the business plan had a line item in it for this. No part of the engineering design package investigated what the operator should do when the process became abnormal and success at coping was at issue. Yes, regulation and safe operation considerations provide for emergency shutdown criteria and the equipment to support that requirement. But operating procedures ignore the subject of activities prior to a challenge to this last line of defense.

The time for remaining silent has passed. As part of any serious alarm improvement project—read that to say operational improvement project—we ask management to visit and decide the guidance that must be given to operators on the aspect of permission to operate. This will require clear, unambiguous understanding of how the decision will be made: under what circumstances will the operator be expected to cease the normal objective of remaining in operation and take affirmative, prerogative steps to move the process to a safe position? That guidance we call "permission to [continue to] operate."

Once management settles the issue of permission to operate, all operating procedures and support technology must be in place, and all affected personnel fully informed and trained to perform appropriately.

5.3 OPERATING SITUATIONS

Not all operating situations and modalities are the same. In this section, we will add some structure to them. This structure will be used later to define situation boundaries to assist in the implementation of the operator's permission to operate.

Operating in Uncertainty

Generally, operators are quite proficient in the day-to-day management of their production units. Production is maintained by the automated controls and other equipment. When variations appear, the operator responds appropriately and normal operation is quickly restored. In this manner, effective process management is maintained by experience, procedures, problem-solving skills, and collaboration. There are a couple of basic ways this comfortable situation becomes problematic.

- Problem incidents begin with a "whimper" or perhaps a small "bang" somewhere, then for a while progress slowly, then take a turn and get complicated, often without the operator knowing, and suddenly get out of control.
- Problem incidents begin with a big "bang" and quickly become (known by the operator or not) out of control.

Unique Events

Most of the seriously damaging incidents never happened before. Nor were they envisioned. For such unique events, experience and specific procedures are of little use. Nor will situation-based training be of much use. Industrial best practices now rely on three key aspects to respond effectively to unprecedented events:

1. Competency-based operator training, where abnormal situation recognition and management are one of the core competencies
2. Abnormal operation detection and identification tools
3. Abnormal situation-management procedures and technology

Explosive Events

Some damaging events seem to happen almost without notice and move extraordinarily quickly into a catastrophic state. In this situation, we must rely on either an operator quickly identifying the situation and initiating an emergency shutdown or the automated activation of emergency shutdown protection equipment. In both cases, it is imperative to ensure that the shutdown procedures and equipment are adequate and effective.

Surprisingly enough, however, such an event almost never arises without some sort of prior symptom. If the right sort of notice were taken, it could likely be effectively managed. Therefore, industrial best practices here fall into key areas:

- Comprehensive and effective equipment health monitoring programs
- Recognition of the role that operating stresses place on both personnel and equipment
- Technology to detect both subtle as well as incipient faults in the process and related equipment

In our discussion of abnormal situations in chapter 2, we presented a figure showing plant states and operational modes and goals (repeated here as fig. 5.3.1). This figure illustrates the three general operational modes all operators face: normal, abnormal, and emergency. For each mode, the operator's goal is identified. For example, if the plant state is normal, then the operator's goal is to keep it normal. In the abnormal

Operational Modes:	Plant States:	Critical Systems:	Operational Goals:	Plant Activities:
Emergency	Disaster	Area Emergency Response System	Minimize Impact	Fire Fighting
	Accident	Site Emergency Response System		First Aid Rescue
Abnormal	Out of Control	Physical and Mechanical Containment System Safety Shutdown, Protective Systems, Hardwired Emergency Alarms	Bring to Safe State	Evacuation
	Abnormal	DCS Alarm System	Return to Normal	Manual Control & Troubleshooting
Normal	Normal	Decision Support System Process Equipment, DCS, Automatic Controls Plant Management Systems	Keep Normal	Preventative Monitoring & Testing

Figure 5.3.1. Plant state versus operational modes and goals

mode, either the plant can be abnormal or it can be out of control. If the plant state is abnormal and out of control, the operator's goal is to bring the plant to a safe state.

By now, you can tell that the real problem occurs when the plant is in the abnormal mode of operation. The basic issue is really simple. The issue centers on how to detect and appreciate the significance of whether the plant state is abnormal or out of control. If all operators had a sign that read "Out of Control" every time their plant was abnormal and out of control, it is unlikely that they would ever try to keep it running in the vain hope of a miracle. That "sign" is probably a bit of a stretch. *As a surrogate, we use the concept of permission to operate to provide similar functionality.*

Definitions

The following definitions are used throughout the remainder of this chapter:

- *Normal region of operation.* The collective operating situation where (a) there are no critical alarms active and (b) the plant critical operating conditions (safety, environmental, and financial) are not violated.
- *Conservative region of operation.* A region of operation that sets limits so that it is unlikely that a manageable upset will go outside of the normal region of operation.
- *Upset.* Any process deviation likely to affect production quality or rate, or place equipment, personnel, or environmental limits at risk.
- *Manageable upset.* Any upset that the operator understands and has every reason to believe he can control and thereby return the process to normal operation in a timely manner and without excessive risk.
- *Unmanageable upset.* Any upset that is not a manageable upset.

5.4 HOW PERMISSION TO OPERATE CAME TO BE

There is an interesting story about how the concept of key operating principles was developed. It starts out like many stories: "Once upon a time. . . . " In this case, a prominent U.S. producer of petrochemical products located in the western United States had been around for a long time. This company's manufacturing plants were better than average, and their personnel were well trained and effective. However, during the normal course of manufacturing, a mistake was made in producing one of their retail products. The product was designed for a niche market use. When it was used, this defective product severely and irreparably damaged users' equipment. The mistake in production was not detected, so the product shipped. Before the general word got out, many millions of dollars of customers' equipment were damaged and needed replacement, which was, of course, paid for by the petrochemical company.

The company's management decided to uncover the root cause of this costly mistake in the full belief that (a) if it were resolved at only the symptom level, a similar problem would repeat again and (b) the understanding of the fundamentals of the failure might lead to important understandings that could have an enterprise-wide impact. Both were true, very true.

What they did was a case study in really good forensic examination. They collected every available record of incidents, accidents, and such, covering about 5 years. If the reports were complete, they recorded causes and recommendations, taking care to ensure that the lessons learned were detailed and sufficiently deep to uncover root causes. If the reports were incomplete, they made attempts to revisit them and complete the investigations. All of this work involved a significant amount of effort. The next, and most crucial, step was to categorize the incidents to look for common causal issues. They found them. They found a list of over a dozen operating actions that almost always were the root cause of serious operating mistakes. If those erroneous actions had not been taken, most of the incidents would not have happened! Permission to operate was the most important way of preventing these erroneous actions and so forms the foundation of this chapter. The others are discussed in section 5.14.

5.5 HOW PERMISSION TO OPERATE WORKS

Operators are empowered to operate the unit(s) under their responsibility by the authority from management that we shall call *permission to operate.* Until permission is withdrawn, the operations staff will employ their best efforts, utilizing plant procedures, training, and their other considerable abilities, to ensure safe, productive, and responsible operation. When permission is withdrawn, current efforts to keep the plant running will cease, and all future efforts will be directed to placing the plant in a safe state. Of course, you know that there will not be a watchman looking over the operator's shoulder and calling the shots: OK to operate; watch out; not OK to operate. Rather, permission to operate will be a structured judgment process that operators, supervisors, and plant staff can understand and apply.

Most plants today have no such understanding. Most plant managers have never thought about this concept. Most operators have no idea how far they should go and that there can be a policy, thought out in advance, that actually provides understandable rules and procedures for this. But now that you have heard about it, does it not sound like just about the only reasonable way to approach this issue?

To implement this concept, management will establish criteria by which operators can unambiguously ascertain whether they have (constructive) permission to operate their unit. Once conditions change so they no longer have permission to operate, the operations team must take all necessary steps to cease operations by moving their units to predetermined safe operating states.

5.6 PERMISSION TO OPERATE

Countless incident postaudits have concluded that most catastrophic incidents happened *after* the operating personnel were unable to ascertain what was going on. To say this another way, once an operator was no longer able to identify what was going on, rarely were his actions successful in restoring normalcy or averting an event. We can now very simply state what permission to operate is:

> The operator has "permission to operate" the plant under his/her control so long as he/she fully understands what the process is doing.

Using this guideline, production management will be able to move from a position of de facto permission to one of providing explicit permission by defining the circumstances under which they believe the operator can be successful. Note that this procedure has a profound effect on the operator's responsibility and authority. So long as the rules give the operator "permission," then he is presumed to have the authority to operate and will use all his skill and knowledge in the pursuance of his task. However, as soon as that permission is withdrawn, then the operator must, and without fault (lest he not do so the next time), cease operating and move the plant to shutdown or alternate safe state. Management therefore assumes total responsibility for this decision. Even if it may be later proved that the operator "should have known what might be happening but did not actually know" or the plant was actually not in upset, the operator is to be commended for his action, since he reasonably thought it to be so and acted under authority of his charter.

5.7 ALTERNATIVE METHODS FOR GRANTING PERMISSION

There are a few basic alternatives that can be used to operationalize the "permission" decision. Management will chose or otherwise specify which, if any, should be used.

De Facto Decisions

- *Carte blanche.* Permission to operate is granted *if and only if the operator is reasonably sure that he understands what is going on and quite confident of his ability to manage.* No bravado, just honest self-awareness. The decision is simple and direct. It is up to the operator to manage.
- *In usual upset.* Permission to operate is granted *if and only if the operator is reasonably certain that the plant is in a usual upset state, regardless of whether or not he understands specifically what the current upset is, and he is confident he will be able to manage it successfully. If anything about this upset is unusual, permission to operate is withdrawn.* Again, simple and direct. It is a bit more subjective than carte blanche, but still clear and unambiguous.
- *Manageable unusual upset only.* Permission to operate is granted *if and only if the operator is reasonably certain that though the plant might be upset, he has demonstrable reasons to know or expect the upset to be manageable. If he determines or begins to suspect that the upset might not be manageable, then permission to operate is withdrawn.* The only manageable upsets are those that are routinely seen and for which procedures that are expected to work exist. This case is quite a bit more subjective and requires substantial operator competence to employ.

Operating Modality Decisions

The plan for predesigning the operator's role based on plant situation is illustrated by Figure 5.3.1. Recall that the figure shows three operational modes: normal, abnormal, and emergency. For each mode, the operator's goal is identified. For example, if the plant state is normal, then the operator's goal is to keep it normal. In the abnormal mode, the plant can either be abnormal or it can be out of control. If the plant state is abnormal or out of control, the operator's goal is to bring the plant to a safe state. Notice that now the critical plant infrastructure systems involve, in order of escalating danger, the PCS alarm system, hardwired emergency alarms, protective systems, safety shutdown equipment, and eventually any physical or mechanical protection and containment.

The following is a list of alternative definitions for granting permission to operate. See Figure 5.7.1. Generally, plant management will select one of these or develop something else that is compliant with their operating philosophy. When the defining rule for granting the permission to operate is violated, the operator must, without delay or need for further decision, supervision, or other authority, move the plant to a safe state or shut down.

- *Normal operation definition 1.* The plant is operating within the conservative region of operation and there are no unmanageable upsets.
- *Normal operation definition 2.* The plant is operating outside of the conservative region of operation but within the normal region of operation and there are no upsets.

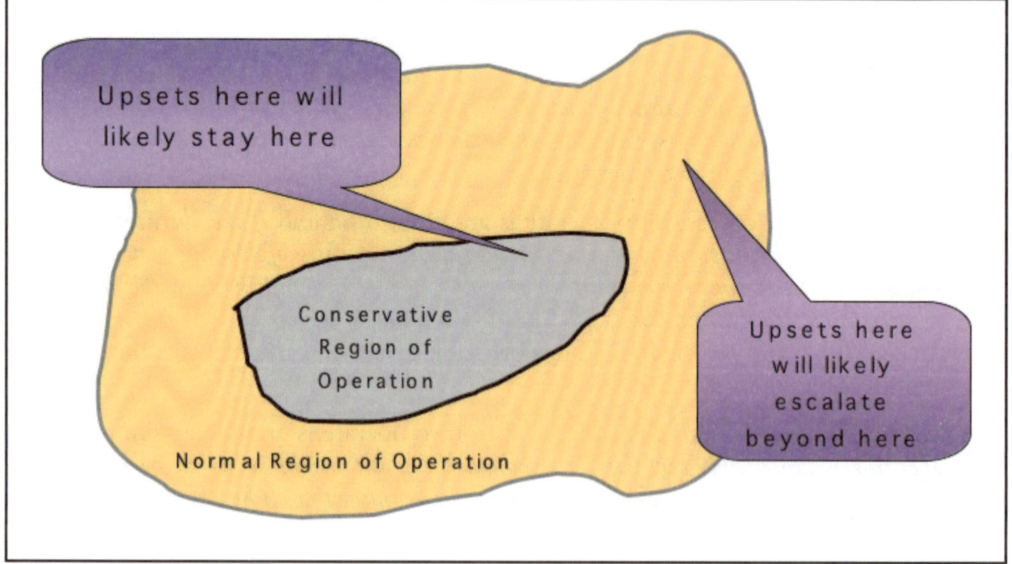

Figure 5.7.1. Regions of operation

For example, if the plant adopts definition 2 for normal operation, whenever the plant moves outside the normal sphere of operation and there is an upset, the operator is required to discontinue normal operation.

5.8 MANAGING THE OPERATOR'S PERMISSION

We now explore how to actually manage this whole idea of permission to operate. For all of this to be useful, you will need ways for the operator to know and understand when to continue and when to stop. These are suggestions, not necessarily requirements. Each plant will develop the specifics as part of an alarm philosophy or other operational guidelines.

In this section, we see some quite specific ways for you to identify situations where a clear decision can be arrived at whether the operational charter for the operator will be maintained or not. Previously, we discussed criteria for the operator to use that identified the plant operating regimes and made a determination from that. However, in practice, this is not enough. We need a second way, a check, if you will, to ensure that the operator's charter to operate is still in force.

Four separate categories are discussed. Any item from any category that withdraws the operator's charter to operate is enough to terminate his permission to operate. Remember that these are examples only. Their purpose is to provide clear, understandable alternatives to you for consideration. It is fine to use as many of the types as needed, or your plant may develop new ones that suit.

Qualifying Abnormal

Let's assume that the plant is now abnormal. The following situations will identify what is meant by continuing to be abnormal. If any one or more of these exist, then we must conclude that the plant remains abnormal, regardless of what opinion the operator might have.

- *Situation 1.* Additional alarms activate in the same or a related plant area.
- *Situation 2.* Few, if any, existing alarms in the same or a related plant area have returned to normal within a reasonable time.
- *Situation 3.* There are one or more items of equipment (valve, pump, sensor, or anything else not trivial) that then fail or are newly discovered as failed.

No Help at Hand

Difficult abnormal situations where there is not a second experienced plant person at the operator's "left hand" in the control room should automatically suggest that the operator is either already in trouble or soon to be. Consider this to be sufficient to withdraw permission to operate.

Observer Evaluation

We continue this discussion by looking through the eyes of a mythical observer in the control room who might be side-by-side with the operator who is making observations about what is seen. Of course, we know that such an observer may not actually be there, so this is a "thought" discussion. The format is question and answer. This methodology will ask the questions. Our observer is responsible for the truth of the answers. The answers will determine whether the operator retains permission or not.

- *Question 1.* Has the operator tried more than one approach to turn the abnormal situation around and it is still abnormal?

  YES—Discontinue operation

  NO—Continue to operate

- *Question 2.* Does the operator appear to be able to explain what is likely to happen to the plant in the next 10 to 20 minutes?

  YES—Continue to operate

  NO—Discontinue operation

- *Question 3.* Is the operator using a predetermined documented approach to work the abnormal situation?

  YES—Continue to operate

  NO—Discontinue operation

- *Question 4.* Has the situation remained quite abnormal (even if it does not appear to be gravely) for longer than a reasonable time?

  YES—Discontinue operation

  NO—Continue to operate

Operator Evaluation

Now we move to the operator himself. This part is self-examination. Each question is expressed in the third person but would be understood as being read and answered by the operator as applying to him personally.

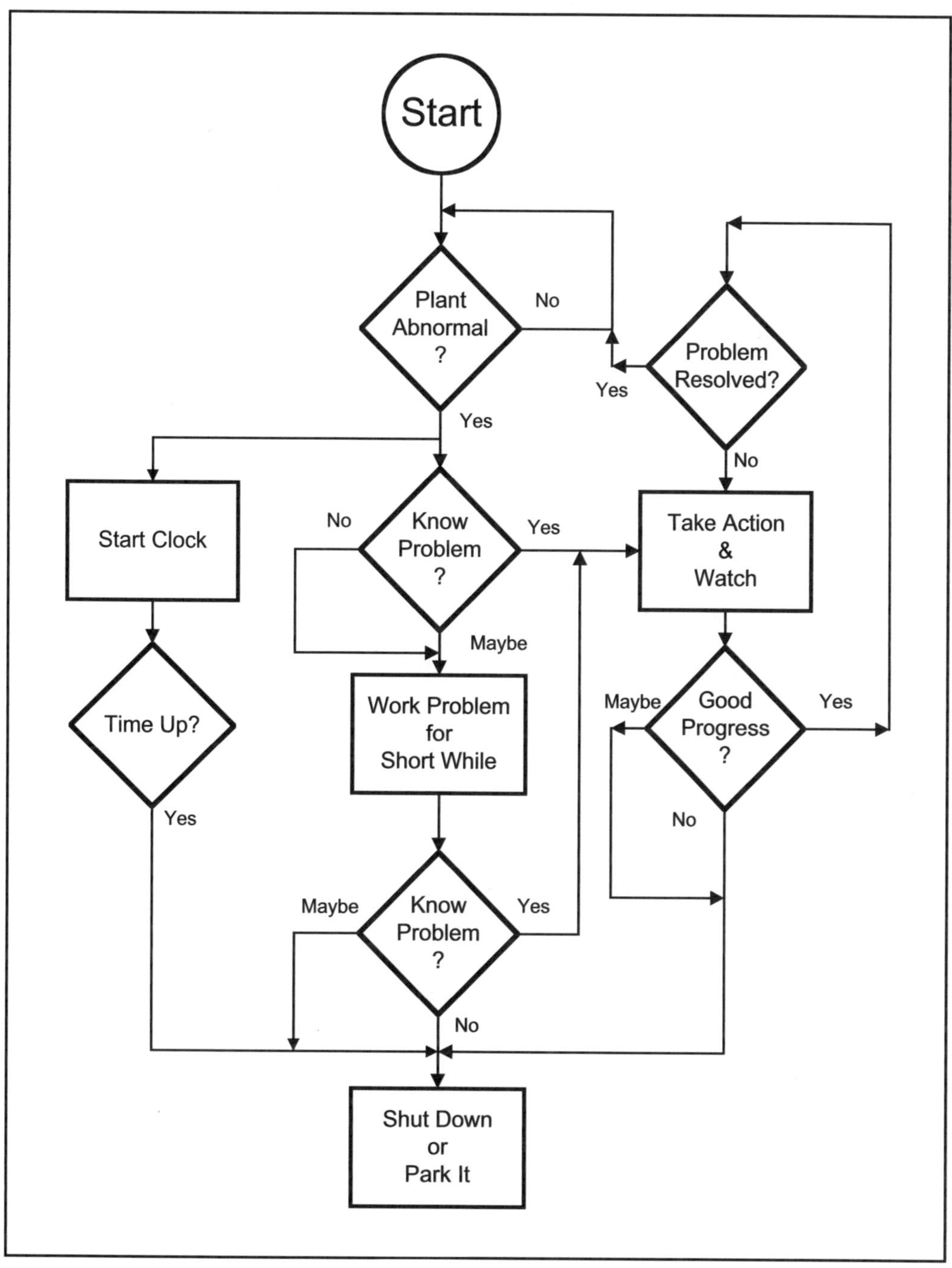

Figure 5.8.1. Permission to operate diagram

- *Question 1.* Does the operator feel too harried to take a careful, methodical approach to understand, evaluate, and manage the situation?

  YES—Discontinue operation

  NO—Continue to operate

- *Question 2.* Does the operator feel that there is enough time to slow down and think things out?

  YES—Continue to operate

  NO—Discontinue operation

- *Question 3.* Has the operator tried the same or similar things to regain control more than once in this event without clear success?

  YES—Discontinue operation

  NO—Continue to operate

Putting It All Together

You get the idea. Where there is uncertainty, either explicitly by the operator not knowing or understanding, or indirectly from what is happening to the operator, the operator should close things down. Figure 5.8.1 illustrates the decision process.

5.9 SHUT DOWN AND SAFE PARK

When permission to operate is withdrawn, the operator must stop operation. Stop means shutting down. Shutting down is harsh. It is time consuming. It is expensive, and it is fraught with opportunity for more trouble with equipment damage and some poor-quality product. What if there were alternatives to shutting down? Safe park is such an alternative. It provides a valuable way station between full operation and shutdown that can buy time for remediation, reduce the risk of unwanted consequences, and permit a smoother and more efficient return to normal operation when it is OK to do so.

For a safe park, a set of process operating states are identified that can be expected to render the process in a place that is safe and somewhat operable, though far from productive. The objective is to stabilize the process and avoid the necessity for complete shutdowns. The constructive result is an ability to forestall the impending consequences likely to come as a result of the abnormal situation in progress—that is, to slow down or freeze time. Recall that time is the most valuable commodity that an operator can have during an abnormal situation. If enough time is available, almost any situation can be managed. Moving to a safe state can be extremely effective when employed for downstream units during an upset in the upstream ones. This process is also quite effective for stabilizing an upset unit itself under certain conditions. These safe states are carefully identified and controls put into place to automatically move to them at operator initiation. Though, in some cases, operators following a carefully scripted procedure can do the transition manually.

Here are some of the practical ways to employ the concept of exploiting this idea.

Operator-Initiated Shutdown

The standard emergency shutdown is familiar to all current industrial manufacturing plants. In one form, OSHA mandates their presence. They are currently designed as a last-resort control measure. These we call emergency shutdowns (ESDs). Operator-initiated shutdown is one small but significant step back from that. For this, much more care is taken to do things in an orderly, controlled, but direct manner. It is a predesigned, preimplemented capability that the operator can initiate from the PCS that will start a largely automated sequence of shutting down the unit. Implementing this capability usually requires the replacement of most manually operated field components with ones that can be operated remotely. A few manual ones can be left, but the "people action" must be carefully coordinated. When implemented correctly, this will ensure that the plant will be in a much less damaged shutdown state than if the ESD system had done it. This should eliminate or significantly reduce equipment damage. As a result, it should facilitate a faster restoration of production when the plant is ready to restart. Of course, those situations that could not be properly managed by this controlled shutdown will be caught by the ESD, thus ensuring process safety integrity foremost.

Automated Shutdown

There is a small chemical plant that faced the problem of managing upsets in a very straightforward, logical, yet unusual way. The management recognized that there might be a very dramatic payout for exploiting the operator load differences between normal operation and upset or abnormal operation. Recall that during normal operation, operators are utilized about 25% of their time. But during upsets, they are loaded to 300% of their ability—since operator load is not elastic, this means that up to two additional operators might be needed during upsets. In a small plant, this flexibility was not economical. Their solution was to invest in additional instrumentation and controls equipment to fully automate a reliable shutdown and start-up.

As soon as the operator determined that the plant was unusually abnormal, rather than attempting to work out what might be amiss and try and fix things, the operator, instead, initiated the automatic shutdown process. The plant was then quickly and properly shut down. The result was minimal production losses with only a rare equipment problem. Any equipment problems that did happen were almost always due to the abnormal situation that caused the situation rather than a flawed shutdown. Once the system was shut down, the operator would attempt to diagnose the problem and affect a repair or call in others to assist. Most problems could be promptly remedied. Once this was done, the operator would initiate the automated start-up process, and the unit was timely back on line.

They did the job well and, at the end of the day, saved the plant quite a bit of money. In the final analysis, the combined savings in manpower and avoided equipment damage far outweighed the expense of the automation and its associated upkeep. Not surprisingly, during normal operation, the plant actually ran better due to the added instrumentation and operational flexibility.

Safe Park

Full automation for start-up and shutdown is normally not considered for a complex processing unit. Moreover, many are highly integrated with neighboring units. Consequently, having an option that is capable of providing additional time for the operator to work problems and at the same time greatly reduce exposure to the danger and damage of continuing to operate would be very valuable. That option is *safe park*. Safe park is a methodology that designs appropriate ways for production units to move to a much safer and more reliable operational mode to buy time and at the same time to avoid the problems of an extensive shutdown.

For each needed safe park unit, a carefully worked-out scenario is developed that permits it to be parked instead shut down. Simple examples are placing distillation towers in total recycle, putting pumps on low power with sufficient kickback to avoid operational problems, and placing a furnace on low fire but in a way that controls thermal stresses and inventoried product problems. The specific details will be governed by the particular unit, of course. Any additional needed instrumentation and controls are added. Operating procedures should be expressly designed for this activity, and the corresponding protocols are in place.

5.10 SPECIAL TECHNOLOGY

By establishing the criterion for understanding the plant's operating status, have we replaced one dilemma with another in the guise of improving operations? Have we simply transferred the problem? The answer is a definite "No." The reason is basic and important. If the operator ever fails to understand what his process is doing—no matter what the reason, no matter what additional technology could have been provided to help him to see what is wrong and find a prompt and effective solution—at this very moment he does not know what is happening. Using the permission to operate model, he is obligated to cease operations and move to safe or shutdown state.

The more effective the operator is, the better job he will do at foreseeing the abnormal and managing it before things get out of hand. There are two key areas where technology can assist in the application and effectiveness of the operator. The first is in the area of detection and warning of abnormal situations. The second is in effectively placing the process in a safe state or shutdown.

Detection and Warning of Abnormal Conditions

Many of these tools are in the development state or are proven in the lab and are awaiting commercial introduction. However, most are in place and providing value in at least one major refiner or chemical manufacturer. They are listed in this summary document to inform where it might be useful for specific problems.

Conditions Related to the Plant

- *Sensor validation.* These are for transmitters to ensure that they are working and the values they forward are reasonable and tracking the process. Early warning of sensor failure (or malfunction) is extremely useful to maintaining production integrity and valuable for ensuring proper function of advanced controls.
- *Control loop validation.* These are constructed around the critical control loops in each production unit. There are several approaches. Some use plausibility calculations on mass and energy balances around a small piece of the unit. Some infer variables from other upstream and downstream variables to calculate the loop variables in question. Where extensively used, they provide an extraordinary measure of early notice.
- *Multivariable state measurements.* This is a very powerful tool to determine whether a process is in control. The theory takes advantage of the fact that regardless of the large number of plant variables, there are actually a small number of degrees of freedom. Therefore it is usually possible to construct a visually understandable view of a process and make an early determination of how normal things are. There are other tools that use process measurements and extract important features to determine whether things are normal. Many of these will likely find their way into industrial plants by the mid-twenty-first century.

Conditions Related to the Operator

- *Operator vigilance.* Most of the time an operator is not in close view of the process. Often there are distractions. There is always the looming fatigue that goes with shift rotations, especially during the quiet shifts. The technology for maintaining operator vigilance involves both identifying the state of alertness of the operator as well as keeping the operator alert and focused on important tasks and ready to monitor for problems.
- *Operator intent.* Due to the complexity of many tasks that an operator might be called on to perform, coupled with the complicated nature of the modern PCS and its associated database support technologies, it is not unusual for an operator to attempt to do something and find it difficult to do. An intent recognition tool attempts to identify what the operator is trying to do and offers to assist with the

task. This is especially important during high-stress situations and highly involved or complex procedures.

5.11 OPERATOR REDEPLOYMENT

Permission to operate is a fundamental way to understand and manage the very tricky business of knowing when to continue to operate and when to stop. And that is important. However, before we consider this matter resolved, there is one more important tool we can add to the operations table: dynamic operator deployment. That is a fancy way to suggest that moving operators into different roles during abnormal situations can have a profound effect on both their understanding of the situation and their ability to manage it effectively.

Let us first take a look at a schematic central control room. Figure 5.11.1 shows a typical plant arrangement with four operators. The basic and primary materials flow is shown progressing from the upper-right operator counterclockwise through all four stations. The primary shipping is done from the plant managed by the operator of the lower-right station.

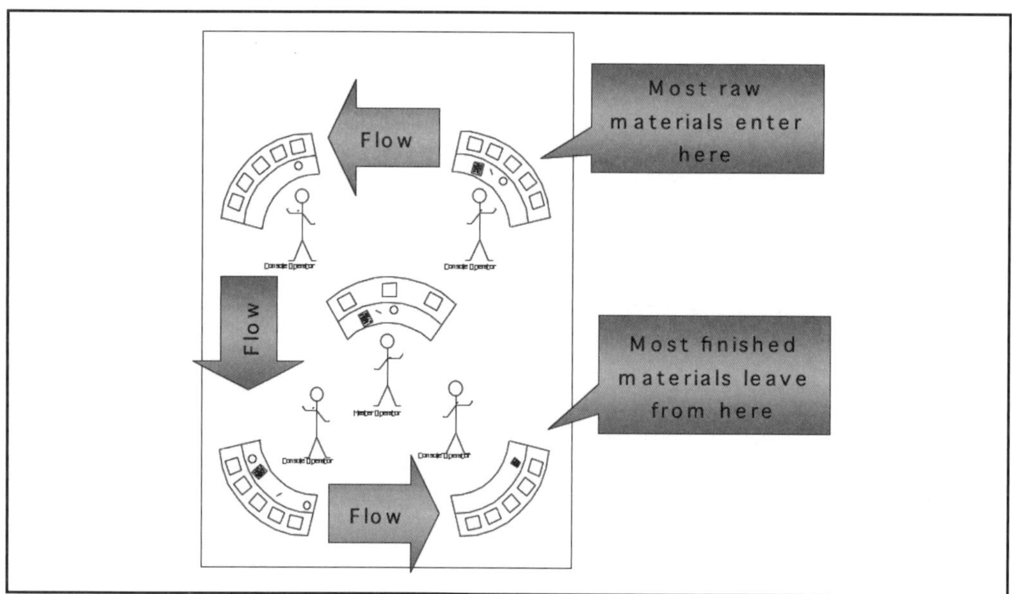

Figure 5.11.1. Typical control room operator arrangement

Let's suppose that the second operator's area has a significant problem (upper left in fig. 5.11.2). Let's term this operator the *console operator.*

At this point several important operational events happen in quick, preplanned, and proceduralized steps:

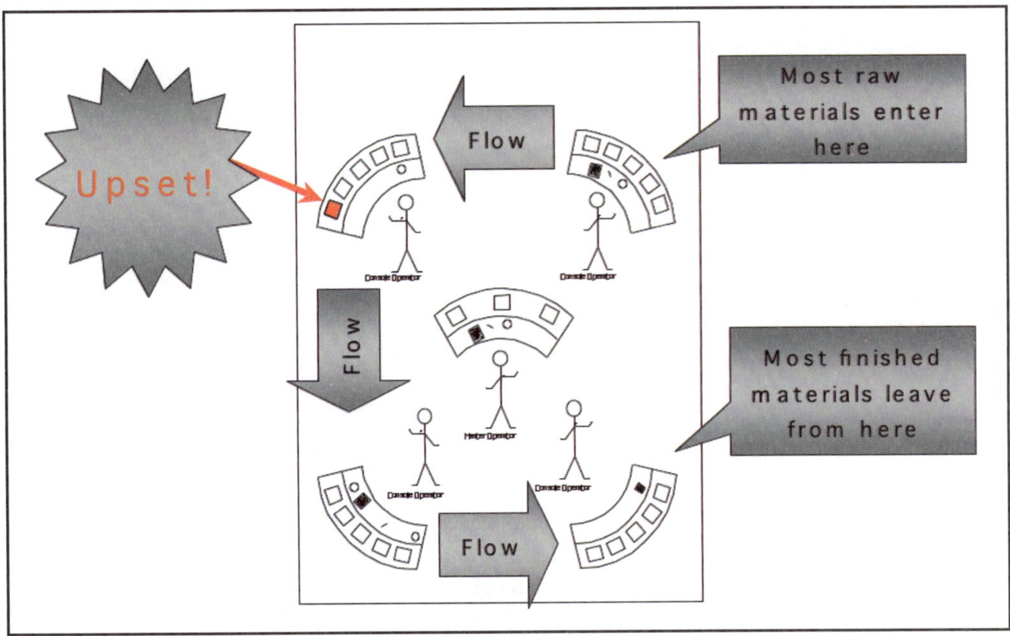

Figure 5.11.2. Second operator develops a significant problem

Step 1. All of the console operator's operational responsibilities pass to the control and management of the nearest upstream operator. In our case, it is passed to the closest operator upstream in the material flow path—upper right.

Step 2. Our console operator remains on station. However, his focus is narrowed to include only the limited portion of his plant that is upset. For the moment, his only task is to locate the appropriate displays and control "handles" that might be needed for management. No independent actions are taken. Only at the direction of the managing operator (explained in step 3) will he perform clear and distinct "hands-on" control and other state changes to plant equipment that are possible from the control room. His counterpart in the field will do the same for field equipment.

Step 3. Another position, that of the *managing operator* (fig. 5.11.3), steps in. This post may be filled by the operating supervisor for the shift, by a senior floating operator, or by any other prequalified operations individual. This managing operator will, for the entire course of the upset, be responsible for the minute-by-minute operational tactics to be used to manage this situation. The managing operator will direct the console operator to perform single tasks, one at a time. Only the current task will be told to the console operator. When that task is complete, he will be told the next task. For example, the managing operator might instruct the console operator to perform the following task sequence:

1. Put the reflux flow controller in manual and set the valve output to 20% flow.
2. Stop the pump on the rundown tank.

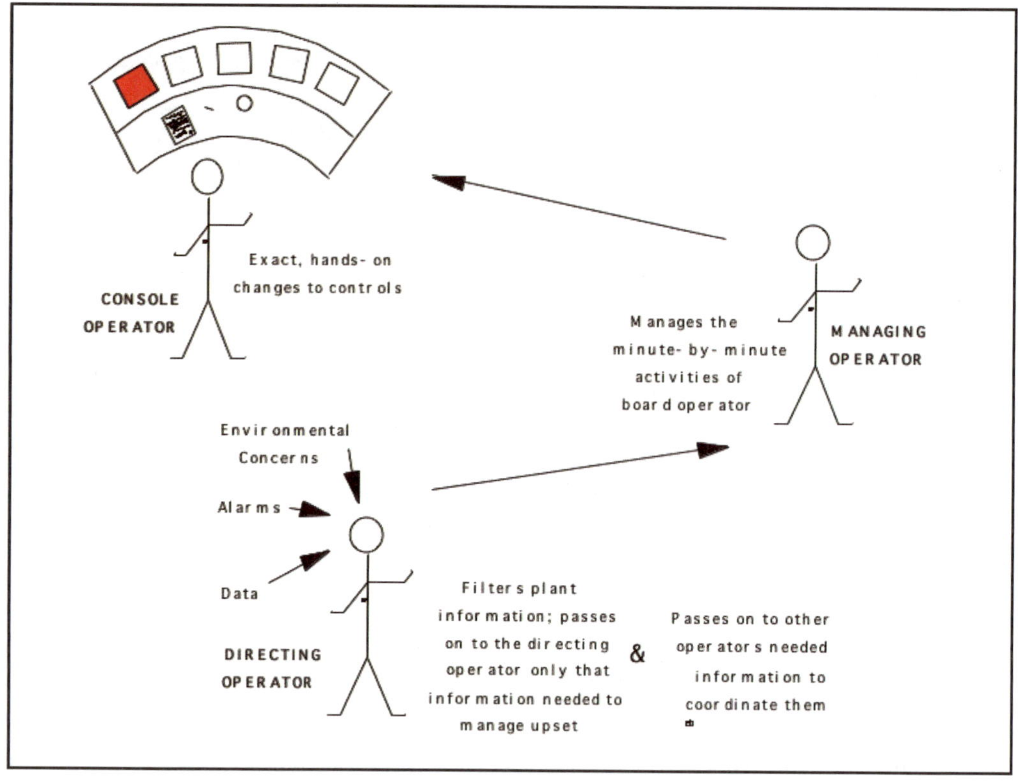

Figure 5.11.3. Operator deployment

3. Reduce the steam flow to the reboiler to keep the temperature in the upper tray area around 165°F, give or take.
4. And so on until the current specific tactical plan of the managing operator has been carried out.

Remember, only one task at a time is on the agenda for the console operator. It is up to the managing operator to decide what to do and in what order.

Step 4. A *directing operator* (fig. 5.11.3) steps in to assist the managing operator by providing all external situational information needed to fully understand the rest of the plant and its state. His job is to be aware of the entire rest of the plant and only pass on to the managing operator what is needed to manage the immediate and specific threat posed by the upset event.

Many plants will quickly suggest that they do in fact use this concept during upsets—not always, but enough to suggest to themselves that this might be not much of anything new. That should be comforting to those of you reading about this for the first time. Good, but not good enough. When those same plants, claiming that this is a practice, are asked if it is part of the formal plant procedures, they will say no. When asked if it is practiced to build skills and competencies, they will again say no. When asked if there are any other ways that this is incorporated into the operational infrastructure, they will again respond no. I leave it to you to draw the lesson.

You might be interested to know that this process for operator redeployment is an actual working procedure for U.S. nuclear power plants to manage upsets. It has been used with considerable success. It is proceduralized. It is practiced. It is part of the fabric of operations.

5.12 PROCESS COMPLEXITY

Moving operators around can provide important strengths to manage abnormal situations. However, not all production processes are the same, and not all operators are interchangeable. Process structural complexity has a part here. It can have a significant impact on the difficulty of managing a production unit as well as operator training and the ability to substitute operators. "Difficulty of managing" means, of course, harder to control and keep working as designed and as required for productive, safe, and reliable operation.

Consider two basic types of complexity—linearly related operating blocks making up a unit and integrated/complex related operating blocks making up a unit. Figure 5.12.1 illustrates a typical linearly related unit. A word about the pictorial nomenclature: the horizontal arrowed lines represent major material or energy transfer conduits. These lines sometimes align vertically to route to the next connection point; however, so long as the arrow enters or leaves blocks horizontally, they represent transfer. The rectangular blocks with letters inside represent a transformation (significant processing event) that converts the infeed coming in at the arrow (at the left) into product at the arrow going out (at the right). While differing processing events (lettered blocks) can and usually do have important differences in action and complexity, for purposes of this discussion, those differences are not of interest. Looking at Figure 5.12.1, we see that the product of block A is supplied to both block B and block F. The product of block B only supplies block C, and so on.

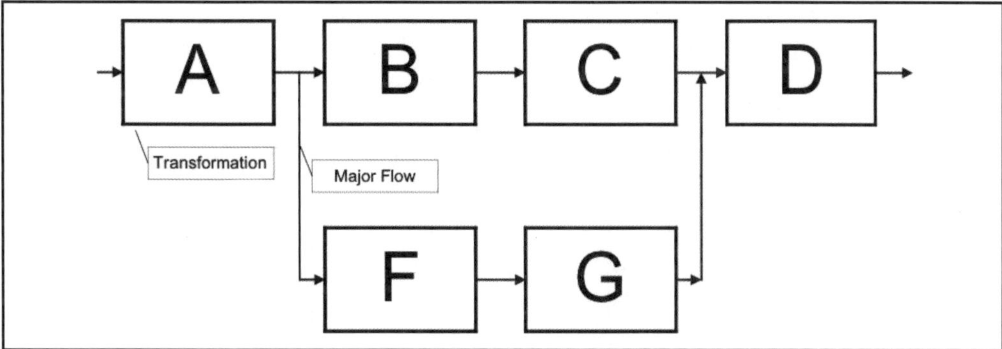

Figure 5.12.1. Linearly related operating block unit

Linearly Related Complexity

We take a closer look at linearly related units to see what operational observations can be made. First, and perhaps most important, they are relatively easy to understand (individuals in the Western world generally think and visualize in this form of relationship). The function of these units is more readily visible. In the event of problems, or just to gain a better understanding of what is going on inside the process as a whole, each processing block can be separated or decomposed from the rest of the process. Once separated, we can readily examine it as if it were the only thing in the world of interest. No other block will affect how this block works. Yes, other blocks might change the material or energy coming into our block, but our block will always work the way we discovered when we examined it when it was (mentally) entirely separated from the others.

Consider the skills and capabilities of operators of linearly related operating units. First, they can, and usually are, generalists. All they require is the ability to understand the concept of input streams and output streams, the knowledge of what makes good feed and good product, and how this block producing the product works. They can readily be cross-trained since the basic skill sets for all similar processing blocks and plants are similar as well. Since the skill sets are similar, they can more easily be substituted for other operators in similar units. To summarize what we've said here, for these types of units, the operators are able to readily understand all parts and could likely replace each other when necessary.

Integrated/Complex Related

We now examine a more complex type of production unit. The integrated/complex related unit will look similar to Figure 5.12.2. Again, a few words about the pictorial

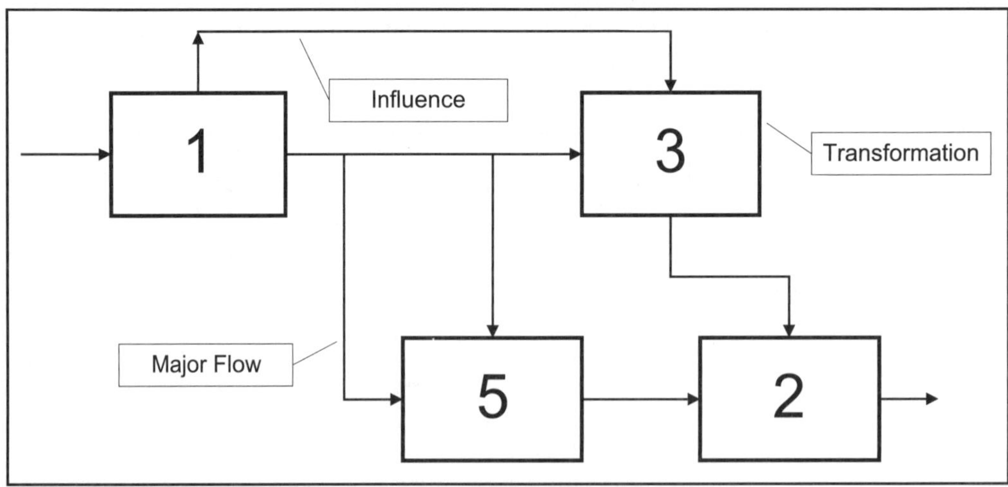

Figure 5.12.2. Integrated/complex related block units

nomenclature: the vertical arrows leaving and entering numbered blocks from and to above represent significant influences not material or energy. These influences affect how the specific block will function. For example, block 1 supplies material/energy directly to blocks 5 and 3 (arrows entering from the left sides of each block). However, block 1 also sends instructions (influences) to blocks 3 and 5 (arrows in from the top) that tell block 3 and block 5 to change what goes on inside the blocks as they separately processes its infeed into product. It matters less exactly what block 1 is changing in block 3 or block 5. What is of importance is that depending on how block 1 influences blocks 3 and 5, they will behave differently, even if the material/energy infeeds were to be unchanged.

Examine the problems of operating and skills and capabilities desired of operators of integrated/complex related operating units. The operation of this type of unit is much harder to understand than the previous linear type. There are more subtle implications and effects. Most problems or issues in the functioning of one part of the plant cannot be separated completely from the rest of the plant. It is as if we need to constantly be looking at the entire, complicated thing just to understand even relatively small irregularities in its parts. As a consequence, the people operating such a plant need to be specialists. Because specialists are required, it is therefore harder for any one individual to keep an eye on anything more than the part they are experts for. Thus, they are not easily interchangeable in the event substitution is needed.

At the end of this entire model stuff is an important concept. Depending on the internal structure and complexity of given production plants, operator requirements and operational requirements can be vastly different. Management must make provisions for this in staffing, training, and policymaking. This is much different from what is general practice in most industrial plants today.

5.13 TRAINING AND SKILLS

If something is rather simple in its construction, most of its aspects are easily gleaned through the exercise of ordinary skills employed using ordinary care. Things that are not simple are not so readily understood. Moreover, when we must understand them, it often requires specialized skills and capabilities. We examine two quite different enterprises that have dramatic training and experience methods. It is hoped that some of the concepts from one might be useful to the other.

Industrial Manufacturing

Industrial manufacturing enterprises, like the ones we consider for alarm improvement, are generally manned by individuals from management, operations, technical support, and administrative support. For operations, we usually find two groups: operators and supervisors. Most operator positions are hourly (often supported by union representation). They are recruited largely from the general labor pool in the community. At times, they are staffed by others transferring to operations. They receive qualifications testing before they are hired, some level of individualized or group formal or semiformal foundation training, and a sizeable amount of on-the-job situational-response training. Their job is to make sure that the production facility produces the required amount and quality of product at an acceptable material and energy consumption for those aspects under their control.

Many supervisors/managers are promoted from the most qualified and experienced operators. The position is usually salaried. There may be qualifications testing before they are promoted, but usually not. There is little if no executive or manager training; although some do offer a general suite of salary-level training that is also offered to the general salary staff of all departments. The job of supervisors/managers is to ensure staffing of the team, resolve scheduling and other daily requirements and differences, assist or advise during emergencies, improve the general performance levels of the team, and satisfy all other administrative requirements.

The overall operations team, the front line of the production enterprise, functions with its workers mostly reacting to the production situation of the moment and its managers making do with an implied model of success (based on experiences as an operator, supplemented by what could be recalled from working with previous supervisors and observed from current peers). There is very little in the way of formal continuing training and experience enhancements, save from what might be gleaned from daily experiences. Finally, and perhaps just as telling, throughout their careers, there is very little real evaluative and strength-probing evaluation done and shared back with the individuals. What little is done tends to be highly variable and generally not followed up with careful coaching and development. There are exceptions, of course.

Military Training

Contrast the above situation with another "purpose-designed" model described below.

Military entities, for all of their historical use and perhaps even misuse and abuse, have personnel categories that are very carefully structured. Again, we have the two functional groups: officers (managers/leaders) and soldiers (workers/technicians).

Soldiers are prepared for duty through an extensive regimen. First, they are fully screened (physically, emotionally, and intellectually). Then they are carefully indoctrinated with their purpose, mission, and functional model. Following that, they are trained by experts in a dual role configuration: First, they are fighting men and women who analyze first-level tactics, perform that role as individuals and a team, and assume limited roles of authority as required in the heat of crisis. Second, they are specialists with specialized fighting skills or technical skills. They are assigned to a daily working team to further develop their performance skills and training. Their training and development will permit them to be reassigned to any similar unit as well as substituted for other positions within such units. Periodically, they receive additional specialized training from formal experts.

Officers are prepared for duty by the two-phased process of careful selection and then extensive training. Their selection comes not from the general pool of soldiers, but from a separate pool. Of course, officers meet the same minimum physical, emotional, and intellectual standards of solders as well. Their training is also done by experts. First, they are carefully indoctrinated with their purpose, mission, and functional model. Interestingly enough, part of this training is done by senior soldiers. This has two effects: it provides clear skills taught by those who know it works because they've done it, and officers are left with a new respect for the capabilities and role of the soldiers who will serve under them when they assume command. Moreover, in order to advance in command responsibilities (rank), they are required to undertake periodic formal advanced training. At the upper levels, this training will be as extensive as any advanced postgraduate degree in the nonmilitary workplace. Within their large sector of operation (infantry, air, armor, etc.), they are fully interchangeable with other officers and often rotate units as they advance significantly.

This discussion is not for a moment going to suggest that industry go out and immediately adopt a military organizational and training model. But, when those concepts are introduced to the industrial manufacturing process, important ideas for change receive consideration. Moreover, the understanding of training strategies and the appreciation of plant complexity are vital to the establishment of rules for granting permission to operate.

5.14 OTHER KEY PRINCIPLES OF OPERATION

This section documents most of the other key operating principles that were discovered at high cost by that major U.S. producer of petrochemicals. Recall the earlier section 5.4, "How Permission to Operate Came to Be" for that story. These key operating principles are grouped into five areas: operating, field, safety, design and inspection, and management. At the heart of these rules is the fundamental edict:

- No situation is more important than obeying these rules for safe operation.
- No one can give anyone permission to violate these rules.

Additional Operating Principles

- *Operate the process only as long as you fully understand it.* This is the gist of permission to operate.
- Never violate a procedure.
- Do not operate for extended periods of time with alarm conditions or repeated alarm activations.
- *Believe your instruments.* Field verify any concerns.
- *Make sure equipment is fit to return to service.* All problems found and corrected; all parts ready; equipment fully functional; equipment aligned for operation; no abnormal conditions; all codes met.

Field Principles

- Never violate formal lock-out tag-out procedures—even one mistake is *two mistakes* too many.
- Always *formally* isolate *blinds, plugs, or double-block with bleed.*
- *Always* assume that there is stored energy in a system before opening it. *All block valves leak, all drains and vents are clogged, and all local instruments are not working.*
- Check-valves leak in *both* directions and are often sealed shut when they should be open.
- *Repair* all broken equipment.

Safety System Principles

- *Verify* all relief systems.
- Do not send liquids to any relief system.
- Do *not* "gag" or disable *any* relief or safety valve—*ever.*

Design and Inspection Principles

- Use only *positively* identified and labeled materials for design, repair, or replacement.
- Have mandatory inspection programs to make sure all materials have not weakened due to *corrosion, erosion, mechanical stress, thermal stress, chemical alteration.*
- When a failure occurs, inspect similar systems for potential failure of similar components.

Management Principles

We've saved these for last, when actually they are the first ones needed. These are for the guy in charge and his direct reports. Here they are:

- Clearly set your own site Rules or Principles. Make sure that everyone knows what they are, why they have been set, and why they cannot be violated.
- Require a MOC process for everything.
- Specify that all design reviews must be conducted by qualified personnel who were *not* involved with the original design.
- Fitness for service (personnel and equipment) must be *proven*, not assumed.

- It is everyone's responsibility to assure that prior site experience *and* relevant industry experience are fully integrated into this site's operation. Recall the wise manager from the "General Lessons from Incidents" in chapter 2, who initiated a program of positive materials identification and saved a serious incident from happening within his own plant after hearing about a tragic incident at another similar plant.

5.15 WHAT IS BEING DONE BY OTHERS

The concepts described in this chapter are being used at production sites by many of the major industrial companies. All are serious efforts aimed at making a crucial improvement in safe plant operations. Technology vendors are also hard at work developing support technology to facilitate an even wider application base. These companies range from the PCS technology vendors, both the major equipment suppliers, as well as providers of third-party controls software packages and knowledge-based technology firms.

Permission to operate. At least one major refiner uses this as their de facto rule of operation. It is been in place for several years and is more effective than anticipated. One major chemical manufacturer used this is as guiding principle in preparing for their Y2K response to problems.

Safe park. This is in use and part of the culture of several major refiners and being seriously considered by others. An interesting sidelight on this is that it has been used to validate management-selected staffing levels. In response to union objections that there was not sufficient manpower to control upsets, the management dictated that the plant has been designed to manage upsets by moving immediately to a safe operating state. This procedure allows maintaining safety at the reduced staffing levels used during normal operation.

Technology in Development

Much of the technology mentioned here is either proprietary, being developed independently by a number of end users, or under development by commercial technology vendors or academic institutions.

5.16 CONCLUSION

For many readers, the topics in this chapter might have seemed to be out of the normal pathways for alarm improvement. By now, it is hoped that you might have a broader view and some alternatives to consider. Alarms are an operator support tool, not the be all and end all. It will be the operator who must use the alarm tool as an augmentation to the job of production management. Good production management means the proper management of risk and the best assessment of what can be done to achieve responsible plant operation. The operational infrastructure can make a difference. The goal of this chapter was to provide alternatives. The examples of what might be done and considered are provided with the intention that plants be able to design effective policy for improved production management.

5.17 NOTES

1. Ian Nimmo, "The Operator as IPL (Independent Protection Layer)," *Hydrocarbon Engineering*, September 2005.
2. U.S. Chemical Safety and Hazard Investigation Board, "Urgent Recommendation [BP Texas City Explosion and Fire, March 2004]," news release, August 17, 2005.
3. Health and Safety Executive, *The Explosion and Fires at the Texaco Refinery, Milford Haven, 24 July 1994* (Sudbury, Suffolk, UK: 1994).

CHAPTER 6

Alarm Philosophy

Commitment is the stuff character is made of; the power to change the face of things.

—Chinese fortune cookie

Effective design is not done on a framework of the haphazard. Robust, functional, and usable systems are based on a clear design philosophy that incorporates the essence of the nature of their function. Alarm systems are no exception. Chapter 6 explains how the alarm philosophy is developed and later used to provide the foundation for alarm system design.

Start with what the dictionaries think about the word *philosophy*:

- Theory or logical analysis of the principles underlying conduct, thought, knowledge
- The study of basic truths . . . logic . . . rationalization
- A particular system of principles

Alarm management philosophy is going to be somewhat broader than the straightforward dictionary entry. Think of alarm philosophy, or alarm management philosophy, as a one-to-one synonym for "the design basis for an alarm system."

This chapter will lay out the process that leads to the development of a site alarm management philosophy. We start by making sure that everyone is comfortable with what the philosophy is (a design specification for alarm improvement), what it looks like, and how to prepare one. We then move on to identifying all the key assumptions that create the foundation on which we construct the philosophy. Building on the assumptions, the discussion progresses to identify all the elements that need decision and specification—which, when done, will be the actual philosophy. The chapter contains many lists of items to consider. It also contains specific guidance regarding how to consider the items and what some of the alternatives might be for the large number of items that need decision. The chapter closes with the detailed description of what a philosophy workshop is and how it is conducted. Examples of a completed philosophy document can be found in appendix 4.

6.1 KEY CONCEPTS

Philosophy purpose	The alarm philosophy document is where all aspects of the alarm design (or redesign) are specified. It will also contain all additional aspects of site infrastructure that are needed for project success. This will serve as the entire design basis for the work. All site personnel, all contractors, and all consultants will rely on it. Incident investigations will use it.
Detailed engineering	The detailed specifics for the work are located in supplementary documents that must comply with the philosophy in all respects.
Participants	All key site personnel will develop it together. They will have been empowered to speak for site management.
Living document	As work progresses, the alarm philosophy document may be changed, but always using the site-approved management of change process.
Approval	Site management will approve this.

6.2 CAVEATS

A Foundation Is at the Bottom

Building something new, if done well, involves laying a foundation, then adding the first level and adding additional levels resting on the levels below until the needed number of levels has been reached. Good construction requires that each level used to support the next level be sufficiently strong to perform all of the necessary requirements. There is a story here, which architects and building designers are fond of sharing with others, that brings the true implications of this concept home.

The story goes somewhat like this. An owner of a building decides after the plans are in a preliminary design stage to add an additional floor to the top of his structure. Let us say, for example, that the structure has ten floors and an eleventh floor is contemplated. The architect comes back to the owner and quotes a price of 20% additional. Whereon the owner replies to the effect that all he is doing is adding a small thing at the top, so why isn't the cost much closer to 10%? The architect's reply captures the essential truth of the design change: "While you think that you are simply adding a floor to the top of your building, you are, in fact adding a floor to the bottom of the building." This means that all of the floors below the new one being added at the top need to be strengthened to accommodate the additional loads. The top floor only needs to support itself. The bottom floor needs to support all floors above and is commensurably more costly.

Think of alarm redesign as the building. The alarm philosophy is the foundation. Every time you want to change or add a seemingly small thing along the way, remember that it will need to conform to your design—the philosophy. Quickly, those small things can require significant modification to a philosophy that was developed along different lines. Everything built on the philosophy depends on it in some way or another. Not only is that dependence during design and construction—it will be for longer. It will be for as long as the alarm system is used.

Owner versus Designer

The entire purpose of an alarm system is to provide effective operator support. At the end of the day, the operator's ability to understand, use, and benefit from any alarm system is the final test. Plant operations therefore owns the alarm system. However, "the ability to understand, use, and benefit from" does not necessarily mean that the owner is the best or even a qualified designer. After all, you are pretty much adept at using your automobile. You can easily decide which brand and model handles to your satisfaction. It is doubtful if you would be as effective at designing the fuel injection system. However, experts at designing fuel injection systems also drive automobiles.

Alarm systems are designed by those qualified and experienced in the technology. Operators have a vital role to play. Some may even lead the activity. Most times, it is a team comprised by operators, engineers, technicians, safety, and management personnel.[1]

Reliance on Philosophy

There might come a time when an operator or engineer may be called on to "justify" (sometimes it is really *defend*) a decision or action. Management should understand that an appropriate reliance on the philosophy must be construed as evidence of good faith and acting with proper responsibility.

Completeness

Items and lists provided in this book are suggestions for consideration. They are not exhaustive, of course. And you do not have to use them. But they should be well worth your consideration. The final list that each site develops will need to represent the best efforts of that site. It should cover those concerns and issues your site feels must be included in the alarm system design basis.

6.3 GETTING STARTED

Let's dig into this. We are at the solution part of the alarm improvement roadmap (fig. 6.3.1). The manufacturing team under the leadership of senior plant management develops the alarm philosophy. It describes the intended alarm design/redesign process from

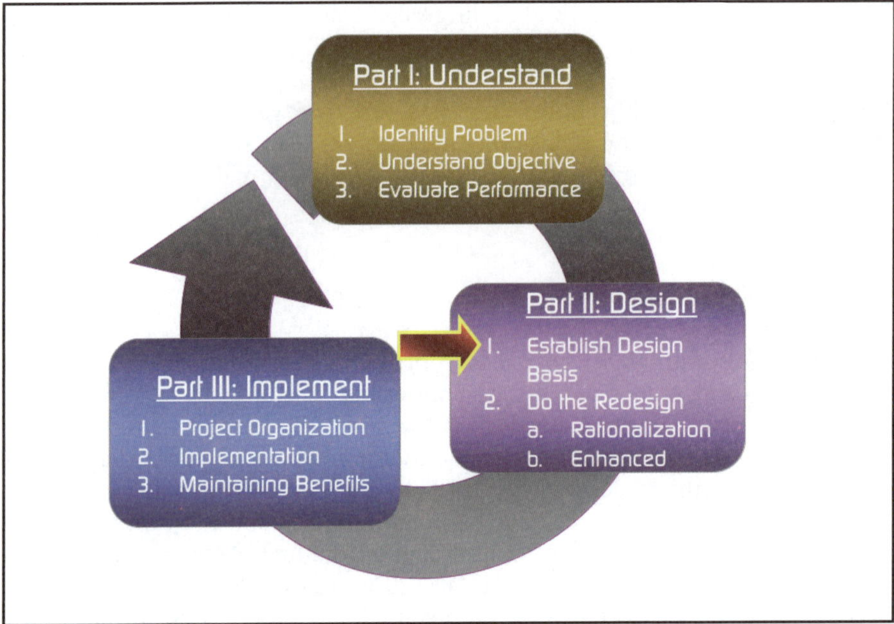

Figure 6.3.1. Roadmap location for philosophy

beginning to end. It starts with the buy-in of site management, progresses through the critical success factors, and includes the complete engineering design requirements to do the job well. It enables the provision for adequate financing, development of realistic schedules, and the inclusion of cooperation and participation from all the other key site players. The list of the detailed engineering design requirements will include the working definition of alarms; their proper response; other details of the alarm system; the integration with maintenance, training, and the remaining plant infrastructure; and the specific path to implementation. This will all provide postimplementation robustness and relevance.

Operator Survey

The alarm system is a primary operator support tool. A good one can really assist the operator. A poor one is what you might have now. Many sites ask the operators how they think their current alarm system is working. There is a formal survey form that EEMUA 191[2] recommends. A revised version is shown in appendix 3.

Advice to the Reader on Timing of This Topic

The discussion of alarm philosophy comes quite naturally at this juncture in the development of effective alarm system designs. However, it is suggested that you delay a bit more before actually preparing one. In other words, please wait until you have read and understood what this book is all about before you set out to do the important task of developing your own alarm design by developing a philosophy.

6.4 SPECIAL ALARM ISSUES

Types of Alarms and Their Recommended Use

Conventional PCS controller loops can configure many types of process alarms. The following are examples:

- High absolute
- Low absolute
- High-high absolute
- Low-low absolute
- High deviation
- Low deviation
- High rate of change
- Low rate of change

All these types are useful. Each provides a certain specialty that will be just right for some situations. Which ones to you will want to use and for what purpose, generally falls into one of the three categories discussed below. There are exceptions, of course.

Normal Process Abnormalities

High and *low* in this discussion refer to the directionality for approaching the abnormal situation, not anything else. The high and low absolute alarms are the workhorse of alarm systems. They are used to signal events that are enough out of the ordinary to require operator interventions. The high-high and low-low absolute alarms are reserved for the unusual situations where the few normal abnormalities can suddenly, and without the usual warning, escalate to the very serious. This escalation would only happen during the normal course of the operator working to resolve the current alarm. Once the situation has escalated, the consequences and operator actions change dramatically. For all other cases, the high-high and low-low alarms would be redundant with the high and low alarms—therefore, they would be unnecessary. Please also refer to "Special Cases of Redundant Alarms" below.

Incipient Process Abnormalities

Incipient abnormal situations are those where the process slowly approaches an abnormal situation. It is the very act of the slow movement that makes the situation difficult for the operator to notice. And since it is slow, the total amount of movement is small. Before an absolute alarm limit could be exceeded, the plant would be in serious trouble. These situations are usually best identified by rate-of-change alarms. As the process moves toward the abnormal, the rates of change usually increase.

Processes that Move a Lot

There are some processes that change their production aspects quite a bit during a normal production cycle. The production requires close attention, but the attention points are always moving about. One way to track the abnormal situations is to watch for abnormal excursions from normal. Deviation alarms are ideally suited for this purpose.

Smart Field Devices

Besides improved functionality and ease of application, smart field devices also incorporate significant diagnostic ability. Most of the diagnostics track device degradation and alarm them long before the device fails. Most, if not all, of these diagnostics are of little value to the operator—but extremely relevant to maintenance personnel. Current practice is not to configure operator alarms for these but to allow the PCS to route them to maintenance personnel directly.

Light Boxes

Light boxes are the colloquial term used to describe those individual alarms that are announced by dedicated electronic hardware separate from the PCS. The usual form is a wall- or panel-mounted illuminated engraved window that will light and sound a hardware horn whenever the point goes into alarm. Some use differing colors for the window or lightbulb to indicate alarm severity. They occasionally use different sounds for the horn to do the same.

If light-box alarms are meant for the operator (our operator in the operator area), then light-box alarms are handled just like any other alarm, with the exception that they are usually considered to be a "required alarm." But even required alarms are examined to ensure that the priority is correct, the activation point is correct, and the appropriate operator support documentation and training is in place.

Special Cases of Redundant Alarms

After all is said and done for a properly designed alarm system, you will notice that an entire class of alarms that might have been used before is not to be found now. You will note that there are no (or very, very, very few) prealarms. You won't very often find a high alarm and then a high-high alarm. Nor will you find low and then a low-low. This is not an accident. It goes back to our fundamental precepts underlying alarms.

Before alarm redesign, the reason that plants used both a prealarm and an alarm were to ensure that a few important situations were not missed by (a) announcing the problem early and then (b) confirming that the problem still existed later on as it persisted. The prealarm (the high or low) was provided to alert the operator that there was a problem. The follow-on alarm (high-high or low-low) was provided to let the operator know that the problem (which might have or might not have been seen or worked on earlier) is still there and getting worse. After the activation of the follow-on alarm, it was expected that operator action take place, usually serious operator action. So far, this all sounds good. The failure becomes obvious when we think about what should be done to configure the prealarm and the follow-on alarm, using our newfound approach to effective alarm design.

Consider the follow-on alarm first. Properly designed, this alarm would require operator action and must alert the operator with enough time remaining for good operation to be restored. During the process of restoring good operation, we expect our operator to keep a continual eye on what's going on and remain aware of how things are progressing. Since, by design, the operator should be able to manage this abnormal situation from the time the alarm activated, there is nothing to be gained by alerting earlier. Nor should we assume that the *process safety time* is any less, nor the *time to manage fault* any longer, for the follow-on alarm than would be the case for any prealarm. If anything, the follow-on alarm would represent a more serious situation. As such, it is reasonable to expect both the time for diagnostic effort and the plant response time to be longer than they might be for a less-serious prestate. Consequently, a prealarm represents nothing more than a redundant alarm. It should not be used. You should very rarely see high-high or low-low alarms.

About Alerts

Prealarms are not such a good idea. To replace them, we might be tempted to provide much the same functionality by using an alert. Chapter 12 presents an illustration of the alert concept. Alerts aren't alarms, so we won't have any redundancy to worry about. As things start to get bad, we just issue an alert for each case. Sounds like a good way to have our way. Or is it?

Alerts are meant to convey messages to the operator without (improperly) having to resort to alarms. But alerts are important in their own right. They convey important information that the operator needs to stay on top of things. And just like any other tool, if it is used in a way not intended and for which it is ill suited, it becomes less able to do what it was intended for. If we were to use alerts to provide prealarm warnings for "important" alarms, we would generate a lot of alerts. In effect, we would be trading alarm overload for alert overload. Overload is overload. Maybe the best thing is to avoid the prealarm conundrum altogether and trust in our alarm design to do what we designed it do to.

Many PCS systems do not provide even a rudimentary messaging or alert type of functionality. In order to remove all the alarms that are present in an unimproved alarm system, it may be necessary to use a workaround. If the PCS alarm system has one or more alarm priority levels that are not required for operator alarms and it permits those alarms to be (a) routed to places other than the operator's station and (b) configured so they will not sound the operator's horn, then those extra priority levels may be used as alerts. None of these alerts is counted as or treated as an alarm in the philosophy or anywhere else.

Classes of Alarms

There is a movement among alarm practitioners to categorize alarms beyond required and ordinary. Typical extra categories are *highly managed* and *safety*. Those practitioners go as far as suggesting that they be configured and operated differently from the required and ordinary ones. Please resist this movement. It is not needed. It is an unnecessary complication that provides a bit of comfort at the expense of good alarm practices.

In a properly configured alarm system, alarm importance is built in. For most plants, abnormal operation can result from missed low-priority alarms as well as from missed high-priority ones, from alarms that derive their importance from financial impacts as well as those that relate to environmental or safety ones. Any attempt to overlay the best practices with additional alarm complication will only serve to confuse the operator and unnecessarily complicate a well-designed alarm improvement process.

6.5 OVERVIEW OF ALARM PHILOSOPHY

The redesign of an alarm system will necessarily impact a number of the entrenched parts of the existing plant infrastructure. A one-size-fits-all approach usually does not work very well here. Every plant has its own history, its own way of doing things, its own problems, its own goals, and its own style. So that all of those diverse aspects can be managed well, the chapter on alarm philosophy has been designed to bring out the really critical ones. The plant will decide what and if to include and how it will be done. With that disclaimer in place, you will find lots of reminder lists of items that usually are considered in most philosophy designs. You should find them quite useful. They help to ensure that something important is not inadvertently left out. As we start this coverage, please consider that the philosophy can only contain what is put into it. All assumptions, all preconditions, and all other parts that you want to think of as being there should not be left up to chance. Lay them out and include them specifically into the document you will prepare.

Philosophy 101

Before you get started with any alarm improvement project, it is important to recognize the fundamental aspects underlying successful process management. These will form the foundation of the design assumptions. They should be included in your document. They should be fully specified and explained, including specific action items that support them. The alarm improvement teams can only suggest the explicit arrangements that will be needed to adequately cover nonalarm items. They are included as part of your alarm redesign philosophy, not only for use of the alarm improvement teams, but also for the entire enterprise. The other steps necessary to ensure that the needed coordination is done should be assigned to others in the plant who normally deal with the particular infrastructure item. Example items include maintenance practices, general training practices (as opposed to alarm redesign training), operating procedure revisions, and the like.

The following items are the key bases to cover. The material illustrated in each part conveys helpful suggestions about what to include as well as what others have found to be a best practice. Your site will use them, modify them to suit, or develop others in their place. Your alarm philosophy will define and clarify each of the items that follow.

Operator-Centric Items

- *Responsible operator.* The alarm system is not intended to take the place of proper operator management of the process or the operator's constant, watchful eye and exercise of insightful judgment.
- *Qualified operator.* Operators will be fully qualified and appropriately trained and monitored (performance, physical health, emotional and psychological health) for the job.
- *Operator ownership.* Alarms and alarm system are for the operator, not maintenance, not the safety department, and not the environmental department.
- *Alarms mean action.* All alarms will be responded to in a timely and appropriate manner.
- *Activations provide sufficient time.* All alarms should provide sufficient time for the operator to manage the process abnormality.
- *Priority guides.* Alarm priority will be used to guide the operator's order of attention to alarm activations.
- *Alarm response information.* Appropriate information will guide the operator to understand the abnormal situation and help decide and implement remediation actions.
- *Appropriate design.* Alarms will not be used to compensate for poor process design, poor equipment, inadequate maintenance, weak or ineffective procedures, and inadequate personnel training and readiness.

Plant-Centric Items

It is assumed that the plant has been designed with due care. Its construction and operation are proper and according to plan. Therefore the alarm system itself should not be used to patch up, accommodate, or otherwise make up for inherent plant design inadequacies, maintenance inadequacies, poor procedures, unduly stressful operations, operation outside of proper design conditions, and the like. If the plant shares this view, then it is important to make sure that the alarm system is designed to be consistent with the primary understandings. Moreover, it is essential that the entire plant team recognizes this and takes the necessary steps to modify practices and standards to ensure that the plant is in condition to take proper advantage of the alarm system.

Broken and missing equipment. All equipment that is part of the plant and used for production shall be replaced if missing and fully repaired if operationally impaired or broken. If certain equipment has ongoing and unavoidable operational problems and therefore cannot be rendered fully operational during production, then specific procedures must be in place to handle this situation. Those procedures must include the operation of the alarm system during this situation. Training and controls must be modified to include proper accommodation of this situation. It must not be left to the alarm system to identify and moderate such situations.

Unusual plant operation. The plant has a design basis and an approved operating envelope. The alarm system will have been designed with that in mind. If the plant will be operated outside proper limits, for whatever reason, proper MOC should be used rather than reliance on the alarm system. While alarms might be important and useful, they have not been explicitly or implicitly designed for such operation. Therefore it is likely that activation points, priority, and other alarm information and performance may not be adequate.

Unusual plant situations. Plant manning and operating procedures have been developed and approved based on the plant practice and needs of operation. Any significant departure from these expectations shall not rely on the alarm system for adequate operation. Unusual situations include the following: operating during severe weather or other unusual natural situations, operating or attempting to operate during severe manpower shortages (perhaps due to illness, weather, or other emergencies), operating or attempting to operate with marginal or otherwise unsuitable or untested raw materials or other resources, and operating with marginal or untrained personnel or management.

Recognition of limitations of alarm system operation. Unlike safety systems and other key infrastructure components that have been designed and implemented to provide unambiguous operation under all conditions and situations, alarm systems are not. Alarm systems must not be relied on to cover all situations and accommodate to all insults. Careful attention should be directed to uncover missing safety systems and other safety infrastructure to avoid any undue implicit reliance on the alarm system to keep the plant and personnel safe.

Alarm System Purpose

The purpose of the alarm system is to bring a potentially abnormal process condition to the attention of the operator in time for appropriate remediation and with appropriate guidance for success. In the event that the operator suspects or is already aware of the possible existence of an abnormal situation or condition, the alarm system shall provide confirmation for those concerns.

One might observe that the alarm system is the presafety shutdown system. Thus the alarm system provides the last clear chance for plant operations to restore good operation before the plant is forced to end operation. As such, it offers a productive way to appropriately support the enterprise.

Philosophy Intent

The alarm philosophy is, in the simplest of explanations, a complete design requirements list of how the alarm system is supposed to be designed and operated. It will *not* explain how to construct a project, how to estimate costs, how to man the work, or which vendors to approach to assist you. Its sole purpose is to describe the end result in as careful a way as possible. Here are some of the key parts to include.

Needs evaluation. The needs evaluation is an examination of the performance of the current plant to identify where it stands against the requirements for proper design and operation. This part must address what measurements are to be made, how they are to be done, what constitutes meeting the desired performance requirements (KPIs), and how to interpret the results in such a way as to provide action guidance and change requirements.

Design. The design will include what the new alarm system should look like, how the detailed redesign specification should be constructed, how changes in the other parts of the plant necessary to support the alarm redesign are coordinated, and how the new alarm system is to be produced.

Implementation. Implementation will cover all aspects of bringing the new alarm system to life and ensure that it fits in with the rest of the site infrastructure. Included are operations practices, procedures, revisions, training, documentation, MOC requirements, and the assignment of responsibility for the entire supporting infrastructure changes into the appropriate hands.

Validation. No new design is done until it is proven to perform as designed. How the new system will be proved shall be specified. Require all inadequacies to be addressed according to procedures laid down in the philosophy. The methodology for handling any uncured or incurable performance is also specified here. *A validated alarm system is one that has been shown to match the design requirements.*

Auditing. No new design is effective unless it is shown to resolve the issue that prompted it. Here is where the performance of the redesigned alarm system will be measured and compared against the requirements. Not only will the project goals be audited, but also the resulting performance must be validated against the project targets. For example, does the alarm system interfere with the operator's understanding of abnormal process operation? Does the alarm system appropriately guide the operator to understand and manage alarms? Is the alarm system a distraction during normal operation? Inadequacies will be addressed and remedied. Here is where all such requirements are specified. *An audited alarm system has been shown to meet the actual needs of the plant, whether or not those needs were adequately expressed in the design requirements.*

Maintaining. What works today may not necessarily work tomorrow. Here is where the plans are laid down and enforced so that the alarm performance will consistently deliver the required benefits into the future. Here is where we make sure that there are appropriate requirements for continued auditing, for auditing after incidents, and for auditing after modifications and any other events likely to impact the design or performance of the alarm system.

Elements in the Philosophy

The following list highlights the key areas that the alarm philosophy should cover to guide design decisions:

- *Alarm design principles.* What specifically defines alarms? How should they be set up? How should they be interpreted and incorporated into the operator's kit?
- *Key performance indicators and critical success factors.* What important requirements need to be met for a proper alarm system design? How will we audit for proper performance?
- *Approved management of change requirements.* Do the modifications, additions, and changes to the existing MOC requirements include alarm management? Is the alarm system as well as the greater controls and operations support infrastructure appropriately secure against reckless and nefarious attacks and other insults?
- *Process for rationalization.* Specifically, what will be the recommended approach to be used to arrive at the requisite number of properly configured alarms?
- *Activation point determination.* Exactly what will be the procedure used to set the alarm activation point values?
- *Priority assignment.* How will priority be used and assigned to each alarm?
- *Alarm presentation.* How will each alarm be shown to the operators? How will operators locate the needed information to effectively deal with the alarm?
- *Enhanced alarming.* What logic and other controls will be put into place to ensure that each alarm activation properly reflects the current plant state and operations need?
- *Operator roles.* What is the operator expected to do before, during, and after alarm activations?
- *Interplay with procedures.* How will procedures reflect the alarm system? How will the proper use of the alarm system be represented in procedures?
- *Training.* How will training be modified to accommodate the new alarm system design? How will the new alarm system design rely on training?

- *Escalation.* When things go wrong, what is the plant's expectation for alerting others, obtaining additional assistance, changing operating goals, and shutting down or "parking" the plant?
- *Maintenance.* What are the maintenance requirements for the new alarm system design, both for the alarm system and for the rest of the plant?
- *Support/technology required in addition to the alarm system.* There are a number of key aspects of operation that need to be recognized and provided for that cannot be provided by the alarm system itself (or which currently depend on the alarm system, but should not). These are very important in their own right. To the extent that alarms will no longer provide information or support for these activities (even though they were done ineffectively), other ways to provide this support must be identified by the philosophy. The short list of these items includes (a) messaging, (b) true operating condition or status of plant, (c) tracking event escalation and migration (to other operator areas), and (d) system alarms for both process-related aspects (sensors, transmitters, etc.) and nonprocess aspects (storage, data highway loading, etc.). Refer to chapter 10 for a more complete discussion.

6.6 ALARM PRIORITY

Important Message

Pay particular attention to the definitions you will use for scoring each alarm in the area of *consequences and severities.* You will be doing the scoring for each alarm—which means thousands will be done. Since the priority distribution of the result will not be known until that work is well underway or complete, you will want to be sure that your definitions are correct. For if they are changed, then all alarms that were categorized before will need to be redone.

With consistent definitions, it is possible to modify priorities en masse simply by adjusting the consequences, severities, and urgency map to priority. This is the power of using tools.

Alarm priority is a clear and powerful way to expose and manage operational risk. Alarm priority encapsulates the urgency and impact effects of what will happen if the abnormal process situation, reflected by the alarm activation, cannot be remedied or cannot be remedied in time. Regardless of the reason for the operator's inability to provide a proper control action or the process's inability to respond in any particular case, priority stands as the single message of the abnormal situation's importance. Current industrial practice uses three or four levels of importance. These simple levels of priority are called on to reflect a potentially broad number of attributes. The attributes most often considered include the following:

- Safety of personnel
- Environmental impact
- Product quality, production rate, production schedules
- Equipment integrity
- Enterprise finances
- Business reputation

A word of caution is given here. It is important to keep the list of impacts as short as possible. While it might seem helpful to include a range of enterprise effects to please a broad constituency, remember that for each consequence you will be required to score or rate each and every alarm. For thousands of alarms, this gets tedious. At some point, the alarm design team will likely become overwhelmed.

Priority Levels

When we examine the configuration possibilities found in PCSs, we often find the ability to configure only a very small number of priorities, say three or four. Or, just as likely, there will be a provision for eight or, for a few others, many dozens of levels. In the first case, the level problem is solved by default: three or four levels available means you will use the three or four. In the last case, we can have too many. We will need to understand what it means to have numerous priority levels.

Priority will be used by the operator to do one single task—decide the importance of alarms. He will use this importance to decide which alarms to work on and in which order. Alarms that activate one at a time and do so very seldom would not need any priority. (Though traditionally, even for this case we used two levels: emergency and other. Remember the dedicated alarm panels?) Actual production situations are quite different. Often alarms will activate close together. If alarms really do reflect issues and problems that differ significantly in their importance, then a way must be found to convey that importance. In an alarm situation, generally the operator will concentrate on understanding and working alarms by selecting the highest priority first. Once those are under good management, the next lower priority level will be examined, and so on until all have been worked. Situations do occur where the higher-level priority alarms consume all available operator process management time. It is entirely possible, even expected, that the lower-priority ones might not be addressed at all. Then the plant will actually take the "hit" of those abnormal situations coming to true incidents. All the more reason to assign the correct priority level to each alarm.

Too Few Priority Levels

The question of how many levels to use is important and must be visited and decided. If we use too few, then we will often face situations where alarms have activated and there are too many in the highest priority to be worked on adequately. "Worked on adequately" really means resolved in time to prevent or at least manage the impacts implied by the process getting into trouble. In addition, we will have a more difficult time matching priority to the actual levels of urgency and consequence. This will show up mostly in those "gray areas" that always seem to appear between carefully established boundaries. A slight miscalculation, or the missing of a small but important aspect, might mean using a lower level when a higher one should have been used or using a higher level when a lower one should have been used. Either way, we mislead the operator.

Three or four priority levels are generally used. The actual number of levels will be decided by the alarm philosophy. Table 6.6.1 summarizes the names for three and four priority levels. Some plants use three levels. Broad experience recommends four. Why four? Suppose we used only three. The top level, emergency, is always reserved for very few situations. These are cases where life, serious equipment damage, major financial impacts, or very detrimental environmental damage are at risk. Therefore we would expect very few alarms to be configured at this level. Recall the configuration metrics from chapter 4, where EEMUA 191 recommends about twenty alarms total. With one priority level taken up this way, we are left with only the two remaining levels to cover all the rest of the cases.

Experience has found that real problems and actual plant situations aren't so cooperative. They usually come in small, medium, and large. Without "medium," we will have to force-fit all the alarm priorities that would normally be in the middle (medium priority) to be at one end or the Other—that is, to be either high or low. Using four levels permits us to reserve a category for the very, very important and urgent situations and still retain the three useful levels for the rest. To summarize this important point, if priority will be used as it is intended, then we should limit the number of levels to a small manageable amount—the recommended number is four.

Priority labels for *three* levels
Emergency
High
Low

Priority labels for *four* levels
Emergency
High
Medium
Low

Table 6.6.1. Three and four priority levels

Too Many Priority Levels

Using too many levels often presents a different problem. While we do not see any one priority level overused, we will see small differences in urgency and consequence evaluations reflected in different priorities. Just as we saw in the case of too few levels, we again find that slight miscalculations might spell the difference between using a lower level when a higher should have been used or using a higher level when it should have been lower. If the operator will truly use priority to triage upset alarm activations, and we expect higher levels to be managed before going to lower levels, then an error in priority might just cause an important "lower-priority" alarm to be delayed too long. That means the priority could and should have been higher.

Priority Names

Why would anyone really worry about the labels for the levels of priority? After all, most of the labels are used so much that they appear standard. They are familiar. We know what they mean. But what if we do not actually know what they really mean? Maybe, just maybe, different plants, different locales, or differing cultures infer different meanings to familiar label names. This common understanding we supposed to be there just might not be. If we are not careful, we might end up with a similar "conflict" that we have with "low" process variable measurement level versus "low" priority. Remember that high alarm on a low level? Label confusion is only a part of the concern. Just as important is what the label connotes.

For example, when a medium priority is assigned to an alarm, it is important that when that alarm activates, the designer and the operator both know how important that alarm is. The operator must know instinctively what medium really means. This is quite different from knowing that all low-priority alarms would indicate less of a plant risk (in the way that the alarm philosophy has defined risk and how it maps into priority) than a single medium alarm. We need the operator to be able to understand each priority name without just consideration of rank order of importance. A simple but easy-to-understand illustration for this need would be to consider any two alarms whose priority differs by only one step. Pick medium and high, for example. If only a medium alarm were to activate and no other alarms were active, what is the operator feeling about the potential problems to his unit were he unable to fix the situation? How would that feeling change if the single alarm were to be high priority instead of medium? There would be a difference. To ensure a reasonable level of consistency between operators, the philosophy should include a clear and useful explanation of what the different levels of priority mean in terms that relate to the operator and his job. This definition process will become extremely valuable during the site testing of the rationalization priority assignment process covered in chapter 7.

This discussion will not attempt to lay out anything close to definitive recommendations on labeling. Rather, we will simply advance a few options for consideration. To keep this simple and grounded, let's presume that our plant has determined that four levels of priority are needed. Table 6.6.2 suggests a few options that plants have historically used for such labels.

Please do not take this discussion as a recommendation or even suggestion that each site should consider using labels different from the conventional ones. However, there have been honest discussions where the conventional labels were standing in the way, in a serious way, of a plant accepting the idea of rationalization. They had to work it out before anything else could be considered. We are probably not able to fully understand the motivation and possible nuances that produced this type of situation. Rather, it is only important to recognize that the alarm improvement ground is newly plowed.

Priority level	Conventional	Alternate 1	Alternate 2
Highest	Emergency	Emergency	Extreme
Middle high	High	Critical	Major
Middle low	Medium	Danger	Serious
Lowest	Low	Warning	Minor

Table 6.6.2. Alternate priority level labels

Planting the proper seeds might take more than the usual amount of give and take. Good places to do so are areas that are unlikely to force compromises in the basic fabric of alarm improvement. Labeling, where the site has assured everyone that it does not confuse, is one of those areas that can be worked out.

Humorous Illustration of Priority

There is a humorous story that comes to mind when we try to provide an illustration of how priority might be used by the operator. You will note that a few of the definitions for alarm priority are more than a bit dated! Here is how it goes:

> Our operator is in the middle of eating a bowl of favorite soup. If the operator, upon hearing the alarm activation, drops the bowl in the haste to address the situation, we call this a priority 1 (the highest) alarm. On the other hand, if after hearing the alarm, the operator puts the bowl aside to manage things, we call this a priority 2 alarm. If after hearing the alarm, the operator pauses just long enough to refer the problem to someone else, it is likely to be a priority 3 alarm. Finally, if after hearing the alarm the operator merely turns it off without missing a spoonful, we have a priority 4 (the lowest) alarm.[3]

Consequence and Severity

This is how plants factor in the important aspects of enterprise integrity. The first task will be to lay out all the key factors of concern. We call these concerns *consequences*. Consequences are normally measured by assigning two or more levels of values called severities. For example, let us say that reputation is an important enterprise value. The enterprise is saying that if its operations go astray and things happen, they want to consider how the going astray and things happening might impact their reputation. To do this, they need to produce a way to measure the effects of a production misfortune on reputation. Then they would use the measure to determine the impact of production misfortunes as a result of any specific abnormal situation on their reputation.

Definitions

Consider misfortunes that cause the plant to fall behind on product deliveries to valued customers. Such a situation might cause the plant to become generally known in the manufacturing community as one that does not reliably meet production promises. Similarly, other production misfortunes (e.g. quality slippage, failure to manage hard-to-fi nd resources, etc.) can be mapped into their impact on reputation. Please note that reputation is not an often-used consequence. It is discussed here as an example that will provoke thought rather than ask the reader to bring previous experience into play. However, it certainly can be and is sometimes used.

The procedure is repeated for the other consequence categories. Generally, plants won't need to use more than three or four categories. The most used categories include safety, environmental, and financial. Once the list is firmed up, the next step is to break the set of categories into a limited number of severity levels. We generally use none, low, medium, high, and emergency. Table 6.6.3 shows a typical blank form. The rows are labeled with the specific consequences the plant decided to use for priority assignment. Three consequences are shown, but any reasonable number can be used. The columns are labeled with the chosen breakdown levels (severities) for each consequence. Five levels of severity for each are shown, though more or less can be used. All these labels should mean the same thing to everyone in the enterprise. An excellent way to start on this is to begin with already-developed risk categories in use at your plant. Usually they can be found and modified to fit into alarm use. Doing so can have important benefits. First, it is easier to build on something already started. Second, and most important, it allows the alarm system to very closely align with the existing site protocols and initiatives—thus reinforcing the usefulness of the site protocols and placing alarm management squarely inside plant policy.

With the consequence categories in place and the severity levels labeled, we now establish a working definition for each severity level of each consequence. For example, a working definition for the severity levels of safety is illustrated in Table 6.6.4. Similarly, the remaining working definitions are produced for each of the other severity levels and consequences.

In this manner, we produce a filled-in definition for each cell of Table 6.6.3. You will need to use these definitions to rate each candidate alarm. For example, consider that a particular high-pressure alarm were to be completely missed by the operator. If the safety consequence of that missed alarm would result in the injury of a worker and subsequent week off of the job, then we would rate that alarm a medium in safety. Once these definitions are in place, the next step will be to add numerical weights.

	None	Low	Medium	High	Emergency
Safety					
Environmental					
Financial					

Table 6.6.3. Blank consequence-severity grid

Consequence	Severity level	Working definition
Safety	None	Negligible
Safety	Low	Minor injury to one or more; perhaps requiring first aid but no lost time
Safety	Medium	Injury to one or more; possibility for limited lost time; all return to work in existing capacity
Safety	High	Serious injury to one or more; likely lost time; possible disability furlough
Safety	Emergency	Very serious, usually permanent injury to one or more; possibly one or more deaths

Table 6.6.4. Example severity categories for safety consequence

Scoring

Now that we have identified the severities for each consequence, the next step is to develop a way of taking the severities to determine the correct value for the risk that they signify. To do this, we will identify a numerical value for each of the empty "boxes" (Table 6.6.3) that reflects the specific severity for the given consequence. For example, we might want to subjectively assign a medium safety impact to our process as 75. (The actual number scale is purely abstract—any convenient scale may be used.) For this example, place "75" in the box (Table 6.6.5) where the row for safety intersects the column labeled medium. Continuing, suppose we consider a high safety consequence to be a third more severe than the medium one. To do that, we place the value 100 (100 = 1.33 × 75) in the safety row under the column heading high. Continue the process until the entire safety row has numbers that you are comfortable with. You are free to use any convenient numerical scale. At this juncture, all that is required is the safety row to have values that adequately reflect the plant's relative severity for each described safety level as it is compared to the other safety levels. The number scale is arbitrary. The relations between the numbers capture all the need.

To continue, use the safety row as a baseline. All the other consequences will be rated in a similar manner. However, there is an added requirement that each of the remaining severity scales be determined not only relative to the other severities in that consequence but also relative to the baseline severity scale. In our case here, it would be the safety row. As an illustration, consider the environmental consequence to be the next one we develop. Using the definitions worked out earlier (see Table 6.6.4), select the single severity definition of which everyone feels the most comfortable. Try to select one in the center area of the severity list. For our example, suppose we pick the high severity level as the starting point. To assign a numerical value, use the definition for environmental high and refer to the numerical number scale you decided earlier for the safety consequences. For our example, recall that the consequence-severity value for safety high was 100. Considering only environmental high for the moment, select a severity number that makes sense for this consequence and severity knowing that you already picked 100 for safety high. It might be that, given the definitions for each severity, the plant would consider an environmental high to be about half again more serious than safety high. Therefore you would enter the value of 150 as the severity for environmental high (150 = 100 × 1.5). Once this first value is picked, the entire scale for the consequence is considered set. From this point on, all the rest of the severity weights for this consequence are picked relative to environmental high. Use only the definitions for environmental severities to rate different severities against the environmental high one (the 150 value) just picked. We do it this way, since there should be no expectation that the differences between severity levels for the different consequences will fall in the same ratio. Each severity is a separate entity. We must be free to score severity levels with appropriate weights that reflect the plant's concerns. In practice, however, the entire table of values usually retains a good relative balance between corresponding severities.

Continue this same way to determine consequence-severity weights for all remaining consequences. You would start with the next consequence and rate this consequence for a severity of high. At this point, all the consequences will be properly proportioned, each against the other, for the high severity. Then taking only that consequence, proceed to fill in the row in a similar manner to what we've done for safety and environmental. Table 6.6.5 shows a completed alarm priority scoring table. This is an example. Your values will be different.

We now have numbers for all consequences at all severity levels. We are ready to decide how these numbers can be combined to a single measure. Once we have a single measure, we are able to compute the recommended priority. But we are getting ahead of ourselves. The discussion continues with urgency.

Urgency

Some alarms must be attended to more quickly than others. Urgency is the descriptive term used by alarm management that keeps track of how promptly (how urgently) the operator needs to attend to a particular alarm. Urgency is factored in using multipliers that adjust the priority score to reflect the time needed to attend to the alarm. The objective of the urgency multipliers is to increase or decrease a score from the consequence-severity rank to include how urgent it is for the operator to find and resolve problems associated with the particular alarm. In general, urgency multipliers are constructed

Severity → ↓ Consequence	None	Low	Medium	High	Emergency
Safety	0	25	75	100	300
Environmental	0	75	100	150	250
Financial	0	25	50	75	150

Table 6.6.5. Example alarm priority scoring table

to increase or decrease the base priority (the determination that you would arrive at without considering urgency) by a predetermined portion of a level for each change in urgency. Again, using scenarios for which you know the priority you want to use, change the urgency by one level and see if the priority changes appropriately. In order to do this properly, you will be using multipliers close to a value of 1.

Not all urgency changes will result in a priority change, so do not be too ambitious with this. Keep it reasonable and real. The real use for this feature will be to adjust alarms whose priority score might be very near, but not over, a breakpoint value. The multiplier will then push it over to the next level, higher or lower, depending on the direction for urgency. In some cases, the "push" might not be over the line. That is fine, since it means that the priority without urgency will have been solidly within its level—which is where it likely should remain. Chapter 7 goes over the process for calibrating these values.

Table 6.6.6 shows a blank urgency table. The first column shows the ranges of time that are used to define different urgencies (from your alarm philosophy). The second column is used for those rare priority assignment protocols (again, from your alarm philosophy) that use only urgency to determine priority. This is not recommended, since it inappropriately places urgency over all other consequences. The third column is used for the urgency multipliers that "move" priority to include urgency. Table 6.6.7 shows an example filled-in table.

Time available (in minutes for effective action)	Priority (urgency-only priority assignment)	Multiplier (multiplies consequence-severity values)
≤ ___		
> ____ but ≤ ____		
> ____ but ≤ ____		
> ____		

Table 6.6.6. Blank urgency table

Time Available (in minutes for effective action)	Priority (urgency-only priority assignment)	Multiplier (multiplies consequence-severity values)
≤ 3	Emergency	1.4
> 3 but ≤ 10	High	1.2
> 10 but ≤ 30	Medium	1.0
> 30	Low	0.9

Table 6.6.7. Example urgency values and multipliers

Priority Assignment

The final step in setting up the assignment of priority is to develop the map that takes a score of urgency combined with impacts and severities and produces a recommendation for alarm priority. Briefly said, all we require is a table that uses the combined score and recommends the appropriate priority. Table 6.6.8 depicts an example for a four-priority system.

To use this table, let's suppose that the overall score with all the urgencies and impacts rated for a given alarm produced a numerical value of 270. Two hundred and seventy falls between 250 and 349 in the table. Therefore the recommended priority would be "medium." Let's see how this table is obtained.

Priority	Breakpoint value
Emergency	From 500 and above
High	From 350 up to 499
Medium	From 250 up to 349
Low	From 100 up to 249
[Might not be an alarm?]	0 up to 99

Table 6.6.8. Example alarm priority break points

Alarm Priority Assignment Setup Review

Setting up the priority rating process is straightforward. First, identify the plant's major concerns and assign definitions and numerical values to them. Next, factor in how important it is for the operator to respond promptly. Following that, identify numerical ranges for each priority level. Finally, take the numerical values from the urgency-weighted consequence severities and map them into an appropriate priority. Let us consider the (recommended) situation where priority will be determined by a weighted combination of the severities augmented by urgency. This process is diagrammed in Figure 6.6.1.

Here is a step-by-step illustration of how it works:

1. Decide the list of consequences (first column in Table 6.6.3).
2. Decide the number of severity levels and their labels for each of the consequences (number of columns after the first in Table 6.6.3).
3. Decide working definitions for each severity level for all consequences (Table 6.6.4 for example of "Safety" consequence).
4. Decide on numerical scoring values for each severity level of each consequence (example in Table 6.6.5).
5. Decide how many urgency (time to respond) levels you will use (example in Tables 6.6.6 and 6.6.7).

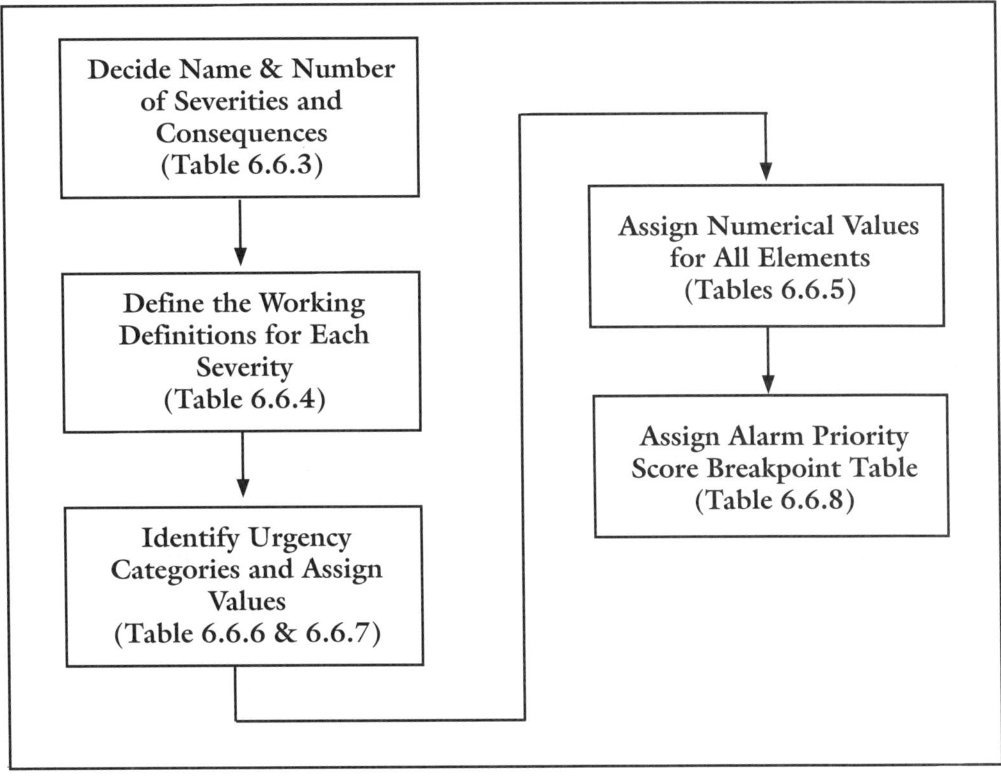

Figure 6.6.1. Procedure to set up rationalization priority matrices

6. Decide what the "urgency only" relationship to priority will be (example in Table 6.6.7) in the (unlikely) event that this will be the method used for assigning priority.
7. Decide the time range, in minutes, for each urgency level (example in Table 6.6.7).
8. Decide on the method to use to combine all the consequences and severities into a single value (example in Table 6.6.8).

These tables will be used for the rationalization discussed more fully in chapter 7.

6.7 ENTERPRISE PHILOSOPHY FRAMEWORK

The alarm philosophy is a site document. Regardless of whether the site is the entire business or the site is only a part of a larger enterprise, the eventual alarm philosophy for a specific project should represent that site. The reasoning for this is simple. Every philosophy must be complete and useful at the site level. Only in this way can the philosophy serve as a proper blueprint for the specific site's alarm system redesign. Your site might be a member of a larger entity, either an extensive plant that might be colocated with others or one of many separate sites in various locations working under the umbrella of an extensive corporate infrastructure.

Overview

We begin our work by assuming that there is a larger enterprise composed of a number of sites. If the philosophy will be done only for a single site, then the process is the same except that the framework document (meant to be the corporate start) will be your actual philosophy. Recall from chapter 3 that we have a well-defined team structure for coordinating sites within the larger corporation. This is repeated as Figure 6.7.1. For an enterprise, the first major step in the production of a working philosophy will be a framework document prepared by the corporate steering team. This team is usually composed of key members of each of the local site steering teams augmented by high-level corporate resources including important representatives from management, finance, and safety and environmental. The framework document is designed to serve as the starting point for each site's individual philosophy.

The job of producing a workable philosophy within a multisite enterprise is an iterative one. Figure 6.7.2 illustrates the way it usually works. The first step is the preparation of a corporate-level shared vision of the philosophy. This is sometimes referred to as the enterprise framework document. This philosophy will serve as the starting point for the sites to develop their own. Next, each site takes the vision and proposes modifications, clarifications, and refinements that it feels will facilitate the framework document to work better. It is important to point out clearly that this first activity by the sites is to improve

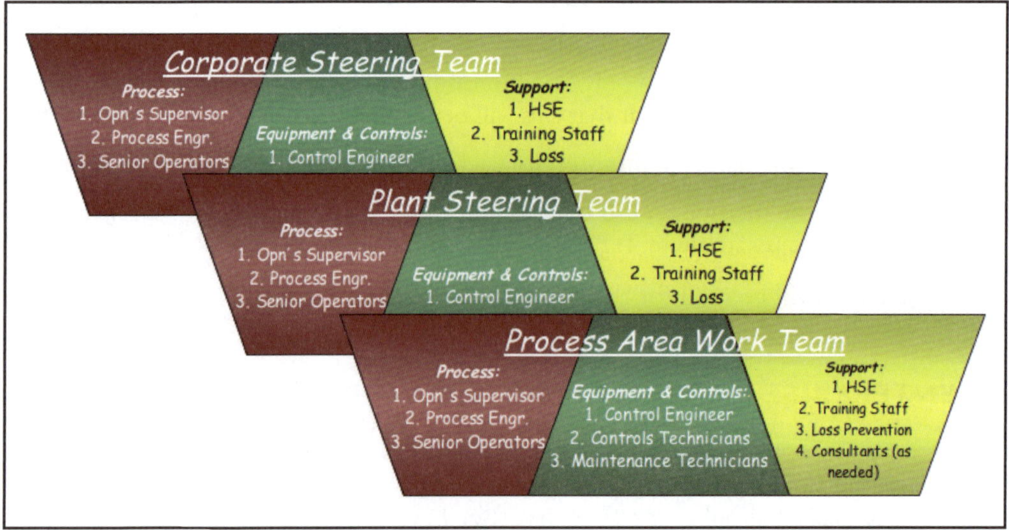

Figure 6.7.1. Enterprise alarm improvement team structure

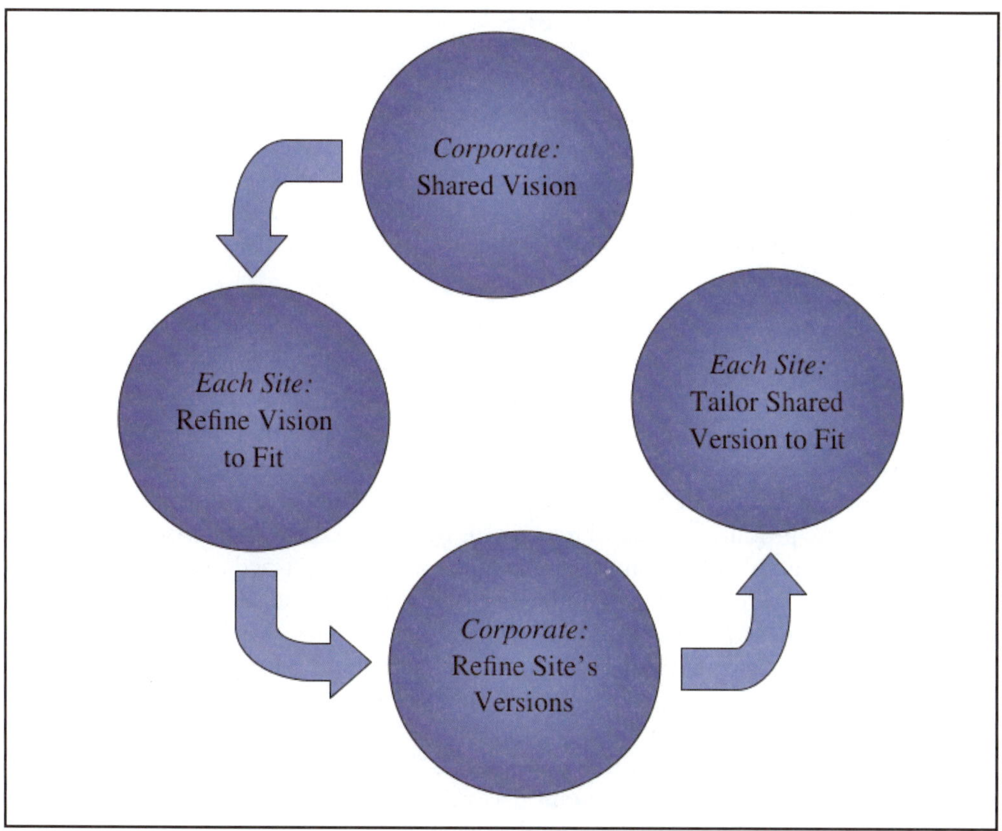

Figure 6.7.2. Iterative philosophy development process

the shared vision—not to move on and produce their own site version. The individual reviews from each site are brought back to the corporate team. Their job is to refine the shared vision based on the sites' suggestions. Where there is a large gap between what the sites suggest and what the corporate vision is shaping up to be, it might be necessary to refer the corporate vision back to the sites for one or two more go-rounds in order to arrive at a consensus document. Again, this document is the shared vision only.

The final step is for the sites to take that shared vision and use it to develop their specific site philosophy. This document may not necessarily require approval from other sites or the corporate steering team. It is recommended that the sites share their individual philosophies with the other participating sites. This is done not so much for approval, because this is generally not required, but for information purposes. There are often items in one's site philosophy that can be useful to other sites. Even if it turns out to be nothing more than better wording or to recall a forgotten item, there is useful benefit.

Framework Philosophy Document

The alarm philosophy framework document contains all aspects of alarm redesign that are required by the enterprise. Many of the requirements will be similar or even identical to those that could have been developed independently by each site. Some may not. The framework structure ensures that none of the expectations of operation that have been set at the top levels of the enterprise will be left unmet at the plant levels. Starting from a framework will eliminate unnecessary duplication and provide the individual sites with a jump-start on their own work.

At the Enterprise Level

It is in the nature of a larger enterprise to have goals and objectives that might not have even been considered at a plant level. For example, let us consider a Plant Beta that is part of Enterprise Alpha. Alpha is a large, well-known producer of gourmet food items. Beta is a local site producing ethanol from corn. Beta was recently purchased by Alpha as part of a corporate diversification program. Alpha has distinguished itself in the public's eye as a high-quality, responsible manufacturer. Plant Beta is a local producer whose site is located in a small farming community in the Midwest. Beta has been known by the local community for having frequent issues with landfill and odor problems. Because it is an important employer in its community, citizens have tended to look the other way. An independently developed philosophy done at Beta might not place much emphasis, especially if it were pricey, on landfill or odor issues. On the other hand, Enterprise Alpha places a very high value on their reputation as a good community citizen. Consequently, a framework document will place each individual plant on notice that including reputation aspects into their local philosophy will be expected.

Some items in the framework may require additional costs or raise troublesome issues for individual sites. However, since the enterprise has decided the items are important, the local sites will have to face the costs since the overall benefit to the enterprise has been judged worth the additional effort and expenditure by the sites.

Charter

Produce a clear understanding of the enterprise charter: why it is in business, the mission, the goals and objective, and the key directives of existence.

Principles of Operation

Understand all essential enterprise-level principles (if they exist explicitly) of operating production sites. For situations where these principles are only implicit or differ from person to person, make a serious attempt to collate them into a form that will provide sufficient guidance for the individual plants. What you are doing here is to attempt to avoid a later problem of having missed something important from the top enterprise level that should have been known at the plant level but was not specifically included.

Also, you will want to document the thinking that went into the plant's decision about this issue.

Applicable Law

Important, but perhaps obscure, legal issues may need to be communicated from the top entity to the plant levels. This isn't meant to be a disclaimer. However, it has happened that legal issues have not been communicated. This can be even more important for multinationals. The significant point here is to ask for and respect any laws that are communicated downward.

Reputation

All products manufactured at the plant level and the plant business practices reflect back on the enterprise name. Here is where the local infrastructure is designed to protect the overall enterprise.

Quality

Ensure that production is designed and operated in order to maintain a consistent quality product. Note that the level of quality (extremely high to ordinary) is an enterprise target. What is addressed here is the consistency aspect. By emphasizing consistency, the enterprise is saying that it does not want a customer to pay for more quality than he receives, nor should he receive a higher quality than he paid for.

Safety

Ensure that employees, visitors, the public at large, and all other stakeholders are adequately protected from harm or injury due to plant operations.

Technology

Ensure that lessons learned at the enterprise level are exploited at the plant levels. It is often the case for large enterprises that differing plants obtain, develop, or identify technology (e.g., process improvements, useful third-party equipment, better understanding of existing equipment or business or technology practices) that can be of important benefit to the other players in the enterprise. It is here that the sharing is explained and the infrastructure for doing so, both at the starting gate and for the life cycle of the operations, is laid down. It is here that some enterprises place requirements on design consistency. This is done to reduce the reinvention of important "wheels" as well as to enable better sharing of new ideas, technology, and personnel. Any added costs have been weighed and found to be well spent when compared to the benefit.

Personality

This is a hard-to-explain issue. Not that it is so technically involved, but rather that it is always somewhat nebulous. Nonetheless, it is significant. Every enterprise has a history.

The founder had a dream; the early partners provided the needed capital or business management acumen, or a brother or other family member disputed things and so spun off a competitor early on. During its operational history, the enterprise entity has developed a way of thinking and doing business. It is important to recognize this and ensure that these values are reflected in the way each individual site does its business.

The alarm system is, in part, a "defense system" that assists operations to manage the plant. In most situations with limited resources, doing that job often requires that trade-offs be made. At the simplest level, consider the common trade-off between spending large sums of money to provide extensive equipment or sophisticated technology versus spending a modest (and perhaps much more appropriate) amount and accepting the limitations attached to that decision. Or consider the choice between purchasing the highest-quality raw materials versus the practice of purchasing raw materials based on market cost—balanced against the desired level of finished-product quality and the pricing structure attached to selling it.

Factoring It All into the Philosophy

You have the idea of what this is all about. You know why it is important. We now explore how everything we've mentioned before is factored into the alarm management process. The first step for the corporate steering team is to identify top-level issues from the corporate infrastructure and ways of doing business that need to be communicated down to the individual site steering teams. There are two places to visit to assist in this job. The first place is the items in the consequence-severity lists that are to be used for rationalization. Consequences are the types of bad things that can happen when a plant goes abnormal; severity is how much each bad thing can hurt if the plant cannot be returned to normal. Typical consequence categories include personnel safety, financial, environmental damage, maintenance and production costs, reputation, and such. If the enterprise has important things to say about any of these categories and how important (or unimportant) they are, the messages are formulated and communicated to the sites here. The second step in determining important enterprise philosophy issues is to obtain a clear understanding about how the enterprise leadership makes the trade-offs between being "perfect" and being "profitable." This, of course, will need a bit of translation for the individual sites so that they will have a workable way to understand and use this information.

6.8 SITE-LEVEL PHILOSOPHY

Begin with the enterprise framework philosophy. Examine each aspect of the framework document: the modifications and reshape it to become a good fit for the site. Item by item, make modifications and adjustments so that it will suit your site. The following list of items should be a good starting point. Each entry will suggest areas and tasks to do that provide the structure and information to start. Remember that while the source for this work is the enterprise framework document, your site is expected make any necessary modifications, additions, and clarifications to it to make the enterprise framework suitable to use for your specific site.

Wording and definitions. Make changes that are necessary to the wording and definitions to ensure that it is understandable and appropriate to your specific location. This is not an "anything goes" activity. The burden of action, so to speak, is on the site to justify changes. Changes must not be done capriciously or as an expedient to eliminate important work or reduce costs by limiting the scope too narrowly.

Work rules. Make changes and adjustments to align the process to the required local work rules and practices. Again, this is not a blank check to eliminate something important and essential. Rather it is to harmonize the framework so that it better fits local practices. If there are essential conflicts between local practices and essential framework requirements, these must be addressed at the highest levels.

Geography and locale. Make clarifications, modifications, and changes that reflect differences between the physical location of your individual plant and the assumptions used for the framework. Typical factors might include the presence of severe weather situations, nominal climate, seaside corrosive atmosphere versus desert dryness, and so on. Take care here to be sure that needed site nuisances are adequately reflected in your philosophy, for you will rely on these later.

Plant production equipment. Modifications may be necessary to ensure that the local design (as reflected by the philosophy) can be implemented in the hardware and other resources available to the plant. For example, different DCSs can have small but important implementation differences; a design that works in other control systems at other sites might not work at your site, or your plant design may not be compatible with certain procedures that work well at other plants of differing designs.

Products. Different products might well call for different ways of managing production. For example, a plant with a highly reactive intermediate product or significantly exothermic reaction may need to utilize very different operating methodology and therefore may need very different alarm support. This is not to say that these different exceptions will permit wildcard exemptions in design requirements, but rather that they may require specialized care to ensure that the spirit of the framework is met at the same time as the demanding requirements of the site are accommodated.

Culture. Very often it is possible to do something in a number of different ways. The culture part of this is to ensure that the ways your site will use for designing, implementing, operating, and maintaining the alarm system be compatible with the plant culture. Again, it is not permitted to eliminate important aspects of the alarm system design or performance just because individuals or departments at the site might not like them. Tailoring things to local culture is not a grant of permission to ignore or underplay necessary parts of the design. New approaches can often lead the way to beneficial changes. Look for them.

Site Personality

Every industrial plant has history. Ask any of the old-timers. If they haven't already told you many, many times before, they'll certainly be accommodating and give you all the early history about the plant in those good old days. Not that anything tangible might remain, but those experiences, dreams, and hang-ups have helped shape the plant into what it is today. Moreover, as new situations present and new players enter the management scene, changes are inevitable. The sum total of this history contributes to what we call *site personality.*

Personality can cut two ways. On the one hand, it is important to include as many of the lessons learned from the past as possible. There is little more embarrassing or more telling than to fail to respect your plant's history lessons that have been paid for at high cost. Be sure to include those lessons. On the other hand, we sometimes use history to prematurely conclude that something is not important. It is very easy to toss aside what is too difficult to understand, do, or maintain and excuse ourselves by saying, "It's not our style." Doing things that way is almost certainly not a good way to run a business. You will want to be very careful not to substitute "not necessary" for "dislike."

The Rest of the "Bases"

- *Charter.* Has the purpose of the plant and how it fits into any larger picture been clearly laid out?
- *Principles of operation.* What is different between the foundation document and your specific plant site? What is it about your plant—in kit, product, or business climate—that needs to be identified and factored into the alarm design (philosophy) for your site?
- *Applicable law.* Are there any special legal requirements or codes to be met that will necessitate changes in the alarm system design or operation?
- *Reputation.* Does the operation of your plant place extra burdens on the enterprise that, while acceptable at the plant level, would unduly damage the enterprise reputation and for which your alarm system operation must be changed to respect?
- *Quality.* What are the plant's quality goals? How does the alarm system need to be shaped to provide adequate operator support to meet your goals?
- *Safety.* What things about your plant place a greater (or lesser) burden on safety that needs to be explicitly accounted for in the design and operation of the alarm system? While the plant might not place a higher expectation, does the community in which the plant is located have higher expectations? If so, how will they be translated into alarm design requirements?
- *Technology.* Are there unusual limitations or advantages to your plant provided by the existing or planned technology infrastructure that should be factored into the alarm designs? If so, how will they be included?

6.9 ALARM DESIGN PRINCIPLES

As is already very apparent, there are a number of ways to achieve good alarm practices. Within each way, there are choices that can be made to tailor a method to suit. What this section is all about is choice. Actually, it is about choosing. That is, from the broad range of what can be done or what might be nice to do, the philosophy for your plant must contain decisions of what you will do. It is like in a restaurant: no one wants to eat everything on the menu. You make selections, asking for special changes where needed, and place your order. Your order is the result of all the choices you make. Likewise, your alarm philosophy will be the result of all the choices you need to make that impact the design and operation of your alarm system. The philosophy will need to be general enough to provide long-term viability while at the same time clear enough and specific enough to provide immediate design and implementation guidance.

Chapter 1 laid out key design principles for an alarm philosophy; here is where the high-level assumptions will need to be "approved," or others substituted for use, throughout the philosophy design process. When decided on, they will pave the way for the detailed design of the new alarm system. The reader is reminded that entries and lists given here are suggestions for consideration. Each plant site will develop the specific list of principles and explain their meaning as well as provide functional and operational definitions for them. There are two categories of principles: fundamental and functional. Fundamental ones are those whose job is to hold up the entire alarm system.

Fundamental Principles

These are so basic and important that they are repeated from chapter 1.

- *Every alarm must require an operator action.* This is the defining attribute of an alarm. For other situations where there is an alarm being considered that does not require an operator action, that potential alarm must not be considered as an alarm—for example, please leave it off the alarm list.
- *Adequate time should be allowed for the operator to carry out the defined (required) response.* As we saw in chapter 2, we use this requirement to determine each alarm activation point. Moreover, this requirement is also used to ensure that the necessary operator support tools are in place and are adequate to do the job. Finally, if it is not possible to meet this objective, then this abnormal situation must be managed in some other appropriate way instead of relying solely or primarily on operator intervention.
- *Each alarm (activation) shall alert, inform, and guide.* As the alerting part of the alarm activation, the event of an alarm activation must also provide information that assists the operator to understand what the alarm is and what it means, and it must also include guidance for the operator to assist in the determination of root cause as well as to suggest the way toward fixing or resolving the problem.
- *Only important things should be alarmed.* Yes, even unimportant things can require an operator response. But, unimportant things do not threaten the process integrity or safety or other important attributes of the enterprise. They must not clog up the alarm system.

Now is the time to point out the true power of these principles. Yes, they are important in their own right. But their vitality and power comes from their ability to guide almost all decisions. To believe the principles is to accept their guidance. Anytime there arises a situation where knowing the "right answer" is important, know that the right answer is always the one that is fully consistent with the fundamental principles! Any decision, plan, or alternative that is not consistent with the principles will likely not work for you.

Functional Principles

- *Every alarm presented to the operator should be useful and relevant.* It is not enough for the alarm event to ask for an operator action. The intended action should be necessary for prudent, effective operation of the plant. For example, suppose a high-level alarm activates at a drum level of 55% for a drum that is used for surge to ensure that upstream flow variations are smoothed out before the flow enters a critical plant operation. If the drum were to fill completely, it would just mean that flow variations would not be damped out for a while. Another example would be a low-pressure alarm on the outlet of a spared pump that is not running.
- *Every alarm system should be explicitly designed to take into account human limitations.* Alarms and alarm systems are not supposed to just scare the operator. The alarm system must be robust enough to provide a sufficient level of redundancy and accommodation to allow the operator to make the usual type of errors or mistakes. This is not to say that the alarm system must be foolproof or that the alarm system must be fail-safe. It means that the alarm system will be constructed in such a way that minor, ordinary types of human error won't result in everything crashing down. A simple example would be to request confirmation for operator actions that have a major impact on the plant if done at the wrong time or in the wrong amount, say missing a setpoint by a factor of 10.
- *The alarm system is a key part of safe operations.* The entire alarm system, including each individual alarm, will receive the attention it requires to ensure that it remains in good working order and that it is properly integrated into the plant infrastructure.

Key Performance Indicators

We now take a good look at what the plant wants the alarm system to do. You will translate those expectations and requirements into a reasonably short list of measurable items, the performance indicators.

Chapter 4 went over the generally used alarm design and performance metrics. From that list, after adding any new ones of your own, you will want to select the few that the plant will choose to provide an overall indication of the successful design and operation of its new alarm system. Yes, they are all important. But you will need the short list of ones to aim for. Some will be useful for design and implementation. Others will measure the real performance of the alarm system in daily operation. The full list would be much longer than most plants will want to keep track of. Remember, the key performance indicators are the ones that will be constantly tracked and reported against. Sure, if they were all consistently great, then everyone can get a daily boost. But Nature does not always cooperate in the way we would like. During upsets, maintenance, or what have you, usually some of the lesser metrics are challenged. Would you want to have to provide "explanations" every time that happens? On the other hand, the key metrics will be more tolerant to the normal noise of operations. Choose the few that you really need.

Select only the metrics and any other indicators that measure the important parts of the alarm system performance. These measurements should show that the alarm system is either in good working order or not. That is it. That's all KPIs are designed for. Certainly, later on, if some of the KPIs are off, then other metrics can be examined to gain more useful information and point you in the direction of a better understanding of what might be going on or going wrong.

Critical Success Factors

Here is where the plant will decide how the alarm improvement project will be judged. Does it meet its intended requirements? It is tempting to simply suggest that the project will be considered a success if all of the KPIs are met or exceeded. Unfortunately, meeting the KPIs is probably necessary for project success, but certainly, and unfortunately, does not guarantee it. A successful alarm system is much more.

The alarm system will be designed to provide important operator support. That support is vital during abnormal situations. We expect the alarm system to enhance the operator's ability to manage abnormal operation. Consequently, the key critical success factors will often center on the following questions:

- Is the alarm system capable of providing the necessary warning to the operator?
- Is this warning provided with enough time to manage the abnormal situation?
- Does the alarm system provide sufficient information for the operator to assess the nature of the abnormal situation?
- Does the alarm system provide appropriate and adequate information to facilitate the operator to manage the upset?
- Does the alarm system avoid interference with the operation of the process during normal, nearly normal, and dramatically abnormal operating situations?

Approved Management of Change Requirements

All plants have an established MOC process and the associated procedures to ensure things are done well and completely. For alarm management, all that is necessary will be to ensure that the plant MOC process is working well and to include any additional alarm management concerns into a revised set of procedures. For example, it is probable that when some mechanical component is changed or upgraded, the plant's existing MOC process will likely not include a visit to the alarms surrounding that equipment or a suggestion that the activation points be checked to ensure that the operator will still have sufficient time to react in the event that the alarms activate. However, this is something that should be done. MOCs need to include these items. Refer to chapter 9 for more details regarding the items usually included in the plant's MOC procedures for alarm management.

Procedure for Rationalization

The procedure to be used to reduce the number of configured alarms will depend on the chosen rationalization method. Chapter 7 provides an in-depth coverage of the approaches. Even if all other criteria and objectives remain fixed, the very nature and process of the different rationalization methodologies will yield differing numbers of configured alarms. Moreover, the alarms that are configured will likely differ in their identity. The plant must review the various approaches and either select one or carefully develop a new one. Once a general methodology is selected, the plant will also need to specify the exact procedures to be used to implement that approach.

Alarm Configuration: Specific Issues

For an abnormal situation to be announced to the operator we can use an alarm. In order for an alarm to activate, it must be configured. Alarms can be configured in quite a number of ways. Here is a list of some of the more important items to consider:

- What type of alarm best captures the abnormal situation (high, low, high-high, low-low, low rate-of-change, etc.)?
- At what configuration location should the alarm be configured (input block, controller block, output block, separate alarm block, etc.)?
- What are the situations (plant operation, alarm system status, etc.) that will affect the alarm configuration (inhibition, priority change, activation point change, etc.)?
- Are there any PCS alarm configuration limitations that require special design or implementation consideration?

Alarm Activation Point Determination

Once an aspect of a process point is selected to be alarmed, the determination of the activation point becomes the next most important decision in the entire alarm management process. The alarm philosophy must clearly specify how the activation points will be set. Improper values result in early alarm activations that produce large numbers of nuisance alarms. Improper values also lead to delayed alarms that activate too late for the operator to have sufficient time to remediate the situation. Chapter 2 discussed the thinking that you will want to use for setting activation points. Chapter 7 shows you how to actually set appropriate values.

Priority Assignment

Priority will be used by the operator to decide the order in which to work competing alarms. Therefore it is essential that the priority of each alarm properly reflect the plant's wisdom and guidance in this area. There are a number of good but differing ways to determine priority. Chapter 7 provides an excellent discussion of alternative methods. Your site will review the priority assignment methods and either pick the one that is most appropriate or modify them to produce the one that will be used.

Alarm Presentation

Decisions will need to be made that determine how alarms will be displayed on operator screens, what screens will be used, what modifications will be necessary, what messages will be sent, what use will be made of alarm horns or tones, and all the other choices that impact how the alarm activation is communicated to the operator. One of the most important results from all the attention to the design of alarm systems has been the recognition of the importance of keeping an overview of the plant in the forefront of the operator's view. Chapter 12 covers this topic in detail.

Operator Roles

It is here that we lay out what is expected from the operator to keep the process running well. You will also want to visit what the operator needs to do to understand the true process conditions without reliance on alarms as a wake-up call. You will decide what the procedure should be for acknowledging alarm activations. You will lay out if and how operators can modify alarm settings. Here you will lay down all the requirements that the operator must follow to maintain his charter to continue operating as opposed to not. Chapter 5 covered the key components of the permission and how it is used. You will want to include that or other alternative specifics into your philosophy.

There are important situations and issues to both recognize and address that impact how the operator conducts business during abnormal operations. They are the following:

- *Functional prioritization.* The prioritization rules and results must be sufficient for the operator to make prompt and accurate grouping of alarm activations. It must always be clear to the operator which sets of active alarms are on the "do me first" list. Moreover, the remainder of the alarm activations should not distract from the ability to provide that list.
- *Manage conflicts.* There must be a clear, straightforward, and effective process in place to handle all these conflicts: (a) multiple alarms, (b) limited operator resources, (c) general operational goals and conflicts that restrict the operator's ability manage the upsets, and (d) any vital limits of operation that may dramatically impact safe operation.

Interplay with Procedures

Operating procedures are an important part of proper and safe manufacturing. Therefore, alarms and alarm-related aspects of operation should be adequately covered by plant procedures. Refer to chapter 9 for more coverage. Alarm issues will be found in both alarm procedures as well as operating procedures. Alarm procedural coverage means providing specific procedures to deal specifically with alarms and alarm events and conditions. These reflect how alarms themselves will be managed. A representative list of alarm-related procedural items includes the following:

- Acknowledging
- Silencing
- Modifying alarm activation points
- Suppressing, inhibiting, bypassing, and other forms of disabling alarms
- Validating alarm conditions
- Using alarm response information to manage upsets

Operating procedural coverage includes how alarm handling and interpretation is incorporated within the usual process-related procedures. A short list of process-related procedural items will include the following:

- Start-up
- Shutdown
- Handling alarm floods
- Operator responsibilities during normal operation that maintain situation awareness without an undue dependency on alarms

Training

The alarm philosophy will decide those special parts of training that need to be ensured as part of the alarm redesign process. This would be a good place to remind you to be cognizant of this process. The purpose of carefully identifying the range of items and issues to cover during the alarm design process is to ensure that alarms and alarm issues are properly folded into the plant infrastructure. Care should be taken to avoid using this opportunity to overstep and take on the larger task of fixing all that ails with the current way things are done at the plant. Therefore, if there were a feeling that overall plant training is lax or that it is done one way and could be much more effective if done another way, this would not be the forum for that change. Yes, repairing the plant training process is very important. But no, it is not a specific component of an alarm philosophy.

Alarm-related training would include specific training techniques and methodology that prepare the operator to use the new alarm system. Topics would include the following:

- Understanding the alarm philosophy
- Alarm design basis
- Performance indicators
- Management of change
- Alarm activation points and priority
- Proper operator roles
- Alarm notifications
- Escalation and operating charter
- Expectations and procedures surrounding maintenance
- Proper use of the alarm information databases

Escalation

Chapter 5 exposed a vital problem that most of us normally never think about. After all, operators have been able to operate without understanding escalation. Well, yes, they have, but often not well. Therefore, it will be important for the plant to work out and then require appropriate procedures for handling situations where the operator should seek help. Moreover, in the unfortunate event that the operator cannot maintain operation, there needs to be guidance into when to cease operation and move the plant to a safe state. Your philosophy is where those understandings are laid down and expected.

Maintenance

There has been considerable discussion regarding plant design and maintenance. Now is the time to decide the expectations and lay down the rules for enforcement. These rules are intended to ensure that proper plant maintenance will be done for all equipment and controls in the plant.

Plant Maintenance

The message is this: fix everything that is broken. Poor or spotty maintenance produces unnecessary challenges to the alarm system. The plant design, in all likelihood, was not done with the expectation that odd parts would be broken. Extra equipment was not provided to aid operations when things were not working, except for a very few critical parts. Plant operating procedures were not written to accommodate lots of nonworking equipment or replacement equipment that might be just limping along. In one way or another, these deficiencies negatively impact operability. Alarms do not know about the maintenance status of the plant; they will activate when they are designed to, no matter what. This makes the operator's job harder and the plant's operation less predictable.

Alarm Maintenance

By the same token, maintenance will also include the specific requirements necessary for keeping the alarm system in good working order. Areas of interest include monitoring the KPIs and other alarm activation and action records for trouble spots. Typical KPIs include the following:

- Alarms that are annoying, stale, or otherwise not operationally necessary or helpful
- Alarm data that are misleading or wrong
- Alarm configuration parameters that seem to be inappropriate (e.g., activation point, priority, automatic disabling, etc.)
- Plant events or situations not adequately covered by alarms that should be covered (according to the philosophy and plant needs)
- Unauthorized and temporary changes or modifications to any aspect of the alarm system that are not appropriately recognized and managed
- Other deficiencies of alarm design or operation that become evident during production

6.10 EXAMPLE PROCEDURE: TO SILENCE OR TO ACKNOWLEDGE

Whether to silence the alarm horn or to acknowledge the alarm is the question—to play on the haunting phrasing of Shakespeare's words in *Hamlet.* This is all about the operator. And these words, rather these choices, are the lead-in to the frontline, real-time managing of alarms. Let's take a look at what the likely events might look like from time to time through the eyes of the operator. Figure 6.10.1 illustrates what might be happening in the control room. Starting (in the upper-left corner) with the process being normal, things can stay there, become abnormal without alarms, go back to normal, or become abnormal and alarms activate.

From the moment the alarm horn sounds until things are back to normal, the operator's life is changed. Until this situation is back in place, nothing else matters very much except dealing with these situations. Before we tackle the subject of the operator and what he might do at this time, we take a brief digression to check in with what the controls platform does with alarms that are active and either silenced or acknowledged. Figure 6.10.2 shows the process control system's conventional alarm behavior. As soon as an alarm activates, the PCS flashes and messages appropriate things, sounds the horn, changes color of appropriate things, and makes symbolic changes, all to capture the operator's attention. If the operator chooses to acknowledge the alarm, the horn turns off, the flash turns off, and there is a visible state change to show that the alarm has been acknowledged. On the other hand, if the operator first chooses to silence the horn, only the horn is silenced. The remaining alarm status indicators remain unchanged.

Now comes the question, should there be rules that must be followed by the operator to silence the horn and to acknowledge the alarm? You are saying to yourself, "What rules? What is this book talking about? I never saw any such thing! I silence the horn to

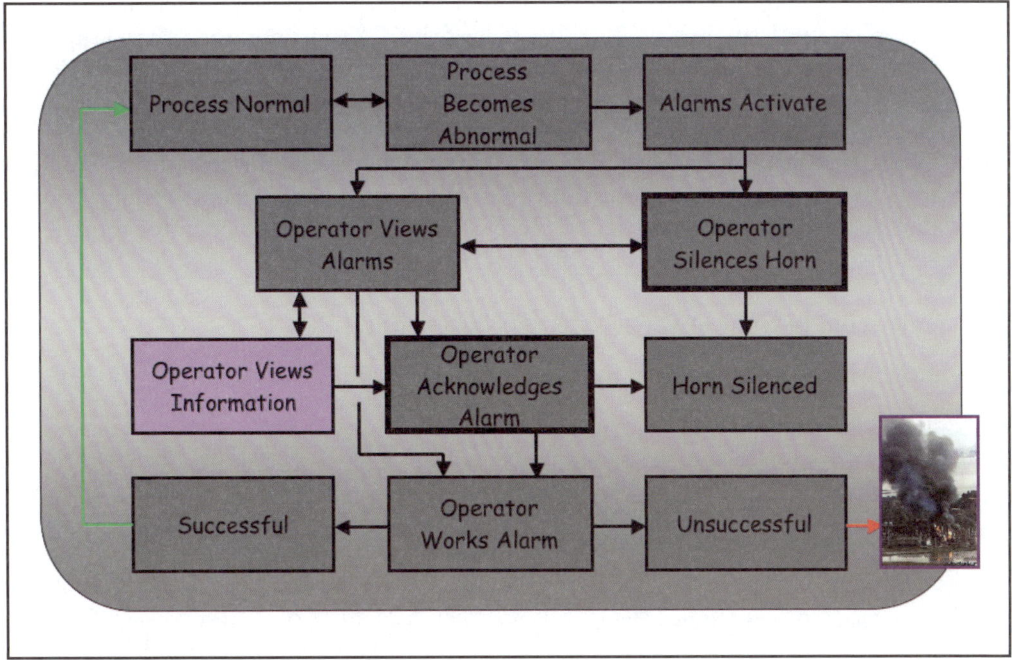

Figure 6.10.1. Process condition and silence and acknowledge actions

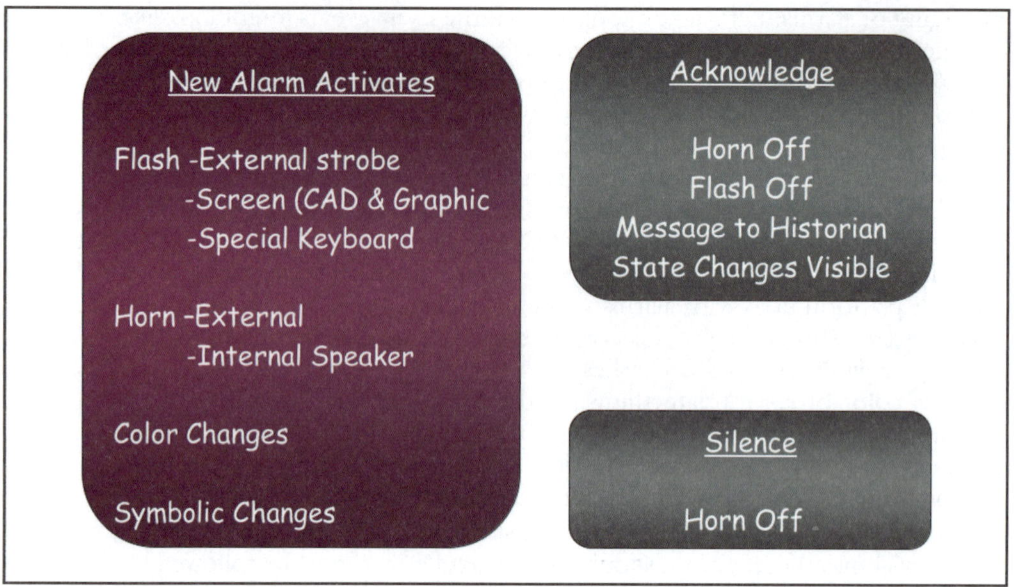

New Alarm Activates
Flash -External strobe
-Screen (CAD & Graphic
-Special Keyboard
Horn -External
-Internal Speaker
Color Changes
Symbolic Changes
Acknowledge
Horn Off
Flash Off
Message to Historian
State Changes Visible
Silence
Horn Off

Figure 6.10.2. Process control system's behavior for alarms and acknowledge or silence

end the annoying racket. I acknowledge the alarm so that when it gets back to normal, it will clear off of my alarm display list. Who needs rules?" You were right—that is, up until now. But from now on, being without rules may not be the way to go. It might not be the way, because having rules will go a long way toward regaining control of the alarm system and turning it into a powerful operator support tool.

Suppose the operator has several alarms that are all active at the same time. We want the alarms to be worked in a successful order. We also want the alarm system to avoid unneeded distractions and annoyances. Here is what we need. The operator must

- know when a new alarm arrives;
- understand how the alarm impacts process integrity;
- know when work still needs to be done on an alarm;
- not be unduly distracted.

So what might these rules be? Our design clues come from the PCS's behavior in response to whether the operator chooses to acknowledge or silence (fig. 6.10.2). Here is what you might come up with for the "rules":

In order to silence the alarm, the operator must do/agree with the following:

1. I see that there is a new alarm.
2. I have a reasonable idea what this new alarm is all about.
3. I promise to get back to this alarm and work it as soon as I have time.

In order to acknowledge each alarm, the operator must do/agree with the following:

1. I see that there is a new alarm.
2. I understand what this alarm is about.
3. I promise that I am going to work this alarm right now.

Silence is the only way the operator can quiet the horn and not actually work on the alarm that caused the horn to sound. In effect, he keeps the bookmark of an unacknowledged alarm to remind him of the alarms yet to be worked. Yes, acknowledge would also silence the horn, but it also will fail to distinguish between alarms that have been acknowledged and worked and those that have only been acknowledged to silence the horn. Sure, this is a simple-minded solution. But it is a powerful simple-minded solution.

6.11 PHILOSOPHY HIT LIST

It is going to take some work to capture all of a plant's possible concerns and issues regarding alarm design since each plant has different management, different business objectives, and different locale. So, the list that follows should be considered suggestive rather than exhaustive. The column termed "module" refers to the specific philosophy workshop module labels of section 6.12 later on in this chapter.

Item	Description	Module
1	Explicit statement of the purpose of the alarm system (in the site's own words)	A
2	Explicit statement of the alarm design principles to be used by the site teams	A
3	How candidate alarms will be identified	A
4	How the alarm activation point will be set	B
5	How the alarm priority will be determined	B
6	Appropriate operator behaviors for (a) Monitoring alarm conditions (b) Interacting with the alarm system (silence, acknowledge) (c) Responding to alarms (d) Escalation and safe operation states (e) Bypassing, inhibiting, and enabling alarms	B
7	Alarm presentation to the operator (visual and acoustic)	B
8	How to achieve situation awareness (a) Without alarms during normal and near-normal operation (b) With alarms and during other abnormal situations	B

9	What rationalization method will be used	C
10	Alarm design metrics	D
11	Alarm performance evaluation metrics	D
12	Validation of new alarm system	D
13	Incorporation of alarm issues and concerns into procedures and other operating practices	E
14	Appropriate training	E
15	Alarm performance assessments (APAs)	E
16	Proper training and other training issues and requirements	E
17	Proper management of change procedures and processes	E
18	Management of DCS infrastructure and other system alarms	E
19	Approved advanced alarm techniques (a) Basic logical management (b) Enhanced knowledge-based management	F
20	Messaging and other operator notifications; any additional non-alarm-related technology and support required	F
21	Handling of enhanced field sensors (smart transmitters) and other advanced process control equipment and software	F
22	Ways to ensure that alarm system infrastructure stays in perspective (avoiding another niche empire)	F

6.12 ALARM PHILOSOPHY WORKSHOP

By now, you are probably thinking, "How can anyone possibly produce a philosophy document that works without spending a lifetime and commandeering half the plant?" Good question. Easy answer. We use alarm philosophy workshops.

Alarm philosophy is finalized at the individual plant level. This is the place where it will be used. This is where it will be implemented. For plants that are operationally independent and not part of a larger business or enterprise configuration, this is fine. For others, the philosophy will be framed at a higher level first and then tailored at each site level afterward. And in those situations, an overall workshop is held first. When the high-level framework is approved, each individual site will hold its own workshop to work out all local issues and finalize the design for themselves.

Workshop Details

The workshop generally takes 3 days to complete. In the rare event that reaching consensus is difficult, one additional half day might be necessary. The workshop is a hands-on activity. There is a lot of ground to cover. The work product will be a document that contains significant information and decisions. The successful philosophy will provide the following:

- Complete understanding of what alarm management is all about
- Comprehensive design basis for the alarm redesign process
- Ability to sell the plan to management

Meeting Facilities

The work will be done both in combined sessions and in smaller breakout work groups. The combined session room should be large enough for the entire group to sit around a central U-shaped seminar-type table that has room to spread out, accommodates lots of documentation, and provides room for taking notes. A whiteboard and several paper easels should be supplied. There needs to be enough wall space to put up completed easel papers and other material important for the group to keep in view. With the view in mind of using material that is file- or computer-based, several projectors and connecting cables should be there. One projector is usually used to show source materials, while the second, which is managed by a notetaker or documenter, is used to show work in progress. Appropriate wall location should be blocked out for the projectors to use, of course.

Each breakout room should contain a large work table, a white board, and an easel or two. Usually the group is small enough that a projector is not necessary, but having one available is a nice touch. Otherwise, everyone just crowds around the PC screen or a secondary flat-panel screen.

Module Identifications

The meeting proceeds according to the following content modules.

Module A: Foundation. Bring everyone up to speed (short background in alarm philosophy, review of operator surveys, etc.). Decide the plan for documenting the workshop. Arrive at specifications for the following alarm terms and processes: definitions of terms, purpose of alarm system, and alarm design principles.

Module B: Key design issues. Decide on policy for setting alarm activation points, alarm priority, alarm presentation, and operator roles.

Module C: Rationalization engine. Determine the preferred rationalization process and any special concerns or issues that need to be taken into consideration during the actual rationalizations.

Module D: Performance. Set configuration and performance metrics, key performance indicators, and the validation process.

Module E: Infrastructure. Develop approaches to modify procedures for training, MOC, and ensuring that the alarm system remains viable down the road.

Module F: Beyond basic alarming. Evaluate and decide how to do advanced alarming, what to do about system alarms, how to handle specialized smart field devices and other infrastructure requirements, and how to ensure that the alarm system remains true to its intended role.

Module G: Arbitrating consensus (optional). In rare circumstances, the group finds itself having considerable difficulty reaching a consensus in one or more critical areas. When a problem does not impact the progression of the workshop and cannot be resolved by extending the module in which it arose, this extra session is held to work out the all of those differences. Its purpose is only to resolve important, difficult issues that remain. If not needed, skip it. However, you should schedule it anyway just to ensure that the group would be available in case it is needed.

Workgroup Configuration

The combined group is divided into two smaller working groups. If the total workshop is attended by five or less, the smaller breakout sessions are not held, and all work is done as a single group. The total workshop participants should normally number less than fifteen. This workshop is a hands-on real working session. Do not pile on the participants. It is difficult for a large number of participants to stay focused, not to mention commit to attending the full session. But do not hesitate to compose the group of the adequate number of plant players and participants.

Divide the overall participants as evenly as possible into the two breakout groups. Avoid concentrations of disciplines and participants from the same areas. You are looking for as diverse a group in each working group as can be had. This is not the venue for the issues and concerns of any site, discipline, individual to take control. What you will be looking for is as complete and representative a discussion as possible.

Working Format

The workshop consists of the six modules previously discussed. A seventh module has been identified, but it is not used unless absolutely necessary. Each module generally runs about 4 hours, with one in the morning and one in the afternoon of each day. However, since it is difficult to predict where any given site might get bogged down, it is customary to begin a new module soon after the previous one is complete. Each module is organized around a central theme, one theme to a module. The modules are designed to provide a general orientation, progress into small-group work, and recombine for group decision making. Here is how they are structured:

1. Orientation and activity definition session (about 1/4 hour) for the whole group.
2. Small-group parallel breakout sessions (1 to 2 hours), during which all module objectives are worked on by each of the smaller groups.
3. Interim consensus session (1 to 1 ½ hours) to unify the first portion of work by the small groups into areas of agreement and areas of nonagreement. Areas of nonagreement should be supported with specifics as follows:

  a. Reasons that go back to one or more existing fundamentals of alarm management that have been agreed to by all but are either interpreted differently or applied differently
  b. Important previous experiences, specifically identified, that do not seem to have a place (as yet) within the alarm management framework process being used
4. Small group parallel breakout sessions (about 1/2 hour) to work out leftover differences. The groups should be prepared to return with reasons for the remaining differences that specifically address, point by point or item by item, the other groups' previously expressed concerns.
5. Final consensus session (about 1/2 hour; unless problems develop) to unify the module's work. Where unification is not possible and resolution of the issues will not stand in the way of continuing to other modules, the building of a consensus is postponed until the optional seventh module. If it is necessary to reach a consensus within the current module before proceeding, the current module is extended.

If consensus cannot be reached after an additional half hour, it could be because of one or more of the following problems or concerns:

1. There is weak facilitation unable to forge the unity.
   a. Improve facilitation, either by changing team makeup or enlisting outside assistance.
2. There are fundamental aspects of alarm management that are not understood and shared by the group as a whole.
   a. Take time out so that they can be identified and understood.
3. One or two individuals are attempting to hijack the process or grandstand.
   a. Privately approach the hijackers and work out an arrangement.
   b. Elicit the whole group to recognize what is going on and attempt to use peer feedback to gain participation from the errant players.
4. The issue is valid, yet participants are struggling. The participants are reasonably confident the additional required work will result in harmony.
   a. Table the issue until the optional module.

Documenting the Workshop

The use of a structured query-answer form or database is extremely useful and is usually available from an experienced consultant or professional alarm management facilitator.

It often will not be possible for one individual to do all the formal note taking and other documentation requirements. Moreover, the process of documentation of these activities will need to take into account a number of difficult-to-track items. There is the usual give-and-take of discussion. Some organizations will want to annotate who made what important statement; others recognize that the credentials of the work are determined by the group doing the work and only keep track of significant items or issues that might need to be reflected back to the author for additional clarification or comment. There will be a great deal of source materials that might be used for reference, to support a major decision, or to define a key issue. And there will likely be a lot of extemporaneous whiteboard and easel use. A useful approach to divide the work is shown here.

Workshop documenter. Select a single individual whose job is to keep track of all of the group work and lead the preparation of the finished document.

Backup documenter. Select a single backup documenter whose job is to keep track of all supporting documents and other key materials that are used by the group as resources. This usually involves making an annotated running inventory so that the materials can be collated or later obtained and incorporated in the record of the workshop. The backup documenter will also fill in for the documenter during any absences or when the documenter is leading or otherwise actively in the middle of the discussion process.

Breakout documenters. Select a documenter for each of the breakout groups. The job will be the same as the workshop documenter, but for the specific breakout group. This individual will also be responsible for working with the workshop documenter to help prepare the workshop work product documents.

Backup breakout documenters. This person fills the same role as workshop backup documenter; there should be one for each breakout group.

Facilitation

An experienced alarm management individual skilled in group facilitation should lead the workshop. There are likely one or more such individuals already on the corporate/enterprise steering team. However, this role might make it difficult to participate as a working member of the team. Most sites prefer to either bring in an experienced corporate facilitator who is adequately briefed and prepared in the discipline to facilitate a technical working activity, hire an outside alarm management consultant to work with an in-house facilitator, or manage the workshop directly.

Preparation

Participant Preparation

It should be clear that all participants in the workshop would need to come ready to work. Therefore they should have had a serious introduction into the entire process of alarm design and alarm management processes. Some sort of formal training is highly recommended. There are excellent comprehensive alarm management workshops available. Often a 2-day alarm workshop is coupled at the beginning with this 3-day philosophy workshop to round out the week.

Exposure to a casual reading of EEMUA 191 is generally considered helpful but certainly *not* sufficient or adequate for serious participation. Earlier attempts at alarm assessments and perhaps even attempts at alarm reduction projects and other remediation of bad actors can be helpful. Keep in mind that a philosophy workshop will make important decisions that determine both what will be done to work out an alarm improvement project and how it will be done. Decisions should be made based on experience and reality, not conceptual expectations.

It is not unusual for some sites to be more advanced in their alarm management thinking than others. One or another individual might have taken an interest in the area, some problems might have developed with the performance of the existing alarm system and they might have been looking into matters, or the plant may have had a more serious event that pointed to the alarm system as a possible contributor to harm. Moreover, there may have been early preliminary studies done of the alarm system performance with data and supporting understandings. It would be helpful for all sites to bring any of this background work done to the workshop. In addition, participants should come with a good understanding of what their operating procedures look like, how effective they are for the operators, and how training is done.

Attendees are urged to bring to the workshop all key materials in their possession that will assist them in the job of participating in the workshop as well as those that might be needed to share with other attendees. Often participants find that while they were very familiar with a particular study, a favorite display on the PCS, or an interesting data study, they did not bring them along because they were so familiar; however, when the issue comes up for discussion, those same participants then wish that they could easily show something related to the others. Plan to bring these materials along. Here is a specific list:

- EEMUA 191 (if they have it)
- Notes from alarm management training workshop (attendance is almost a requirement)
- Workshop guide (this book, of course)
- All other items needed from the lists in the earlier paragraphs above

Site Preparation

There is also some work that the individual site teams will want to do in order to prepare for the workshop. If possible, sites should conduct APAs using one of the many software tools on the market. In addition, it is recommended that sites conduct an operator survey of the alarm system. Refer to appendix 3 for a sample survey. Surveys will enable the participants to base their concerns in areas that actually impact their specific sites. They are also useful to provide a proper overall grounding into the significance of alarm performance issues on plant operations. They will benefit from understanding how their control systems displays and navigation are designed and used. Finally, the area of operator's responsibility to manage operations and engage in proper escalation of issues and ceasing operation should be understood and addressed at the site level before the philosophy is developed. Care is needed to ensure that the alarm philosophy and project is not governed by the survey. Good alarm design is not responsive to an approach designed to remedy performance symptoms.

The specific list of items the workshop host might bring along would include the following:

- General site statistics
- Number of operators
- Approximate number of loops per operator
- Approximate number of configured alarms per operator
- Some idea of alarm activations during normal operation and abnormal operations
- Results of site operator survey, if taken, otherwise sample operator survey form (appendix 3)
- Example alarm response sheet (appendix 6)
- Summary of alarm management metrics (appendix 7)
- ISA 18 Standard for Alarm System Design and Operation[4]
- Example philosophy developed by the 1998 Honeywell European Users' Group (appendix 4)

6.13 ENTERPRISE PHILOSOPHY FRAMEWORK

For plants that are part of a larger enterprise, the first philosophy workshop normally would be done at the enterprise level. The process and procedures are the same as described in this chapter, except the participants are drawn from the enterprise. Instead of producing a site-specific document, the work product will be a framework to provide the blueprint for the individual site philosophies. After the framework is prepared, it is carefully reviewed at each site (usually by the site steering team) to ensure that it is complete and useful. Note that the feedback from the site teams is directed toward the framework structure, not the site's specific needs. It is not until the framework is complete that the sites then take it on as their starting point. Then each site works it as described here to finalize it for their site.

Most site finalizations take about the same time as the enterprise one to produce. For one thing, the participants are mostly new. For another, there are just lots of materials to decide on and review and affirm or modify. It all takes some time. Since this is producing the entire framework for the alarm system design, it will be time well spent.

6.14 CONCLUSION

The alarm system is at the front line of the operator support tools list. As the plant works, the operator will use all his knowledge, skill, and insight to keeping the plant running efficiently, safely, and economically. To help ensure that important, unusual, or abnormal situations are not inadvertently missed, an alarm system is provided. In order for this alarm system to deliver on its mission, it needs to be designed to fit the plant and provide for the plant's needs. The philosophy guides the way it is done.

This chapter is the backbone of successful alarm management. Without a clear blueprint and roadmap, it will not be possible for plants and enterprises to design, develop, and implement good working solutions to their alarm management problems. Many sites have attempted to attack the symptoms of poor alarm management. They have achieved various levels of success at it. Many are proud of how many of the bad actors they have eliminated. And operations is quick to agree and praise those efforts. But in the end, only annoying alarms have been accounted for. The more important concern of providing the operator with an effective tool for better managing the plant has been missed. Moreover, since it appears to all that alarms have been improved, there is even less incentive to work on the real issues.

We continue with the next chapter on rationalization. Rationalization is the engine that produces the improved alarm system configuration.

6.15 NOTES

1. W. L. Frank, J. E. Giffin, and D. C. Hendershot, "Team Make-up—An Essential Element of a Successful Process Hazard Analysis," *CCPS International Process Safety Management Conference and Workshop* (San Francisco, September 1993).
2. Engineering Equipment Materials Users' Association, *Alarm Systems—A Guide to Design, Management and Procurement*, EEMUA Publication No. 191 (London: EEMUA, 2007). http://www.eemua.co.uk.
3. InTech, Instrumentation, Systems, and Automation Society (now the International Society of Automation) (Research Triangle Park, NC: 2005).
4. Instrumentation, Systems and Automation Society, *Management of Alarm Systems for the Process Industries*, ISA 18.02 (Research Triangle Park, NC: ISA, 2008), http://www.isa.org.

CHAPTER 7

Rationalization

Courage is resistance to fear, mastery of fear—not absence of fear.

—Mark Twain (Samuel Langhorne Clemens)

Rationalization is the heart of alarm improvement. It is the process and technology by which each needed alarm is selected and the configuration design and supporting information for the task is built. In this chapter, you will learn the objectives of alarm rationalization, how to select the working teams, and the steps to complete a rationalization. Rationalization is a structured process. It includes deciding which points to alarm (including calculated and imputed variables), determining the alarm activation point, the setting of priority, and all other remaining alarm response information including potential causes, appropriate operator responses, and likely consequences of error.

An objective of alarm rationalization is to create a PCS configuration with the correct number and configuration of alarms. When properly and carefully done, the new configuration should result in significantly fewer alarm activations. Additionally, those alarms that do activate will be more important and provide more useful operator guidance.

7.1 KEY CONCEPTS

Rationalization requirements	1. Specify only those points that require alarming. 2. Determine the alarm activation point and priority. 3. Identify all necessary documentation to provide to the operator to work the abnormal situation. 4. Arrange for the rest of the plant infrastructure to complement the alarm system design.

Rationalization process	Built using the fundamentals. Starting from the poor alarm performance statistics and using a program of fixing each one that appears to be annoying the operator is NOT a recommended rationalization work process.
Alarm priority	The sole purpose of alarm priority is to specify how operational risk is managed by the operator.
Final test	The final test for a good alarm system is the following: 1. Alarms designed to meet the four fundamentals of alarms 2. Alarm activation rate does not exceed the operator's ability to manage them 3. Alarm system provides a full measure of operator support

7.2 INTRODUCTION

The objectives of this chapter include the following:

- Fully understanding the objectives of alarm rationalization
- Understanding how the alarm rationalization process works
- Knowing how to form the rationalization teams
- Examine typical rationalization results

If this is the first chapter in the book that you have turned to in the hope of taking a shortcut into getting to the "meat" of your job of alarm improvement, let me suggest a few words of advice. The first word is "philosophy." Rationalization requires a plan. Without the philosophy, rationalization becomes an exercise to improve skills, not a way to produce actual useful results. The second word is "data." If you are ready to rationalize, you must already have a plant with alarm problems. You will need to fully analyze your plant's performance to really understand what is important and what is not in order to place the proper emphasis on which alarms might be important and which might not. The third and final word is "perspective." Rationalization, as the roadmap shows, is a waypoint of a carefully constructed trip. Without at least having visited or previewed the other waypoints, it will be difficult to appreciate the fundamental requirements, what a proper rationalization will produce, and what needs to be provided by other parts of the alarm improvement process so that it all will work out well. Figure 7.2.1 reminds us where we are in the alarm improvement roadmap.

Let's be sure we start out on the right foot. Rationalization will take significant effort. Each and every alarm you end up with for your plant will have to be justified and configured and documented. The last one you work on is just as important to the operator as the first one. All the intervening ones count as well. Your critical objective is to achieve an alarm activation rate that your operators can work with. The only way to do that is to configure the least possible number of alarms to do the job. When they are configured, you want them to be correctly done. The choices presented in this chapter

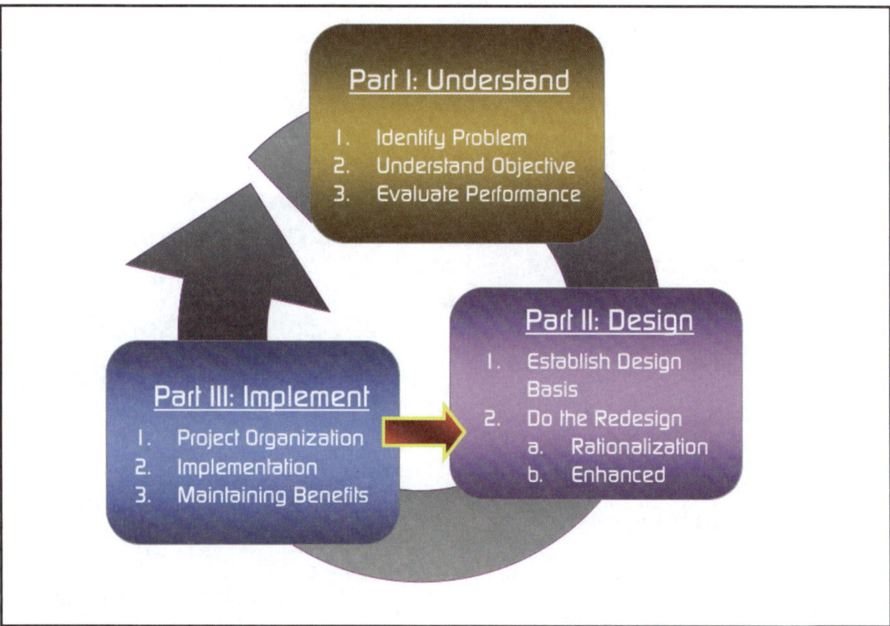

Figure 7.2.1. Rationalization on the alarm improvement roadmap

are real choices. Your plant should take the time to understand them and decide which will work best. Please approach this from the vantage point of a new way of doing alarm design rather than finding out which way is the closest to the way your plant may be used to doing things in the past. Where some choices are much better than others, this book will suggest important considerations to guide you. Figure 7.2.2 illustrates the rationale for rationalization.

Basic Approaches

There are two basic approaches to rationalization:

1. Start from where you are (the existing configuration).
2. Start from zero.

Starting from where you are is really starting from where the plant is now. It involves a complete review of each and every existing alarm. During this review, the decision will be made to keep the alarm as is, keep the alarm but modify some of its configuration parameters, or eliminate the alarm completely. This process also includes the identification of any new alarms that are needed for completeness.

Starting from zero begins with a blank slate (or blank sheet of paper, the "white sheet," if you will) by initially assuming that the entire plant has no configured alarms whatsoever. The plant is divided into its primal set of smaller subsystems, both for ease

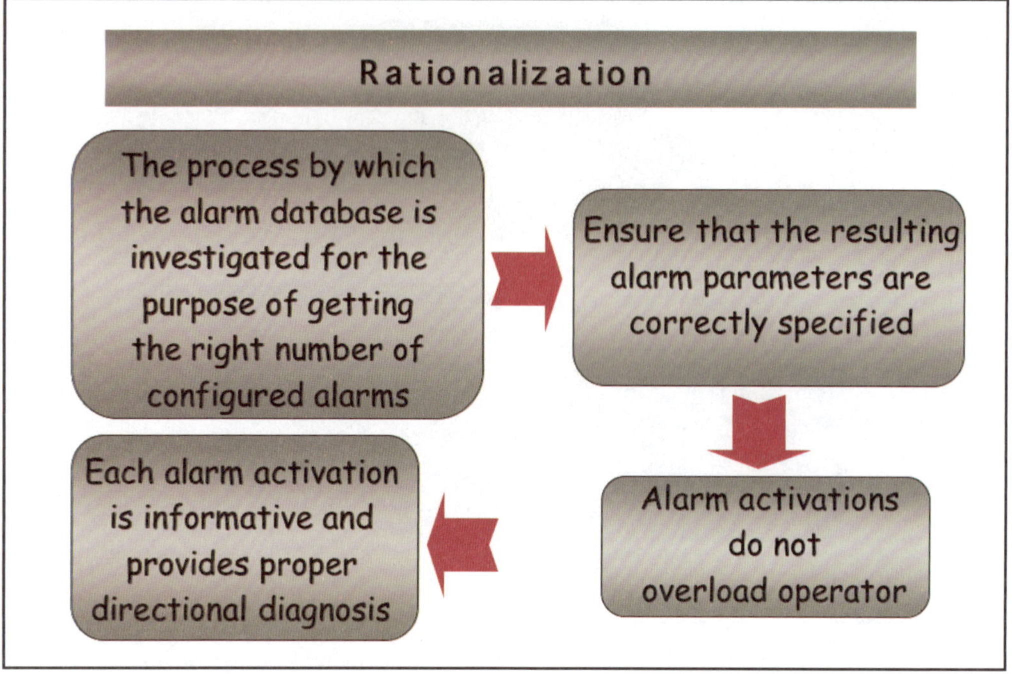

Figure 7.2.2. Rationale for rationalization

of understanding and to facilitate the design process. For each subsystem, an analysis is conducted to decide the minimum number and type of alarms needed to manage it properly.

Regardless of which approach is employed, the result is to make sure that each configured alarm will be understandable, prioritized, relevant, unique, and timely. For most plants, a successful rationalization results in a significant reduction in the number of configured alarms—on the order of five- to tenfold, which results in a reduction in alarm activations during operation. It is not unusual to completely eliminate alarms during normal operations, to have very few alarms during abnormal operations, and to have a manageable number of alarms during upsets.

Cornerstone Concepts of Alarm Management

Be careful. This is important. The literature on alarm management and materials provided by the leading vendors in alarm analysis and management software suggest a wide variety of different approaches to how to do alarm rationalization. Some differences reflect style. Others base their validity and relevance on the presumed practical aspects of efforts expended against results obtained. Some even go as far as recommending only doing as much as a limited budget permits. Even more confusing, most of the differing methods suggest approaches that are consistent neither with themselves nor with each other—that is, if one were to follow one recommended approach, correct decisions made there would be incorrect if another approach were to be followed. All this might suggest that alarm management is still in its formative stage, too early to be taken seriously or invested in responsibly. This is certainly not the case!

As we do often throughout this book, let's remind ourselves of the four fundamental precepts of what defines an alarm:

1. Operator must act.
2. Provide time to act effectively.
3. Provide adequate guidance.
4. Alarm only important conditions.

For each candidate alarm, when this alarm activates, the operator must be required and expected to take an action. Moreover, the alarm must be activated in such a way that provides enough time and guidance for the operator to act effectively before the alarm condition progresses to the point of serious impact. Add the requirement to guide the operator and not to distract him with unneeded conditions and you have all that is needed to understand alarm management. These are the cornerstones of alarm redesign. Every—and read that as strongly as you might—every decision necessary to provide a responsible alarm system will be based on those four precepts. Every potential rationalization plan, every suggested approach, and each individual step along the way must agree with and not conflict with those precepts. This book is about providing the broad understanding and effective methods for reaching a proper alarm system—one that is straightforward, consistent, and highly effective.

There is no dearth of discussion and no limited number of opinions about what is acceptable as results of a properly rationalized alarm system. You found metrics (or targets) for acceptability explained in chapter 4 on alarm performance review. You selected the ones the plant will use in the philosophy in chapter 6. Now is the time to understand the fundamental basis for the setting of those metrics. There are two classes of metrics: primary and secondary. The primary metrics are based on the limitations of operators to manage the abnormal situations announced by the alarms. What are the tasks our operator must do after an alarm activates? Secondary metrics are those whose value we estimate and if met, are likely to provide good results for the primary metrics. Secondary metrics exist simply because it is not possible to control enough of the primary ones in order to get the job done. By managing the secondary ones, we expect to get close to the primary metric targets.

Once again, we are face-to-face with our fundamental precepts. So when we put forward that an operator must not face more than five or six alarm activations in an hour, we mean that if an alarm system produces more than that on a regular basis, it is not likely that the operator can successfully manage things. So long as the rationalization process can end up with alarms that satisfy this requirement, we are fine. This whole book is dedicated to providing you, the reader, with the thinking and methodology to efficiently end up with a useful alarm system. Its experience is offered as a way to save many of the false starts and inefficiencies that usually accompany early efforts. Its examples and approaches are provided as a way to encourage efficient understanding. Its completeness is offered in the hopes that it will be able to provide that measure of courage and perseverance for you to succeed.

7.3 ABOUT THE WORD "RATIONALIZATION"

How did "rationalization" come to be the key name for this alarm improvement step? No one really knows for sure. This gives us liberty to explore a likely genesis. Examine the construction of the word RATIO-N-ALIZATION. From this word, two other words "ratio" and "ration" appear.

Ratio. The "ratio" part of the word rationalization now takes on an important message. Since we will have only a rationed few number of configured alarms, we must take care that the ones we use are constructed in the proper ratio: more in the more difficult parts of the plant and only use important alarms (higher priority ones) to denote important situations.

Ration. Improving alarm systems means making a serious reduction in the total number of alarm activations. Without much thought, we come up with two very simple solutions. Either we provide for very, very few potential (read that to mean "configured") alarms so that even if they all activate at the same time, there will not be many points in alarm, or we somehow make sure that the alarms we already have (and it does not matter now how many we have configured) simply do not get activated. The latter sounds just fine, since using it will mean that we do not have to look at the configuration at all, just keep the process from ever going astray far enough to activate more than a few alarms at a time. However, if we could control the process that well, we would not need alarms in the first place. At the end of the day, it is clear to see that our only significant chance to manage the number of activations is to make the configured alarms all essential and as few in number as possible. Consequently, we "ration" the limited few of them to uses that are important and useful.

Rationale. Go back to fundamental reasons for the methodology to improve alarms. Using the fundamental concepts, rationalization now takes on a purposeful design with results that can be traced to important concepts of process operation.

Now that we are more comfortable with the terminology, we can turn to what happens when we do a successful rationalization. The major accomplishment is that by doing it we gain about 90% of the benefits of improved alarm performance. That is a bit better than Pareto (remember the 80-20 rule?), but we had to do more work for it as well.

7.4 CHECKLIST

There are a number of methods to conduct rationalization. If you have already decided the approach and have it fully specified in your alarm philosophy, then that is the method you will use. Simply work your way through this chapter, picking out the appropriate

Item	Section	Notes
Rationalization approach	7.7	Start from zero, start from where you are, or use structured combination of both
Alarm activation point	7.15	Use method shown here or another methodology just as good
Alarm priority assignment	7.16	Method might be limited to the capability of plant rationalization tool
Alarm documentation	7.6	Alarm response information for each alarm configured; tracks the rationalization process

Table 7.4.1. Rationalization alternatives checklist

methods and procedures. If you have not fully specified this approach, you might want to pay particular attention to the wide range of alternatives. Some are equally effective. Others can make a big difference in both the end quality of the work and the difficulty in getting there. The following is a list of the basic decisions you will be making as you go along.

7.5 GETTING READY TO RATIONALIZE

This is the place where the job gets started. We want to end up with only the needed alarms. It is a simple objective. To do it will take specialized procedures and hard work—not necessarily hard labor, just relentless attention to the task, details, and keeping the plan in the forefront. The approaches described here work. One or the other, and sometimes in a bit of a combination of the two, is used by almost every serious alarm improvement project.

Housekeeping

Housekeeping is actually not a part of rationalization. It is the step just before we start and was introduced in chapter 3. Its importance is vital. The purpose of doing the housekeeping is to remove as many of the challenges to effective process operation as possible that are due to poor equipment operation or malfunction. Your alarm system should not be asked to accommodate or supplant missing, defective, broken, or otherwise improper equipment. More to the point, your plant was designed, built, documented, and operators trained on a plan that included all equipment and all equipment operating according to preestablished requirements. Anything not there, not working, or otherwise not doing its intended part places good operation in jeopardy. The original design was neither planned nor implemented with the expectation that things would not work as planned. Procedures were not written for how to operate with lots of odd pieces of nonworking equipment. MOCs were not written to account for each piece of underperforming equipment. In fact, with the possible exception of in-place spared equipment, there is in all likelihood no infrastructure that makes it okay to have things not working properly. So we fix it all first.

A short list, in addition to fixing broken items, includes the following:

- Ensure that sensors and transmitters are properly selected for their specific application, use primary measurements (delta pressure) rather than separate instruments and a calculation, and ensure that the instrument range is adequate for the operating range, not just a standard range to make it easier for instrument configuration and repair.
- Ensure that sensors and transmitters are properly located, properly mounted, and properly configured to minimize erroneous readings; ensure proper electronic or mechanical transmission of signals, and that they are adequately protected from interference from other nearby equipment and surrounding operations.
- Ensure that controllers (control loops) are properly configured and tuned (this is a big one!).
- Ensure that proper signal filtering is built into the instrument selection, added at the instrument interface with the control equipment, or configured in the processing blocks of the control system.
- Add any specialized suppression or processing to eliminate spurious alarm conditions resulting from signals that are intrinsically noisy or variable and for which no other method of management is feasible.

Bad Actors

Attack and conquer the bad actors in the plant is one of the first "kind" acts that rationalization teams can do early on in their work. Make sure that you follow the plant's MOC procedures for all alarm removals and "repairs." Bad actors include the following:

- Nuisance alarms including chattering, repeating, related, consequential, items out-of-service, and the like
- Alarms with no response including alarms that activate and for which the operator does nothing or has no idea of what needs to be done (and therefore again, does nothing)
- Stale alarms that have been active and on display screens for weeks, months, or even since the plant was commissioned
- Functionally disabled or never-used alarms

Do not lose sight of the fact that cleaning up the bad actors is a very small part of alarm system improvement. Bad actors can, however, be the most visible portion of poor alarm system behavior. Please keep in the forefront of your mind that this cleaning is just part of the staging of a serious project. Doing it well will endear you to the operators and probably the process engineers who have been bothered by them. Success here is mostly good will. You will not be in any better position vis-à-vis good alarm guidance, appropriate alarm activation points and priorities, and certainly no better for alarm flood situations. That is why we do rationalization.

Filters and Deadbands

There are a few aspects of the signal and alarm-processing infrastructure of the PCS itself that may need attention.

Filters

To improve the chances that a process measurement (or other real-time signal) will produce an alarm condition whenever it crosses the boundary between normal and alarm, it is necessary to ensure that normal process noise be appropriately managed. There are two built-in tools that address this situation. The first is the filter mechanism that can be used to preprocess analog measurements in the signal flow path from analog-to-digital converter to the PCS computational database. The filters are usually controlled by a parameter called a time constant or a filter constant. Since their configuration and time-based definition vary from manufacturer to manufacturer, no specific numerical values will be recommended here. It is suggested that those values properly balance the desire to minimize the propagation of process or measurement noise, against the requirement to keep the data current with actual physical conditions. Too "heavy" a filter will quite effectively remove or minimize noise. At the same time, that "heavy" filter will delay the visibility of fast-moving process changes. Consequently, filters should be set by good engineering practices that are appropriate for proper operation of each individual control loop. For signals that are not part of a control loop, the control loop settings are a good choice there as well.

Alarm Deadbands

The alarm-processing portion of the PCS contains a specific parameter called deadband. Occasionally, it is called by its classical name of "lock up." Both refer to the same thing. The parameter is usually expressed as a percentage. They are often shown as a percentage of full measurement span. The deadband is used in the alarm clearing computations. Refer to Figure 7.5.1 for an illustration of how this works.

The value of the deadband is the difference between the setting for the alarm activation point and the deadband line. In the illustration, the alarm starts out in a clear or not active state. At the first vertical line, the signal exceeds the alarm activation point and the

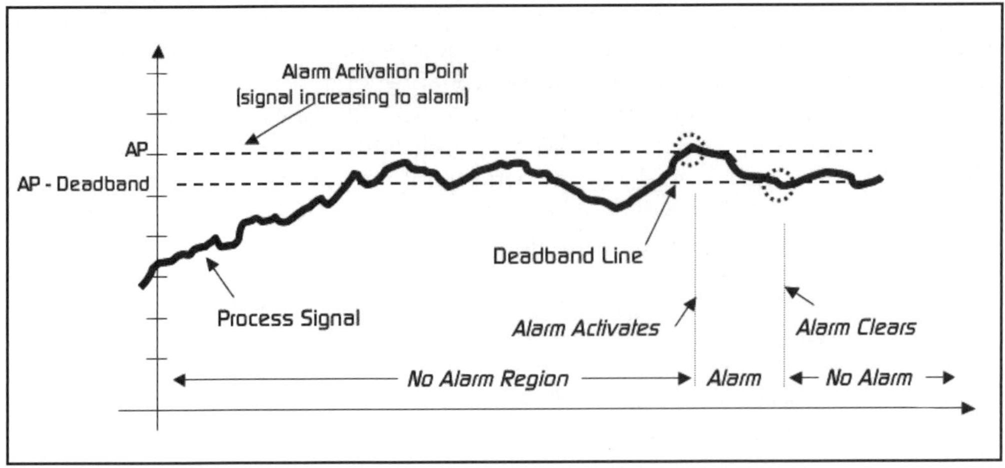

Figure 7.5.1. Illustration of alarm deadband operation

alarm activates. The alarm stays in the active state until the second vertical line where the process value has decreased below the deadband line. In this way, noise from the process does not result in spurious alarm activations and alarm clears.

The recommended values for deadbands vary with process signal type. For rapidly changing signals like pressure and flow, the bands are wide. For slower signals like temperature and level, they are narrower. Analyzers have a signal to noise ratio that can be quite low. New readings are likely to be widely separated in time due to sampling loop lag and analyzer processing time. For this class of signal, we use a very wide deadband.

As far as determining the actual values for the deadbands, the best and most effective approach is to conduct a small noise study for representative flow, pressure, temperature, and level variables. Then set the deadbands for each class about 20% larger than the normal noise values. However, it is recognized that not all sites have the patience and/or time available for such a study. Table 7.5.1 shows generic recommended starting points for these values.

Signal class	Recommended deadband (percent of nominal signal span)
Analyzer	10
Flow	5
Pressure	5
Level	2
Temperature	1

Table 7.5.1. Recommended alarm deadbands

The Data

The key steps for rationalization are pick the variable to alarm, decide the activation point, assign priority, and develop the alarm response information. You will need plant information to do that properly. The list includes the following:

- Complete PCS configuration in machine-readable format
- Real-time alarm and event data in machine-readable format
- HazOp data
- Operating procedures
- Piping and instrument drawings
- Shutdown logic
- Incident reports
- Any special operating or performance issues and considerations

As the alarm improvement teams do their work, the information from the list above is used extensively as resource material to better understand the alarms and the plant. For example, reading through the HazOp information, we might come across a condition where the plant, through a small fault in a shared utility, causes a reaction to accelerate dramatically. So far, operators caught the situation, but to ensure that they do not miss it, a special alarm was added on an otherwise nondescript filter. For some reason, all the operating procedures were not fully updated, so this alarm, while known by most operators, was not included in the normal operating procedures. Ordinarily, during a rationalization, this filter alarm would likely have been either deleted or modified. Access to the HazOp avoided this problem. The alarm is retained.

During the alarm improvement process, new things will be discovered or rediscovered. Often new ways of understanding the operation and inner workings of the plant come out. Therefore the team is advised to be aware that existing documentation should serve as a guide, not a specification. Your work should recognize this potential and plan for ways to use it. Some of the ways to make productive use will be facilitated by having safety and training members on the team. They will quickly recognize what needs to be done as far as improving the existing infrastructure and hopefully make plans to get it done. At the very least, recording the concerns and issues for later use in MOC and follow-up would be a good idea.

Alarm Documentation and Rationalization Tools

There is so much data to enter during the rationalization phase that you will want to use a purpose-built tool to assist you. Refer to chapter 4 for a more detailed discussion. The review of the commercial tools is repeated here for completeness. They are listed alphabetically.

- *AMO-Rt (PAS, Inc., Houston, TX).* AMO-Rt is a comprehensive alarm management solution, covering all the key aspects of alarm management, including documenting and defining customer alarm strategies relative to best practices (such as EEMUA guidelines), as well as the necessary tools for implementing, enforcing, and auditing these strategies in real time.
- *LogMate from TiPS (Georgetown, TX).* LogMate includes an alarm analysis engine, a rationalization utility, a configuration enforcement tool, and an alarm key performance indicator (KPI) engine. The analysis engine supports standard EEMUA-based statistics that can be generated contemporaneously or e-mailed on a scheduled basis.
- *ProcessGuard and Alarm MOCCA from Matrikon (Edmonton, Alberta, Canada).* ProcessGuard addresses real-time alarm assessment. It connects to any PCS or SOE platform and consolidates the data within SQL or Oracle. The diagnostic package provides a thin-client Web interface that automates qualitative alarm load assessment.

Rationalization Is Not Just About Numbers

This a fine place to bring this up. All of this time we have been recommending that a good target for rationalization would be a reduction in the number of configured alarms by a factor greater than four and probably closer to five or even more, and this is generally true. Most plants are overalarmed by a very large amount. These reduction factors are purposely high since it is necessary to impress practitioners that a good rationalization will likely remove a great number of not-needed alarms. This is to reinforce that no one can expect that just tinkering and a "by guess and by gosh" approach will do the job. It will not even be close. The anchor point for our thinking is that the alarm system must produce alarms at a rate and presentation that is useful to the operator. The only explicit, manageable way to do that is to manage the configuration of alarms. We manage the configuration based largely on the configuration targets found by previous "travelers" on the alarm management pathway.

The number of configured alarms is not the really important part. Their proper design is. So long as abnormal situations are properly announced and responding to alarm activations does not unduly overload the operator, the alarm system is working. Consider the case where a plant only has a very few number of configured alarms that, by those numbers, appeared to not need rationalization. Yet that same plant has an alarm system that does not assist the operator to manage plant upsets. We do not have a paradox here. We have not uncovered a major inconsistency in the alarm analysis and design process; rather, the opposite is true. This situation is a good example that conveniently amplifies the very fundamentals that make up the rationalization process. Here is why.

Recall that the end product of rationalization is to do the following:

- Specify only those points that required alarming.
- Determine the activation point and priority.

- Identify all necessary documentation to provide to the operator.
- Arrange for all other complementary infrastructure to match and complement the alarm system.

Even for a plant that might start out with about the right number of configured alarms, we would use exactly the same rationalization methodology. This means that we use the same process for developing the alarm philosophy: the alarm response information, understanding the consequences and such, and providing all the other operator support information and guidance, and so on. In this case, however, it usually makes best sense to do the rationalization using the method of "starting from where you are."

7.6 ALARM RESPONSE MANUAL

The very first benefit that early alarm improvement projects produced was a completely documented record of all configured alarms. It did not start out to be anything important. It was one of those things that just happened. It turned out to be one of the more significant early justifications of the validity of alarm improvement. Here is what it is all about.

We start with an alarm response (or alarm summary) sheet. And here is the punch line, right up front: the alarm response manual is simply the collection of all of the alarm response sheets. At first, it was an actual paper sheet. Now, it is a virtual sheet, usually in a structured database format. The sheet is a synopsis of all alarm information we know. It contains alarm identification information, alarm configuration information, alarm causes, alarm consequences, ways to validate that an alarm condition exits, what to do to fix or reduce the alarm condition, any links to other automated events, and finally, a record of testing and other links to special procedures or situations. Figure 7.6.1 shows a typical example sheet. The sheet will be discussed section by section. To facilitate understanding, after the descriptor information is mentioned in the explanatory text, the actual data from Figure 7.6.1 will be shown in braces {} to connect our discussion back to the alarm response sheet.

Our example alarm response sheet will be explained below. The first section is the header.

Header Information

The header block contains all the descriptor information and most of the configuration information. The first item is the Tag ID {FI-2009}. Next, the point descriptor, which is the actual wording, is used for identifying the tag in most text fields within the PCS {PB Feedwater Flow}. There is a place for general Comments {blank in our example}. The nominal control Setpoint is shown {195} and Units {KLbHr}. The Alarm Priority is shown {High}. Finally, the location of the point within the PCS point architecture, Operating Group {12}, and Process Area {PB}.

Configuration Data

This section contains the range values for the analog signal and the alarm activation points for all configured alarms. In our example, the tag FI-2009 has two alarms configured, a high flow {High Alarm} with an Alarm Point {200}, documented in column one; and a low flow {Low Alarm} with an Alarm Point {100}, documented in column two. The upper Range Limit {275} and lower Range Limit {0} is also shown.

Next, we will examine the remainder of the sheet for the high (flow) alarm only.

Causes

This section documents all reasonable causes that would likely lead to the alarm activating—that is, would cause a high-priority, high-flow alarm for FI-2009. In our example, there are the following four: {1. Failure of LVC/LIC-2011; 2. Plugged flow element; 3. Incorrect steam flow measurement; and 4. Instrument failure}. Any one of these failures, if not corrected, would lead to the high-flow alarm activating. This list was compiled by interviewing operators and reviewing operating procedures and all incident reports. Incident reports examined include those relating to the primary concern, FI-2009, as well as all of the "cause" instrument issues or failures.

Alarm Summary Data

Tag ID: FI-2009	Alarm Priority: High
Point Descriptor: PB Feedwater Flow	Alarm Status: Enabled
Comments:	Operating Group: 12
Setpoint: 195 Units: KLbHr	Process Area: PB

HIGH ALARM	LOW ALARM	
Alarm Point: 200	Alarm Point: 100	Alarm Point:
Range Limit: 275	Range Limit: 0	Maximum:
Causes:	**Causes:**	**Causes:**
1. Failure of LVC/LIC 2011 2. Plugged flow element 3. Incorrect steam flow measurement 4. Instrument failure	1. Low boiler feedwater header pressure 2. Control valve failure 3. High steam drum level 4. Piping failure in exchanger 5. BFW pump failure 6. Mechanical failure	

(continued on following page)

Confirmatory Actions:	**Confirmatory Actions:**	**Confirmatory Actions:**
1. Check steam drum level LIC 2011 2. Check strip charts, check "periscope" 3. Check BFW header pressure FI 2078/2077, DEA Pressure PI 2098/2099 4. Check BFW pumps 5. Check flow control valve FCV 2009	1. Check steam drum level LIC 2011, strip charts, periscope and level control valve 2. Check BFW header pressure PI 2078 3. Check for piping failures 4. Check BFW pumps	
Consequences of Not Acting:	**Consequences of Not Acting:**	**Consequences of Not Acting:**
1. Boiler could overheat and seriously damage tubes 2. Insufficient steam flow, high press	1. Boiler could overpressure and rupture 2. Insufficient steam flow, low press	
Automatic Actions:	**Automatic Actions:**	**Automatic Actions:**
None	None	None
Manual Corrective Actions:	**Manual Corrective Actions:**	**Manual Corrective Actions:**
1. If steam drum level and steam pressure are high, put feedwater valve in manual and begin to close it 2. If steam drum level is low, manually increase opening of feedwater valve	1. Start another pump 2. Put feedwater valve in manual and open 3. If drum level is high, check level control valve, wait out upset; may have a "swell" if load INCREASED, be ready to INCREASE BFW flow 4. If problem with exchangers, check PDIC 2075, open bypass FDCV 2075	
Safety Related; Testing Requirements		
None	HazOp 12-4453; Yearly Testing	

Figure 7.6.1. Example alarm response sheet

Here and in the next parts of the alarm response information, you can observe how the seemingly long list of documentation items now becomes important and useful.

Confirmatory Actions

Before we ask the operator to work this alarm, we must take the additional step of ensuring that the alarm is a true representation of a physical situation gone abnormal. The fact that our alarm has activated does not mean, for sure, that the physical flow (boiler feedwater) measured by instrument (called FI-2009) is actually flowing at too high a value. A transmitter might have failed, but the actual flow remains correct.

There are five confirmatory actions in our example {1. Check steam drum level LIC-2011; 2. Check strip charts, check "periscope" [strip charts refer to an existing analog pen recorder; periscope refers to a special level detector installed on the process]; 3. Check BFW [boiler feedwater] header pressure FI-2078/2077, DEA [deaerator] Pressure PI-2098/2099; 4. Check BFW pumps; 5. Check flow control valve FCV-2009}. Any one or a combination of items from this list would confirm that our actual flow is too high. If none are out of the ordinary, the operator should consider that the flow is probably okay, but some instrument might be in error.

Yes, we did say that one of the "rules" for responsible operating (see chapter 5, "Permission to Operate") is "trust your instruments." Nothing will be said here to contradict that guiding principle. However, trust is not always good if blind. Experience has shown that a brief time spent confirming the alarm condition can prevent a great deal of wrong focus in abnormal situation management. Moreover, the simple process of checking and confirming will often uncover many of the direct causes for the alarm.

Consequences of Not Acting

Consequences are important to bring out here inasmuch as they play a large part in determining the priority. Just as important, providing them here enhances the impact of the issue for the operator and assists in understanding the underlying process operation. Not acting is a dual concern. First, we expect the operator to see this alarm after activation and start to work the issue. Of course, if there are other tasks of higher urgency, then those will probably come first. This alarm is a high priority. Unless there are emergency ones activating or there are several other high priority ones at the same time, our operator is expected to direct his efforts here. Second, even if our operator is quickly focused on this alarm, the consequences are meant to convey what will likely happen if efforts to "fix" the problem are not successful in time.

Returning to our example, there are two consequences: {1. Boiler could overheat and seriously damage tubes; 2. Insufficient steam flow, high pressure}.

Automatic Actions

Alarm activations are sometimes used as digital switches to automatically initiate a process operation. Sometimes they are used to automate a blow down or start an overlevel protection pump or other odd equipment. On the other hand, there are times when the alarm condition happens so quickly that there will likely never be enough time for the operator to react and the process recover as a result of operator actions. In these cases, an alarm event might set in motion automatic actions to control the situation. Those are documented here.

For our example, there are no {None} automatic actions.

Manual Corrective Actions

Manual corrective actions are the bread and butter of the operator's handling of alarm conditions. Often all that is needed to start the process heading back toward where it should be is to put the problem controller in manual. This will stabilize the process if process or controller variability was the likely cause of the alarm. Other times, another related part of the process will be manually changed. At times, our operator may direct that the position or alignment of certain manual valves in the field be modified. Still other times, our operator might shut down one part or another in response to broken or malfunctioning equipment.

In our example, there are two recommended manual actions: {1. If steam drum level and steam pressure are [both] too high, put the feedwater valve in manual and begin to close it [thus diminishing the flow until it is back into an acceptable region]; 2. If steam drum level is low, manually increase opening of feedwater valve [to add more feedwater]}.

Safety-Related Testing Requirements

Some alarms are used to recognize and manage more serious operating conditions. Many of these might not be so apparent on the surface. Consequently, our alarm might have been added due to a specialized hazard condition. Alternatively, it might be present as part of a safety warning system but does not require a safety integrity level (SIL) rating of its own. Moreover, the instrument itself might be prone to certain degradations that must be checked. In any event, this is where any of those unusual conditions are documented.

For our example, there are no items in this list {None}.

Example Online Alarm Response Sheet

It is all well and good to have collected this information. Its true value, however, becomes realized when it can be made available to the operator when needed. Some plants choose to integrate links to the alarm response sheets. Figure 7.6.2 illustrates what this might look like as an online operator support tool. This figure contains mock information, so please do not examine the actual text too closely for content.

A control loop faceplate is depicted on the left of the figure. The button at the lower right (with the check mark) is the link to check the alarm response information. The right portion of the figure is the information that it is linked to. There is a box that depicts all alarms for the selected tag. Eight alarms are depicted; they represent all possible alarms for this tag, however, in actual practice, you will see only one or two used. The two dialogue boxes at the bottom contain the response information as well as additional point descriptors and other configuration information. The example was prepared for use on an Emerson DeltaV DCS system. It links through a standard configuration parameter "ALM_Help."

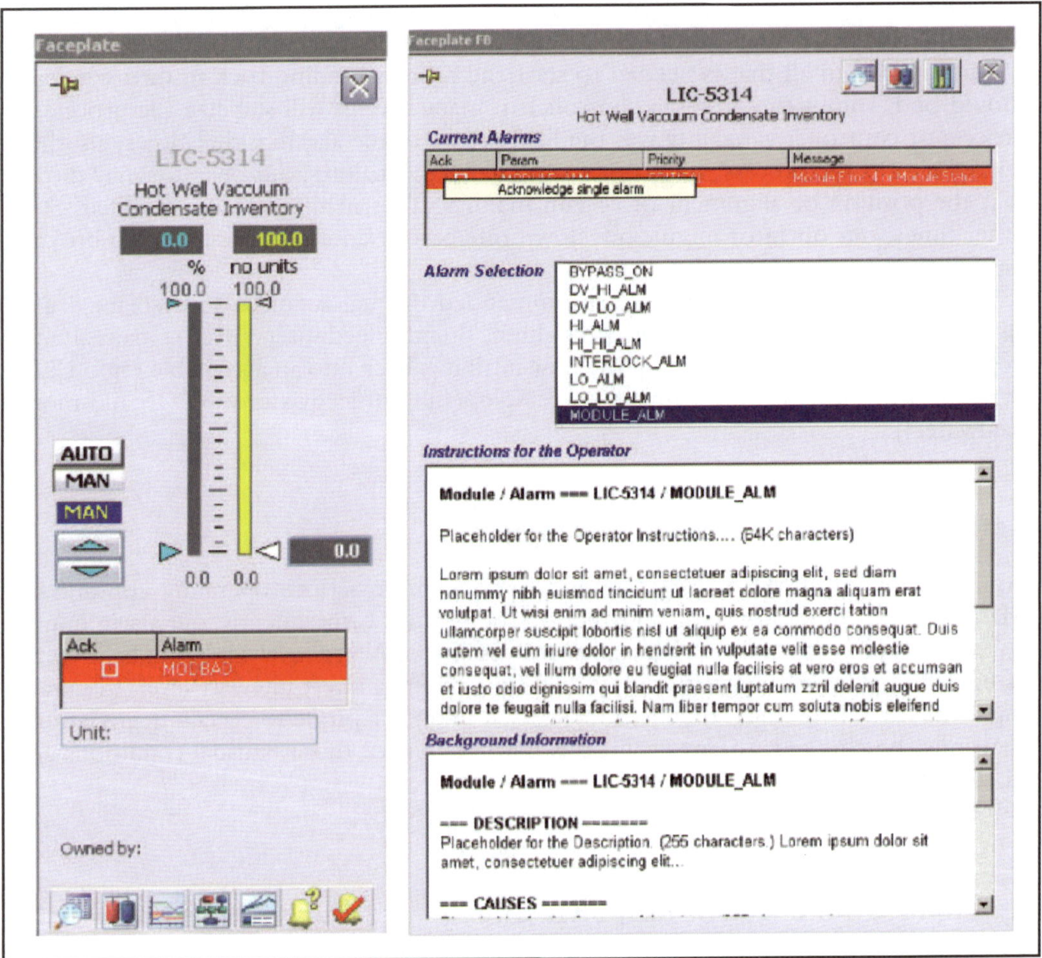

Figure 7.6.2. Example of an online linked alarm response sheet

Additional Items

While not on our example alarm response sheet, a few other items will need to be determined and documented as part of the complete understanding of this alarm. These items come from the data needed to properly determine the alarm activation point. The list includes the following: the likely rate of change of this measurement (FI-2009), the normal variability of the measurement, how difficult it is to manage this flow successfully, and how much time the operator will need to manage the situation if a problem develops in this control loop. Refer to section 7.15, "The Alarm Activation Point," for a complete discussion.

7.7 RATIONALIZATION METHODS

Alarms Are Not the Important Part

I would like to suggest an important concept for your consideration. It has to do with what alarms are all about. It has been introduced in several places so far, but now is the time to drive it home:

> The alarm system, and all alarms that are created, is only to notify the operator that there are abnormal situations within the plant that require timely attention to manage.

An alarm is only a means to an end. Alarms and alarm systems are not at all about finding where to set the alarm activation point for process variables that need alarming. It is the abnormal situation that must drive all aspects of the alarm creation and design.

For example, consider a plant that has a process flow, a primary reactant that joins others into a continuous stirred tank reactor. Do not worry about understanding reactors—I do not much either. It is just a convenient example to play with. If the reactant flow is too low, the reactor will go exothermic and the whole thing will heat up and explode. If the reactant flow gets too high, the reaction will slow down, but the reactor is most likely to overfill. The reactor has a level monitor and a temperature monitor, among others. These are enough to start with. Start with the flow. Looks like a good idea to have both a high and a low flow alarm. The low flow will guard against the reaction running away; the high flow will keep the reactor from overfilling. The low flow is important, since a reactor can run away quickly and we do not want that. The high flow is not needed. If the level grows too much, we have a high-level alarm. We need the level alarm since there are a number of ways the volume balance in the reactor can go amiss.

Let's return to the example above, but from the abnormal situation perspective. For our reactor, the two important abnormal situations we want the operator to know about are a reactor runaway and the reactor overfilling. We then look for the two best ways to identify each condition to warn the operator. OK, I know that those of you who understand reactors better than I already have your eyebrows raised. I oversimplified this, but I am sure that you can think of another example that is well grounded in chemical engineering or paper or cheese that will do better. The message remains.

Rationalization Approaches

There are the two basic rationalization methods: (1) start from where you are and (2) start from zero. The process is very different for each approach. The results are usually different as well (fig. 7.7.1). The very nature of starting from where you are requires a great deal of "letting go," so to speak. Unfortunately, many sites find it difficult to overcome their attachment to whatever "comfort" level their old alarm system might have provided. They are unwilling to cut the ties too readily. Consequently, for most projects, doing it that way reduces the number of configured alarms by only 10%–35% the first time through. This really means that you will be retaining 65%–90% of the original alarms. On the other hand, the results of *starting from zero* are a reduction of a two-and-a-half-fold up to a tenfold reduction. This method requires the use of a more holistic approach. In the final analysis, either approach, when seriously utilized and worked in a dedicated, professional manner, will produce a workable alarm system.

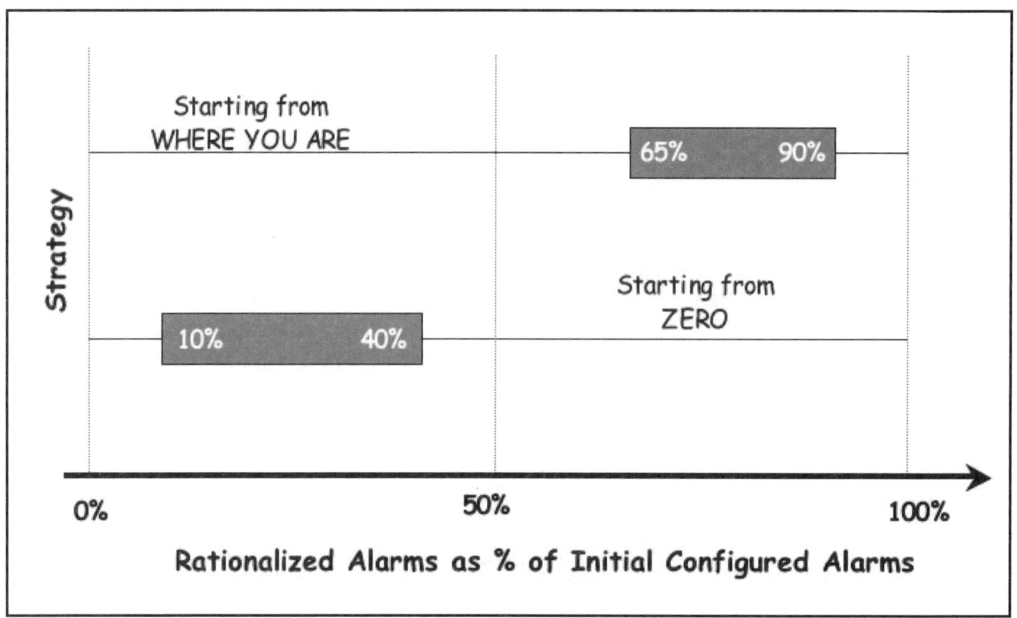

Figure 7.7.1. Rationalization numerical results depend on rationalization approach used

"Starting from Where You Are" Rationalization

Starting from where you are means just that. Capture the existing alarm configuration database, alarm activation points, priorities, and all the rest of the PCS and alarm system configuration. Then, using whatever strategy might appear to be useful (one by one, by classes, by physical entities, etc.), decide whether this alarm should stay (and, if so, what its new activation point and priority should be and so on) or whether it should be eliminated completely. Also, as you go along, you will identify any alarms that are not presently configured but needed. Do this until the entire configured database has been reviewed.

There is a practical problem with doing it this way. There is no way to take advantage of the work until it is completely done. There are no halfway benefits, no partial advantages.

"Starting from Zero" Rationalization

Starting from zero means your work starts with no configured alarms. The design team must add each alarm that will eventually end up in the configuration. We introduce the motivation for this concept by a story.

The Loss-of-View Lesson

This is a true story.

The cornerstone for starting from zero derived from a situation that many of the early users of PCS hardware had the horror to experience. We call it "loss of view." Loss of view is a technical term used for systems that use electronic displays to show information and such to the operator. Loss of view is used to describe the situation where all view screens or displays are suddenly blank, blue, appear "frozen," or otherwise unable to provide any useful information to the operator. At this moment, and for as long as the situation lasts, the operator is functionally unable to obtain any information about what the process is doing. Nor are there any "controls" by which the operator can make any interventions (auto/manual changes, start or stop a pump, etc.).

Let's get personal. Imagine that you are in the control room when a loss of view occurs. Well, when it first happened to a major refiner, it made company news and led to a fundamental new understanding of how to handle this in the future. But we are getting ahead of the story. Let's look over their shoulders during the event. There is a control room with a number of operators and suddenly everyone sees nothing. As far as they can tell, the screens are the only equipment affected, but who can tell what! After a short time, our control room fills up a bit. Visitors come, from the operations superintendent to plant manager and some in between. Eventually, the senior operations team decides that the situation is too out of control to do nothing and orders the plant shut down.

So without the aid of any board-mounted controls (read that PCS controllers accessible by screens), operators were sent into the plant and tasked with shutting it down. They manually open and close valves. They start and stop pumps and fans. They trip out package units. They bring everything they find in the plant to a shutdown status. They managed to do it without much personal injury but when things finally cooled, there was a lot of equipment damage. And there was lost production as well. The whole thing cost a very large amount of money—so much, in fact, that a major company investigation was launched to understand what really happened and decide how to prevent it in the future. Here is what they found:

1. When view was lost, the electronic displays were totally unavailable and no work-around existed that would have restored the view quickly.
2. There was no secondary method of monitoring any important aspect of the process; the operators were totally blind.
3. All other PCS functions were found to be working. This means that all control loops were operating and all valves and other outputs were being appropriately managed by the control system. All alarms were functional. All PCS interlocks and other logic were fully functioning.
4. All safety-related shutdown systems were fully operational; all physical level protections (i.e., overpressure relief valves, dikes surrounding tanks, etc.) were in place and functional.

Based on these findings, here is what they concluded:

1. Their plant had been carefully designed to operate, the PCS controls were fully operating, all safety interlocks were in place, and all physical containment and all code requirements for safe operation were in place and functional.

   The best thing to do was leave things alone and let the control system do its job, then just wait for view to be restored. The worst thing to do was to do what they did—try to shut down without the eyes and hands of the control system and operator.
2. Even with the existing PCS, the operators lacked an effective overview of the essential process operation. In the "old days," anyone could come into the control room, walk to the panel board, see at a glance the essentials of the process, and determine what was going on.

   Therefore they required that each process unit be examined and the four to eight fundamental variables identified. Once identified, they placed electronic analog chart recorders for each one next to their appropriate PCS electronic display screens. It was then possible for the operator to know at a glance that his process was basically sound or, if something was amiss, to know where it was and where to look for a resolution.

The loss-of-view lesson for alarm management is this: *it is possible to identify key fundamental aspects of a process, which, if monitored, would provide a robust measure of the health of the process.* This robust list of variables is not usually extensive. Taken together, they will provide a sufficiently complete picture. The real lesson from the loss-of-view story is how it reinforces the concept that fewer, but fundamental, alarms can do the job. This is why starting from zero works.

7.8 REQUIRED ALARMS AND COMMON ELEMENTS

Before we jump into the working details of alarm rationalization, there are two important items to nail down: identifying required alarms (sometimes referred to as "highly managed alarms") and identifying and handling common elements.

Required Alarms

A funny thing happened along the way to rationalizing plants in the early days of alarm improvement. At the end of these first projects, plants were shocked to find that a small number of important alarms had been deleted. These alarms had originally been configured to accommodate special concerns of the plant or perhaps to comply with perceived or actual regulatory or code requirements. When these cases were examined after the rationalization work, it was clear what had happened. What happened was that those "important" alarms were not important enough to satisfy the new standards of proper alarms. According to how the plant decided to configure alarms, those had not met the test. Since there was nothing else to go on, they were deleted. No, this situation does not serve to identify flaws in either the alarm philosophy or the way it was interpreted and used for rationalization. Nor does this serve to suggest that those "important" or required alarms are unnecessary. To the contrary, all of them should be retained and properly examined for appropriate configuration and documentation—but not deleted.

Proper rationalization calls for identifying all such required alarms and ensuring that they are kept. We do this at the very beginning of rationalization. The reason is that all alarms that have been prespecified (by code, law, or overriding plant practices and/or requirements) are important and must be retained. We explicitly now modify the alarm rationalization process to identify them and ensure that they are retained. In these cases, the plant is strongly encouraged to review the alarm configuration parameters and documentation to ensure that not only do the alarms conform to the requirements, but also that they will activate when needed and provide the operator with appropriate guidance.

Common Elements

All plants have parts that are unique and represent special cases. Most plants also have a number of common parts or elements. For example, in a chemical facility, there might be lots of pumps, heat exchangers, tanks, separators, and the like. The pumps, for example, might be classed as small, large, and unique. For the nonunique ones, it has been found to be very efficient to design alarms for them as a class. Then whenever a similar pump comes up in the plant, we will use the design that has been prespecified. While there might be several classes of each type, doing it this way can significantly enhance uniformity as well as efficiency in the work process. Do the same for the other common elements.

By the way, it is often useful to work the common elements using the approach of starting from zero. This approach has been found to be more robust and result in a lower number of alarms. A lower number of alarms for a plant component that is repeated at a number of plant locations produces a multiplied reduction effect for the total number of configured alarms.

7.9 "STARTING FROM WHERE YOU ARE" RATIONALIZATION

Alarm improvement can start with a complete list of each and every configured alarm in the entire plant. The work will be to simply go through the list, cut every unnecessary alarm, and add any alarms that are needed. We end up with a new alarm system. Unfortunately, we often end up with too many configured alarms. People working generally in an environment where the default—albeit unconscious—bias is in favor of the existing alarm will find it hard to remove that bias. This has been the traditional approach for alarm improvement for established sites. It makes complete sense. After all, one of the key alarm problems is caused by too many alarm activations. If we reduce the number of configured alarms, there are fewer that can activate. There is nothing wrong with that thinking, at least on the surface.

What is mostly wrong with starting from where you are is that the starting point can have a very, very large number of existing alarms. For a single operator area, it is not unusual to find 5000 or even 15,000 or more alarms in the operational PCS database. Lest we lose sight of the effect, if all the alarms were to activate at the same time (really impossible, but let's just consider it for the shock effect), one could actually see the 5000 (or 15,000) activations occur. No, of course, that is not where this point is going to be made. The point of this is that if we start from where we are, we will have to somehow go through all 5000 (or 15,000) alarms, one at a time, and see whether we want or need them. We cannot just read the tag and make a decision. No, we will have to know what the alarm does and ask ourselves, or an experienced operator, is it needed, and if so, why; or if it is not needed, why not? After it is all done, we want to end up with only a 1000 or so.

Well, this process usually just cuts the number by 25% on average. Start with 5000 and end up with a bit fewer than 4000. Do it again and you end up with about 3000. You will have to do it a total of four times to get it down to around 1500. And by this time, all the members of the alarm improvement team will want to jump out of the nearest high window. But it could be done that way.

There is one other wrinkle in this procedure: it is difficult to really appreciate deleting alarms one by one when there are lots of them still left. It is hard to see the true nature of which alarm might be important and which one is not when one has to look at many other potentially redundant ones. Which one is best? Does one show a nuance that the other does not? And each time one is deleted, you will need to revisit nearly the same scene again and again. But this can work. It has been a tried-and-true way to do it, so far.

Work Process

The basic work process is depicted by Figure 7.9.1.

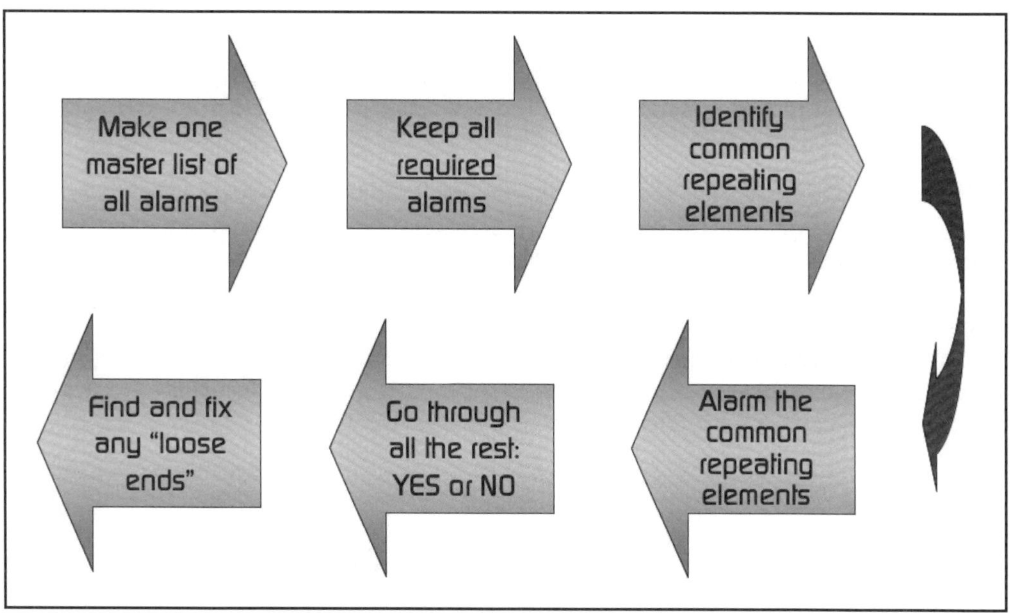

Figure 7.9.1. Starting from where you are workflow

Make a Master Alarm List

We get the master alarm list from "exporting" the PCS configuration files and then "importing" them into a documentation and rationalization (D&R) tool. Simple to say and simple to do.

Keep All Required Alarms

Go through the list of configured alarms and flag all the following alarms for keeping:

1. Alarms that are configured due to regulatory requirements and other requirements of lawful plant operation
2. Alarms that have been explicitly specified by plant HazOp reviews and other explicit safe-operation requirements
3. Alarms that have been explicitly added as a result of incident reviews and resulting investigations and added to the plant operating procedures and training protocol
4. Alarms that are present to support specific plant operations that have been explicitly identified to be vital to plant operations and are incorporated within operating procedures and training protocol

This is not to say that these alarms will be kept regardless. But any consideration of their removal must trigger a plant MOC and otherwise rise to the top of a "watch and review" list for those responsible for validating alarm rationalization results.

Identify All Standard Process Elements

Most plants contain a significant amount of standard process elements: pumps, heat exchangers, demisters, slug-catchers, and the like. Make a list of them.

Alarm the Standard Process Elements

Start with the list of standard process elements like pumps, heat exchangers, demisters, slug-catchers, and the like. Continue the alarm process by considering alarming them in a standard way. The process is simple. Take each process element type and work out the simplest method for alarming them. You might need a few cases here. But they will be obvious. If you are having trouble starting, consider using the start from zero approach for one of the elements (see section 7.11). Once you complete the list, simply locate each of them in the alarm tag list and modify their alarms accordingly.

Go through the Alarm List and Say "Yes" or "No"

Now is the time to start working though the alarm list one at a time from start to finish. Use either the Method of Flows or Method of Elements discussed in Section 7.14. Each configured alarm will be either affirmed, and hence kept, or denied, and therefore removed. For each alarm kept, we do a full documentation (see section 7.6, "Alarm Response Manual"). The information is also entered or computed to provide for a proper activation point and priority.

Find and Fix Any Loose Ends

You are nearly at the end of the process. At this point, it will be necessary to visit all alarms that have not been evaluated and evaluate them. At the conclusion, the D&R process is complete.

7.10 "STARTING FROM ZERO" RATIONALIZATION

In recent years, after companies have had one or two rationalizations completed, the results were not what they had expected. The process quickly became repetitive and less productive. And they ended up with many more configured alarms than they had wanted. They began looking for a more efficient way to reduce the number of configured alarms. One of the early techniques to emerge was the use of classes of subsystems. When they identified certain subsystem classes that repeated again and again, they began to try to understand their similarities and differences. They wanted to see if there was sufficient commonality to make it worthwhile working out alarm conventions as a class and apply that result everywhere the subsystem was present. This worked surprisingly well. They began looking for more similarities and working out more conventions for more classes.

Eventually, the process had evolved far enough along that it became natural to consider a radically different approach to rationalization. Starting from zero was born—or, as some refer to it, the "white sheet" approach, or "Carte Blanche" in French. Surprisingly enough, after working out the kinks, the approach turns out to be more indicative of actual process abnormalities. It does so in a better, focused way than the old methods. To say it another way, the variables that are important, and hence should be watched by the operator (recall the loss-of-view story), give a very good picture of the true operating situation, and there we place our alarms.

This is a good time to think about how suited this method might be for new (grassroots) plants. New plants sometimes come with a recommended set of alarms from a licensor or perhaps from another similar plant elsewhere in the enterprise. Since the new plant is not in operation, prospective operators and supervisors as well as engineers will likely have little stake in keeping the starting set of alarms. Moreover, there will probably be very few who are experienced with the plant. All the more important to start from the basics: Starting from zero is the preferred method here. Not only will it produce an excellent alarm system directly, but also the process usually produces better operating manuals and better controls. All these benefits derive from the fact that starting-from-zero involves going back to basics. This fosters better understanding and more appreciation for the choices that can be made in operation. It goes without saying that this process enhances the safety and overall operability of the new plant.

Work Process

The work process is illustrated by Figure 7.10.1.

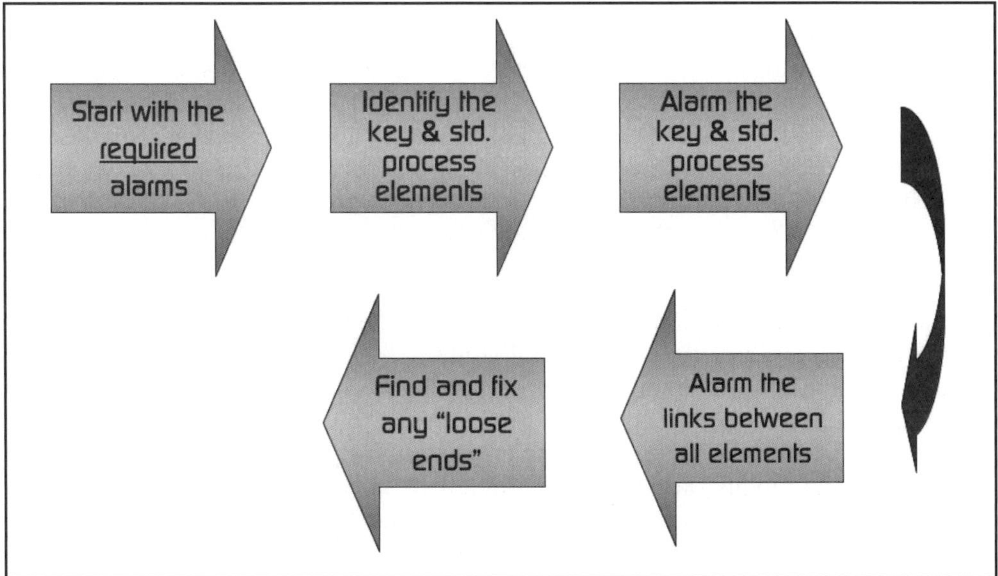

Figure 7.10.1. Starting from zero workflow

Keep All Required Alarms

This is the same step as the "starting from where you are" approach. It is repeated here for completeness. Go through the list of configured alarms and flag for keeping all the following alarms:

1. Alarms that are configured due to regulatory requirements and other requirements of lawful plant operation
2. Alarms that have been explicitly specified by plant HazOp reviews and other explicit safe-operation requirements
3. Alarms that have been explicitly added as a result of incident reviews and resulting investigations and added to the plant operating procedures and training protocol
4. Alarms that are present to support specific plant operations that have been explicitly identified to be vital to plant operations and are incorporated within operating procedures and training protocol

This is not to say that these alarms will be kept regardless. But any consideration of their removal must trigger a plant MOC and otherwise rise to the top of a "watch and review list" for those responsible for validating alarm rationalization results.

Identify Key and Common Process Elements and Subsystems

This step is broader than the similar one outlined in the start from where you are approach. Most plants contain a significant amount of common process elements: pumps, heat exchangers, demisters, slug-catchers, and the like. As in the previous method, make a list of them. Now the new part. Make a list of the key, important process subsystems. This list will likely include items like reactors and separators, for process industries; paper stock machines and web processes for paper making; steam boilers and turbines for power industry; and the like. These are all of them, not just the ones that repeat a number of times. Use the Method of Flows or the Method of Elements to work your way through the plant. That process is described in section 7.14.

Alarm the Key and the Common Process Elements and Subsystems

Earlier, we put aside the common process elements like pumps, heat exchangers, demisters, slug-catchers, and the like. We will continue the alarm process by considering alarming them in a standard way. The process is simple. Take each process element type and work out the simplest method for alarming them. You might need a few cases here. But they will be obvious. Once you complete the list, simply locate each of them in the alarm tag list and modify their alarms accordingly.

Now that you are warmed up, you are ready to tackle the process subsystems. Use the approach of "only four alarms" (section 7.11) to work your way through each subsystem. Unlike the standard process elements, these are likely to be one or two of a kind. Do this for all the other individual subsystems you have.

Alarm the Links Between All Elements and Subsystems

This should pretty much include all the small pieces (plant components) that connect the subsystems together. In general, this category amounts to less than 5% of all alarms that the plant will configure. You will want to use P&IDs and process flow diagrams (PFD) to locate and keep track of these parts. At times, the odd point might also show up on a procedure or other operating or maintenance instruction.

Find and Fix Any Loose Ends

You are nearly at the end of the process. At this point, it will be necessary to visit all abnormal situations that have not been evaluated and evaluate them. Step back and view the entire operator area. Check to make sure that nothing important is left out. HazOps and incidents are reviewed as a final check to ensure that all is covered. Lest we forget, by this point, we will have identified the alarms, the activation point, the priority, and all of the other documentation (alarm response sheets, etc.) needed. At the conclusion, the D&R process is complete.

Wrap-Up

The process works. Fortune 500 petrochemical companies use it with success. Moreover, when they examined the total effort to do it this way with what it took them to do a similar plant starting from the existing alarms, they found that it was very close to the same. But doing it this way, they ended up with significantly fewer alarms (a reduction by a factor of four, five, and even ten). In addition, they felt that they had a much better understanding of the fundamentals of operation than before.

7.11 ONLY FOUR ALARMS

A good way to understand how the whole idea of starting from zero can work and build acceptance of the process and demonstrate the process is to do the "only four alarms exercise" with a group of process experts. Here is how it works. Start by selecting a fairly conventional process element or subsystem. Typical selections include heat exchangers, reactors, distillation columns, and compressors. Figure 7.11.1 illustrates a typical one. So, pick one. For the selected element, develop a full P&ID. Keep it real. Keep it generic.

Compose two equally qualified, informed working teams. It will work just fine with one, but two makes the process more interesting and often provides additional insight. Each team will work independently at first and come up with its own design. The first task will be to understand the fundamentals of operation for this process unit. Using the results from that part, move on to identify the key, overarching variables that govern its successful operation. Finally, considering those key variables, they are allowed to design

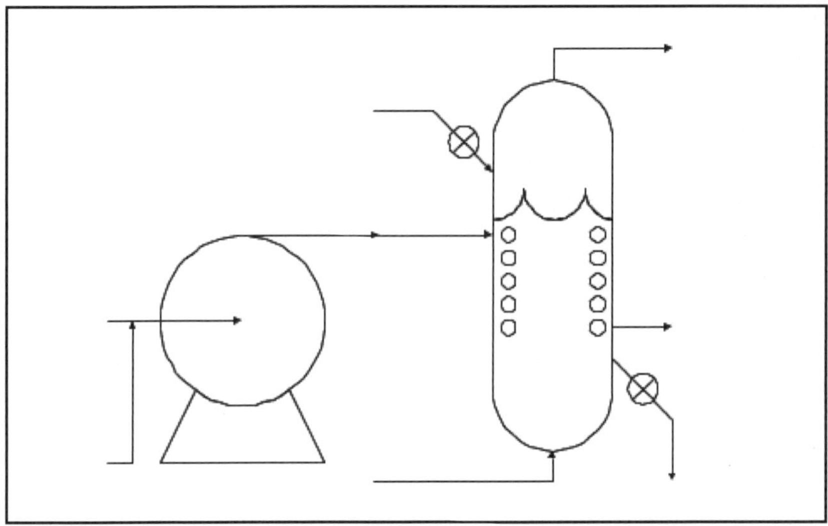

Figure 7.11.1. Pump and tower example

only four alarms. Once the team picks their alarms, they also need to explain why they chose as they did. This part commonly takes an hour or two, but sometimes a bit longer.

After time is up, it is often discovered that the teams may have failed to do it with just four alarms. During questioning, we find that perhaps one team's minimum is five. The second team needs six. No team should need many more. You have success! The exercise produced exactly the results that were sought. Four alarms is an arbitrary number. There is no magic or other hidden benefit to that number. The point is that the job could be done using a very small number of alarms. In the end, each team will have been able to alarm a basic processing element using only a fraction of the usual number of alarms in contrast to a more conventional approach. Additionally, by starting from zero, they become better able to understand the intrinsic aspects of normal and abnormal operation.

Recall that two teams are recommended to work the exercise. Using two teams working on the same exercise provided the opportunity to pause the process about two-thirds of the way along. During the pause, the teams share information. Usually, there will be considerable agreement but each might have taken a bit different approach to get there. After comparing notes, they go back to work and finish up. In an actual plant situation, we would ask the team to meet again and resolve all differences.

A variation of this exercise that has also proven successful is to give the problem to a number of individuals working alone. After they obtain results, all results are shared with each participant. Each participant is then asked to take the combined feedback and work out a resolved solution. Convergence is usually very good. One or two large differences are almost always the result of personality rather than fundamental disagreement. Refer to section 7.13, "'Starting from Zero' Examples."

7.12 IDENTIFYING SUBSYSTEM BOUNDARIES

Things are taken apart in order to work on the smaller pieces in an easier and more complete way than possible when whole. It is really easy to take anything apart. Any scheme will work. Anyone can do it. On the other hand, it is extremely difficult to take things apart in a way that permits it to be perfectly put back together afterward. How often have you heard the story of an item taken apart and then reassembled only to find three screws left over? The art and science of taking things apart in a way that permits reassembly is called decomposition.

Decomposition

Decomposition means to take apart. To start from zero, we will need effective ways to take a large plant and divide it down, decompose it into manageably small parts. If it were just a matter of breaking it apart, then any forceful way would do. We could be as haphazard as we wanted. But breaking it apart has a purpose. We will want to be able to understand what is happening in each of these pieces in a way that will make sense. Random pieces do not lend themselves to planned processing. Once the work of understanding each separate piece is complete, we put things back together again into a functioning whole.

An effective way to break something large into smaller pieces is to take advantage of boundaries. Boundaries can be natural, as land is divided by rivers, lakes, mountains, forests, and the like. Boundaries can be functional like political boundaries. Figure 7.12.1 illustrates typical boundaries.

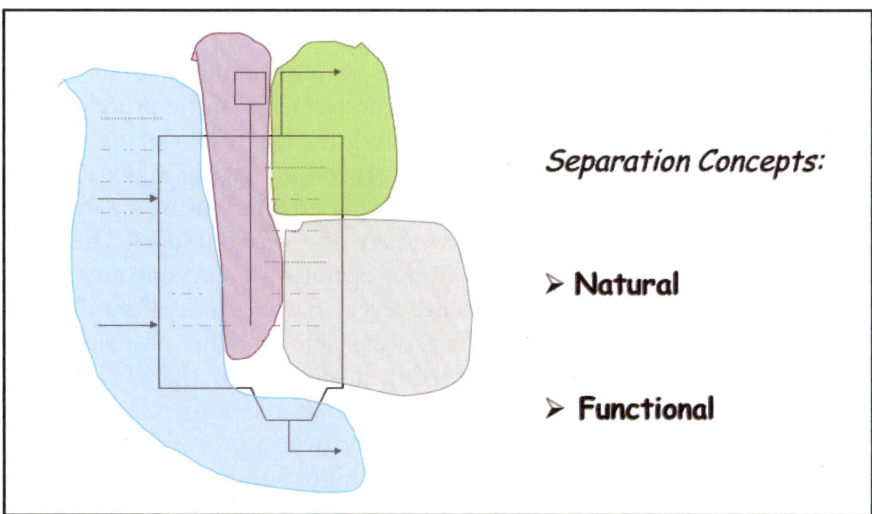

Figure 7.12.1. Decomposition boundaries

Notice that this example has four identified regions or subcomponents. We were able to identify the boundaries between the regions by finding places within which most of something was, and that something was not much elsewhere. Or, we found the boundaries by looking for places where the adjacent piece(s) were less connected than elsewhere. Fortunately, there are some very good techniques we can use to help identify places where separation can be best made. One of the more important is called Transformational Analysis. Figure 7.12.2 shows what one looks like.

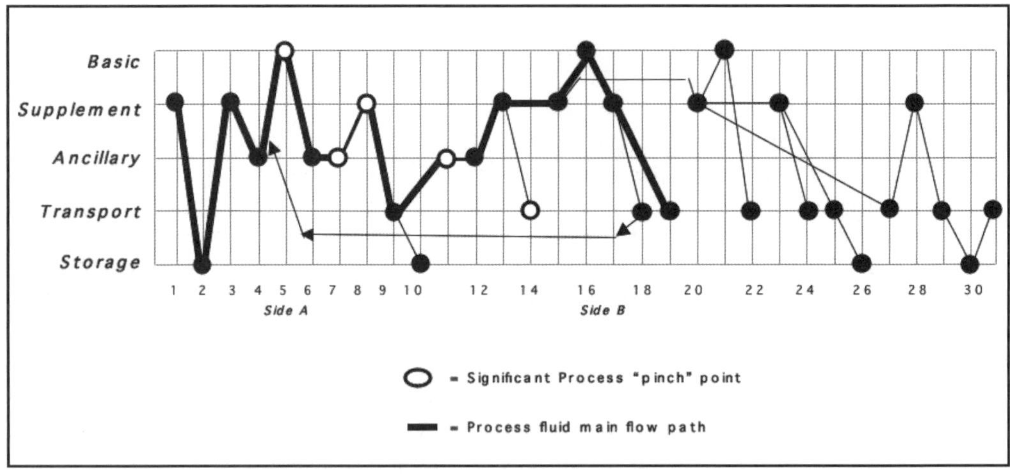

Figure 7.12.2. Transformational analysis diagram

Transformational Analysis

The diagram depicts an operator's entire process, from beginning to end. On the horizontal axis, at each block along the way, are single elements of the process. At the left are the "raw" materials, either true basic materials or the end results of upstream process areas that come to our operator for additional processing. At the right are "finished" products or the end result of our processing that is passed on to the next part of the plant for additional processing. Our example shows thirty-one steps or blocks. This means that there are thirty-one process-type elements in our operator's processing area.

The vertical axis of our diagram shows five levels. Each level depicts a different level of sophistication and/or complication of a process element. The most simple is at the bottom, the most complex is at the top. Starting at the bottom, the STORAGE level is just that, a place that material is being stored. That could be a tank or a warehouse. The next level up is TRANSPORT. Transport is exactly that, a process component whose only job is to move product from one part of the unit to another. That could be a pipeline or an over-the-road vehicle. The next level is ANCILLARY. At this level, simple processing is done to our material. Examples are heating, cooling, pressurizing, aerating, stirring, and so on. The important distinction is that some of the gross properties of the material are being altered in a very simple and basic way. The next higher level is SUPPLEMENT. Here is where our material is altered in a way that changes its intrinsic makeup. The mixing of several materials together or the separation of a mixture into components, say by distillation or centrifuge, are example processes at this level. While perhaps costly, it is generally possible to reverse what is done at this level. At the top level, we find BASIC. Here fundamental changes in the material are made. A good example is a reaction. Another is an alloying of metals. The product at this level does not resemble any of the constituent materials that went into it. It is not generally possible to reverse what is done at this level.

The level of processing for each of the thirty-one steps is indicated on the diagram by either solid or outline "dots." Outline "dots" are used to identify steps that are known (by those who understand this process very well) to be places of unusual difficulty to manage. Solid lines connect the "dots" to permit our eyes to follow the diagram. Heavy lines are for main material progression. Light lines are for minor (sometimes feedback) amounts of material flow. Our example process shows the material entering at progression Step 1 and proceeding to progression Step 31. Notice that most of the material stops at progression Step 19. Starting with Step 1, we see that the first thing our process does is a SUPPLEMENT. The very next step, the product goes to STORAGE. Next, it comes out of storage and goes directly to another SUPPLEMENT step. Then, at Step 4, it receives an ANCILLARY type of processing. Immediately after that, it goes into a BASIC processing, and so on until it is done.

Now that we have a good idea how the diagram works, it is time to see how it might help understand where good places exist for decomposing the process into smaller parts. By now, it must be occurring to you that the steps that place lowest in the diagram are probably the better places. This is best illustrated by first thinking about what might result if we selected a step at the top. Suppose we decided to decompose this process at Step 5, a BASIC step location. First, let's remember what it means to break up the process. Breaking it apart means that we will want to consider the part before the break as one part and the part after the break as a separate part. The first part will be Steps 1–4. The second part will be, say, Steps 5–9. But we already know that Step 5 is very complicated. To do it correctly, Step 5 must be able to know and manage as much of the inflowing energy and raw materials as possible. But that part is mostly Step 4 and earlier. And those steps are in another part. Therefore Step 5 either must do without this information and management of what is coming in, require extra effort by the part containing Steps 1–4 to do its job correctly, or require extra information from the part containing Steps 1–4 so Step 5 might try and compensate for inadequacies at Step 4 as best as it can.

On the other hand, if we were to break the process apart at Steps 2 or 9, we will already know most if not all of the properties of its inflow material, since all that happens before is either STORAGE, for the case of Step 3, or TRANSPORT, in the case of Steps 10 and 11.

Let's test our hypothesis of whether it is more difficult to manage the higher levels of processing (ANCILLARY, SUPPLEMENT, and BASIC) as opposed to the lower ones (STORAGE and TRANSPORT). The test we will use is to go to the operating records for this process and determine where the incidents and accidents happened to see if they correlate with level of difficulty. Figure 7.12.3 shows incident statistics overlaying the transformational analysis diagram.

The incidents shown in Figure 7.12.3 were compiled over 6 months. For this process unit, there were fifty-four incidents recorded by the site. The most, twenty-nine, were at the most complicated BASIC Step 5. Total incidents for all SUPPLEMENT levels were sixteen. Total incidents for ANCILLARY levels were eight. Incidents for TRANSPORT amounted to one. There were no STORAGE incidents. A cursory examination indicates that incidents fall off by a factor of two at each descending level. It is also interesting to note the breakdown of incidents into their contributing factors. Shown are the contributing factors of *people* (in blue, from the 12 o'clock position clockwise), next *equipment* (in deep magenta), and then *process* (in yellow). Contributing factors were determined by investigation and represent the one that was the most likely cause of the incident. At Step 5, the breakdown is as follows: people, 12; process, 1; and equipment, 16, for a total of 29 incidents.

We are left with strong evidence that using a transformational analysis diagram can be useful in process unit decomposition.

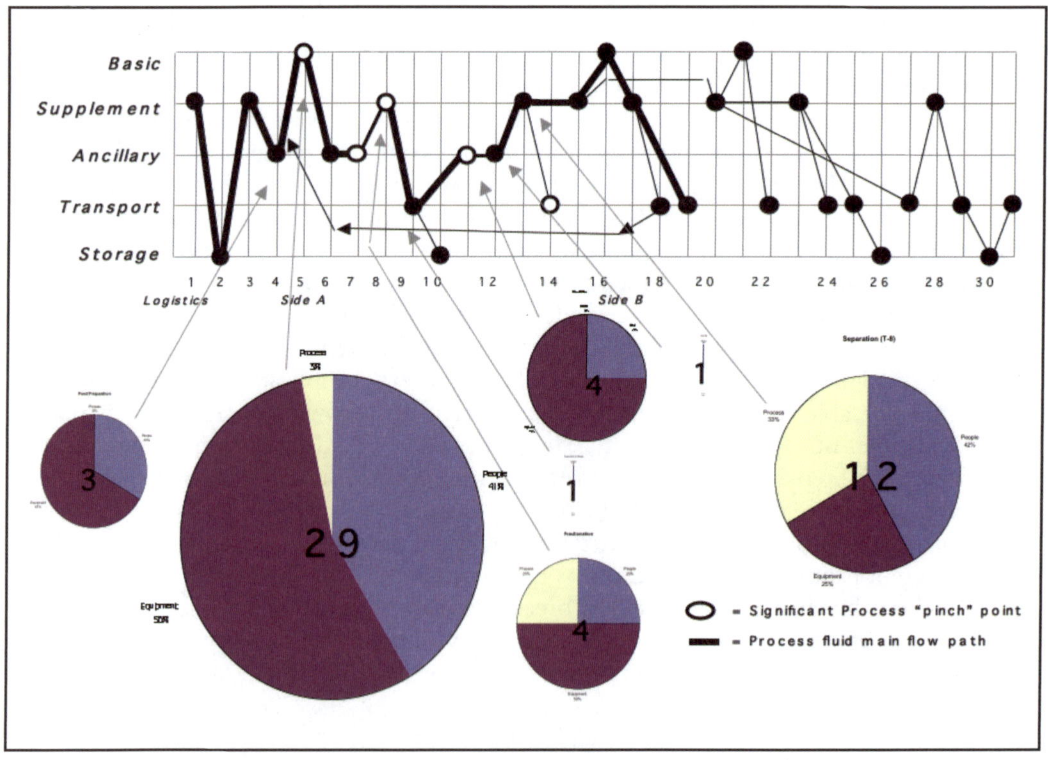

Figure 7.12.3. Transformational analysis diagram with incidents

Subsystem Boundary Attributes

With our process decomposed into workable subsystems (or pieces as we sometimes refer to them), we can turn our attention to the task of finding the key influential aspects between the pieces. We find the first influence at the boundary. Refer to Figure 7.12.4 for our example.

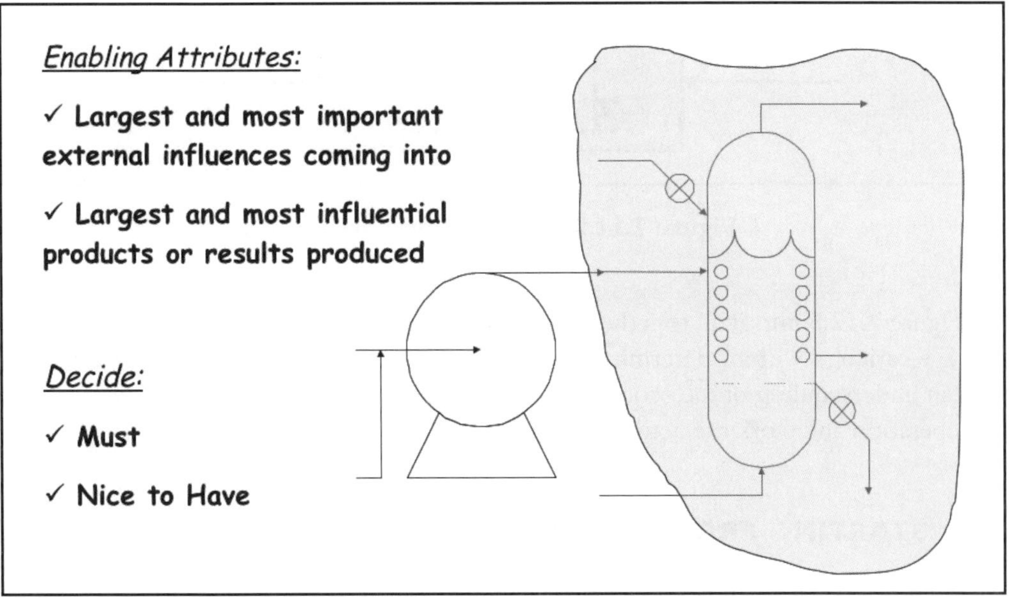

Enabling Attributes:
✓ Largest and most important external influences coming into
✓ Largest and most influential products or results produced
Decide:
✓ Must
✓ Nice to Have

Figure 7.12.4. Boundary attributes

Our boundary has been drawn around a tower. The tower and two coupled heat exchangers are inside; the pump and everything else are outside. The first thing to do is to list all the items (e.g., mass flows, energy flows, information flows) that cross our subsystem's boundary. Make a list of the largest and most influential items that come into our subsystem. Make a similar list of the largest and most influential products or outflows leaving our subsystem. Categorize these influences as (a) vital to know or (b) important but not essential. These influences provide a way to start building a candidate alarm list.

Subsystem Internal Attributes

We now turn our attention to the interior of our subsystems. Identify all key variables that have a fundamental role in the operation of the subsystem. Remember the "loss-of-view" discussion? There should only be a very few key, fundamental variables that indicate its health.

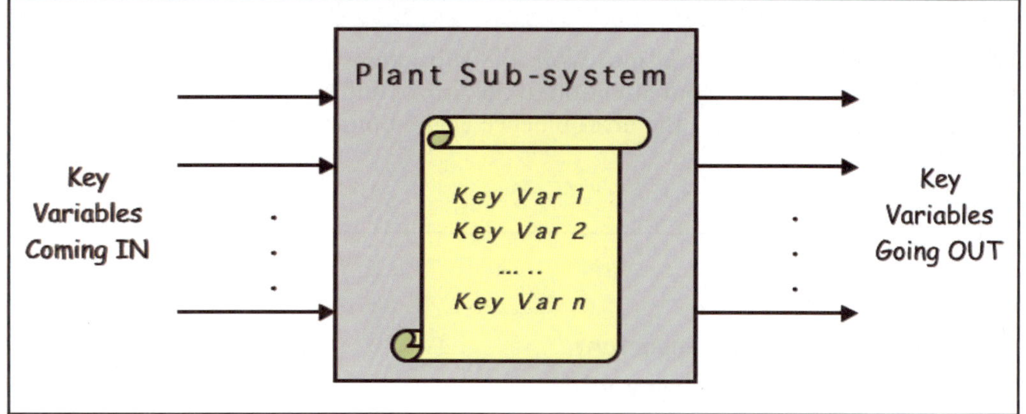

Figure 7.12.5. Key subsystem influences

Figure 7.12.5 puts it all together. Using the list of key variables in, key variables out, and key variables within, determine the smallest subset that can provide the operator with an understanding of the process. We want this list to be sufficient to identify abnormal operation and work the activity of correction.

7.13 "STARTING FROM ZERO" EXAMPLES

We examine several examples to see how starting from zero can actually work.

Furnace

Consider a furnace as an example of an appropriate subsystem for determining a minimal alarm set. The furnace in question is one of several, and the group is part of a larger production system at a major chemical facility. The furnace is shown in Figure 7.13.1. We will not go into detail with this. Basically, there are a number of burners, each shown with gas fuel lines, control valves, the firing box, combustion air system, purge system, and stack and emissions control. Here is how the alarm candidates worked out:

- *Key inlet variables.* Low gas pressure, low combustion airflow
- *Key outlet variables.* Excess oxygen, steam temperature, flue gas temperature
- *Key internal variables.* Fire-eye
- *Other variables to watch.* Energy balance, efficiency, water temperature, liquid level in upper drum, comparing furnace curve to calculated pressure curve

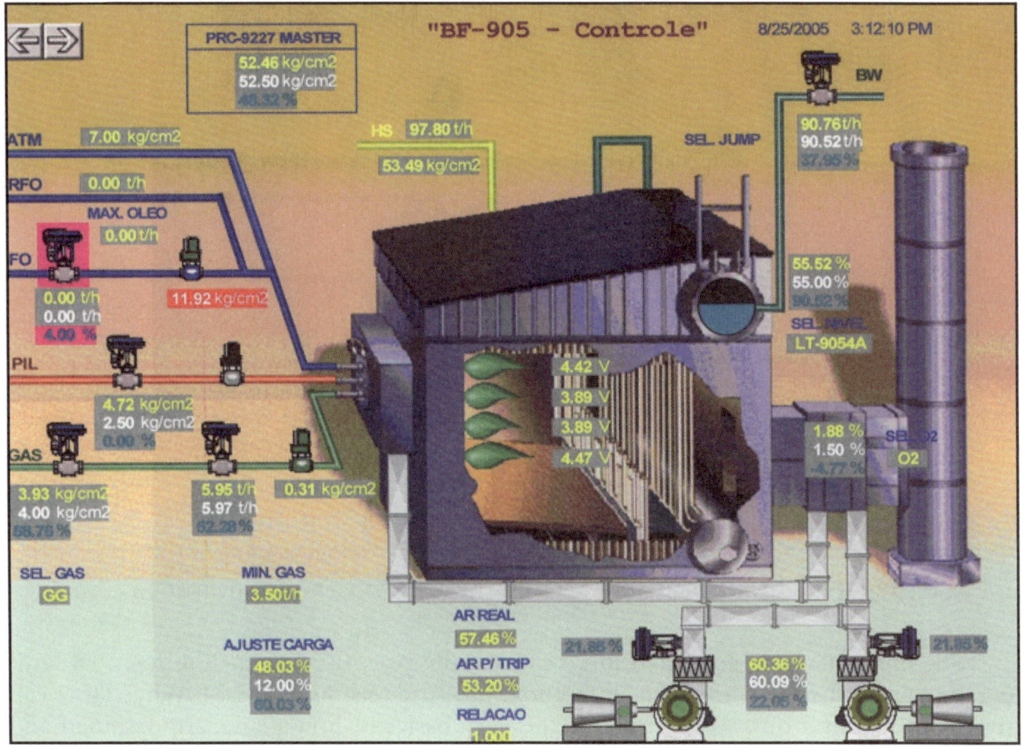

Figure 7.13.1. Example furnace for alarming

For this example, the working group ended up with six alarms. There were other interesting aspects (see the list under "other variables to watch"), but at the end of the day, the operators felt confident that with the six alarms, they would not miss anything important in the system. Sure, as soon as any alarm activated, they would be having a good look inside to begin the process of diagnosis and management. However, the six alarms were sufficient.

Heat Exchanger

Consider a very simple shell and tube heat exchanger. Figure 7.13.2 depicts such a process element. The use of this heat exchanger is to produce an outlet temperature on the tube side of the desired value. There is one medium flowing through the shell side. There is another medium flowing through the tube side.

Our exercise is to determine the minimal alarm strategy for this device. After some brainstorming, it is easy to come up with a list of all possible alarm possibilities. Such a list would include tube flow, tube inlet temperature, tube outlet temperature, tube differential pressure, shell flow, shell inlet temperature, shell outlet temperature, and shell differential pressure. All in all, we have come up with eight variables.

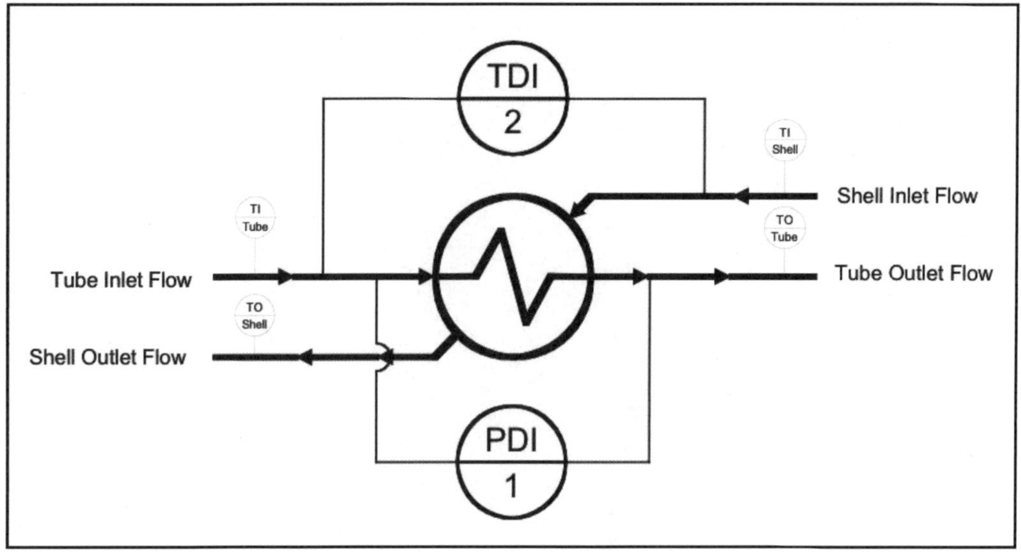

Figure 7.13.2. Simple shell-and-tube heat exchanger instrumentation

The minimal alarm configuration requires only two: tube-to-shell differential temperature and tube inlet-to-outlet differential pressure (see fig. 7.13.3). Why? How can this work? To find out, we go back to the very basics of how plants are designed and controls implemented. First and foremost, it is not the heat exchanger's job to make the shell outlet flow arrive at any specific temperature. That is the job of a temperature controller located elsewhere. Nor is it the heat exchanger's job to maintain any specific flow; that is up to the flow control loop(s) also located elsewhere. The important point of this is to identify only the intrinsic functionality of the heat exchanger. Knowing this will be all we will need to keep track of any abnormal operation. We need to know, how can this go amiss? We want to know, what is the best way to keep track of it when it does?

The real job of our exchanger is to move heat in the desired direction—either away from the shell side to the tube side or from the tube side to the shell side. The exchanger can neither work nor work well under two basic conditions:

- If there is fouling in the tubes, a tube leaks, or there is some other problem that causes an excessive pressure drop
- If there is some other interference that gets in the way of the heat transfer between the shell and the tubes

The differential pressure looks after the tube mechanical issues (except for leaks); the differential temperature looks after the thermodynamics issues.

You might ask, "What happens if there is a shell-side leak?" This is a legitimate concern. Leaks like this can happen. But unless the leak is dramatic or significant, it most likely will not affect the heat exchanger function. Are there alarms that might be needed for this? There might be. In the case where the shell liquid is benign, say cooling water,

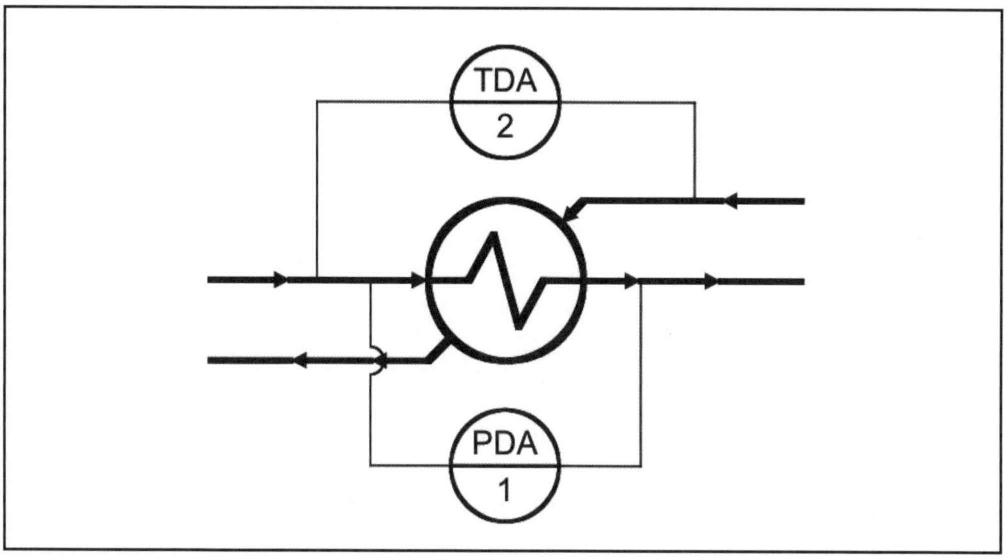

Figure 7.13.3. Simplified alarming for shell-and-tube heat exchanger

or by-product hot water, the plant might just leave it up to normal visual inspections to uncover such a leak. If the shell fluid is another process stream or a fluid that might create a personnel, economic, or environmental problem, then the plant will certainly want to know about this leak. However, important leaks occur in more than just the heat exchanger. A leak in other piping feeding the shell is just as likely. Consequently, a single mass balance test for the entire portion containing the shell fluid (shell itself plus inlet and outlet piping) would be used. Or if a fluid is costly or dangerous, then we have a special case and should treat it as such.

You get the idea here.

7.14 WORKING THROUGH THE DATABASE

The power of a structured approach is that each progressive alarm or abnormal situation that is rationalized is related to either the previous alarm or abnormal situation, or to the following alarm or abnormal situation. What was considered and decided for the earlier situation naturally leads into considerations and decisions for the following situation.

A plant has a lot of alarms or potential alarms. We are talking about thousands for typical processes. We need a methodology for working our way through the extensive number of existing (or potential) alarms that is efficient and provides insight. Instead of using an alphabetical or random list, we use a structured process. There are two straightforward ones: (1) the Method of Flows and (2) the Method of Elements. The utility of these approaches is that they provide a very efficient working procedure that can build on the experience of handling the alarm that was rationalized just before the current one in the list. The types of situations and abnormalities in the process that will be discussed in the current alarm are quite related to the situations and abnormalities that were reviewed for the alarms just finished. Everyone is ready.

Method of Flows

Using the Method of Flows means to start with the major input flow to a unit and progress down the flow line until the outlet. For each tag, identify the alarms to keep and those to delete. Once a tag has been completed and before moving on to the next major flow tag in the chain, investigate all of the "related tags" for alarms. We do this since the thinking that went into the alarm above will likely be close to the thinking in understanding the related alarms. Recall that this list comes from the information on the alarm response sheet shown in Figure 7.6.1. So, we are already familiar with the understanding needed to rationalize. But only go down one level in the process of following the "related alarms." Once the related tags are done, return to the major flow and move down that chain. If you do not, you will quickly find yourself hopelessly distracted. Continue the process with all other major flows.

For example, in Figure 7.14.1, start at the broad arrow number 1. Start with Flow 1 (F-1). Is an alarm needed on F-1? After F-1 is rationalized (alarmed or not), we ask what are the related tags. Those tags are the ones that are referred to in the alarm response information as "related," "confirmatory," or "manually adjusted" process variables that related to F-1. For each of these variables, we ask whether to have alarms configured for them. If so, we configure them. Note that we do not check on potential alarms for

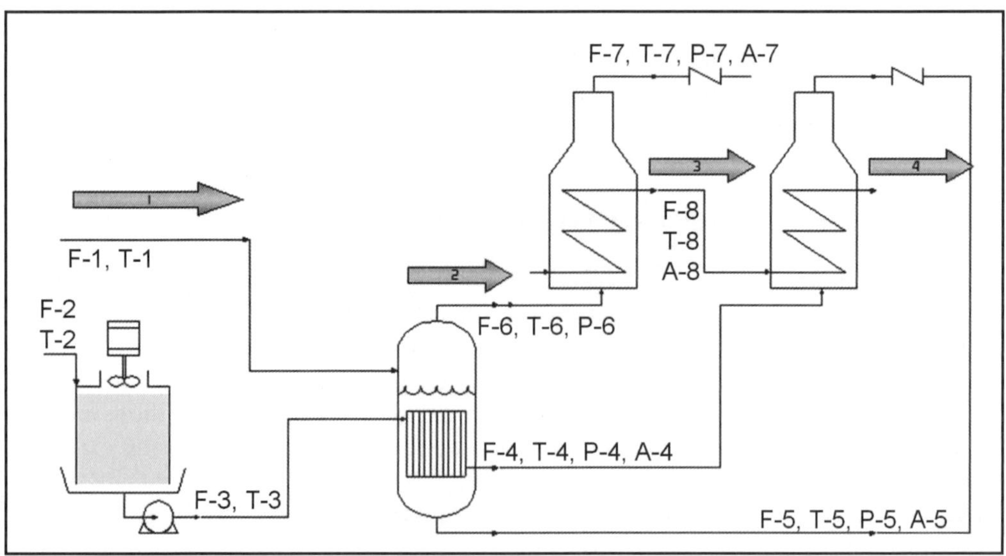

Figure 7.14.1. Method of flows rationalization progression

variables that are on a list of "confirmatory" variables for "confirmatory" Variables—that is, as the major flows are followed, the method only goes one level deep. Any of the other variables close by are either picked up by another variable needing it to be related, confirmatory, or manual corrective actioned, or they are picked up on another major flow. Finally, if they are not picked up by following the flows, they are rationalized when the loose ends are handled.

If T-1 has not yet been considered, we consider it now. Repeat the method of alarming that was done for F-1, now focusing on T-1. Once T-1 is complete (whether alarmed or not), according to the Method of Flows, move on to the broad arrow number 2. The same process is followed as before. Work with the Temperature 6 (T-6), Pressure 6 (P-6), and Flow 6 (F-6). Then move on to broad arrow 3 and so on.

Method of Elements

Using the Method of Elements means to start with the first key element or processing element (yes, in the chain of key flows, if you choose) and work yourself around the element in a causal way. As in the method of following the key flows discussed above, investigate the related tags after the primary tag. Then move back to the element and follow it around until done. Then move down the key flow to the next element. Referring to Figure 7.14.2 as an illustration, focus on the heated reactor tank in the center. To use the Method of Elements, each alarm or potential alarm for this reactor is examined until every abnormal situation for the reactor is covered. This examination is figuratively

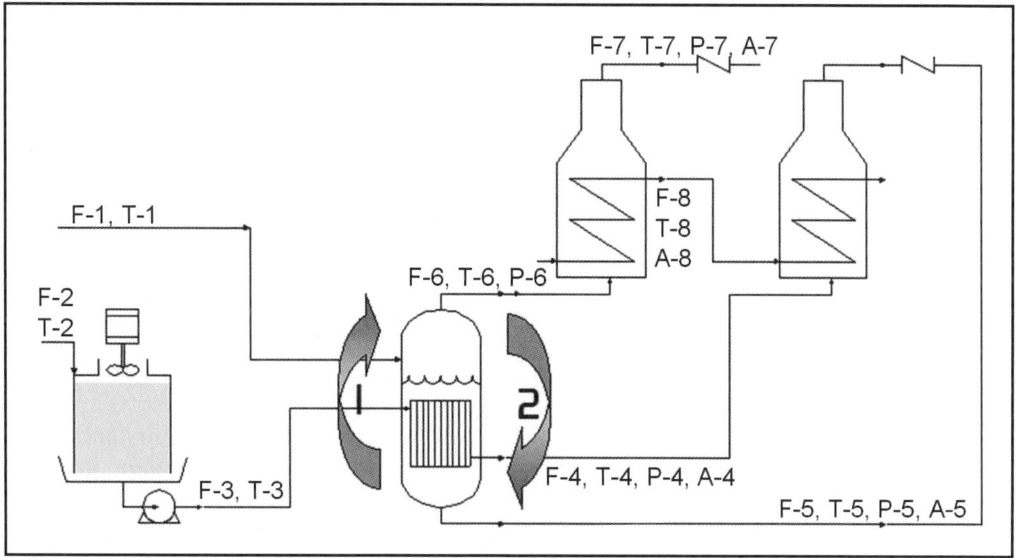

Figure 7.14.2. Method of elements rationalization progression

illustrated by the two curved arrows (labeled 1 and 2) that surround the reactor. Once this reactor is examined in full, another element is selected for examination.

Begin with the reactor bordered by the two curved arrows. The following streams connect this element with the rest of the plant: 1, 3, 4, 5, and 6. Stream 1 involves F-1 and T-1. Stream 3 involves F-2 and T-2. Stream 6 involves F-6, T-6, A-6, and P-6, and so on. In addition, there are a number of internal streams or flows and such within the reactor. Start with any convenient one and rationalize it. Use the process discussed above in the Method of Flows for picking related and such alarms. Again, only go one level deep. Then pick up another reactor item and rationalize it. Continue until all the alarms in the reactor are rationalized. Then move on to the next element and continue the rationalization process just as was done here. Any missing alarms are picked up at the end when loose ends are considered.

Choosing a Method

Both methods are very useful. Recall the two rationalization approaches: starting from where you are and starting from zero. In both methods, one of the early steps was to alarm key common elements. For this, we should use a starting from zero approach. Yes, a part of starting from where you are starts, in fact, with a technique from starting from zero. Once the key common elements are done, it is time to select a rationalization progression method.

If your plant will rationalize by starting from where they are, the Method of Flows is a good one. Use a PFD or a P&ID as your roadmap. Remember that not all variables that can be alarmed are on these diagrams, so you will need an exhaustive list of variables that can be alarmed together with a way to identify within which part of the plant they reside.

If your plant will use the method of starting from zero, then the Method of Elements is much preferred. There is a strong reason for this choice. Starting from zero requires thinking about each part of the plant as if it were in isolation from the others. Most large elements, by their nature, are separate pieces. Yes, they are certainly integrated though heat recovery and chemistry, but those relations are somewhat secondary to the physical geography.

Most Powerful Approach

Experience has shown that a very powerful way to sequence the rationalization process is to combine the two approaches. It is very simple. Use the Method of Flows to start. As soon as this method comes to an element, use the Method of Elements and work around the entire element encountered. When the entire element is complete, go back and use the Method of Flows to find the next element in the "flow" path. Rationalize that element fully. And so on, until the entire plant is completed. Again, the loose ends are picked up at the end.

7.15 THE ALARM ACTIVATION POINT

To open this "door," be ready to really change "business as usual!" Alarm settings may have been determined by any number of ad hoc ways in the past. No longer will they be done as before. It is likely that some of the settings that surround important process variables may have been adjusted a bit over time to provide at least a measurable level of protection. Probably most have not. It is rare that any substantial effort will have been devoted to examining alarm settings except, unfortunately, by operators bent on providing more "wake-up" types of notification.

We will now examine how to determine appropriate alarm activation point settings. The proper setting of alarm activation points will be one of the more valuable parts of the knowledge this book has to provide. These procedures will enable you to determine alarm activation points that ensure that the timing of alarm activations will have a reasonable chance to assist the operator to maintain production integrity. In chapter 2, we learned how important it is to be able to give the operator enough time to respond to a process upset. Sufficient time was necessary for the operator to detect a problem, understand its nature and extent, decide what to do about the problem, implement the changes decided on, and wait for the process to return to a satisfactory status. We called this *fault tolerance time* or *process safety time.* Figure 7.15.1 repeats the figure from chapter 2.

The bottom line message is that in order for the operator to have a chance at managing process upsets, we must arrange for the warning of abnormality sufficiently early to be able to correct things. For that to happen, the Time to Manage Fault must be less

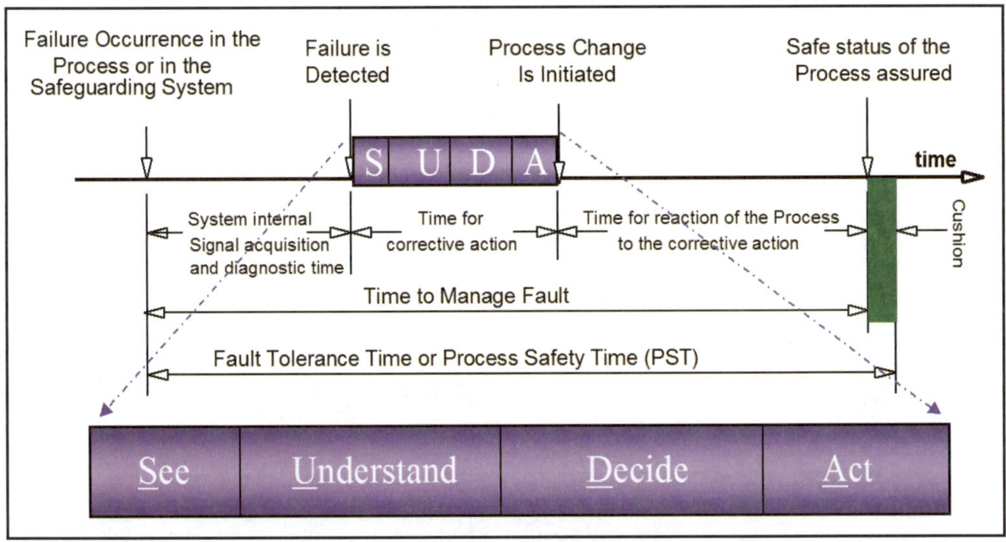

Figure 7.15.1. Process safety time

than the Process Safety Time. If this is so, then the operator and process should have sufficient time to recover. *We use this concept to determine the alarm activation point*—that is, alarm points are set not near the level of unsatisfactory behavior but early enough to provide enough time to manage.

Alarm Activation Point Determination

Here is where we show how to produce an appropriate alarm activation point.[1] We start with a few examples.

Examples

All this can sound deceptively easy. But you know it is not. Here is the rub. Operators need time to manage. But alarm systems do not know about time. They only know about values like pressure, flow, and the like. Thus our job, the way to get this to work, is to find a way to transform the *time values* the operator needs into *process values* the alarm system needs. Consider the data and calculations shown in Figure 7.15.2. We show a temperature variable. The plant has determined that most temperature variables can change about 6°F per minute for fast ones, about 2°F per minute for normal ones, and about 0.5°F per minute for slow ones. Also, in general, it takes operators about 1 minute to manage easy loops and about 4 minutes to manage hard loops. *Manage* includes

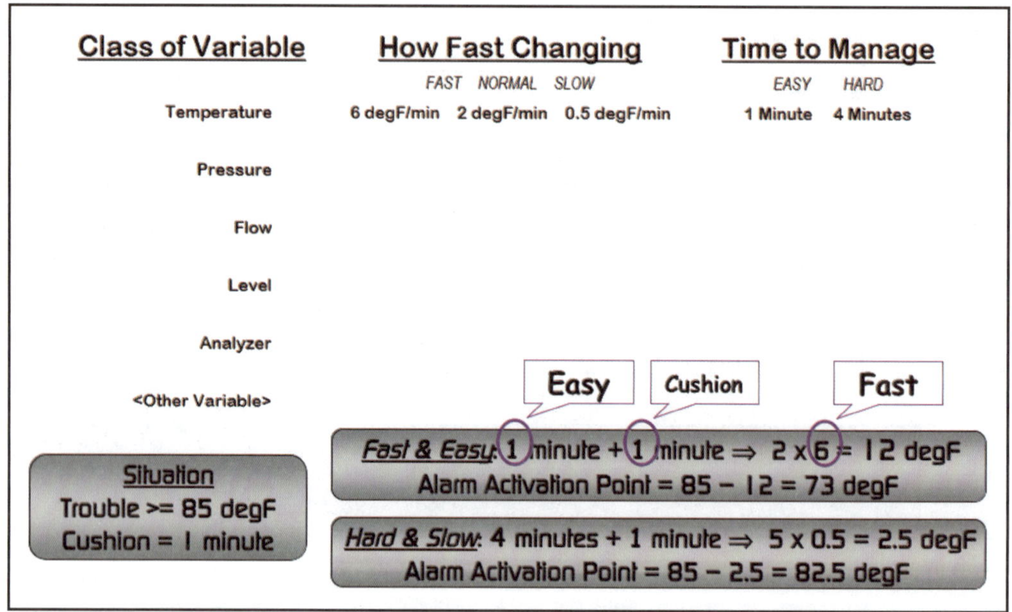

Class of Variable
How Fast Changing
FAST NORMAL SLOW
Time to Manage
EASY HARD
Temperature
6 degF/min 2 degF/min 0.5 degF/min
1 Minute 4 Minutes
Pressure
Flow
Level
Analyzer
<Other Variable>
Easy
Cushion
Fast
Fast & Easy: 1 minute + 1 minute ⇒ 2 x 6 = 12 degF
Alarm Activation Point = 85 − 12 = 73 degF
Situation
Trouble >= 85 degF
Cushion = 1 minute
Hard & Slow: 4 minutes + 1 minute ⇒ 5 x 0.5 = 2.5 degF
Alarm Activation Point = 85 − 2.5 = 82.5 degF

Figure 7.15.2. Illustrations of alarm activation point calculation

SUDA and Response times. Their philosophy has set a cushion at 1 minute (but you will more often use a percentage).

For this particular temperature, the process will get into trouble at or above 85°F. If the temperature loop were "fast" and "easy" to manage, the proper alarm activation point would be 73°F. If the temperature loop were to be different, say slow and hard to manage, then the alarm activation point should be set at 82.5°F (actually, in practice, you would use 82°F).

While not shown in the figure, if the temperature loop were to be hard to manage and fast, the proper alarm activation point would be 55°F.

$$55 = (85 - [4+1][6])$$

Equation 7.15.1.

Basic Equations for Alarm Activation Point

The basic equation for a high directional alarm is

$$\text{Alarm Activation Point} = \left(\text{Process Trouble Point}\right) - \left\{\left(\text{Time to Manage}\right) + \left(\text{Cushion}\right)\right\} \times \left(\text{Process Change Rate}\right)$$

Equation 7.15.2. Alarm activation point calculation

In practice, this is a very simple concept. First, determine the SUDA time. Next, determine (or estimate) the time for reaction of the process. Note that the reaction time is not the full time to return to normal; it is the time estimated for the process to turn around and be able to avoid the unwanted consequences of abnormal operation! Finally, add the SUDA time to the Reaction time and add the *cushion* (which is the extra time that the designers of the alarm system felt necessary in order to provide for all the uncertainties in process management that is normal for production operations). The cushion is normally computed as a percentage of the time to manage (SUDA time plus the Reaction time). Normal values are 25%–50% with a minimum of 2 to 3 minutes.

Equation 7.15.3 illustrates the procedure for calculating the alarm activation point where the cushion is specified as a percentage of the time to respond instead of a prespecified time.

$$\text{Alarm Activation Point} = \left(\text{Process Trouble Point}\right) - \left\{\left(\text{Time to Manage}\right) \times \left(\frac{1 + \text{Cushion}\ \surd}{100}\right)\right\} \times \left(\text{Process Change Rate}\right)$$

Equation 7.15.3. Alarm activation point calculation when cushion is percentage of time to manage

If the cushion for the earlier example were to be 50% instead of 1 minute, the proper alarm activation point would be 49°F.

$$49 = (85 - [\{4\}\{1 + 50/100\}\{6\}])$$

Equation 7.15.4.

A Digression in Setting Alarm Activation Points

Consider a reactor where the pressure should be kept below 34 bar to ensure that the reaction proceeds properly. The normal working pressure of the process is established at 23.5 bar. If an upset occurs and the pressure starts to climb, at what value should the *pressure alarm high* be set? Suppose experience has indicated that the pressure in the reactor generally increases at a rate of about 0.5 bar per minute. Further, let's say that experience has also taught us that it usually takes 10 minutes for the operator to implement a correction likely to restore proper pressure to the reactor. And let us say that the reactor can respond to such a process change within about 2 minutes. Then the *Time to Manage Fault* is equal to 12 minutes. Let us also say that the alarm philosophy has set the minimum difference (cushion) between *Process Safety Time* and *Time to Manage Fault* at 2 minutes. This means we need a total of 14 minutes. Therefore, the *pressure alarm high* alarm activation point should be set at 27 bar. The calculation for this setting using Equation 7.15.2 is shown in Equation 7.15.5.

$$27.0 = 34 - (10 + 2 + 2) \times 0.5$$

Equation 7.15.5.

Key (in order of appearance in the computation):	
27 (bar)	The result of our calculation
34 (bar)	The maximum process operating pressure for control purposes
10 (minutes)	Time the operator needs to orient and implement solution
2 (minutes)	Nominal time for the process to respond to most control changes
2 (minutes)	"Cushion" (minutes) required by the alarm philosophy
0.5 (bar/minute)	Nominal maximum rate of change of the process

Before we examined setting alarm activation points using the earlier method, it was entirely possible that an alarm value of 30 bar might have sounded reasonable. Alternatively, perhaps one might have used even 31 or 29. Or, if the process engineer was very conservative in his concerns for process integrity, a setting of 26 might have been picked. Let's take a look at each of the cases:

Activation Point = 30 bar. This would mean that under normal circumstances, the operator must resolve all upset problems in only 4 minutes instead of the usual 10 minutes studies have found is required. As a consequence, it would be a very rare event for the operator to ever be able to manage this abnormal situation. Plant safety and integrity would only be assured by the presence and proper functioning of interlocks and other physical devices operating automatically. Had they not been present or not properly functioning, an incident would result.

Activation Point = 31 bar; = 29 bar. This would mean that under normal circumstances, the operator must resolve all upset problems in only 2 minutes (6 minutes for 29 bar) instead of the usual 10 minutes studies have found is required. This would be the same consequence as above, except worse. Here it would be an almost certainty that the operator would not be able to affect the outcome at all.

Activation Point = 25 bar. This would mean that under normal circumstances, the operator would have 4 additional minutes more than the minimum necessary to resolve the upset problem. That would be an advantage, would it not? It might seem so at first; however, setting the alarm activation point this low puts us quite close to the normal operating pressure of 23.5 bar. Suppose that the normal process variations are +/– 2 bar. Then an alarm activation point of 25 is actually within the normal variability of the process. Setting the alarm activation point here will produce nuisance alarms.

The Limit of Alarm Limits

As interesting as the above calculations might be, there are limits to the discretionary ability to establish alarm activation point values. For some alarms, when we use the guidelines to establish proper alarm activation points, it is entirely possible to find out that there is no alarm activation point that will allow enough time. To be clear, this means that the limit-of-operation value for a given process signal is so close to the actual operating level (taking into account normal variability) that any calculated alarm activation point will lie on the wrong side of the actual operating level. Refer to Figure 7.15.3 for the cases where the alarm activation point (ap) is below the control loop setpoint (sp). What does good alarm management practice recommend for this situation?

You are now at the point of test. The test is really quite simple. If you believe that all the work that went into the construction of the alarm philosophy was proper, if you believe that your foundation of alarm response is such that the operator must be given enough time to meet the responsibility to manage the upset, and if your understanding of the process is sufficient to accept your calculations of response times, then you must conclude that there is no way that any alarm will work!

The plant really does not care whether any alarm will work. Alarms are not used by the process in any way. As far as the actual operation of the plant is concerned, alarms are a constructive fabrication of our minds. Nonetheless, if we have a desire to operate this process safely and appropriately, the process upset for which we cannot find an appropriate alarm activation point is still important. It still must be managed. It is just that now we know that it cannot be managed in any conventional way that uses operator intervention. There are two cases to consider.

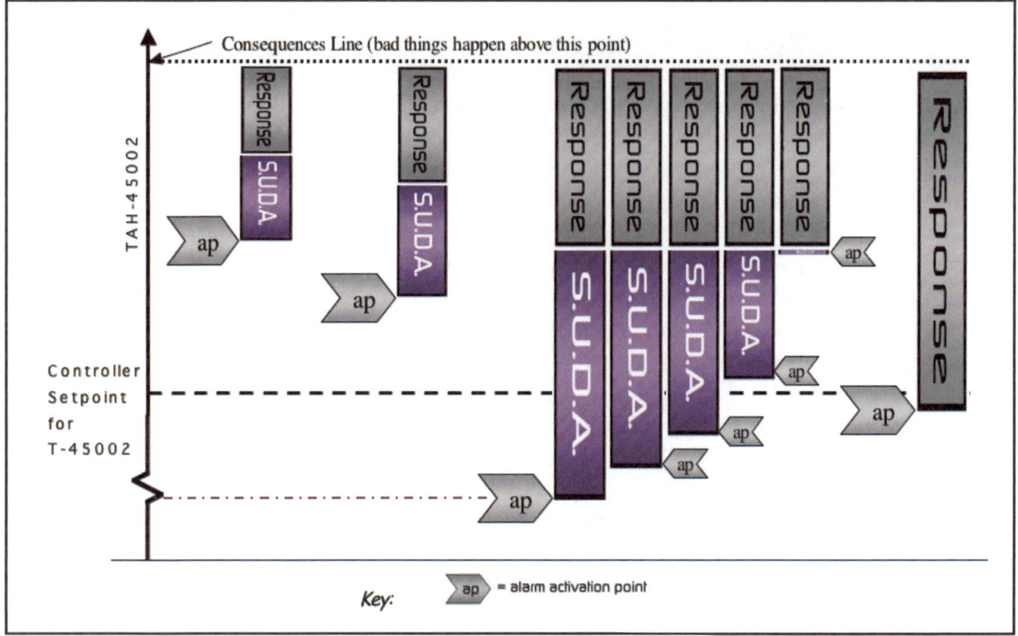

Figure 7.15.3. Alarm activation points compared to controller setpoint

Case 1: After an upset is detected, there is enough time for the process to respond to a corrective action but not nearly enough time for the operator to figure it out and implement it. [e.g., fig. 7.15.3, the alarm activation point "ap" is well below the controller setpoint "sp" and "response time" is above "sp."]

Solution: Either provide the operator with aids designed to dramatically reduce the SUDA time or remove the operator entirely from this situation by designing automatic responses to the upset. When the upset is detected, the automated response is activated without operator intervention.

Case 2: after an upset is detected, not only is there not enough time for the operator to figure things out and implement them, but there is not even enough time for the process to respond to the right action. [e.g., fig. 7.15.3, last case to the right]

Solution: This process is in an unsafe operating state. Either there is a failure in process design or some part of the normal process operational safeguards is not working. The process must be either shut down or redesigned in order to be safe to operate. There are no other alternatives.

Summary

Alarms are to assist the operator to act. If there is not enough time to do the job, then the operator cannot be assisted. In these cases, alarms should not be used as the means to maintain good operation. Some other way must be employed. Automated response and plant redesign must be considered.

Generalizing Alarm Activation Point Calculations

As the earlier examples illustrate, if we know reasonably well the desired operating limits and we have a good idea of how quickly the process moves and what the operator might need in the way of working time, we should be able to set alarm activation points. To do this well, we will want to know some of the basic information about the operator's task and the process behavior. We could estimate for each alarm and potential alarm the values we need. After a moment's reflection, no one would really do it this way. There are simply too many alarms to examine. The situation is resolved by obtaining class estimates for all necessary information. Divide the alarmed variables into classes. For each class, construct nominal values for the parameters needed. Next, adopt general rules for using all these parameters and values correctly. Tables 7.15.1 and 7.15.2 illustrate the needed information. The data you provide there will enable the computation of alarm activation points without the need to conduct special evaluations or computations for each candidate. Some rationalization "tools" have this feature built in.

For example, flow variables are classed as being either a "traditional flow" (in the sense of a flow used to simply move material from one place to another) or a "flow impacting process" (a flow where variations in this flow will have important effects on other parts of the process). For each type of flow, we come up with reasonable values for a likely rate of change. This will be used to estimate process response time to control changes (or other operator management input) as well as how fast the problem will likely

Signal classification	**Likely rate of change FAST** ***(% of range per minute)***	**Likely rate of change NORMAL** ***(% of range per minute)***	**Likely rate of change SLOW** ***(% of range per minute)***
Traditional flow			
Flow impacting process			
Traditional temperature			
Temperature impacting process			
Pressure			
Pressure impacting process			
Level			
Analyzer values			
Analyzer values impacting process			

Table 7.15.1. Alarm activation point calculations—process parameters

Signal classification	Time to manage *(easy loop)*	Time to manage *(traditional loop)*	Time to manage *(difficult loop)*
Traditional flow			
Flow impacting process			
Traditional temperature			
Temperature impacting process			
Pressure			
Pressure impacting process			
Level			
Analyzer values			
Analyzer values impacting process			

Table 7.15.2. Alarm activation point calculations—operational parameters

escalate. For each type of flow, we come up with reasonable values for how difficult it is to understand and manage during upset conditions. We express this in a multiplier value that we will use to ratio the "time to manage" values between normal and difficult. The ratios are done with "traditional" being equal to one and "difficult" equal to a multiplier value selected for the particular type of loop.

The remainder of the variable types are similarly analyzed, estimated, and entered into the tables for later use. Alarm activation points are then calculated in the same way that the earlier examples were done.

Too Much Time; Just Enough Time

Remember the opening thought to this chapter: even the best alarm design cannot overcome nature and design limitations. By now, you might be sensing a paradoxical dilemma with the suggested practical way to determine alarm activation points. On the one hand, you are probably convinced that the method can provide a pretty good setting. It should provide enough time. But, and this is a really big "but," the whole approach seems to be directed to providing just enough time. Not much more. Yes, of course, there is the "cushion" that is built into the calculations, and that will provide a bit more time. However, that time is really meant to be a hedge against unknown uncertainties rather than really more time that one can count on.

Are we unduly reducing the operator's alarm management time? Suppose we start with the alarm activation point that the SUDA approach provides. Then we compare it with the normal plant operating value. Suppose the alarm activation point is a lot closer to the plant trouble point than the normal operating value. Would it be a good idea to give the operator more time by moving the alarm activation point closer to the normal operating value? Sound like a good idea? Not so fast. It is not that obvious. We need to consider variability. Variability generally comes from two places. It comes from the normal disturbances and such that a control loop normally manages. But more importantly, it comes from external disturbances in the larger process that have made their way down to this particular loop or variable. By moving the alarm activation point closer to the normal operating value, there is the distinct opportunity of crossing the line into process variability. To avoid most nuisance alarm activations described above, we want to set the alarm activation point closer to the SUDA one.

Special Case

There is one special-case situation that deserves consideration. It is the unusual situation where it might be possible to determine a value for the process variable being alarmed, which, once crossed, is almost always assured of progressing to trouble. For this situation, and this situation alone, it is recommended that the alarm activation point be set at that value, assuming that the value provides at least enough time required by the SUDA analysis approach. The most likely places to come across this situation are where a boundary is crossed between endothermic and exothermic reactions or failures of mass or energy balances to close. There are others as well.

Alarm "Pick-Up" Order

Well, dear reader, you have been in this very long chapter for a while now. Let me ask for your forbearance to add another important, interesting, and practical topic. I start with a question. What if several alarms with the same priority activate around the same time? Which should the operator choose to work first? Does he need to choose? If so, then how should he choose?

Begin with the situation where alarms activate close to each other, but not at the same time (fig. 7.15.4). Note that the S-1 refers to the SUDA time for alarm 1; R-1 refers to the Reaction time for alarm 1 and so on for alarms 2 and 3.

At Time 1, Alarm 1 activates; at Time 2, Alarm 2 activates; and at Time 3, Alarm 3 activates. Assuming there are no other active alarms, soon after Time 1, our operator will begin to work Alarm 1. After all, that is one of the fundamental concepts of alarm management. But once Alarm 2 activates, what should our operator do? There are two nontrivial choices: (1) stay with Alarm 1 until S-1 is done and then move to Alarm 2 or (2) drop Alarm 1 and start on Alarm 2 until S-2 is done and then return to Alarm 1. The short answer is to drop Alarm 1 until Alarm 2 is handled and then return to Alarm 1. If our operator stays with Alarm 1 until the end of S-1, there will not be enough time for S-2 plus R-2, even if the operator were to get lucky and be able to find a shortcut to

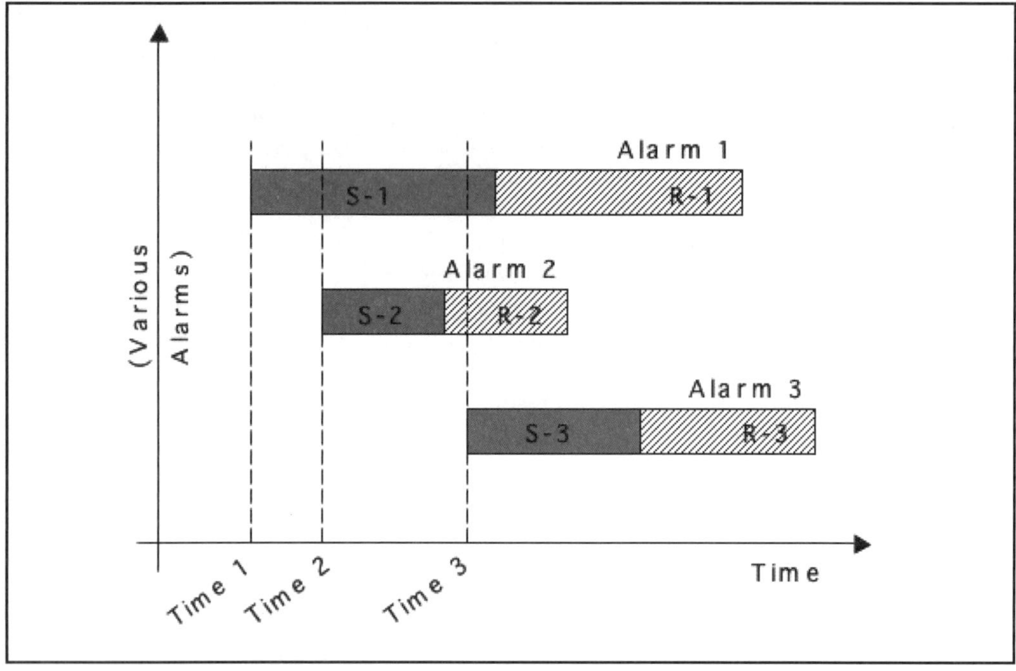

Figure 7.15.4. Several alarm activations around the same time

saving the situation. We hope that by the time Alarm 3 comes in, and if our operator was able to finish Alarm 2 and Alarm 1, there just might be enough time to do some good there as well. No strategy is perfect. Please remember that even a perfectly designed alarm system will not ensure that all alarms can be remedied. Nature's dues will be paid. All alarm management is about trying to avoid overpayment.

How is this discussion useful? Well, the easiest answer comes from the fact that there are two distinct aspects to responding to alarms. The first is the SUDA time (S-x; where "x" stands for alarm with tag x). The second is the response time (R-x). Full operator attention is needed for the former. Just keeping an eye out is OK for the latter. And during the latter, the operator can work on other pressing issues. There is another important aspect. Once the operator has taken an action, even if that action is not the final one needed for sufficient management of the abnormal situation, so long as it is in the appropriate direction of solution, it can change the process safety time for that abnormal situation (refer to fig. 7.15.5). Therefore picking up alarms in the way suggested here might just have even more robustness than first thought. Before intervention, our process was headed for trouble to occur at time T2. After the intervention at time T1, the new trouble time is time T3. We have managed to add more time for the operator to manage.

Consider another example situation. What if our three alarms actually came in at the same time, not staggered? Refer to Figure 7.15.6. For this case, there is no obvious first alarm to start with. The general approach would be to work the alarms in an order

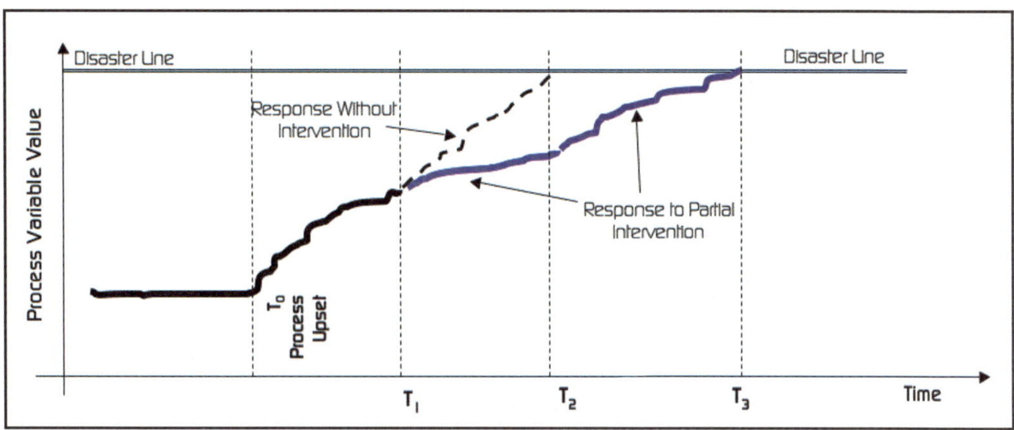

Figure 7.15.5. Change in process safety time due to operator intervention

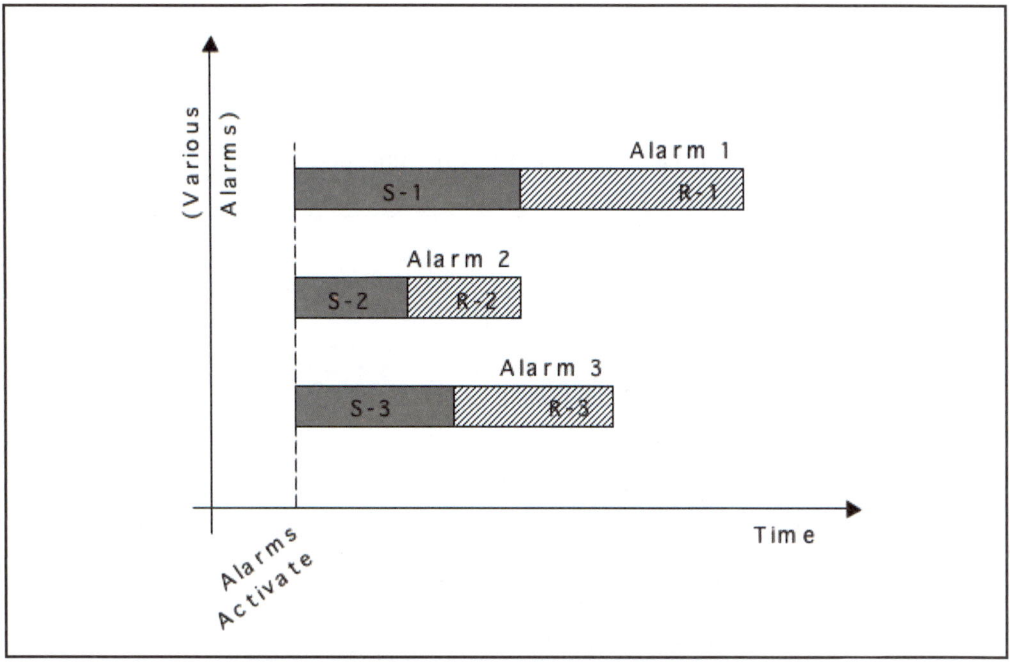

Figure 7.15.6. Several alarm activations at the same time

starting with the shortest SUDA time to the longest. Figure 7.15.7 illustrates such a choice. Note that this leads to the observation that Alarm 2 and Alarm 3 might be managed OK, but certainly Alarm 1 might not. However, that would be the best we could do. What if the alarms were selected differently, say, as shown in Figure 7.15.8?

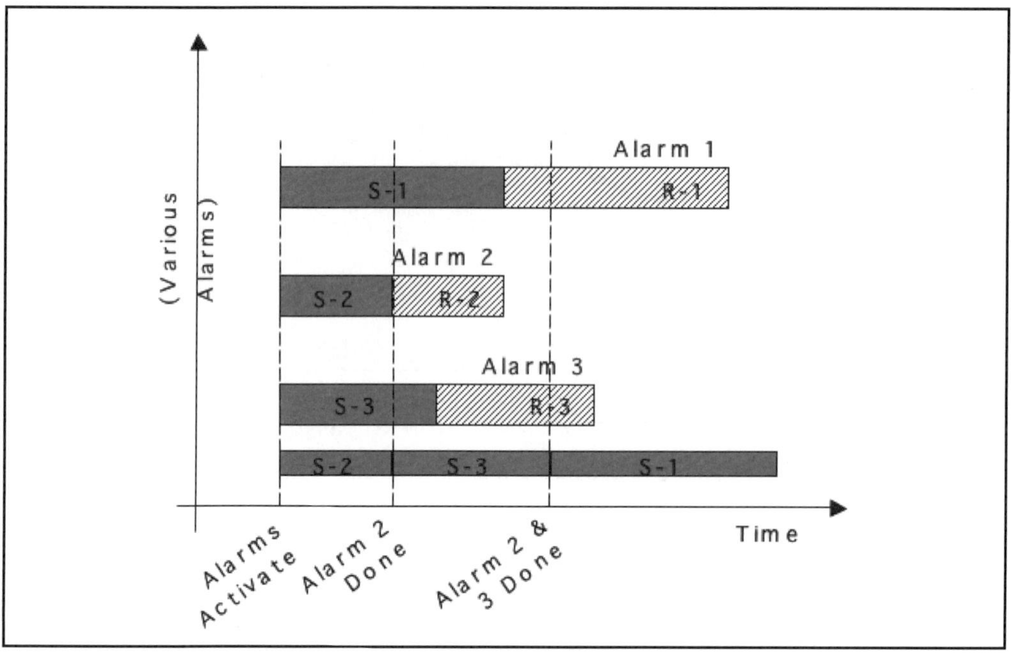

Figure 7.15.7. Several alarm activations at the same time and picked up in proper order

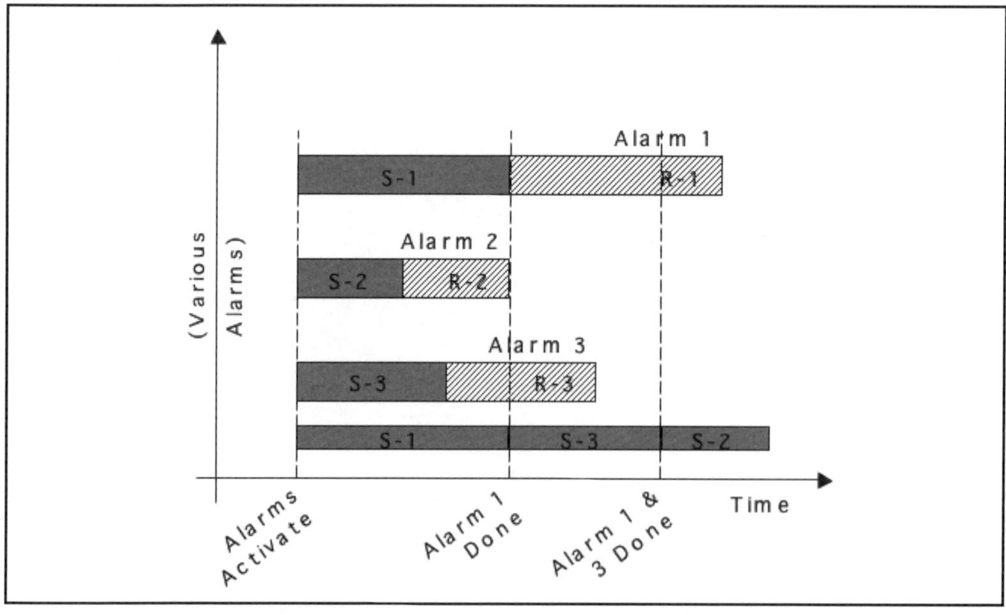

Figure 7.15.8. Several alarm activations at the same time but worked in reverse order of S-time

For this case, only Alarm 1 would likely to have been worked satisfactorily. The others would probably not have been, so the alarm pickup order matters. And now you know why "Time to Manage" is sometimes used to nudge priority assignments in a direction that encourages faster moving situations to be worked first. No guarantees—just another tool in the box.

7.16 DETERMINING ALARM PRIORITY

In chapter 6, as part of the alarm philosophy, you prepared an alarm priority definitions table and a scoring table. The definitions table is used to identify the specific severity of the consequences for each alarm. Using these severities, a scoring table is used to build up a total combined score (risk) for the alarm. An example scoring table is repeated here as Table 7.16.1. In general, the higher the score, the higher its likely priority will be. We have waited until now to decide how to take a numerical score and produce an appropriate alarm priority.

Max severity → / ↓ Consequence	None	Low	Medium	High	Emergency
Safety	0	25	75	100	300
Environmental	0	75	100	150	250
Financial	0	25	50	75	150

Table 7.16.1. Example alarm priority scoring table

In addition to the alarm scoring table, we need to decide the alarm urgency multiplier table. This table was introduced in chapter 6 and is repeated here as Table 7.16.2.

The priority breakpoint table is used to take the score (from Table 7.16.1) multiplied by the appropriate urgency (from Table 7.16.2) and map it into a recommended priority. Table 7.16.3 shows a blank alarm priority breakpoint table.

Time Available (in minutes for effective action)	Priority (urgency-only priority assignment)	Multiplier (multiplies consequence-severity values)
≤ 3	Emergency	1.4
> 3 but ≤ 10	High	1.2
> 10 but ≤ 30	Medium	1.0
> 30	Low	0.9

Table 7.16.2. Example urgency multiplier table

Priority	Breakpoint value
Emergency	From ____ and above
High	From ____ up to ____
Medium	From ____ up to ____
Low	From ____ up to ____
(Might not be an alarm?)	0 up to ____

Table 7.16.3. Blank priority breakpoint table

An example completed table is shown by Table 7.16.4. The next part will go through the procedure to obtain the actual values for this table.

Priority	Breakpoint value
Emergency	From 500 and above
High	From 350 up to 499
Medium	From 250 up to 349
Low	From 100 up to 249
(Might not be an alarm?)	0 up to 99

Table 7.16.4. Example priority breakpoint table

Assigning Priority

The process of determining priority is outlined in Figure 7.16.1.

There are several typically used approaches to decide how to rate an alarm for a recommended priority. Commercial alarm management tools support most. We will cover them next.

- *Sum of all severities.* Simply sum the severity numbers, one for each consequence, of course (Table 7.16.1), and get a total. Use that number to obtain a priority (Table 7.16.4)
- *Sum of all severities modified by urgency.* Sum the severity numbers (Table 7.16.1), modify the sum by the urgency (Table 7.16.2), and get a total. Use that number to obtain a priority (Table 7.16.4). This is the recommended method. It provides an excellent way for the combined risk to the enterprise. There is less tendency for the alarm priority to be artificially inflated or deflated. And it offers a wide latitude to adjust things to produce results that are in line with requirements and expectations.

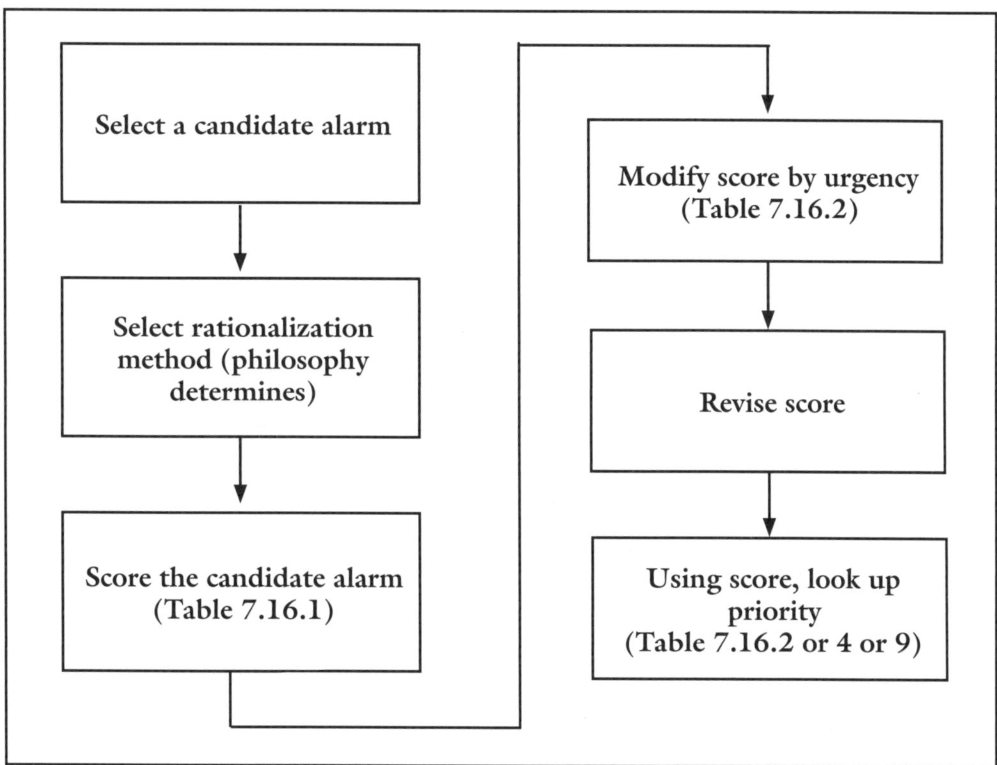

Figure 7.16.1. Procedure to assign alarm priority

- *Maximum severity.* Pick only the score for the maximum severity (Table 7.16.1). Use that number to obtain a priority (Table 7.16.4). Note that this method requires an adjustment to the method. To use the existing tables, it will be necessary to adjust for use with only one score. Either take the score multiplied by the number of severities and use the table, or modify the table entries by dividing by the number of severities.
- *Maximum severity modified by urgency.* Pick only the maximum severity (by label) and the urgency (by label). Use them to obtain a priority (Table 7.16.9). Note that this method reduces the subtlety of urgency. Urgency coupled directly with maximum severity work as a single unit. Without careful attention to the priority table (Table 7.16.9), it will be difficult to meet the priority distribution target set out by the alarm philosophy.
- *Urgency only.* Using the urgency, simply read the priority (Table 7.16.2).
- *Qualitative risk.* Uses a variation of maximum severity modified by urgency. This is an unusual method. See appendix 9 for a discussion.

Choose a method based on your alarm philosophy. It is acceptable that some alarms will have priority assigned by one method and others by another, but this is not usual. Just follow your overall plan. Now you know how this works. Let's run through a few examples, using our illustrated tables, to see what we get.

Calibrating the Alarm Priority Assignment Process

So far, all that you have seen are examples for the alarm priority assignment tables. It is now time to go over the way to determine specific values for your rationalization team to use. There are two anchor points that will help you do this. First, you will use as much as you can obtain of the collective experience of your plant people. Second, you will use the guidelines that were determined in the alarm philosophy to evaluate the final priority distribution of your rationalized alarms. Recall that EEMUA 191 suggested 20, 5%, 15%, and 80% for the priorities starting from Emergency and going to Low. Yours might be different, but here is where they will be used. The following illustration assumes that the rationalization method you are using will be Sum of All Severities Modified by Urgency. The other methods are calibrated in a very similar way.

Setting up the three rationalization tables is a coordinated, iterative process. Plan on going through the three sets of tables (Table 7.16.1, Table 7.16.2, and Table 7.16.4) several times until you get it working the way you would like it to. Since you will be using a rationalization tool to do the final assignment, there is little penalty for getting part way into the process, discovering that you would like to make a change in some of the rationalization matrices, making the change, and redoing what has been done before. After all, the tool is doing the work. It is not much different from writing a document and midway through it deciding to substitute the word "complicated" for the word "intricate." Just use the tool. But remember, all of this seriously depends on how carefully you worded the definitions for each of the consequences and severities in chapter 6. For if you change any of these definitions midway during the project, you will have to go back and redo every alarm you have already done to make sure that it is scored properly according to the changed definitions.

"Test" Alarms

Select about twenty alarms from the plant area to be rationalized. Try to ensure that they are representative, because your alarm priority mapping results will be used across the entire plant. About ten of the alarms should have an urgency that should not impact the priority. Of the other ten, five of the alarms should have an urgency likely to increase the priority and five likely to reduce the priority. Refer to the right column of the procedure in Figure 7.16.2, diamond A. Have your experienced team set the priority for all twenty alarms. Be sure and keep notes to refer to later on as to why the alarms were decided the way they were. Without reference to any scoring tables (because, at this point, they are not settled), ask your experienced team to rate each test alarm and decide, from their understanding of the alarm and the process, what they feel the priority of each alarm should be. You will be using the definitions of priority to guide this process. You

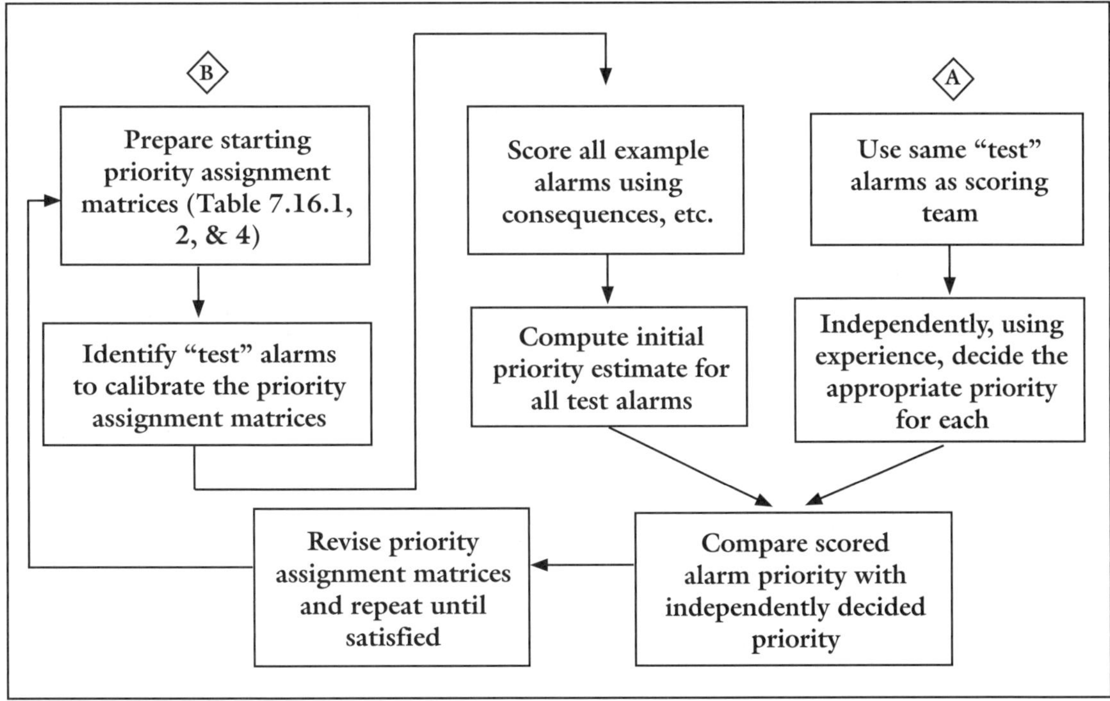

Figure 7.16.2. Iterative procedure to build alarm priority breakpoint table

may consider the definitions (see chapter 6) for impacts and severities for guidance, but no numerical values, please. After each alarm has been prioritized, put this work aside. It will remain your benchmark test during the iterative process. Again, a word of advice: keep in mind that it might be necessary to "negotiate" the priority of some of these test alarms if the team feels that they need it, but that will likely happen down the road a bit.

Breakpoint Table

Select the ten test alarms that have urgency that is not likely to impact priority. Follow the left column (diamond B) in Figure 7.16.2. Use the Alarm Scoring Table with numerical values and the initial estimate values in the Alarm Priority Breakpoint Table to set the priority for each test alarm. As a reminder, this table would have already been defined in your alarm philosophy document. Here we are refining it for use by your project team. Here is how it is done. Pick priority for each using the chosen rationalization methodology from the philosophy. Since you now know both the score and the priority from this approach, compare each, alarm by alarm, with the same priorities that the team first independently concluded were needed. Refine the working break points (Table 7.16.4) until you are satisfied that all ten test alarms score as needed. Note it might be necessary to adjust the priority of a test alarm that was used as reference, if the team feels that it might have been incorrectly estimated.

Table 7.16.5 shows an example comparison of priorities from the first iteration of a hypothetical calibration. An exact match is not necessary. Make some adjustments and do the calculations again. Use these results to adjust values in the break points to achieve as close a match as you deem useful. Remember that this process is a tool. Exact matches are not as important as whether everyone feels that the computed results are close enough for the process to work for the plant. Keep in mind also that each alarm will have a priority computed, but it is always important for the rationalization team to ensure that that priority is correct. During the rationalization work itself, the team will override any that are not.

Alarm number	Alarm priority by experts	Alarm priority using tools	Urgency effect
2	Emergency	High	None
3	Medium	Medium	Down a bit
5	Emergency	High	None
8	Low	Low	Up a bit
12	Low	Medium	None
13	Medium	Low	Down a bit
20	Emergency	Emergency	Up a bit

Table 7.16.5. Results of hypothetical priority calibration work

Add Urgency

Now that you are comfortable with the priority breakpoint table, it is time to factor in urgency. For this part, you will be using the ten alarms that you suspect urgency might have an impact on priority. Start with the five alarms that have high urgency. Using the urgency multiplier table (Table 7.16.2) adjust the alarm score from Table 7.16.1 to obtain an urgency weighted score. Using the alarm breakpoint table you have just calibrated (Table 7.16.4), determine the alarm priority for each of the five alarms. Compare this with the alarm priority the initial experienced group determined. Adjust the multipliers in Table 7.16.2 to align the computed priorities with the estimated ones. It might be necessary to slightly adjust the alarm scoring values in Table 7.16.1, but be careful not to upset all the good work you have so far in calibrating alarms.

Repeat this process for the five test alarms that have an urgency that is likely to reduce priority. Once you are settled, the alarm priority process is settled.

Nonweighted Maximum Severity with Urgency Direct to Priority

This method represents a special case. It is not preferred for assigning priority. It is discussed for completeness and to accommodate rationalization tools that do not have the flexibility to perform the other methods. The limitations of this method are significant. Since the method will select only the single maximum severity:

- Regardless of how many impacts are evaluated and no matter how many of the impacts have that maximum severity, the alarm priority will tend to be biased toward the higher end of the alarm priority range.
- Extreme care must be exercised to ensure that the priority implication of each similar severity cell is equivalent to every other severity at that same level. Hence, each cell in a given column (which means down the list of all impacts) of Table 7.16.3 must be of equivalent priority value. This often means that some of the cells may be left blank inasmuch as the range of the severities might not be as equally broad for all impacts.
- The only adjustments that can be made to align the total numbers of alarms in the priority levels (to conform to the alarm philosophy; see chapter 6) will be to the priority assignment map matrix (Table 7.16.9).

Start with a blank single maximum severity with urgency matrix (Table 7.16.6).

Max severity → ↓ Urgency	None	Low	Medium	High	Emergency
Least time					
Some time					
More time					
Most time					

Table 7.16.6. Maximum severity with urgency blank priority matrix

This will be the method of obtaining priority. For our example, this table has sixteen cells that must be completed (ignoring the severity of "none"). We will use the letters A to P to rank order each of the cells in the table. To use this approach, recognize that the upper-right cell representing the most severe maximum severity for which the operator has the least amount of time to remedy. Therefore this must represent alarms that should have the highest priority. Place an "A" in that cell. In like manner, the cell representing "low" maximum severity and "most time" time should represent the least priority. Place a "P" in that cell. Table 7.16.7 shows our progress.

Now, giving careful consideration to the definitions for each of the severity columns from Table 7.16.2, place the remaining letters (B through O) in the cells of Table 7.16.7. Each letter down the alphabet represents slightly less priority than the one before. Table 7.16.8 illustrates a typical assignment. Notice how the order proceeds in the table. There is no pattern. The only thing that matters is that the team finds that the priority decreases as the alphabet progresses.

↓ Max severity→ Urgency	None	Low	Medium	High	Emergency
Least time					A
Some time					
More time					
Most time		P			

Table 7.16.7. Example maximum severity with urgency starting matrix

↓ Max severity→ Urgency	None	Low	Medium	High	Emergency
Least time		K	H	C	A
Some time		M	J	E	B
More time		O	L	G	D
Most time		P	N	I	F

Table 7.16.8. Example maximum severity with urgency completed rank order

The final step will be to establish cutoff values to map our rank-ordered list into actual discrete plant alarm priority levels. Consider a plant with four priority levels (1 to 4) with 1 being the highest. Using our rank-order list, decide the boundary between alarm priority 1 and alarm priority 2. Then decide the boundary between alarm priority 2 and alarm priority 3. Do the same for priority 3 to 4. You will do this by BOTH considering your definitions for the severities as well as tracking how many of each priority alarms you are configuring during the rationalization. For our illustration, consider priority 1 alarms to be ranks A through C, priority 2 alarms to be D through G, priority 3 alarms to be H through L, and priority 4 alarms to be M through P. Table 7.16.9 shows the final maximum severity with urgency to priority assignment matrix.

↓ Max severity→ Urgency	None	Low	Medium	High	Emergency
≤ 3		3	3	1	1
> 3 but ≤ 10		4	3	2	1
> 10 but ≤ 30		4	3	2	2
> 30		4	4	3	2

Table 7.16.9. Example maximum severity with urgency to priority completed

Remark about Example Matrices

Generally, the rationalization team will use these early tables for a while. After a hundred or so alarms are completed, check the priority distribution. If the priorities are out of balance to the degree that you feel a change is needed, return to Table 7.16.8 and adjust the priority cutoffs appropriately. This will produce a new Table 7.16.9. Please remember that the numerical values shown here in the example tables are just that, examples. Their purpose is to provide illustrations of the concepts and to assist in the understanding and use of the results. They are not necessarily meant to be adopted or used as "starting points" for doing it yourself. They should, however, give you the idea of what they look like and how to do it.

7.17 ALARM PRIORITY ASSIGNMENT EXAMPLES

The following examples illustrate both how each individual alarm can have its priority determined, as well as contrast the results of using the differing priority assignment methods. Each of the aforementioned methods for assigning priority will be used. At the conclusion, the results of all of the methods will be compared. Three separate illustrative process tags have been chosen for this exercise. One tag is used twice with two different urgencies as the only change. All three tags will have been assessed by the alarm redesign team and assigned severities for each of three consequences and urgencies. See Table 7.17.1 for the scores. All calculations will use the previous example tables (7.16.1, 7.16.2, 7.16.4, or 7.16.9). Remember, these examples are going to follow a procedure similar to that in Figure 7.16.1.

Sum of All Severities

For sum of all severities, urgency does not matter, and so it is ignored. Table 7.17.2 summarizes the results. Here is how we score our examples. First, we sum the severities

Tag	Severities			Urgency
	Safety	Environmental	Financial	
F-2301A (urgent)	High = 100	High = 150	Emergency = 150	2 minutes
F-2301B (not urgent)	High = 100	High = 150	Emergency = 150	12 minutes
T-3004	None = 0	Medium = 100	High = 75	45 minutes
A-0562	None = 0	Emergency = 250	Medium = 50	10 minutes

Table 7.17.1. Example tags with consequences, severities, and urgencies

Tag	Score
F-2301A (urgent)	400
F-2301B (not urgent)	400
T-3004	175
A-0562	300

Table 7.17.2. Example tags with sum of the severities score

for each tag. For F-2301A (the more urgent one), the sum of the severities (from Table 7.17.4 and results shown in Table 7.17.2) is 400 (where 400 = 100 + 150 + 150). For F-2301B (the less urgent one), the sum of the severities is 400 (where 400 = 100 + 150 + 150). This is the same as F-2301A since we are not using urgency in this calculation for priority. The remaining sums of severities are computed in a similar way.

Next, use the calculated sum of all severities to obtain a priority recommendation. Use the scores (from Table 7.17.2 into Table 7.16.4) to produce the desired alarm priority simply by looking it up. We obtain a "High Priority" for F-2301A and F-2301B, a priority of "Low" for T-3004, and a priority of "Medium" for A-0562. Table 7.17.3 shows the results.

Tag	Severities			Urgency	Priority
	Safety	Environmental	Financial		
F-2301A (urgent)	High = 100	High = 150	Critical = 150	2 minutes	High = 400
F-2301B (not urgent)	High = 100	High = 150	Critical = 150	12 minutes	High = 400
T-3004	None = 0	Medium = 100	High = 75	45 minutes	Low = 175
A-0562	None = 0	Critical = 250	Medium = 50	10 minutes	Medium = 300

Table 7.17.3. Priority assigned by sum of the severities

Sum of All Severities Weighted by Urgency

Do the sum of the severities calculations exactly the same as for the sum of the severities case above (see Table 7.17.2). Then multiply that sum by the urgency multiplier (Table 7.16.2) to get the final score. For example, T-3004 scores out at 157.5 (where 157.5 = {0 + 100 + 75} × 0.9), which maps to a low priority. Table 7.17.4 summarizes these results.

Tag	Severities			Urgency	Priority
	Safety	Environmental	Financial		
F-2301A (urgent)	High = 100	High = 150	Critical = 150	2 minutes	560 = Emergency
F-2301B (not urgent)	High = 100	High = 150	Critical = 150	12 minutes	400 = High
T-3004	None = 0	Medium = 100	High = 75	45 minutes	157.5 = Low
A-0562	None = 0	Critical = 250	Medium = 50	10 minutes	360 = High

Table 7.17.4. Priority assigned by sum of severities with urgency

Maximum Severity

Maximum severity is not a usual method. It can be used for plants where there are significant differences in the risk exposure for differing parts of the facility. For example, a portion of a plant might be quite unrelated to the main production enterprise and therefore have little impact on product quality or production rate. This portion of the plant may have been designed to eliminate a serious environmental problem. But this problem might not have much financial or safety impacts. Left to its own computations, very low scores under financial and safety would tend to dilute a high score for environmental. Using the maximum severity, we will more likely arrive at a proper priority for this portion of the plant.

For maximum severity, we again ignore urgency. Notice how this method really responds to a single high item. Our tag A-0562 was given a medium priority using the sum of the severities but jumped to emergency when maximum severity is used. Here is how we calculate. A-0562 scored 750 (where 750 = 250 × 3). We use the × 3 multiplier since we use only one of the three possible consequences. A score of 750 maps into a priority level of emergency. Using similar computations, notice that F-2301 stayed the same and T-3004 moved up one level. The results are summarized in Table 7.17.5.

Urgency Only

Using urgency only, all we need to do is to take the information from the urgency column of the table and just look up the corresponding priority (Table 7.16.2). In our example, T-3004 has a priority of low. We identify it to be low by noting that the urgency is 45 minutes. Forty-five minutes translates into a low priority. The results are shown in Table 7.17.6.

Tag	Severities			Urgency	Priority
	Safety	Environmental	Financial		
F-2301A (urgent)	High = 100	High = 150	Emergency = 150	2 minutes	450 = High
F-2301B (not urgent)	High = 100	High = 150	Emergency = 150	12 minutes	450 = High
T-3004	None = 0	Medium = 100	High = 75	45 minutes	300 = Medium
A-0562	None = 0	Emergency = 250	Medium = 50	10 minutes	750 = Emergency

Table 7.17.5. Priority assigned by maximum severity

Tag	Severities			Urgency	Priority
	Safety	Environmental	Financial		
F-2301A (urgent)					Emergency
F-2301B (not urgent)					Medium
T-3004					Low
A-0562					High

Table 7.17.6. Priority assigned by urgency only

Maximum Severity Weighted by Urgency

The case of maximum severity weighted by urgency is used for the same situations that maximum severity is used but where it has been decided that prompt response is very important. This means that for some severe cases, there might be much more time to respond than others. We use Table 7.16.9 directly to obtain priority. The results are summarized in Table 7.17.7.

Summary of Examples

Let's put all of the example results together and see how the variations in priority assignment methods affected the recommended priority. Table 7.17.8 shows these results.

In Table 7.17.8, we see that the method of choosing how to calculate alarm priority can make a significant difference for some tags. The priority for A-0562 varies from medium to high to emergency depending on the method used. Each of the other tags

Tag	Severities			Urgency	Priority
	Safety	Environmental	Financial		
F-2301A (urgent)	High	High	Emergency	2 minutes	Emergency
F-2301B (not urgent)	High	High	Emergency	12 minutes	High
T-3004	None	Medium	High	45 minutes	Medium
A-0562	None	Emergency	Medium	10 minutes	Emergency

Table 7.17.7. Maximum severity with urgency

Tag	Sum of the severities	Sum of the severities with urgency	Maximum severity	Urgency only	Maximum severity with urgency
F-2301A (urgent)	High	Emergency	High	Emergency	Emergency
F-2301B (not urgent)	High	High	High	Medium	High
T-3004	Low	Low	Medium	Low	Medium
A-0562	Medium	High	Emergency	High	Emergency

Table 7.17.8. Summary of rationalization methods

changed priority one step for more than one method. This methodology is a tool. It is responsive. You will need to fully understand the implications before you recommend one method over another. The message here is not that there is consistency, that is a given, but that the chosen method will discriminate. Choose the method to have priority selected based on plant needs. Not all parts of a large enterprise need to use the same method. Your philosophy will have covered why the selected method is responsive enough to accommodate the plant's risk needs.

7.18 RATIONALIZATION WORKING SESSIONS

This section will review the actual working process the rationalization teams would use.

Teams

Recall the formation of the alarm improvement teams from chapter 6. Figures 7.18.1 and 7.18.2 review the team organization and composition.

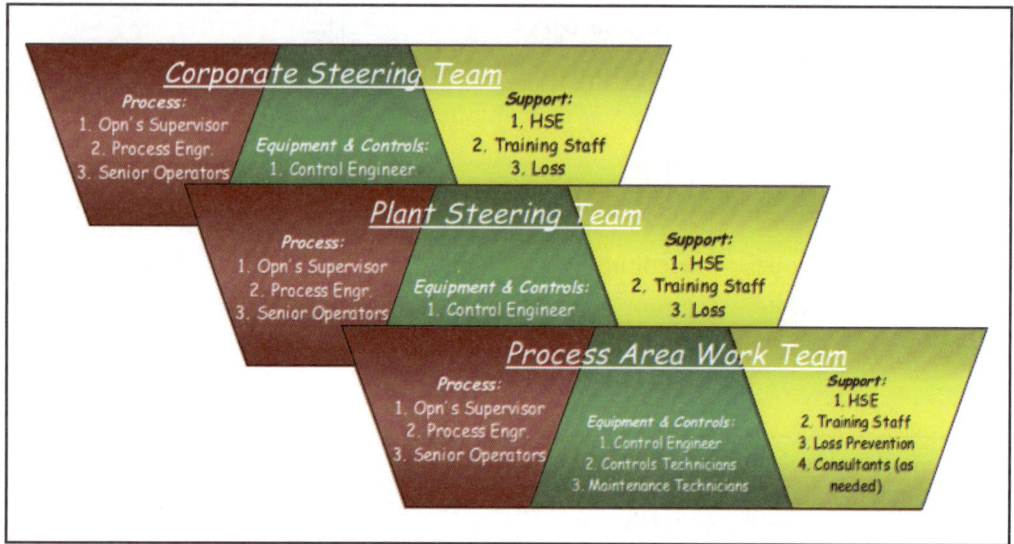

Figure 7.18.1. Enterprise alarm improvement teams

Figure 7.18.2. Rationalization team

At the enterprise level, we have a top-level steering team. Their job is to assist the individual site steering teams to get organized, bring local issues to the corporate table, and to ensure the appropriate corporate messages get back to the sites. The site steering teams, one for each site, have much the same role as the corporate team, but this time at the local level. They help the local process area work teams get organized, ensure that the site understands local area concerns, and facilitate site consistency in alarm redesign and implementation.

The individual plant-area rationalization team is multidisciplinary.[2] There are process experts, equipment and controls experts, and enterprise support experts. Their job is to take the master alarm redesign plan and implement it for their plant area. Just to be sure, plant area is meant to be a single operator position. Not all team members will be present all the time. A core team will do most of the work. Generally, the team consists of three to five individuals. Weekly checkpoint sessions are usually held where all attend. Other times, other members of the team or other resources outside of the team are brought in to assist on this or that particular issue.

There is a daily working team: these are the individuals who sit down and work through the alarm database of tags (for "starting from where you are" method) or equipment (for "starting from zero" method) to identify alarms and specify their knowledge, activation point, priority, and such. This usually consists of an experienced operator, another unit resource (such as a process engineer, knowledgeable control engineer, or, for parts of the plant that are heavily mechanical, a maintenance technician or engineer), and a rationalization facilitator. Individuals may come and go here, as their expertise and schedule permits. However, care must be taken to ensure continuity and enough participation to avoid always having to come up to speed every time you start work for the day. It is not unusual for a plant to dedicate specific members to the team for the duration of the entire operator area. Replacements are planned in case the primary participants are not available due to illness, vacation, or limited plant emergencies.

Participant Preparation

Rationalization participants from the plant need to be broadly experienced in the aspect of the plant in which they are participating. They will be providing information and assisting with judgments that will be integral to redesigning the plant alarm system. At the same time, rationalization is going to be a very different activity for them. They will need to fully understand why their alarm system needs improvement, how the new designs were arrived at, and what the work process to do their part is. Preparation for this part is vital. Experience shows that participants who have attended specially designed workshops on alarm redesign are significantly better prepared and work more productively. They are more motivated and rarely get lost on aspects that are unneeded and unproductive. All in all, spending a few days to understand alarm redesign more than pays for the continual delay and reduced productivity during the work sessions. Remember that any reduced productivity really means that not only is one person spending more time, but the entire team of two or three is working at the same reduced speed.

Work Areas

The first important business is to set aside a physical area for everyone to work. Remember that this is mostly a team effort. There will be great deal of information, drawings, manuals, and so on, all being used at the same time. The group will be meeting for most of the work day, and much will be done using open discussion and discovery. Once the work gets going, it will progress better if the working environment helps everything to go smoothly and efficiently. Provide filing facilities (e.g., file cabinets, book cases, etc.) to store resource materials as well as work in progress. Access to a photocopier is a must.

A fax machine would be nice. One or more computer printers are essential. Many sites use documentation from file-based archives. Having one, and often two, digital projectors will facilitate enough display to be able to use materials from several resources at the same time. Moreover, teams often use one to display the data entry as it is being added (via your software tool) and the other to show the odd resource document. It is also not unusual for efficient teams to be adding rationalization data simultaneously. One will be adding the base information; the other will be adding the results of the ongoing discussions and decisions. You will also want plenty of wall space to tape up items of interest, several paper easels (using the new Post-it style paper is a nice touch), and all the other office tools you usually want.

Note that we did not list telephones and the like. While outgoing calls are the rule, most of the time the team needs to concentrate on the tasks at hand—not being on call for each and every plant issue and concern. Yes, real emergencies and items that will be seriously impacted should not wait. All else is a distraction that will really slow down work and contribute to a significant level of demoralization. Try to avoid it.

Work Sessions

The most important part of work sessions is that they are structured. Structure will help in the beginning where everyone is still learning how this all works and how to do their part. It works later on when the team finds their time and motivation wearing thin. There is no intrinsic blessing to structure as a concept. Not just any structure will do, unfortunately. You will need to set the right structure for your team and site. Some observations regarding other teams' experiences might be helpful.

Set a working schedule that recognizes that the alarm improvement project is important but does not "own" the participants. This means that your team members all have other duties, allegiances, and responsibilities. Schedule so that they can at least keep a respectable participation in them. For example, if an operating group has weekly Friday morning meetings for an hour, delay your Friday work session for the hour. If Mondays are a day where a number of regular site administrative things occur, maybe taking Mondays off to let everyone catch up with things might be a good idea.

Expect participation from all when they are scheduled to work. Work here is largely a team participation event. Even one member missing will upset activities. Every plant has things going on that are out of the usual. Alarm improvement is an important site objective. It should take precedence over daily problems and minor issues. Yes, we are all important, but sometimes importance is best respected when it is not always available at every beck and call.

Structure work sessions to balance "pushing ahead" against an individual's "ability to stay focused no matter how long it takes." Stay balanced between what has been done and what needs to be done. Vary the workday to keep everyone interested and productive. Regardless of the rationalization approach, many teams find it useful to pursue two main threads concurrently—not to be crass, but somewhat like soap operas do. Part of the time is spent on thread one, and then they shift to thread two. For example, one thread might be going through the particular unit, working the alarm rationalization according to plan. The other thread might be to collect ideas to improve situation awareness—the operator's ability to find out what might be wrong without alarms. They use the rationalization results from the previous several hours' work as insight into the situation awareness issue. Sometimes the second thread will be worked for an hour or 2, sometimes for 5 minutes. It depends.

Do no work without documentation. The work product of rationalization is lots and lots of small bits of information, all correct, all in the correct place. There is little room for anything done without recording it. Also, do not lose sight of the reality that some of what is done every day is filling in missing or needed information into the rationalization database, and some is important insight and discovery. Ensure a way to capture the insight and discovery. The rationalization tool you are using adequately covers the other.

Keep statistics. You must know by now that there is a lot of detailed production-type work that must be done. Really, when you look hard at it all, much of it can be very dry—boring, if you will. Teams like to know how well they are doing, especially in this type of situation. Without making the numbers more important than the results, use statistics to keep motivation up and a sense of accomplishment going. Daily averages are helpful. Set a daily production goal that is both reasonable and reachable. Each day it is met is a day well spent. Each day the team feels well spent is one more day they will want to continue.

Keep everyone in the loop. Rationalization reaches deep into the fabric of our understanding and management of the plant. Do not keep everything exclusively in the team. As you find out interesting things, test it on others. If it works, then you have gained something and the other guys get to play a part in it as well. If it was just exuberant illusion, then you have a good way to help the other guys know that they are important in what you are doing and that you respect their knowledge. Do not worry, honest insight that turns out not to be as correct as we might wish is not anybody's misfortune or cause for ridicule.

Keep to plan. The original plan, as set down by the philosophy, was what the plant has asked the rationalization team to do. All decisions should be reflected back to that. In case the philosophy was either wrong or silent on a particular concern, procedure, or what have you, revisit it formally. Do not assume that the rationalization team has the authority to modify the basics on the fly. Changes in operation require a MOC against the philosophy and initial charter.

Work out all other working arrangements as a team, at the beginning and every time they need modification. No one yet has written an exhaustive, step-by-step way to do rationalization that works for everyone and every time. Nor would it be possible. Thank goodness. So have the team work out how they think they want to do things, try them, modify the ones that need it, and keep at it.

Events Schedule

There are three main events: daily work sessions, periodic checkpoint sessions, and work review sessions.

Daily Work Sessions

This work is full time. Most plants adopt a schedule mirroring the normal plant workday. Operations participants generally shift to this schedule for the duration of their participation. Just like the philosophy development activity, there should be minimal interruptions and regular attendance. Where some participants travel to site to facilitate or otherwise participate, they occasionally adopt a workweek of four 10-hour days with alternate weeks starting on Mondays and Tuesdays. This provides for more travel days together every other week. They usually stay near site the weekends in between. Though keep in mind that a 10-hour day, doing the type of work the teams need to do, can be very tiring.

Periodic Checkpoint Sessions

Weekly or biweekly meetings are vital to keep focus and ensure that neither the excitement nor the drudgery of working alarms is distracting the work from its broader focus of production improvement and adherence to good site practice. The meetings need to be carefully planned and efficiently run. Participants include the entire process area work team *plus* supervisors, when available. The plan for the meeting is to review progress, bring up and resolve technical issues or administrative issues that are getting in the way of the work, and to ensure continual buy-in for the project.

A checklist for the meetings generally includes the following:

1. *Project administration aspects.* Schedule, budget, who is doing the work, and so on.
2. *Plant activities that need to be planned for or accommodated.* How doing the rationalization might impact plant activities, how any changes in plant activities might impact the rationalization work schedule or participants.
3. *Technical issues.* Anything that happened to operations that needs to be considered by the alarm redesign team, anything that the rationalization team uncovered that led to a different understanding (either better or not) of the plant and/or how it was being operated or maintained, any issue or problem that needs more work from others to understand and/or resolve, and so on.
4. *"Political" issues.* Improve the communication of alarm redesign back into the plant infrastructure; handle doubts or concerns about why the work is being done, why it takes so long, or when supervisors will get back their operators, and so on. Middle managers can get especially critical of what might seem to be distracting work getting in their way.

Experience has often shown that these meetings quickly became routine and were therefore too short and failed to bring up important issues that needed work. They also failed to adequately inform local management of the nature and implications of what the new alarm system would look like and how it would work. Down the road, they are often surprised at what came out, even though they attended many of the checkpoint meetings.

Work Review Sessions

With the exception of a final approval of the work, this is the forum for any constructive issues to be raised and resolved or resolution planned for.

Work review sessions are attended by the process area work teams and the plant steering team. Managers are also a must. These sessions are held only when needed and for the express purpose of communicating major milestone results and obtaining approvals to continue as is or modify accordingly. They must fulfill the requirements of effective management stewardship. Managers, engineers, and operators must be aware that work covered during these sessions will be considered to be of sufficient quality to meet the design requirements and are acceptable to the plant.

7.19 PARTIAL RATIONALIZATIONS

An interesting twist to the rationalization process has come up. It is termed a partial rationalization here. It also goes by a number of other generic and proprietary names. The assertion made by the proponents is that a plant need not go to the trouble and expense of rationalizing an entire operator's area. It suggests taking only a part; do less rationalization, but do it cleverly. This approach is fraught with dangerous shortcomings. It will expose your plant to operational risk. It is discussed here to inform you. It must not be considered by your plant.

A partial rationalization means just that—a plan to rationalize only a portion of an operator's area alarm configuration. A number of ways have been suggested. Some focus on examining an operator's area and flagging the more important portions. Others recommend using the historical alarm database and only address the ones that have activated in a selected time frame. While these approaches might have some appeal on the surface, any attempts to actually use them will produce results that are inconsistent with the basic concepts and structure of alarm improvement.

Concepts and Experience

To be sure we are all on the same page when we talk about a partial rationalization, we are referring to the act of implementing a specific rationalization result. This is a walkaway implementation. The operator will use this system as is. Observe that we are not talking about a way to decompose the engineering process of reaching a rationalized alarm database, where this is just one step and this step will be followed by other steps eventually leading to a complete rationalization that is then implemented. A partial rationalization is a walkaway implementation.

By now you understand the importance of presenting the operator with a consistent picture. You know the vital nature of providing the operator with a set of alarms that fully satisfy the site requirements. Consistency means that there are no differences in presentation, nomenclature, and construction of the alarms that will get in the way of operators' trust and reliance on the alarm system.

Let's talk about the deception of "comfort." The number of practitioners doing alarm improvement work is increasing. They are gathering valuable experience. More and more are becoming familiar with EEMUA 191 and other guides. This exposure provides a measure of comfort rather than expertise. This is precisely why standards are being developed. Just because someone thought something might be a good idea does not make it so. Comfort must not be confused with good practice.

In the final analysis, the test of a successful partial rationalization must be the following: if any partial rationalization were an adequate design, then it should be possible to completely remove all alarms that were not touched by that procedure. If the answer to this is *true*, then that process works. If not, then it does not.

We now look at some of the partial rationalization approaches that have been suggested. None should be used in place of a standard rationalization.

Bad Actors

Recall from section 7.5 that bad actors can be a real annoyance and a distraction to the operator. Fixing them can be thought of as a very good idea. This is the one partial rationalization method that can be implemented early, without the necessary concern of being complete. However, the reader is cautioned that this approach treats only the symptoms. Doing this too early can easily distract from the true evaluation of alarm system deficiencies, as well as rob a project of the needed energy to do the job correctly. It will not provide a workable, effective, and safe alarm system. This is not to say that it has no part. Cleaning up the bad actors can be a very effective way to demonstrate good will and competence. It is often done as the first step of a full rationalization. It is not *the* rationalization, and that is the point of all of this.

Rationalize Only Important Parts of the Operator's Area

The assumption implied here is that parts of the area are more important than others. More important means (a) more likely to cause trouble, (b) more dangerous if it becomes abnormal, (c) harder to manage, (d) more difficult to understand, (e) more expensive to fix if it breaks, or (f) more costly in product or production rates if it gets disturbed, or any other number of similar definitions of "more important." It is likely that portions of the operator's area do differ. That is true. The differences may be important. That is also true. However, the leap from that plausible argument to the act of only understanding and managing the more important alarms is an unjustified one. And at the end of the day, doing it this way will not have saved much effort. Nor will it provide much benefit. Here is why.

Suppose that this method were used. In order to identify and separate the more important parts of the operator's area, it is necessary to fully understand the entire area. Doing that properly takes a lot of time and effort, though it will be less than that required by a comprehensive look at the full operator area. Once a list of "important" parts has been made, it is unlikely that the number of potential alarms to be examined will be less than 25%–50% of all alarms in the area. That can mean needing to do only one-half to one-quarter of the work of doing a complete job. So we might do it that way. Now take a look at the situation. We have a percentage of alarms for our operator that have been fully examined and redesigned according to our design principles. All those alarms obey the rules for a good alarm and are properly configured and documented; the operator is properly trained, and procedures rewritten are in agreement with the design. Now we look at the alarms that have not been rationalized. They have not been processed at all. For them, nothing has been done to improve either the alarm settings or the documentation and training. So our operator ends up with some alarms that are redesigned and some that are not.

Now suppose that an alarm activates. That alarm will be or will not be one that had been rationalized. Nor will our operator know whether that alarm is one that was rationalized. Therefore our operator will not know for this alarm whether the alarm setting is appropriate, whether the priority is proper, whether the procedures are helpful, or whether the alarm information is useful. Even more basic, the alarm that activated might have been an old one for which no action is required. This alarm would have been removed had it been rationalized. Or the alarm might be one that requires an action and must be accurate and timely, but the operator thinks it is one that was not rationalized, and hence he remembers (incorrectly) that it is "advisory" only. After all, a lot of the old alarms were not useful. Therefore he does not act. What level of trust will the operator have in the alarm system in this situation?

You get the picture. Let's put this in perspective. The sole benefit of having a portion of the alarms properly done is more than offset by the detriment of having an uneven alarm system design and the operator not being able to take advantage of most of the improvements. Little is consistent. Little can be trusted.

Rationalize Only Alarms that Activate

This approach takes historical alarm activation data and uses that to make a list of all alarms that our operator has faced. Since these were the ones that occurred in the past, it is likely that these will be the ones that will occur in the future. And to some extent, that is probably true. But this truth does not lead to benefit. This is because past history is a very poor predictor of future events. History teaches a lesson that we must heed. The study of abnormal situations leading to dangerous operation and on to serious incidents or accidents illustrates time and time again one important lesson: most very bad things do not repeat the same way. Other different bad things occur instead. That is why bad plants do not self-cure. That is why OSHA and other safety responsible government entities require an understanding of the root causes of incidents, not the proximate causes or symptoms. That is where the power of understanding abnormal situations is derived from. And that is why using only previous alarms will not provide sufficient guidance for future situations.

Bottom Line

Let's start the process of understanding the problems of partial rationalization by reminding ourselves of the defining components of a good alarm system. The site's alarm philosophy will have addressed all four of the alarm design fundamentals:

1. Definition of an alarm: requires appropriate operator action
2. Activation point: set to allow enough time for the operator to act
3. Information provided: will guide the operator to appropriate understanding and action
4. Only alarm the important: fewer alarms configured, all alarms needed, and fewer alarm overload situations

Now we compare these fundamental requirements with any operator's area that had only been partially rationalized. It will not matter what partial rationalization approach has been used. We look at the alarm database. It divides into two classes of alarms: (1) alarms that have been rationalized and (2) alarms that have not been rationalized.

Alarms that were rationalized. This set of alarms will be assumed to have been properly designed, documented, represented in operator displays, and trained for. All alarms should meet the requirements of the alarm philosophy. They can be expected to deliver the proper capability envisioned by the plant management and alarm design teams. There is no issue here. Any limitations will be caused by either defects, limitations in the philosophy, or by unavoidably extreme abnormal situations. In summary, these limitations are common to all complete rationalizations.

Alarms that were passed over and not rationalized. This set of alarms may or may not represent alarms that would result from following the plant rationalization plan. The degree of alignment depends solely on whether they, by happenstance (certainly not design), were properly designed before the plant determined what a proper design was, which is unlikely.

Any operator area that is composed of alarms where some are proper and some are not will have the following defects:

1. No uniform form and feel for the operator
2. No compliance with the alarm design requirements
3. No predictable, dependable operation
4. No ability to meet any good-practice standard; no ability to meet any governmental regulatory requirement

To summarize, and this is the important lesson: no partial rationalization meets any good-practice design. Nor will it be possible for a partially rationalized alarm system to meet any present or future regulatory requirements for proper functioning of an alarm system. Partial rationalization places plants in danger. Partial rationalization is to be avoided!

7.20 CONCLUSION

There you have it. In this chapter we have reviewed all the basic requirements of a successful rationalization project. It was done in some detail, especially those aspects like computing the alarm activation point values, setting up the rationalization matrices, and actually calculating the priority. Much of the discussion of the two key rationalization approaches may have been new to you. It should have been helpful. Certainly, there must have been a considerable amount of skepticism about the whole idea of the starting from zero rationalization. But by now, at the end of this chapter, you should feel more comfortable with what it is, how to do it, and whether it will be useful. And that is really the intent of this book: provide new insight, new skills, and new plans to assist you in your work to improve your plant's alarm system.

7.21 NOTES AND ADDITIONAL READING

Notes

1. Patent pending.
2. W. L. Frank, J. E. Giffin, and D. C. Hendershot, "Team Make-up—An Essential Element of a Successful Process Hazard Analysis," *CCPS International Process Safety Management Conference and Workshop* (San Francisco, September 1993).

Recommended Additional Reading

Alarm Management. NAMUR Publication NA 102. Potsdam, Germany: 2005. http://www.namur.de.

Alarm Management and Annunciator Applications Guidelines. Palo Alto, CA: Electric Power Research Institute. http://www.epri.com.

Apple, Steve. *A Practical Approach to Alarm Management Parts 1 and 2.* TiPS TechDoc White Paper. Georgetown, TX: TiPS, Inc., 2008.

Campbell-Brown, D. "Alarm Management: A Problem Worth Taking Seriously." *Control,* July 1999, 52–56; August 1999, 62–66.

Engineering Equipment Materials Users' Association. *Alarm Systems—A Guide to Design, Management and Procurement,* EEMUA Publication No. 191. London: EEMUA, 2007. http://www.eemua.co.uk.

Instrumentation, Systems and Automation Society. *Management of Alarm Systems for the Process Industries.* ISA 18.02. Research Triangle Park, NC: 2008. http://www.isa.org.

Matrikon. http://www.matrikon.com.

PAS. http://www.pas.com.

TiPS. http://www.tipsweb.com.

CHAPTER 8

Enhanced Alarm Methods

We should be careful to get out of an experience only the wisdom that is in it—and stop there; lest we be like the cat that sits down on a hot stove-lid. She will never sit down on a hot stove-lid again—and that is well; but also she will never sit down on a cold one anymore.

—Mark Twain (Samuel Langhorne Clemens)

It might appear that most of the alarm design work should be done. By now, you have put considerable time and effort into understanding how poor and ineffective the old system was. One of the first tasks you did was to buy good third-party alarm analysis software. You went to the trouble of thoroughly examining the PCS alarm configuration data. You spent even more time working with data analysis tools to derive as much information from the data as you could. You put energy into understanding the data and deciding what to do. Finally, you learned how to approach alarm improvement in an organized way. You examined the philosophy, formed working teams, and finished a complete rationalization. So why are you not done? Why can you not just take all of that work, implement it, and turn everything over to operations?

The short answer is that you could stop now. A lot of plants consider it. You could stop here and go on to all of those other things that got put off so the alarm redesign project could get done. But you really do not want to. Before you pack up and move on to other work, you will want to take a look at the limitations of what you have so far. When you do, you will find that it will be a very good idea to spend a bit more effort. Doing this will provide serious added value to the practical usefulness of your alarm system. The justification for this extra work derives directly from two of the original fundamental requirements for alarms: (1) require action and (2) provide time to do it effectively. Without this extra enhanced technology, operators will find that they cannot count on the alarm system to be consistent with these requirements. An inconsistent alarm system will fail to provide proper alarm support to the operator.

This chapter is all about the reasons why you should continue the work of redesigning your alarm system.

8.1 KEY CONCEPTS

Enhanced alarming is necessary to maintain alarm relevancy	All alarms must be actioned by the operator. This means that no alarm should activate that, due to current situation, should not be in alarm. The only way to do this is to use enhanced alarm methods.
Enhanced alarming	Permits an alarm to activate when normally prohibited or inhibits an alarm that is normally permitted. Modifies alarm configuration parameters to keep current with the situation. Provides situationally based operator guidance.
Enhanced alarming controls alarm flood	Without controls on alarm activations or management of activated alarms, high alarm activation rates cannot be managed or accommodated.
Alarms are part of the safety infrastructure	Alarm integrity is essential. No alarm should be modified without operator knowledge and a clear way to manage these modifications.

8.2 BEGINNING

In chapter 7, "Rationalization," you picked the specific plant abnormal situations to alarm. You decided on the proper activation point and the appropriate priority. Each and every time the alarm activation point was passed, that alarm would activate. Each alarm is separate and distinct from every other one. The alarm will activate regardless of whether that variable is still useful, regardless of whether it is in a part of the operation that was spare, shutdown, or being devastated by severe weather or other massive upset. It was not designed to take into account the activation of any other alarm. The alarm system itself takes no notice of what the plant is actually doing. No attention is paid to what the plant is making or what outside conditions are influencing it. Little notice is taken of the wealth of accumulated process and alarm knowledge. Each alarm stands alone.

Recognize that each and every alarm activation, by definition, requires an operator action. You know that an operator would not act on any alarm on a pump that is spare at the moment. But you know that a large number of alarms that might occur during a plant upset might not be useful, nor acted on by the operator. You also know that when several alarms representing the same problem activate, the operator should act on only one of the alarms. When we have are extra alarms that are not useful to the operator, those extra alarms take time and thought away from managing the necessary alarms that are important. A proper alarm system must include appropriate recognition of all of these situations. This chapter covers these situations.

Figure 8.2.1 reminds us where we are in the alarm improvement process roadmap.

Enhanced alarming will take advantage of being able to recognize specific plant situations to improve alarm presentation and handling. This is also where we provide ways to utilize the information gleaned during the earlier documentation and rationalization phases of our work. This chapter covers the additional supportive technology that can be added to a basic alarm system design to ensure that each alarm activation is worthy of operator action—*in all plant operating states* and *for all operating conditions.* Enhanced alarm management (sometimes referred to as logical processing, enhanced techniques, or advanced alarm processing) uses additional alarm information, relevant plant variables, and current operating and plant situations to reach a decision to permit or not to permit a potential alarm to be activated and thus be presented to the operator as an actual point-in-alarm. Enhanced processing is implemented using logic and other computations and tools.

There are four basic categories of enhanced alarm methods (see fig. 8.2.2):

1. *Operator-controlled suppression* (sometimes referred to as standard techniques), which includes shelving, release, out-of-service plant, and equipment under test

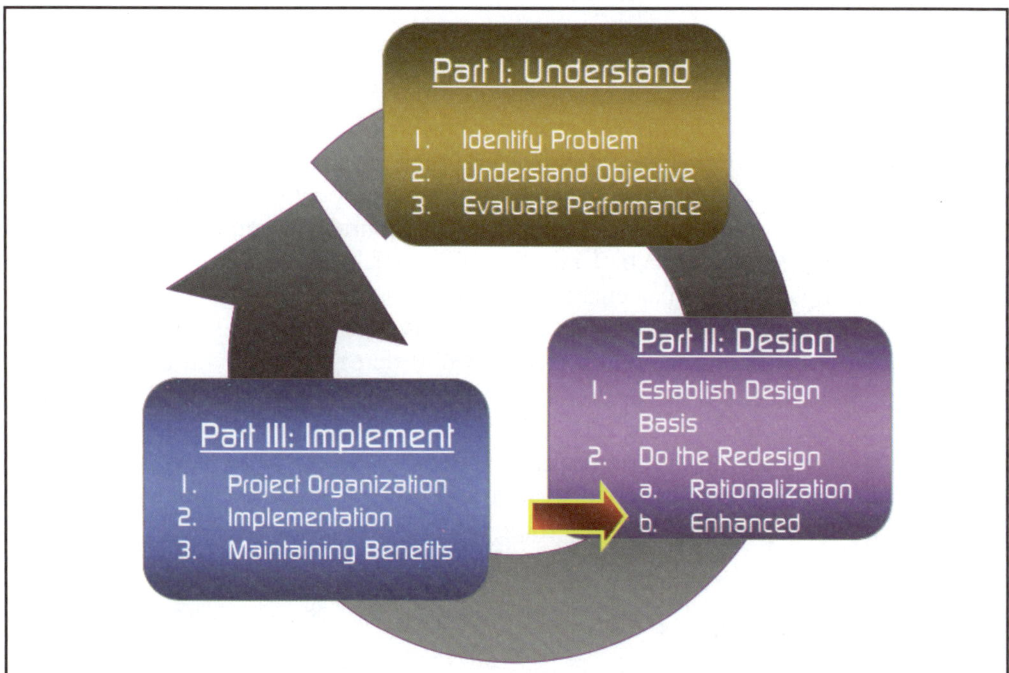

Figure 8.2.1. Enhanced techniques on the alarm improvement roadmap

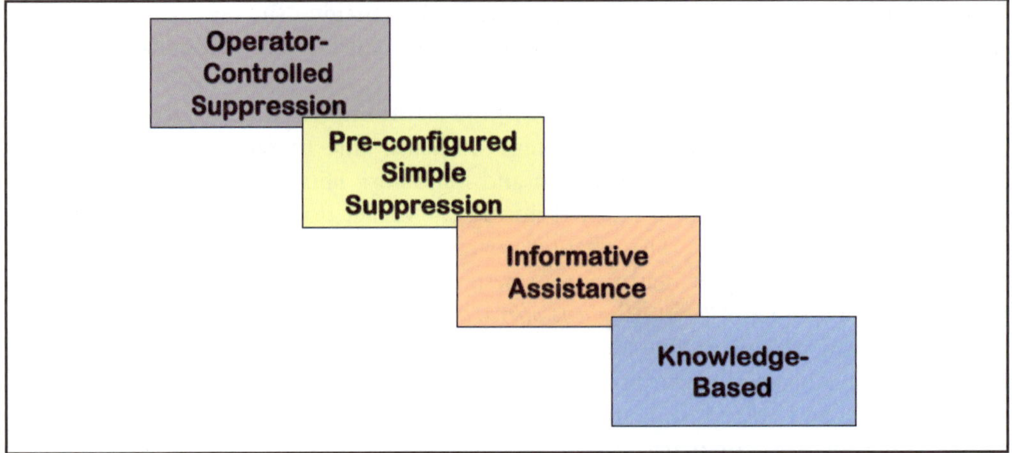

Figure 8.2.2. Enhanced alarm methods

2. *Preconfigured simple suppression* (also referred to as standard techniques), which includes logging of repeat alarms, redundancy, eclipsing, auto shelving, operating mode, and major event
3. *Informative assistance*, which includes providing guidance, grouping, and more
4. *Knowledge-based*, which includes pattern recognition, neural networks, fuzzy logic, and other knowledge-based methods

Readers who are familiar with other discussions on advanced alarm management will notice that the list above is quite different than they are used to. This is by intention. That long list of "conventional" advanced alarming methods is heavy in nomenclature to the detriment of insight and real, situationally linked understanding. Here we employ a more direct and useful approach. To do the job, specialized processing capabilities are designed and implemented that enable the alarm system to work efficiently, monitor how it is working, and ensure that everything is consistent with good plant practices.

One other point before we go on. You will find that the better you are able to get your alarm system to work, the more visible some problems can become. This is not to say that things are worse. Usually, they are not. But it might appear so. If it does, some might want you to back off. Human nature can work this way. Please continue the program. The appearances of getting worse are often just the encouragement you need to show that enhanced alarming is a good thing to do.

8.3 THE SITUATION

After fixing the bad actors and doing a good rationalization, the alarm design so far includes the following:

- A working sitewide plan for how the alarm system should be designed, implemented, and maintained
- All alarms have been configured properly—that is, only needed ones are selected, the correct activation point is determined, the correct priority is assigned, appropriate documentation is produced, and everything is properly integrated into the site infrastructure

At the same time, here is what we do not have but really need for proper operator support:

- Activation of alarms only when they are useful
- Guidance for the operator where specific plant situations strongly affect the proper course of alarm response (e.g., same alarm activations are handled very differently depending on what the plant is doing)
- Ability of the operator to manage abnormal situations when the number of alarm activations far exceeds the operator's ability to accept, understand, and properly respond to every alarm

Enhanced alarming provides the additional infrastructure to accommodate these three missing requirements.

8.4 SAFETY NOTICE

This chapter provides a broad discussion into the reasons why operator-initiated and dynamic, logic-initiated modification of alarm parameters are done. Using these methods judiciously and effectively can make the difference between operations that are barely supported by their alarm systems and operations that provide robust operator support during abnormal situations. However, changes need to be carefully made. This section discusses the important bases to touch.

Operator Awareness

Whenever there is the possibility of adding an alarm activation that might not have been there before or suppressing an alarm activation that usually is expected, the operator might be confused. Consequently, we must be vigilant to ensure that the operator is fully aware of when and where these changes are likely to occur. We must also be careful that the engineers and technicians understand the nature of these conditional or situational alarm controls.

Following plant rules (as established in the alarm philosophy and other plant procedures), the operator may make appropriate alarm system modifications. Embedded logic may also be making behind-the-scene alarm system modifications. These modifications may, at times, be large in number and span more than one operations shift. Therefore it is unlikely that any single operator, using his own devices, will be able to keep at the forefront the exact status of each alarm in his operating area. Explicit capability must be in place to keep the operator aware of the alarm status of all alarms that are not in their normal state. Where the modifications do not involve complete suppression, it is enough to provide an efficient place for the operator to query in the event he desires to know more. For alarms that are suppressed, more is needed. A conventional way is to place "flags" at the places that the alarm appears to indicate that it is deactivated. Other site-approved methods are OK as well.

Monitoring

It is not expected that the list of alarms that have been modified or suppressed will grow day by day. However, it is entirely possible that the number might change from day to day and add up over time. It is possible for these changes to persist for a long time and, in some cases, well beyond the need. An effective method must be employed to keep track of the alarm modifications, together with normal values and reasons for modification. This information will be useful for operators to verify changes and to come up to speed during shift changeover. It will also form the database for plant enforcement of alarm configurations. Section 10.6 covers enforcement in detail.

Unsafe Operations

Dynamic alarm modifications, including but not limited to suppression, must not be employed where the result will disable the notice to the operator of a plant condition that must be made known and acted on. Therefore even though a piece of equipment might be spare and not operating at the moment, if that equipment enters a state that should be known and acted on by the operator, an appropriately configured alarm must still activate.

As an example, consider a spare reactor that has not been in operation for several weeks. Consider a situation where an internal temperature then becomes very high. It might happen. Even though the reactor is considered shutdown, if there can exist a condition that still might be abnormal during shutdown situations, it must be alarmed. For most reactors, such a situation is highly unlikely. But it is possible for pyrophoric residues to remain behind even after cleaning. Even though steps may have been taken to seal the vessel from intrusive air, the possibility for residue and leakage may remain. The possibility for a fire may also exist. Therefore some alarms are appropriate and should remain active. They must not be unduly modified or suppressed.

8.5 ENHANCED ALARM FUNCTIONS

Enhanced alarming performs one or more of these three functions (see fig. 8.5.1):

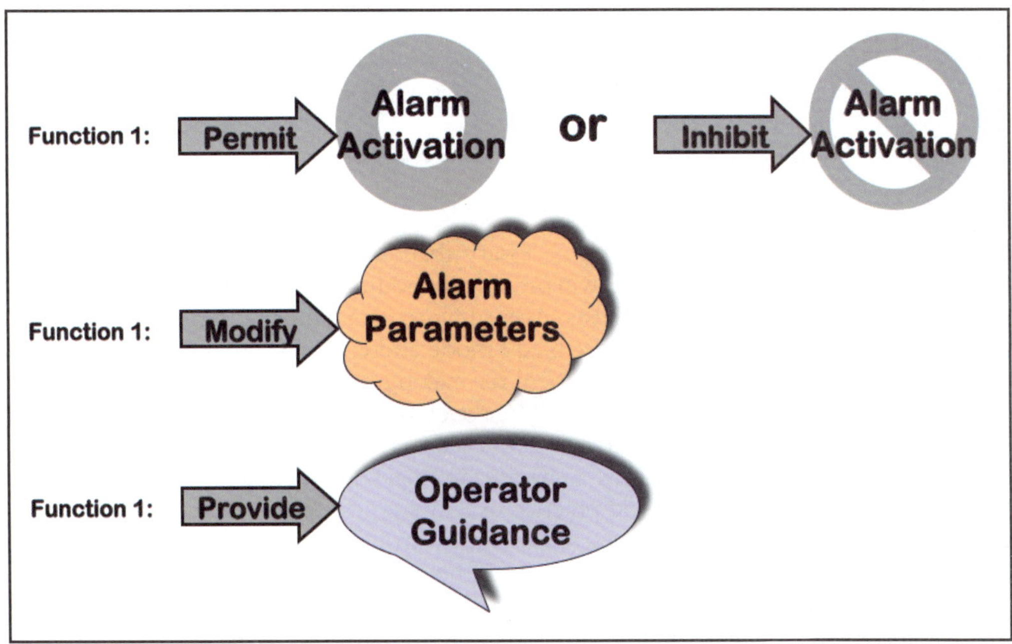

Figure 8.5.1. Function modes for enhanced alarming

1. Permits the alarm to activate, when normally prohibited, or inhibits the alarm where otherwise permitted
2. Modifies alarm configuration parameters to ensure relevance to current operating regime (activation point, priority, deadband, etc.)
3. Provides guidance to the operator to better understand and manage the alarms that have activated

That is it—nothing more, nothing less.

The enhanced alarm methods should be designed in such a way that they are easy to understand and the possibilities of error are extremely low. This usually means that the standard logic software embedded in the control platform infrastructure should be used whenever possible. Special purpose software and other custom programming should be employed very sparingly and only when the embedded infrastructure is insufficient.

Proper management of this functionality will require an inventory of all alarms that can be affected by enhanced alarming. This would involve creating fully annotated, dynamic records of all suppressed or otherwise modified alarms so that this information can be conveyed from operator to operator as well as known to engineering, maintenance, and safety personnel. Moreover, plants may wish to add controls that manage when, by who, and for how long alarms can be modified. Those controls include the ability to determine and accommodate changes in the equipment and configuration that might render some of the logic indeterminate. Indeterminate logic will fail to function.

Special care must be taken to ensure that all logic-based alarming is fully functional and correct despite equipment readiness status and the availability of key parameters and data.

8.6 ENHANCED ALARM INFRASTRUCTURE

General Considerations

To use enhanced methods, it will be necessary that alarm-modifying tools be used to manage this functionality. The tools will provide for adding processed alarms (by tag and other identification) to appropriate lists. According to plan, and at the appropriate times, the tools will place the alarm in a "disabled" mode and ensure that all logs and displays indicate this new status. The exact protocol for making the transition from enabled (the normal mode for alarms) to disabled shall be set by the alarm philosophy. The protocol should address what should be done if an active alarm (i.e., the alarm has been activated) is suppressed. It must deal with how to handle an unacknowledged alarm as well as one that has been acknowledged. Good practice usually suggests that active but unacknowledged alarms remain on operating screens and alarm displays until explicitly acknowledged. After acknowledgment, they may be removed from those screens and displays, according to established plant practice.

The tools will also maintain lists and corresponding displays to indicate to the operator (and others) all alarms that have been inactivated together with alarm status and time of addition to the list. Other tools may be assigned the task of monitoring these managed alarms and notifying appropriate plant personnel about their status. Finally, there are tools that may automatically reactivate alarms based on plant practice.

Alarm Processors

All alarms are handled within the PCS infrastructure (hardware, software, and firmware). The ground-floor alarm processing involves activating the alarm whenever the activation point is crossed, handling acknowledgment, handling the horn, and handling the messages and other indications of alarm condition. This is done by the PCS. This chapter discusses alarm handling that goes beyond this ground floor. There is the basic part that resides in the PCS. Some enhanced capabilities may be there as well; however, much will likely be obtained from a third party and integrated back into the PCS.

Basic Infrastructure

This infrastructure is the PCS alarm processor. Its inner working are not user-configurable. The only flexibility users have is in the selection of the alarm type (e.g., high absolute, low rate-of-change, out-of-range, etc.), the alarm activation point, and the alarm priority. Once these parameters are entered into the PCS configuration, the alarm processor takes over and runs the show. PCS alarm processors also have various levels of capability to control the parametric aspects of the alarm activation. Examples of ways it can control include the ability to suppress or inhibit, the ability to modify the alarm activation point, and the ability to modify the priority.

Enhanced Infrastructure

PCS alarm processors are designed to provide various levels of information about the alarm system itself. However, this information is very limited and usually requires supplementation with purpose-built alarm processing software. The alarm tools discussed in chapter 4 have some of this. Other software is available from third-party suppliers. It is the entire capability of the basic PCS alarming and the additional software from tool makers or custom designers that forms the enhanced alarm processor discussed in this chapter.

Alarm Integrity Monitoring

Suppression can also provide a unique alarm maintenance capability. To gain this benefit, we monitor all alarms that would ordinarily have activated, due to the plant or piece of equipment being shut down, but did not activate. This is important. For in every such case, we have an event that should trigger an alarm (now suppressed or inhibited). If the alarm condition did not show "in alarm but suppressed," then something did not happen that should have. Something is malfunctioning. Be it hardware, software, or documentation, something is not proper. The enhanced alarm processor should inventory and identify all such suspect items. This list should identify problems.

The above tools will be used for all other operator-controlled alarm methods and most automated ones as well. Each method will have its specific features.

8.7 OPERATOR CONSENT

No one needs to be reminded that the operator is in charge of the process area. That is why we have operators. They have the usual tasks to perform. They are expected to promptly and effectively respond to abnormal situations. We rely on their constant exercise of vigilance and good judgment. Therefore it is understandable that most automatic intervention must have a protocol that respects the operator's position. Some enhanced alarm methods may initiate without operator involvement. Some may not. We therefore turn our attention briefly to the conventional practices in the automation industry to obtain explicit operator involvement regarding any actions taken or proposed by the automation system itself.

Most enhanced alarm methods will be fully preplanned and predesigned. Most involve straightforward logic and can be implemented automatically as part of a basic advanced alarm methodology. However, some cannot. There are four ways to evoke the operator's consent to proposed and automatic actions. They are illustrated in Figure 8.7.1 and summarized in order of decreasing automation:

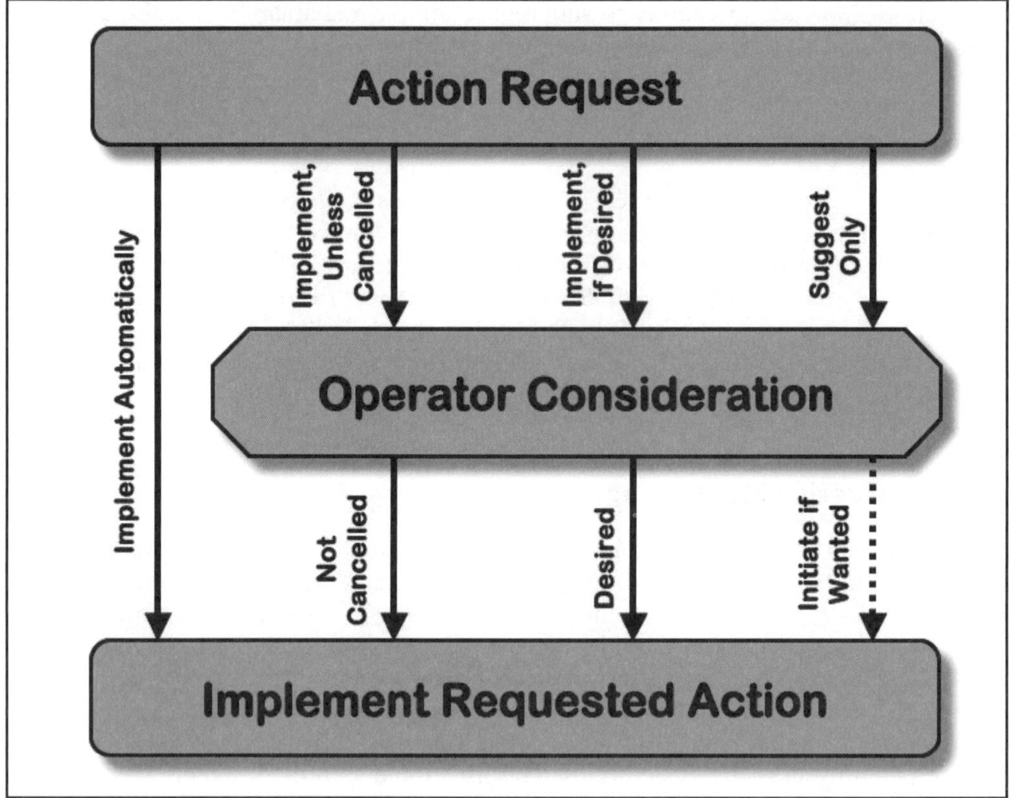

Figure 8.7.1. Operator consent hierarchy

- Implement automatically
- Implement unless cancelled
- Suggest with positive response required for action
- Suggest only

Implement Automatically

The automation system will act automatically to make the preplanned changes in the predetermined situation; it will do this all the time and without the need for any operator involvement. This is the null case, no permission required. *Implement Automatically* is used for all cases where no operator involvement or consent is needed. These are situations where the plant design clearly calls for such an action or a situation that is recognized by a plant integrity survey analysis of an event and so is part of standard plant practice.

Implement Unless Cancelled

The automation system or other support or modification system will provide a message or other indication to the operator that the system will automatically initiate an action unless the operator denies permission within a relatively short period of time. This action will have been fully configured and prepared in advance. It will start automatically unless the operator withdraws consent by taking an overt action to cancel. If the operator fails to cancel, the action will initiate. There is full monitoring of whether the operator cancels the action. If cancelled, it is not suggested again unless the preplanned method detects the need to do so, even given that the operator had earlier cancelled. The *Implement Unless Cancelled* mode is employed in situations where the system has clear and pressing reasons for acting. Often, almost immediate action is required to avoid situations that carry substantial risk of loss. This situation is so serious that the consequences of acting are much less than the risk of not acting, even if the current situation does not progress to the serious state predicted.

Examples include the following:

- Alarm counts indicate that the catalyst regenerator plant is in alarm flood condition. In order not to miss important alarms, the alarm flood view will be initiated in 15 seconds unless cancelled.
- Batch temperature profile and certain temperature rates of change indicate that the reaction is very near the transition to exothermic. Appropriate alarm limits and priorities will be implemented for this phase of operation in 30 seconds unless cancelled. After 30 seconds and until 300 seconds, they can be reset to previous values if requested by the operator.

Suggest with Positive Response Required

The automation system or other support or monitoring system will provide a message or other indication to the operator to initiate an action. This action has been fully configured and prepared in advance. Activation requires a simple consent. A message will ask for the operator's consent. If not granted, the suggested action will not be performed by the system. The system may or may not remind the operator. After a certain time, or when the action is no longer appropriate, it is withdrawn. The *Suggest with Positive Response* mode is employed where the system has strong justification for recommending the action, and the recommended action is the most reasonable one to take. However, operator consent is required, since there might be other important issues to take into account. In addition, it may be difficult for these issues to be known to the automation system. In these situations, the consequences of not acting (not accepting the system's suggestion) are less serious than the consequences of improperly acting. With the success bias in favor of action, but the risk bias in the direction of not acting, the situation clearly recommends that the operator proceed from the position of knowing what to do and being ready to do it, but doing it only after consideration.

Suggest Only

The automation system or other support or monitoring system will provide a message or other indication suggesting that the operator perform an action. There may or may not be any monitoring of whether the operator does, in fact, perform the recommended action or a substitute action. It is up to the operator to take or not take the suggested action. If there is monitoring, the system may or may not remind the operator of the suggested action if not initiated. The *Suggest Only* mode is normally used in cases where the system cannot make a specific recommendation. One case is where the system has several alternate actions to suggest for the operator to consider, and the operator is needed to decide the most appropriate one. Another case is where the system might have only one action to suggest, but the reasoning or justification for the action may be either weak or complicated. Either way, we need the operator to think about what is going on and then decide what to do, if anything.

Examples include the following:

- Batch elapsed time indicates that the reaction might transition to exothermic; do you wish to reset alarm limits and priorities for this phase of operation?
- There are over six stale alarms in the second heat exchanger; are you ready to declare that the exchanger is out of service?

8.8 OPERATOR-CONTROLLED SUPPRESSION TECHNIQUES

Suppression means that an alarm (or set of alarms) will be rendered not operable—that is, even though the alarm activation point might have been crossed, that alarm will not actually go into alarm. The techniques discussed here are designed for the operator to initiate. They are used for situations that occur rarely, or ones that might be difficult to predetermine and thus not readily implementable by logic. This capability provides the operator with an important tool to ensure that the alarms he feels are not relevant do not activate and therefore will not bother or distract him from managing the plant. It is the operator's way to exercise prerogative over any alarm (or in special cases, almost all alarms) to prevent activation.

Suppression is not a casual capability. Operators will suppress alarms that, in their informed opinion, are not needed in the current operating regime or state of upset *and* if permitted to activate would detract from the operator's ability to focus on the proper management of his plant.

Here are the common methods:

Shelving. Shelving allows the operator to disable an alarm in a preconfigured, premanaged way. This is also called disabling or inhibiting; however, actual functionality of these features can be directly linked to specific controls platform functionality. Shelving is used to provide a way for operators to disable any particular alarm properly with the alarm status appropriately monitored and controlled.

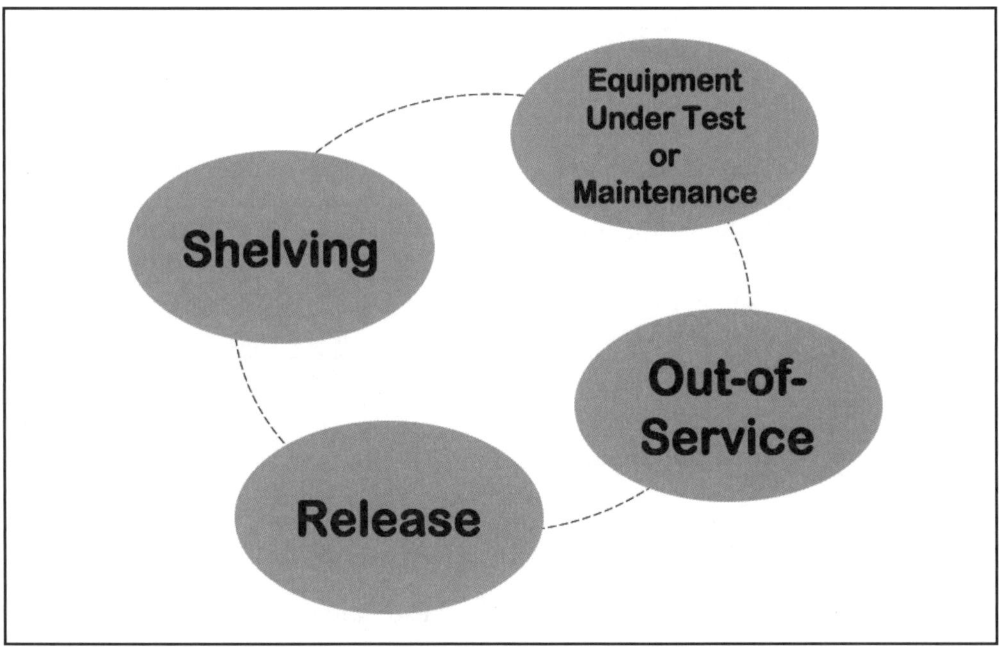

Figure 8.8.1. Operator-controlled suppression categories

Release. Release allows the operator to disable an active alarm in a preconfigured, premanaged way for a long time. Release is used to provide a way for the operator to do long-term shelving without the need to reactivate it manually once it clears.

The release process operates to disable the alarm in much the same way as shelving does, but with important differences. Whereas shelving requires the operator to unshelve the point, release sets up an automatic way for the point to be reenabled. First, the alarm in question must be active. Second, after being placed on the release list, the alarm will automatically be acknowledged and remain disabled until the alarm condition has cleared. Once cleared, the alarm will automatically become reenabled (or automatically unshelved).

Out-of-service. Out-of-service blocks all alarm activations for the equipment for the entire time that equipment is out of service. It is used for installed spared equipment or situations where production rates idle parts of a plant or in parts of a plant idle due to significant maintenance or reconstruction.

Each plant area will be identified in such a way that the operator can declare it to be either in service (the usual state), and therefore all alarms follow their configured plan; or out of service, and therefore all relevant alarms are shelved. Note that not all alarms within a given plant will be shelved, even if the plant itself is out of service. Within out-of-service plants, rarely are all portions completely devoid of hazards to the point of not requiring any alarm monitoring. Examples include portions that retain energy (chemical, thermal, electrical, or mechanical), portions within the physical boundary of an out-of-service plant area but nonetheless are currently in use by other in-service plants, and portions that retain a potentially hazardous inventory, say, of significant mass, possessing chemical or radiation dangers, and the like.

Equipment under test or maintenance. This blocks all alarms from displaying to the operator for the duration of the test and sends a message to a journal or otherwise logs the fact that all alarm conditions that would have alarmed if the test were not in progress would activate. This method is used for equipment that needs to undergo routine testing where alarm activations during the test should be monitored but not passed on to the operator.

Very similar to out-of-service plants, we identify important equipment that are subject to regular testing, either as part of a conscious maintenance program or as required by statute. Then, in bulk, just as in out-of-service plants, the equipment alarms are either enabled or suppressed.

8.9 PRECONFIGURED, SIMPLIFIED SUPPRESSION TECHNIQUES

These techniques are implemented with standard logic without any knowledge engines or anything complicated. They are activated in any appropriate way using a previously explained operator "consent" mechanism.

Suppress. Using preconfigured logic, this prevents the activation of alarms. This technique is used for alarms that fail to have relevance under present operating conditions. All suppressed alarms will not activate and ordinarily do not appear in any log except the list of suppressed alarms.

Included suppression categories are all of the following alarm modes: *bypass*, *cutout*, *disable*, and *inhibit*. They are considered to be equivalent to suppress and so are to be treated as being identical. All further use of these terms should be interpreted as being identical to suppress. Refer to section 8.6 for a discussion of the requirements of design construction for this capability.

The implementation of this method would be along the same lines as operator-initiated suppression; however, instead of the operator doing the decision work, it will be done automatically by preprogrammed logic. Operator consent is usually not required here inasmuch as this activity is no different than any other engineering and process management decision except that it impacts one or more alarms. The only difference is that here it is important that the operator know the situations that lead to alarms being suppressed. The usual method is to maintain a summary display of all suppressed alarms. It might be implemented as a static list or include live data with an indication of whether or not the alarm activation point was reached (hence the alarm would have activated had it not been suppressed).

Logging of repeating alarms. This inhibits redisplay and horn activation of an alarm after it has first activated and then starts to cycle repeatedly. A log will be maintained that periodically logs the number of repeat activations per 15-minute time period so long as the repeating alarm satisfies the repeating definition (see below). Logging of repeating alarms is used for alarms that have a tendency to cycle or chatter, but usual conventional methods of management fail or cannot be used.

This capability is usually designed to initiate using a preconstructed list of alarms that tend to repeat often and cannot be managed any other way. They are usually placed on the repeating list as soon as the third repeating activation occurs within a 15-minute time period. It is also designed to automatically detect repeating alarms and process them in the same way.

Here is what the construction would look like for alarms that are not in the predefined repeating alarm list:

- First, a detection module watches the alarm activations and identifies any alarm that is repeating. Metrics for repeating alarms differ, but the usual one is ten or more alarm activations and clearings (whether or not any are acknowledged) within any given 15-minute period.
- Once triggered, the logging mode continues as long as the initial criterion is met, but no shorter than 1 hour, and more usually, a full shift.
- After the quiet period has passed (where the number of activations falls below five in a 15-minute time period), each alarm is reenabled. It now must requalify as repeating to be logged as such again.

Redundancy. This permits the first alarm to activate and suppresses all the rest. It is used where multiple measurements of the same process variable are present but none can be eliminated by rationalization. This is list-driven. For each redundant set, once any alarm in the set activates, all the rest are automatically shelved. When the active alarm clears, everything is reset and ready for the next time.

Auto shelving. Auto shelving allows the alarm system to automatically shelve an alarm that starts to be repeating (see definition for shelving in preceding section). This is the technique used for alarms that normally are well behaved but for some reason have started to abnormally cycle at a rate that cannot reflect any reasonable process condition and, if allowed to persist, can become a significant annoyance for the operator.

This operates exactly the same way that logging of repeat alarms does but is used for those alarms that were not in a prequalified list. A plant can either use the "logging of repeat alarms" for alarm points that are not in a prequalified list or disable the automatic discovery of logging and use auto shelving. The threshold for auto shelving is usually the same as the one for repeating alarms (ten or more activations within a 15-minute period). The unshelving may either be automatic (at the end of any active period, but not less than 1 hour) or manual and done by operator action.

Operating mode. This modifies alarms by (a) suppressing them, (b) changing activation points, (c) changing priority, and/or (d) changing the list of proper actions to take in response. Operating mode is used when plants undergo transitions to different operating conditions during the normal course of production.

This is implemented in the same way that an out-of-service and/or equipment under test is done. It can either be fully automatic, in that the scheme knows exactly which mode the plant is operating in and sets the alarm parameters accordingly, or managed in any of the usual operator "consent" modes. Either way, it is utilized by employing preconfigured tables of alarm parameters, one for each mode.

Major event. This suppresses less important alarms during serious events. It is used to reduce the impact of alarm floods and permit the operator to focus on the more important alarms. It is implemented using a "major event" detector. The detector must be capable of identifying the events (either from a list or by class). For each such event, there will be a set of alarm parameters or alarm performance behaviors to be deployed. Example events might include the following:

- Partial or complete plant trip or shutdown
- Major environmental upset (spill, release, or serious weather emergency)
- Alarm overload

Consider alarm flood as an example. Suppression would utilize a flood threshold detector. Two of the usual alarm flood threshold detectors are (1) 10 or more alarms within a 10-minute or less time period or (2) 100 or more alarms within a 10-minute or less time period. Once flood is detected, a plant might suppress all alarms of a lower priority than high and automatically silence the alarm horn as soon as any single alarm is acknowledged. Other alarms might be enabled, disabled, or have their alarm configuration values modified.

The important thing to note is that during such a situation, the alarm load must be reduced to a level that the operator can still use alarms to help manage things. Therefore changes are made to ensure that the load is reduced without undue risk of missing anything serious and, at the same time, to provide all the useful information to guide the operator as much as possible, under the circumstances.

8.10 INFORMATIVE ASSISTANCE

Informative assistance is a fancy name for setting up a way to use all the information that you have about the alarm, why it activates, and what to do about it when it does. Rather than have the knowledge sequestered in obscure files or notebooks, we provide a way to have it available to the operator when needed. We also ensure that it is formatted and expressed in a way that encourages the operator's understanding and use. This is the time and place to use all the relevant alarm information available.

When Informative Assistance Is Useful

The process of rationalization produced a lot of very useful information that is tailor-made to assist the operator. Each alarm is backed up by its alarm response sheet that details information including (a) a complete description of the alarm variable, (b) ways to confirm that the alarm activation actually reflects an underlying process problem, (c) what will happen if the alarm is not managed, and (d) what are the likely steps to take to handle the abnormal situation. Informative assistance provides this and other related information to the operator. This information can be presented either before alarm activation or after alarm activation.

- *Before alarm activation.* Link the online alarm summary information to all process graphics where the tag in question appears.

  Anytime the operator has a question about the integrity or proper performance of this variable, provide a mechanism to see additional relevant information. At a glance, the operator will be able to ascertain the normal limits for the process variable and the other associated variables to monitor and evaluate anything unusual.

- *After alarm activation.* Link the online alarm summary information or other pre-prepared operator advice to all process graphics and other locations where the tag appears when it is in alarm.

  At this point, the alarm has activated; therefore, all the information about causes, confirmatory actions, and remediation becomes highly relevant. The operator will use this information to analyze the alarm situation and decide on the most appropriate manner to take remedial steps to improve the situation.

All that is needed to provide this information to the operator is an alarm knowledge database that is structured so that each individual alarm and tag can be recognized and linked to the appropriate places in the main control and alarm infrastructure of the human-machine interface. Additionally, the formatting and presentation of this information to the operator must be managed so that the operator can understand what is presented and navigate to other important information. Typically, links are included on process graphics, alarm summaries, and tag control faceplates. You may also want to link to an event log or other form of operator log so that the operator can record events and actions for later recall and for use in training, reporting, and unfortunately incident investigation.

At this juncture in the platform development of industrial controls, very few PCS vendors actually provide such a capability. Moreover, most of the useful data (alarm summary items, etc.) reside in third-party software or, in some rare, to-be-avoided instances, in homegrown spreadsheets or other odd databases. For third-party software, this is the tool kit you used to monitor alarm performance, check the configurations, and store the rationalization information and results. Consequently, you will use the inherent capability of these tools to link to the primary controls platform. If your third-party tool does not support linking, you will need to change tools or export the knowledge and structure to another suitable platform.

How to Do It

The information produced for each alarm during a rationalization comprises the alarm response sheet. Note that we include the word "sheet" in the phrase to make it clear that we are not referring to the standard PCS display called the alarm summary. Not only is the information extensive, but it is also structured (refer to chapter 7, "Rationalization," for examples). This structure provides the operator with the right information at just the right time.

Examples

Looking at a few examples should help clarify how this all might actually work.

Eclipsing Example

This is the situation where all related alarms are allowed to activate, but the advice provided to the operator deals only with the most important one(s). Those are the alarms that need to be managed.

Eclipsing. In a related group of alarms, eclipsing permits the most important alarm, or the alarm whose response would be the most appropriate for the current situation, to be highlighted for the operator.

More than one alarm will activate at the same time, but the alarms can have significantly different priority, and therefore the operator must focus on the most important one, not necessarily the first one.

As related (or consequential) alarms activate, those that have been preidentified as belonging to an eclipse group are examined. At each point in time, the most important or most informative alarm is identified to the operator as such.

For example, let us suppose that we have an eclipse group around a heat exchanger consisting of four temperatures, two flows, two pressures, and three analyzers. Figure 8.10.1 illustrates this exchanger. Note that this is *tremendously more complicated* than it needs be from what we know about a proper rationalization; however, it will better serve to illustrate our point. Let's suppose that there is a rapidly growing crack due to high thermal stresses in one of the heat exchanger tubes. The first alarm might be a rate of change temperature indicating the potential for thermal stress (TDXA-1). There is an absolute temperature alarm (TAH-6) to keep track of the heat transfer (which we know to be some serious leakage out of our split tube, but the operator does not yet know this). There is also an analyzer alarm on the outlet stream of the shell side of the heat exchanger (AAH-5).

The eclipse processor would immediately promote the analyzer alarm as being the most relevant, indicating a serious leak. Seeing this alarm as being suggestive of the most serious of the problems should facilitate the operator with managing this situation. Depending on the alarm philosophy, all eclipsed alarms can be inhibited. Those that have been inhibited but are already in alarm might be removed from the current alarm status

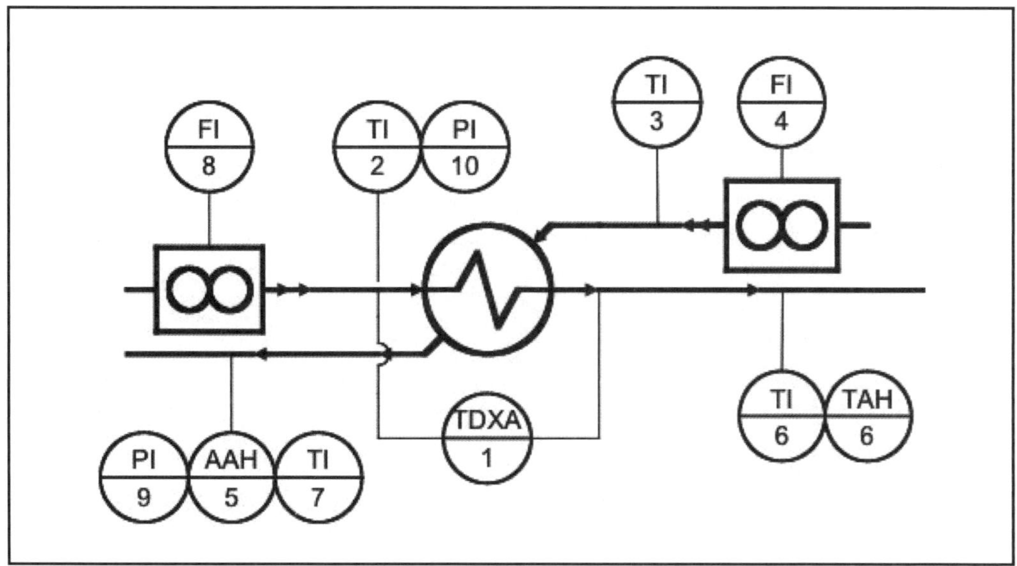

Figure 8.10.1. Heat exchanger for eclipsing example

Grouping Example

Grouping. Grouping permits all alarms to activate but provides guidance to the operator based on the overall event the abnormal situation represents. It is used to collect alarms that, while individually useful, have generally the same operator response.

Decide whether or not there are any natural groups of alarms. Often, there are not, so do not be confused into thinking that you might have done something wrong if you cannot find lots of groups. Likely places for groups are those parts of a plant where the equipment is replicated to achieve a production need. Often this is where capacity requirements are more than any individual piece of equipment can provide and it is more practical to duplicate equipment rather than build larger capacity equipment. Other areas are where process measurements are made in more places than those required for basic controls. Arrays of temperature measurements arranged to provide profiles are a good example of this form of a group.

Regardless of how large the groups are or how many groups you might have, implementing this is rather simple. For each group, do the following:

1. List the alarms that are in the group.
2. Determine the group response—the information you will provide the operator to recognize and manage the abnormal situation within the group.
3. Build the logic that (a) provides the operator with the recommended information and (b) ensures that any other alarm activation from a member of the group, while the first alarm is still active, will reference the first

alarm's information and not provide any additional confusing information or requirements for the operator.

More Examples

Let's look at a few different examples and see how other methods of providing information might work. The data for these examples is taken from the alarm response sheet (see chapter 7). That control loop example is illustrated by Figure 8.10.2.

Consequences Example

- *Background.* Let us suppose that FAH-2009 (the high-flow alarm for tag FI-2009, the PB Feedwater Flow) has activated. This means that the value for FI-2009 has probably exceeded 200 KLb/Hr. The alarm is configured to be high priority.
- *Situation.* All the operator knows for now is that the flow might be too high and, if so, this would be an important situation to remedy. Given that the alarm has been properly configured, the operator should also be confident that if his work is done promptly and accurately, the situation could be remedied.
- *Use of knowledge-linked data.* We link the operator immediately to the consequence information: unless this situation is cured successfully and promptly, either the power boiler (PB) will overheat and possibly rupture or otherwise damage the tubes or the boiler will not be able to produce enough steam flow and could develop very high pressures that might cause damage to downstream equipment.

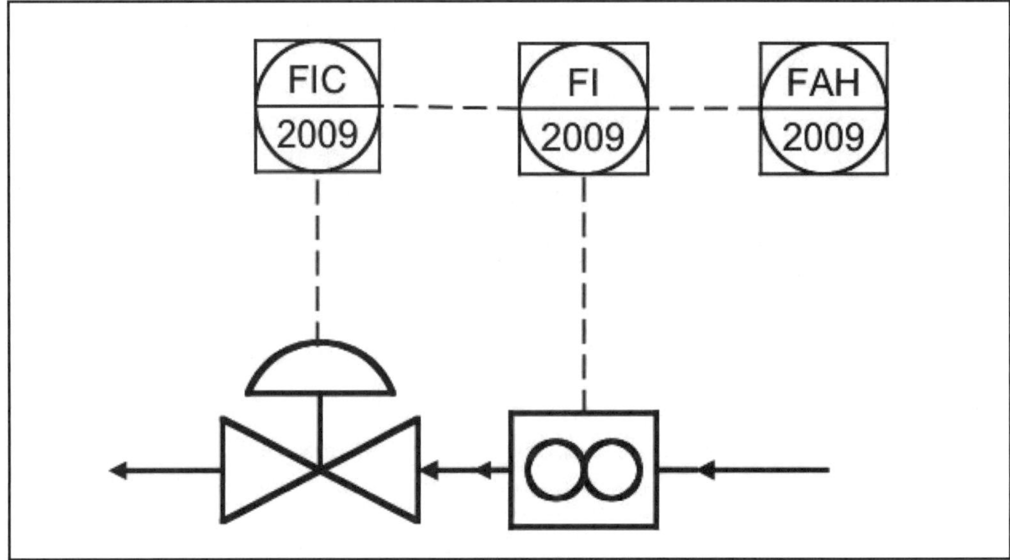

Figure 8.10.2. Boiler feedwater flow example

Without being able to link to this information, the operator would have to remember it, explicitly try to find it on his own, or do without. I leave it up to your judgment to determine the value of having this information readily available.

Remedial Actions Example

- *Background.* Let us suppose that FAH-2009 (the high-flow alarm for tag FI-2009, the PB Feedwater Flow) has activated again.
- *Situation.* Again, all the operator knows for now is that the flow is likely to be too high and that this is an important abnormal situation to remedy.
- *Use of knowledge-linked data.* We link the operator immediately to the information describing what is required to fix this situation (e.g., remedial actions). Moreover, since our causes and corrective actions have been structured, we know the relationships between them. The two capabilities below illustrate what might be done without and with the ability to adequately check the process database and make logical tests and computations.
  - *Capability 1.* No other logical testing is done, and no plant information is used: all that this enhanced technique does is to present the operator with the list of causes and the list of what to try to remedy each. It is up to the operator to check the plant and try to determine which might be the cause for the alarm. Once that is done, the related remedial action can be looked up.
  - *Capability 2.* Additional logic is used to determine which of the conditional situations might be the actual problem: the logic attempts to go through the list of causes and "test" the real-time data in the plant database to see if it is able to uncover any additional information that would lead the operator closer to a primary cause. If there were a mass balance computation being done, it might be possible to (a) suggest that the steam flow (Cause #2) as indicated by the flow meter is likely very different from what it should be or (b) suggest that the level control (Cause #1) is not working (since the liquid mass balance is off and the level is not correct). Once a likely cause is offered, the linked corrective action is suggested to the operator.

As before, without being able to link to this information, the operator would have to remember it, explicitly try to find it on his own, or do without.

8.11 KNOWLEDGE-BASED

The previous categories of enhanced alarming methods have one thing in common: they can be implemented using simple logic. They will do their job without needing much more than a few on-off "switches." Unless the required number of switches gets out of hand, that is about all there would be to using those techniques. For example, we can use real-time alarm activation information to decide whether alarms are activating so fast that the operator cannot manage the plant. When the activation rate exceeds the threshold, the alarm system can be set (switched) to show only critical alarms for the next 5 minutes or so. Other situations might use plant data to switch alarm presentation modes.[1]

Alas, the world, even the defined territory within our plant boundaries, is not always so accommodating. Simple things can often become quite complex. Complex things sometimes become extraordinarily daunting. Situationally based alarm support is one of them. Enter the realm of knowledge-supported decision making. For simplicity, we call these knowledge-based techniques. The list includes pattern recognition, neural networks, fuzzy logic, knowledge-based reasoning, and model-based reasoning. There are others, but these are the most common. However, their use is not necessarily very common. Using them is a surgical-like practice. They should be used only when necessary and only when the operator will actually be assisted well beyond the effort required to understand and use the methods. There are commercial providers of this specialized technology.[2]

Pattern Recognition

A pattern is an arrangement or form, a model or plan. In our case, we take advantage of a repeatable situation to provide knowledge. To use this, we observe that certain patterns of alarm activations not only announce a set of individual problems but, when taken as a group, can also suggest more complex problems with clarity. As an example, consider the situation illustrated in Figure 8.11.1.[3]

The compressor in our plant has a set of nine alarms. Each alarm is labeled with a number from 1 to 9 across the top of the table. For example, alarm 2 is an overtemperature alarm for a cooling chamber, alarm 4 is a high-leakage current into the system B ground circuit, alarm 5 is a common undervoltage alarm, and alarm 9 is a high rate-of-change alarm on a decreasing flow signal. These are the individual alarms. If any were to activate, the operator would identify what might be wrong and then proceed to figure out what to do to get the situation back to normal.

Fault	Alarms								
	1	2	3	4	5	6	7	8	9
Supply A failed	✓		✓		✓				
Pump A failed			✓		✓				
Supply B failed		✓		✓	✓				
Pump B failed				✓	✓				
Filter DP high						✓	✓	✓	✓
Bearing 1 oil flow lost						✓		✓	
Bearing 2 oil flow lost							✓		✓

Figure 8.11.1. Pattern recognition example (from EEMUA 191)

Also shown in the pattern diagram are seven faults along the left side of the table. "Supply B failed" is one. "Bearing 1 oil flow lost" is another. In these cases, there is no single alarm that can identify any of our faults directly. However, we know from experience that if alarm 4 (high-leakage current) and alarm 5 (common undervoltage alarm) occur at very close to the same time, it almost always means that pump B has failed in a way that is well understood. Similarly, other patterns of alarms can point to other well-understood faults. The ability to identify these faults and communicate that information to the operator as soon as they occur is a valuable operator assistance tool.

Neural Networks

You will seldom, perhaps never, come across the need to develop a model of a fault process that requires using neural net technology. But you might. So let's take a quick look. Neural nets are used to describe situations where there are a lot of data linking problems to eventual faults. Unfortunately for most of us, looking at this data and trying to make some sense of what it means is beyond normal abilities, even trained ones. Since we do have a lot of data, there is a way to use it when the need to know is important enough. Typical situations might be a complicated reactor runaway or a batch pharmaceutical run that uses very expensive raw materials and needs to be closely watched.

Figure 8.11.2 shows a typical illustration for a neural net. Do not worry, we will not go too far into this—just enough to give you the idea. There are two basic elements to the net: nodes (the "dots" in the diagram) and the links (the lines connecting nodes in the diagram). The nodes on the left (the dots with the funny looped "hats" on them)

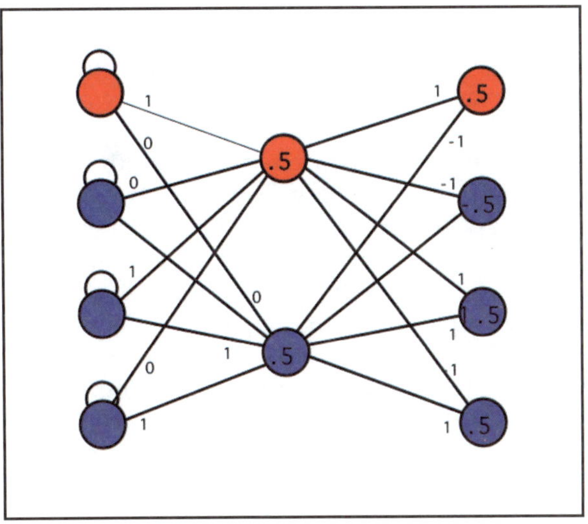

Figure 8.11.2. Neural net illustration

each stand for something we know that can occur as a "cause." In our case, think of them as alarm activations. We have one node for each alarm we want to have in our model. The nodes to the far right stand for things that we know might be the reasons why the alarms occur. Think of these as the underlying faults. Our data then consist of lots of sets of alarms and, for each set, what has been found (or what specialists think should be found) to be the underlying fault pointed to by each specific set of alarms. Using special neural net–building procedures and lots of data, we can build a model. The model is all the connections and intermediate nodes that link the extreme left nodes to the extreme right nodes. This infrastructure is not useful to us. It is only needed to complete the model. Our value comes from using the model to take new alarms (sets of input conditions) and map them to likely advice for the operator (the output conditions).

The useful part is that the sets of data used to build the model do not always repeat the same way. If they did, then all we need to do is to use the pattern model in the earlier section. So, even though our data situations might look more like something statistical, we can still build a useful model to capture the sense of it. To use this tool, all we do is continually feed alarm patterns into the prebuilt neural net model and wait for it to identify a fault. Usually when it does, it is the correct one. Usually when it does not, there is no underlying fault; therefore the alarms themselves should point to the problem that needs fixing, not something else more vital or more obscure.

Fuzzy Logic

As in the case of neural nets, you may never come across the need to develop a model of a fault process that requires using fuzzy logic technology. But, on the other hand, you might. So let's take a look at what it is all about.

Fuzzy logic is exactly what comes to mind when you examine the two words, fuzzy and logic. Normal logic is something that is always the same. Fuzzy logic is used for cases where we have to use the words "often" or "many times," but not "always," to describe something that happens in a variety of ways.

Let's consider a simple temperature control example. For conventional logic controllers, all it does is compare the current temperature with the desired temperature to know whether to heat, cool, or do nothing. Every time we do the comparison and the result is the same, we do exactly the same thing. On the other hand, a fuzzy logic controller (illustrated by fig. 8.11.3) might not turn on the cooling when the room appears to be too warm without checking with other criteria, like when was the last time the cooling was on. If the cooling had been off for a long period, turning it on will produce unpleasant odors for a while. If it is likely that something else will cool things instead, the fuzzy logic might conclude that it is best just to wait and not start the cooling.

A fuzzy logic alarm controller might be used to determine situations where an alarm should be turned off or inhibited for a while. It might be used to modify alarm activation points during somewhat vaguely defined situations. You get the idea. And I did tell you that this is not one of the frontline tools?

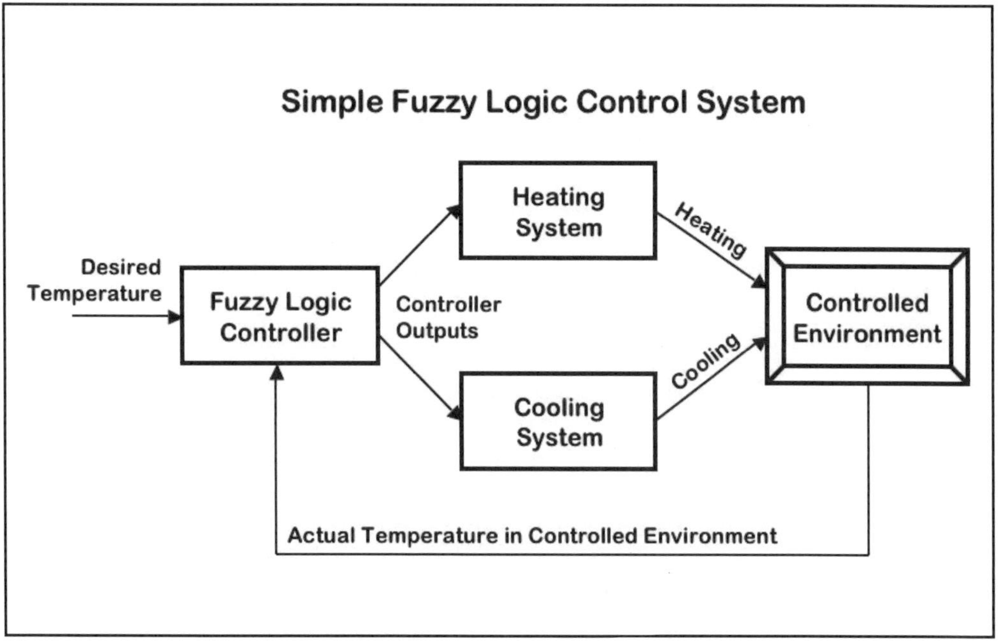

Figure 8.11.3. Fuzzy logic illustration

Knowledge-Based Reasoning

When people work in a plant, even a rather complicated one, it is common for them to learn quite a bit about it and its inner workings, quirks, and the like. Rather than all this casual knowledge gathering itself together and forming anything like a coherent picture, most of it remains a collection of this and that. Each piece is known and understood, but nothing more. Nonetheless, there is important knowledge both within the "pieces" and from the collection of the pieces as a whole. To use this knowledge, we use a knowledge engine.

There are commercial knowledge-utilization engines available in the marketplace. They permit the intake of a large number of seemingly disconnected pieces of information (knowledge chunks). In the jargon of knowledge engines, these pieces are called rules. They can number into even the tens of thousands. They can come from one individual or an entire collection of individuals. Using these knowledge pieces, or rules, the engine searches and connects related pieces to produce a coherent picture of some aspect of the situation in question. It is this coherent picture that is conveyed to the operator as information to assist in the operation of the plant.

Here is a very simple example:

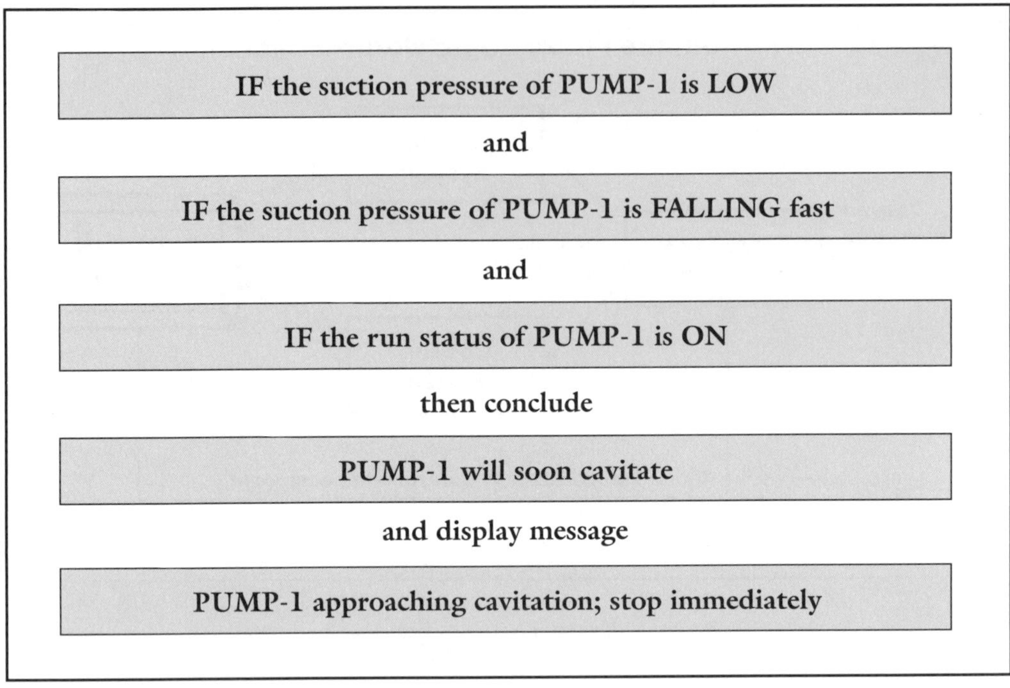

IF the suction pressure of PUMP-1 is LOW
and
IF the suction pressure of PUMP-1 is FALLING fast
and
IF the run status of PUMP-1 is ON
then conclude
PUMP-1 will soon cavitate
and display message
PUMP-1 approaching cavitation; stop immediately

Figure 8.11.4. Example knowledge-based reasoning

There are three equipment states being tracked: (1) suction pressure, (2) rate of change of suction pressure, and (3) run status. None of the three equipment states, by themselves, constitute an alarm condition. Yes, a low suction pressure might have been alarmed, but in this case, a low pressure is not at all so unusual, inasmuch as this system operates always at a low pressure. It is the combination of the three items that portends trouble. This rule captures it.

Knowledge-based reasoning can also be used to interpret alarm patterns and other aspects of plant operation.

Model-Based Reasoning

Plants and their associated major pieces of equipment can complicate things. This complication can arise due to size, the nature of its pieces, or simply because what happens inside them is hard to see, or hard to understand, even if it could be seen. But suppose there was a way to see and understand everything. Suppose one could build a model, a replica, of that plant or piece of equipment according to our own rules of understanding. While there are a large number of ways to build models, let's suppose ours is built

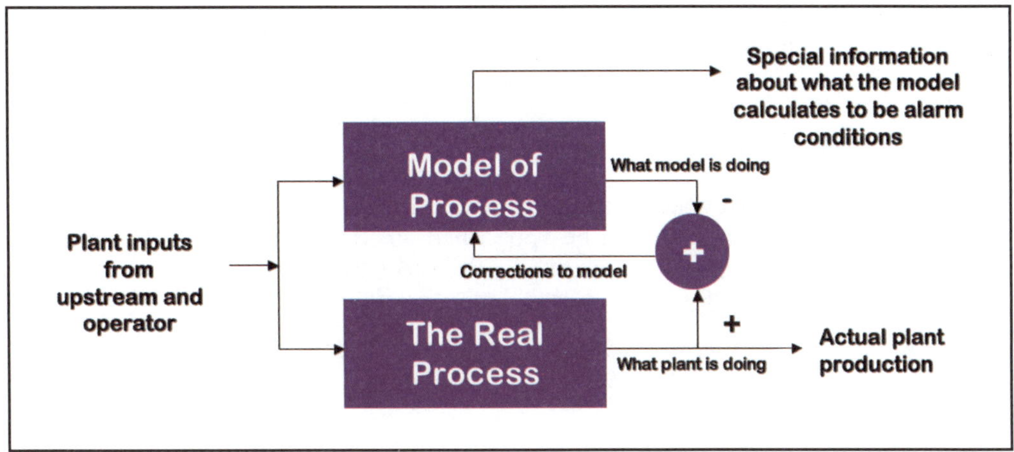

Figure 8.11.5. Model-based reasoning diagram

of mathematical equations. And suppose this model is "tuned" so that it behaved closely enough to the real thing.

Because we constructed this model, it is built of pieces we understand. It is easy to look inside it to see what could not be seen in the original. Moreover, we can also use our model and make it do all sorts of things, even things that might damage actual equipment. It permits us to see what might be going on inside of the actual thing. Knowing what is going on inside permits us to draw conclusions and make decisions. We have described the process of model-based reasoning. Figure 8.11.5 illustrates this construction.

The actual process is the lower box. The model is the upper box. The real plant inputs (e.g., streams, energy, etc.) go into the process. We also tell the model mathematically exactly what inputs go into the plant. The plant produces results (products). The model calculates its results. Both results are compared, and the model is adjusted to bring the model results as close as we can to the plant. Once that is done, our model matches the plant. If we were to look inside (e.g., for reaction products or internal temperatures), we can know them from the model even if we cannot measure them in the actual plant. This information can be used to advise the operator. Such advice can be used to prevent upsets or better manage those that occur.

8.12 KEEPING TRACK OF PLANT STATE

If the alarm system were able to know exactly what the plant operation was at any given moment, this information could be used to enable all needed alarms; disable all unnecessary alarms; configure the needed alarms with the appropriate priority, activation point, and annotations; and provide tailored operator support information to manage alarm events. Knowing the current plant state is far from trivial. Few DCS-based control systems explicitly identify this situation. Some PLC-based control systems implicitly identify this but fail to exploit its benefits by comprehensively managing alarms.

Explicit Plant States

Whenever it is possible for the controls infrastructure to explicitly identify the current plant state, this information should be "published" to the operator and the enhanced alarm management processor. It can then be used to modify alarming. Typical explicit plant states include the following:

- Setup
- Start-up
- Shutdown
- Fault in progress (upset)
- Standby (safely parked)
- Product change
- Grade change
- Testing and maintenance
- Severe weather operation
- Out of service (semipermanent shutdown)

Implicit Plant States

Implicit plant states are the dirty little secrets that few who promise to provide state estimators will talk about. The reason is simple. Implicit states are largely vague and very hard to identify. Yet they represent real conditions and situations. Moreover, these states often represent precursor conditions to the explicit ones—that is, plants are often in an abnormal situation before it is possible to identify that it is abnormal, and certainly well before it is likely that anyone or any agent can declare (explicitly) that it is. Operators may or may not know the actual plant state at any given time. Here is a sample list of the implicit plant states.

- Abnormal operation (nonspecific, general situation)
- Fault in progress (unusual plant states)
- Raw materials changes
- Energy availability changes
- Equipment degradation (both minor and near incipient failure)

Any linking of alarm operation with implicit plant states cannot be done directly, of course. For situations where it is possible to suggest to the operator that an abnormal situation (of the implicit type) is in progress, implicit states can be recognized. Then, after appropriate investigation and agreement by the operator, alarm operation can be adjusted. At that point, we consider the plant state to be explicitly known, so it is used directly to modify the alarm system and advice to the operator. Until implicit states can be identified, they cannot be used for operator advice or any other purpose to manage or modify the alarm system.

8.13 ALARM INFORMATION WITHOUT ALARM ACTIVATION

As we progress in our thinking about how wonderful all these ideas are and how useful they might be for the operators, there is an important reality to face. Getting enhanced methods to work is not only about the "how" of implementation; it is also about the "when." Remember, once an alarm has activated, it is too late to take it back. Once the alarm is activated, in a matter of fractions of a second, the control platform infrastructure will pass the fact to all the agents that sound alerts and warning tones, change colors of displays, generate messages, blink text and graphics, and do all of the other things we know are important to get the operator's attention.

This means that any scheme for implementing useful alarm controls must be constructed to function in a way that will not require taking away an already activated alarm. Therefore it will be necessary to prevent the alarm from activating in the first place until it is known that it should. For example, consider the situation where several alarms are "related" together in a group. For these, we decided earlier that once one activates, we should temporarily disable or inhibit the remainder of the alarms in that group. Getting this to work this way in your controls platform can be tricky. If we could always rely on the alarm processing system to be fast enough to identify the first alarm in the related group and then inhibit the rest, we are OK. But most systems operate on a heartbeat cycle where calculations are done first and then the control implications of that calculation are evaluated afterward. Therefore it is usually not the case that we can always find the first-to-activate alarm and then block the rest. A more dependable method must be used.

Before we go too far into our expectations for implementation, please accept that most equipment vendors have a fairly simplistic alarm control structure. Most of our important schemes may not fit with the alarm-processing infrastructure found there. All provide alarm activations when variables exceed a preset limit. Very few provide anything else in the way of useful enhanced alarm handling. To make matters even worse, few include appropriate links and other enabling tools to do it yourself. So to make all of this work, you will need ways to perform alarm condition calculations, and only then activate the alarm, if it is in a permitted status.

Plant Area Model

The first necessity to efficiently modify or affect the alarm system function requires a rich plant area model and the necessary logic to exploit that situation. Figure 8.13.1 shows the general structure of a model. The structure must be sufficiently rich to support as many levels of aggregation as are present in most industrial plants. This usually means down to the individual piece of equipment. This provides the ability to group alarms into a wide variety of preidentified plant areas so that the enabling or disabling of them can

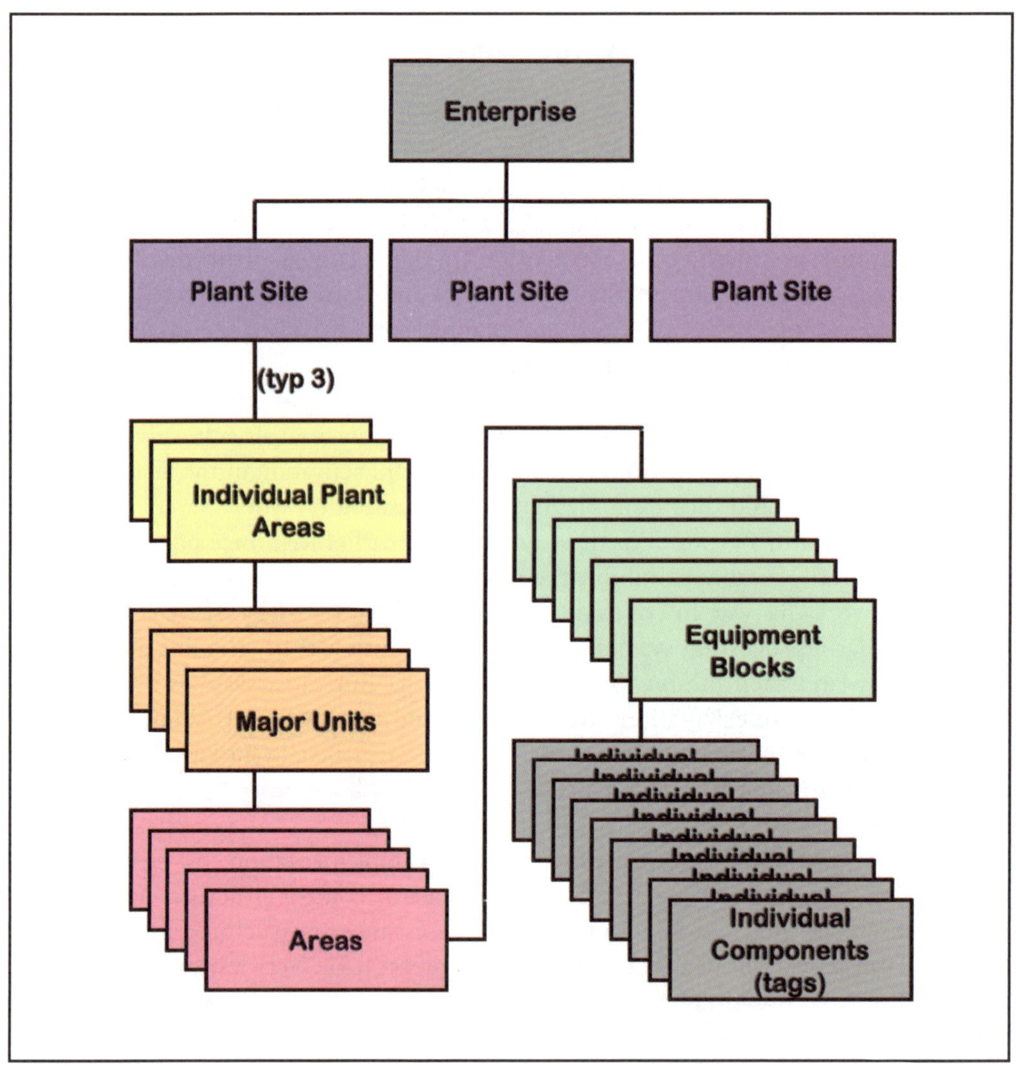

Figure 8.13.1. Plant area model structure

be done efficiently, without resorting to large lists, tables, or other explicit enumeration methods. Chapter 3 introduced the idea. Now we can take advantage of its usefulness.

Begin by examining a typical advanced alarm suppression scenario. Start with a description. Consider a plant with three almost identical processing trains. This could be an ethylene plant that, over time, increased capacity to meet a growing market demand. Very simplistically, a typical plant would contain the following components: a hot side (feed preparation, cracking, fractionation, and drying) followed by a cold side (cryogenic separation in stages) with a recycle stream for the intermediate products that need to be further separated but must return to the main processing stream to do so. Figure 8.13.2 illustrates the general arrangement for a single train. There are four parts to the hot side with clearly referenced labels "AH" prefix and commonly identified parts. The "A" is for Train A, the "H" is for the hot side. Similarly, for the cold side we use "AC," where the "C" stands for cold side.

If one wanted to turn off all alarms for the compression portion (hot side), it would not be necessary to have a list of the individual tags there, but merely to reference all tags in the subset identified by AH Compress. Let's "grow" the model to include multiple trains. Figure 8.13.3 illustrates this situation.

As before, if the plant needed to shut down Train B, all that would be necessary to suppress alarms in Train B would be to reference all alarms for all tags prefixed with BH and BC. Just to stretch the concept, to suppress all alarms for all hot sides, just reference all tags *H (where "*" is the general placeholder for "any and all"). This presents a very efficient and clearly traceable tool.

Conditional Alarming Facilitators

Conditional alarming is the ability to reach into the alarm structure of the PCS for alarm control purposes. It utilizes information about the alarm system itself to modify alarms.

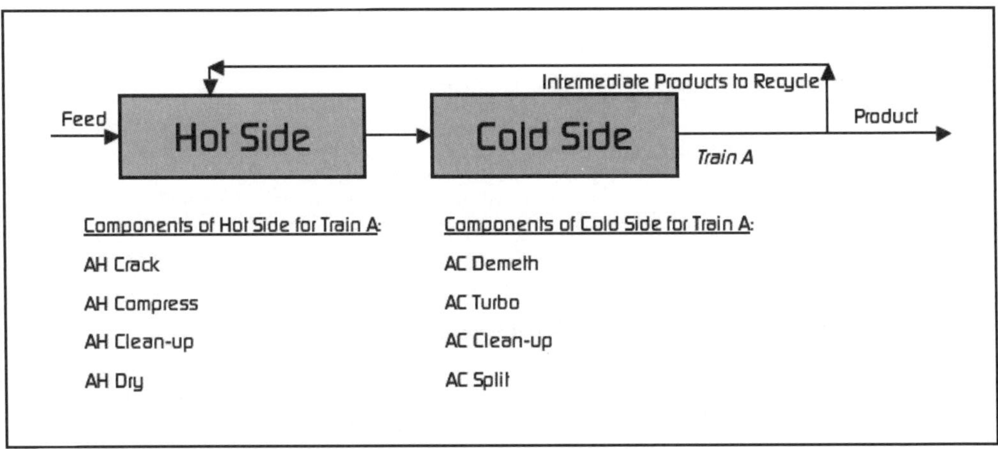

Figure 8.13.2. Example train for ethylene plant

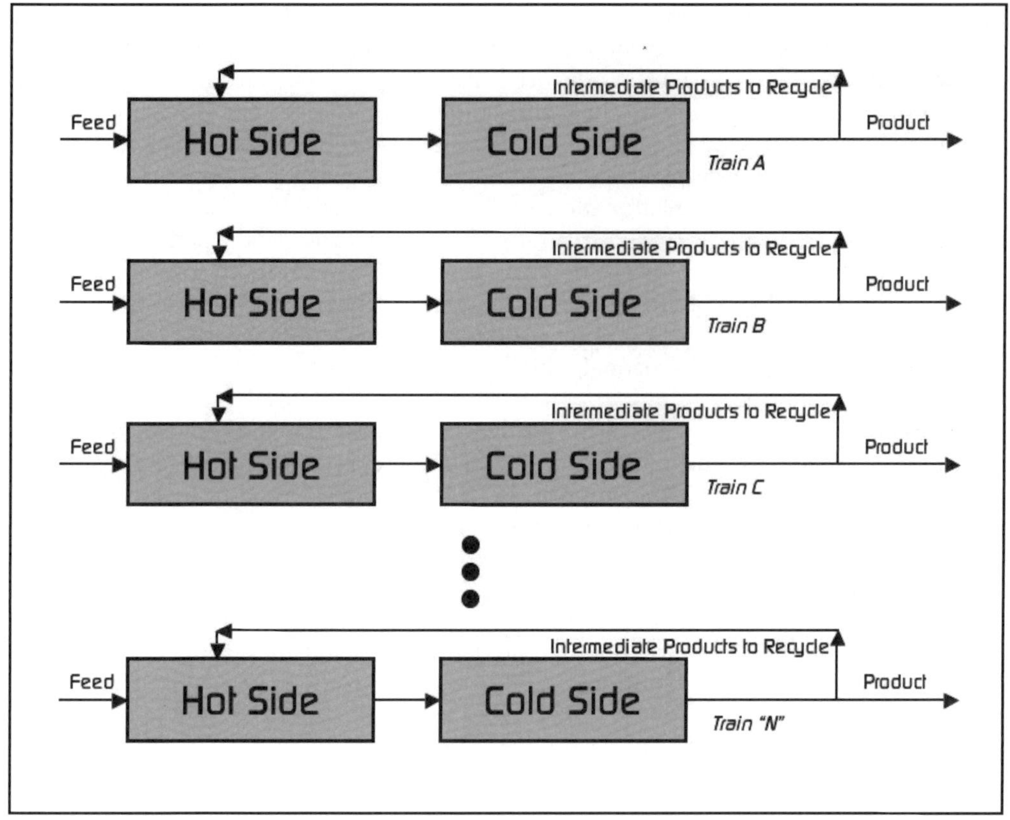

Figure 8.13.3. Example parallel train plant

Therefore it must be possible to have full knowledge of the entire state of the alarm system. Conditional alarming includes the ability to do the following:

- Inhibit or permit activation upon activation point crossing
- Change activation point value
- Change priority. This capability is managed by tools within the native infrastructure, if they exist, or by tools externally (third-party)

8.14 ALARM ACTIVATION PERMISSIONS

This section discusses how we might structure alarm activation permissions to facilitate implementation of our enhanced technologies. The distinctions and categories shown in Figure 8.14.1 are important. Being able to know when an alarm *should* activate without actually *having it* activate is key to being able to control activation and inhibition.

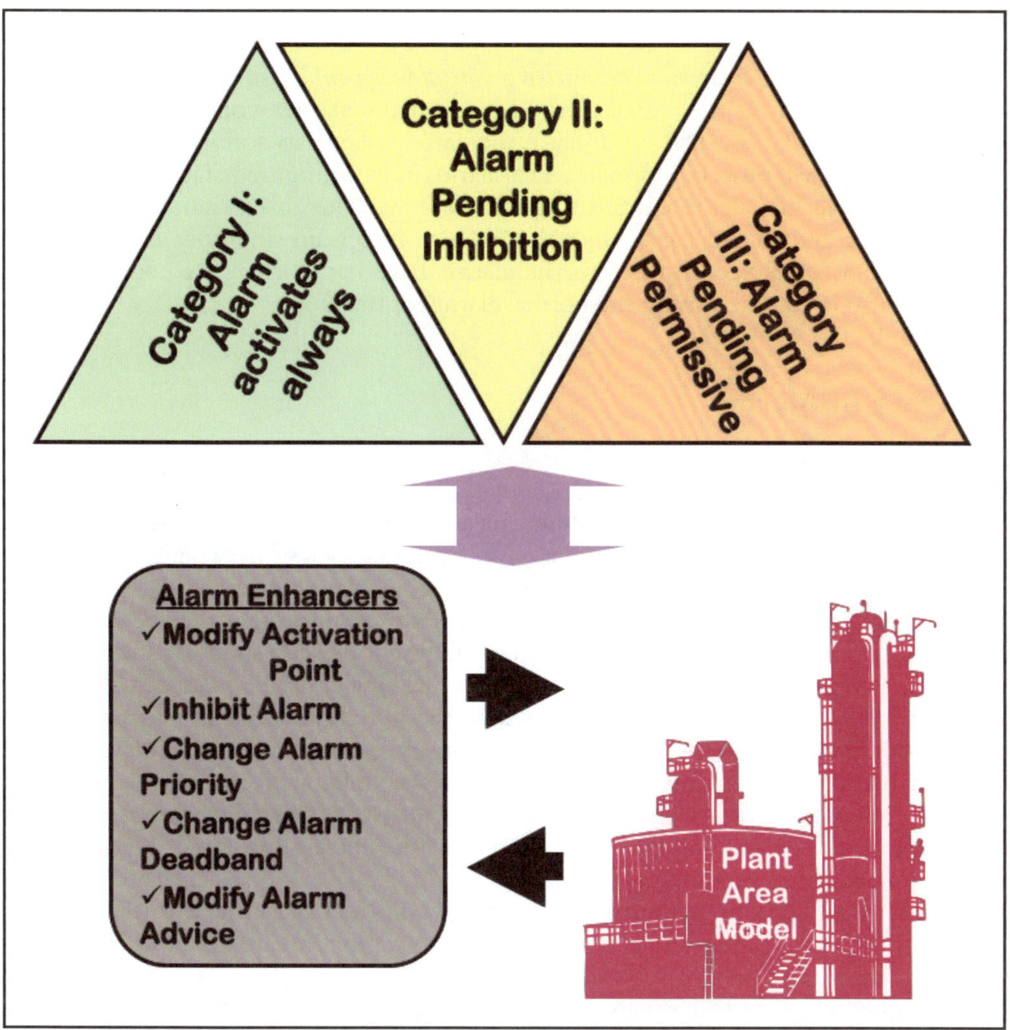

Figure 8.14.1. Conditional control of alarm functionality

Category I Alarms

Category I: Alarm activation is without restriction

For Category I alarms, the normal operation of the alarm system is perfect. Nothing more than a correctly configured alarm is needed. Alarm calculations will be done and alarm activation proceeds normally. These alarms, in their configured state, are expected to activate whenever the alarm activation condition is met, and they are not controlled by other logic except systemwide controls using plant area identifiers. Of course, as needed, the operator can still manually inhibit or suppress alarms in this category.

Category II Alarms

Category II: Alarm activation is permitted pending being inhibited
For Category II alarms, the usual method is to let the alarms work normally up to a point. Alarm calculations will be done, results are available to software logic, but the alarm activation will proceed normally. This alarm, in its configured state, is expected to activate whenever the alarm activation condition is met, but this situation is controlled by other logic and so the alarm might be disabled in certain situations or conditions. When conditions change and before the alarms have had a chance to activate, each affected alarm is inhibited using the normal alarm control logic.

Category III Alarms

Category III: Alarm activation is inhibited pending a permissive.
For Category III alarms, alarm calculations will be done and results available to software logic, but no alarm activations will result unless explicit permissives are active—that is, these alarms, in their configured state, are not expected to activate whenever the alarm activation condition is met. Any activation will be controlled by other logic.

Alarm logic for duplicate and consequential alarms requires that they be considered Category III alarms. If the alarm processing equipment directly supports Category III, we are fine. If not, this level of alarm state identification must be managed a different way. For this situation, the control logic would set all redundant alarms to suppressed. Shadow computations are performed to determine when the alarm would have become activated had it not been suppressed. To do this, we check with the alarm controlling logic (i.e., simulating the calculation to see if the alarm has passed its activation point) to determine if the alarm should be activated had it not been suppressed. If the alarm is eligible for activation (i.e., no others who have a prior call on being active are active), it is set to active. The normal control infrastructure will then cause the alarm to activate. On the other hand, if the alarm is not eligible for activation, regardless of whether it has passed its activation point, it is not activated.

One does not want to have any important alarm being missed due to errors or failures of these calculations and controls. This must be carefully done.

8.15 CONCLUSION

When we think about ways to improve operators' ability to better manage in the control room these days, better management of alarms appears at the top of most lists. It is among the fastest growing areas of opportunity. Alarm activations can occur so often during seemingly normal operation that operators are noticeably stressed. At times, they are completely overwhelmed. Even when the alarm rates are modest, some activations fail to require action and are without sufficient guidance for the operator in order that correct actions can be routinely identified and executed in time. Current best practices in alarm design focus primarily on the task of reducing the total number of configured alarms to the point that only the necessary ones are used, and each one used has a carefully determined activation point and priority. Only after this is in place can we look to additional methodology and technology to improve the operator's chances of successful process management.

Like building a house, unless this first story is sound, it is unwise, even dangerous, to add a second on top. The additional methodology described in this chapter is used to filter competing information for the operator, thus providing useful information even in confusing situations. It is also used to better organize existing knowledge to provide operators with better guidance for managing upset situations. Properly deployed, the additional operator support tools can be extremely effective in facilitating the understanding of abnormal operation as well as more effective management of them.

Like most situations, there is another hand. The other hand is that some sites think that it might be OK to do a minimal job on the basics of rationalization. This can appear to save a lot of time and money. Then they find an advanced alarm package that purports to handle alarm floods. At this point, they call it "good enough." It is not. Without the full rationalization, there is no assurance that the needed alarms are there, there is no assurance that the alarms occur at the right plant condition, there is no assurance that the priority is correct, and there is no assurance that the appropriate alarm response information is provided. In short, all they have is a shaky structure to place an expensive "band-aid" on.

By now, you are ready to shout, "Information overload. I am way too informed. Do I really have to do any of this?" Well, you do not have to do any of this. But if you do take a pass, your alarm system might not be able to fully deliver. It is that simple. Your job is to figure out what is the least complicated and most robust way of ensuring that your alarm design will provide the level of operator support your process and business needs.

8.16 NOTES AND ADDITIONAL READING

Notes

1. Jan Eric Larsson, et al., "A Revival of the Alarm System: Making the Alarm List Useful During Incidents," *Proceedings of the Fifth Topical Meeting on Nuclear Plant Instrumentation, Control, and Human Interface Technology* (Albuquerque, NM: 2006).
2. See Ureason, http://www.ureason.com, and GoalArt, http://www.goalart.com.
3. Engineering Equipment Materials Users' Association, *Alarm Systems—A Guide to Design, Management and Procurement*, EEMUA Publication No. 191 (London: EEMUA, 2007), 91. http://www.eemua.co.uk.

Recommended Additional Reading

EPRI Advanced Alarm Management. Project 39.001.

PART III

Implementing Alarm Management

CHAPTER 9

Implementation

Even though you are on the right track, you will get run over if you just sit there.

—Will Rogers

This chapter is where all the previous alarm improvement work is made real. Here is where the goods are delivered to the operator. Without careful implementation, none of the carefully-done and dearly-paid-for work would be of any use. No design ever ran a plant. No report ever enabled the operator to do a better job. No redesigned alarm system ever worked better than its predecessor unless and until it was actually put into practice—implemented. So, please, take the time to read this and prepare and execute a really good implementation plan.

Plants already have implementation policies and procedures that guide work like this. You will use them, of course. This is not at all about a rehash of project execution guides. Rather, it covers the alarm improvement parts of those procedures that you will use. Let's hit the high spots so we can move on to finishing the job of getting alarm improvement working for your operators.

9.1 KEY CONCEPTS

Scope	Must implement a complete operator area at the same time.
Timing	Implementation can be timed for plant shutdowns or done "on the fly."
Use your simulator	Plants with reasonably good simulators have the ability to both test the alarm system performance *and* train operators on the new alarm system itself.
Implementation is more than changing alarms	All of the plant infrastructure that changes to support the new alarm system is included. The short list includes graphics, procedures, and training.

9.2 BEGINNING

Once the rationalization is complete and all the needed enhanced methods are designed, the job of alarm improvement is nearly ready to proceed to installation. Before the new alarm system can be installed, we must ensure that the plant infrastructure is ready for it. Chapter 9 covers the additional parts that are necessary to turn a good alarm system design into a good alarm system in operation. Included are the management-of-change issues and the importance of follow-up studies to ensure proper operation.

At this point, the new PCS alarm system is only on paper. The essential technical personnel are onboard and standing by ready to implement the design. It is important to select an implementation sequence that is both compatible with the PCS architecture (data highway segmentation and such) as well as consistent with operator responsibilities. No plan should modify just a portion of an operator's span of control. So all of this applies to one or more complete operator areas of responsibility.

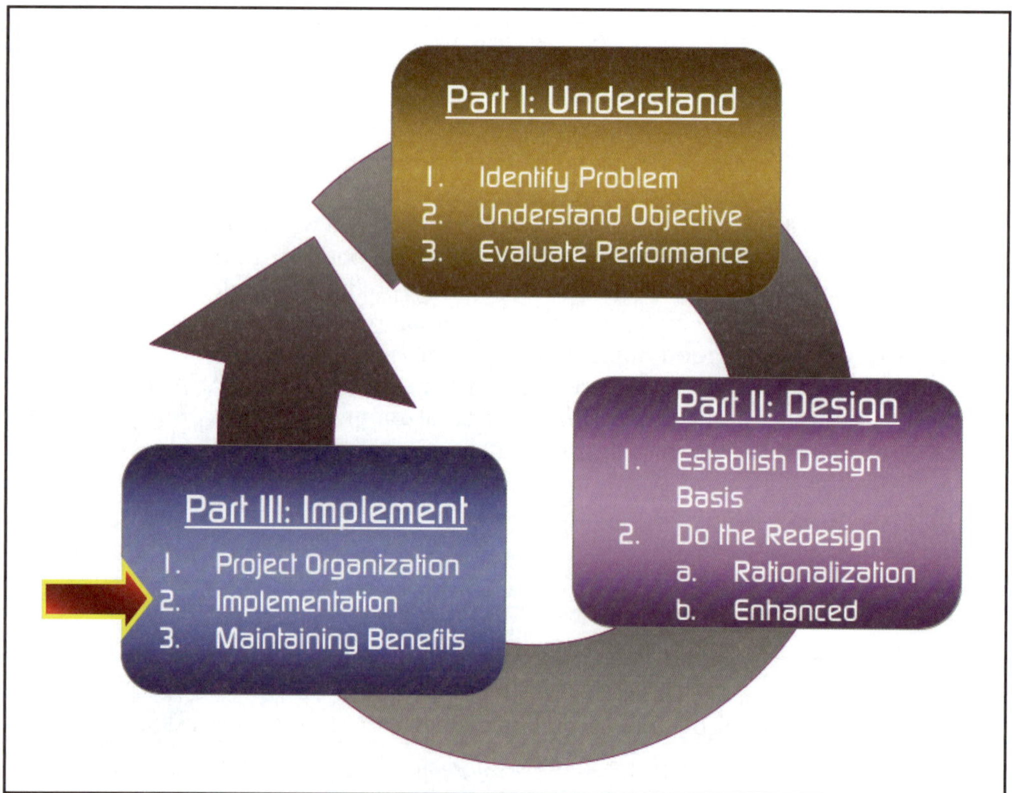

Figure 9.2.1. Alarm redesign roadmap location for implementation

9.3 IMPLEMENTATION STEPS

The steps in the implementation process are really quite straightforward. If you have done project work, you can readily see that they are consistent with your past experiences. The plan to be used for implementation will include the following steps. They are not necessarily in the order you might follow, but they should be in the ballpark.

1. Obtain all of the management of change (MOC) authority to proceed.
2. Translate the design for the alarms into the proper PCS configuration code ready for download.
3. Prepare the enhanced alarm capabilities.
4. Prepare additions and modifications to the process graphics.
5. Prepare and review training materials.
6. Train operators and other personnel.
7. Update all plant documentation.
8. Update and put into practice all other complementary plant infrastructure changes such as maintenance, incident investigations, and so on.
9. Download and activate the new configuration and graphics changes.
10. Replace old procedures and guidelines with the new ones.
11. Download and activate all remaining changes.
12. Verify all of the changes.
13. Review the operability to ensure things are working as designed and that your "as-designed" fulfills the enterprise requirements for good and safe operation.
14. Obtain all MOC approvals to put new design into service.

The order of implementation will be left up to plant project schedulers to make a good fit into the daily plans and planned and unplanned operational interruptions that are part of the enterprise.

Approvals

We started alarm improvement with an alarm philosophy (see chapter 6). As a first step, it needed plant approval. Even though the philosophy had the approval of the plant decision makers, that approval was conditional upon a proper and timely completion. The advanced approvals could not and should not have anticipated all the nuances and special cases that your work uncovered. Moreover, once the work was done and the new design fleshed out, it is likely that it will look different from everyone's initial, and largely conceptual, understanding. To bring it all back into a proper focus, this is the time to formally obtain the approvals. You will use the same approval and sign-off processes that are used for all other plant projects and changes. You will include any special aspects unique to an alarm improvement project.

The first item of business is to develop a way of communicating the entire work to management. This often starts with a report that outlines all the work to be done. From the problem definition (documented by that initial APA, and hopefully not supplemented by a disaster or worse that was experienced along the way) through the final list of alarms and their rationale, this report outlines why and what was designed. It includes a copy of the philosophy and full lists of the rationalization results. The graphic design changes, training changes, and enhanced alarm logic are all there. A detailed "cutover" plan shows that you have clearly thought it all through.

You are now asking the appropriate management for permission to implement. They will want to know that the plan is complete. They will need to understand and accept that in order to fix the alarm system, the fixes will also affect a broad area of the enterprise. There are even policies there for when the operator must shut down (or safely park) his plant. For some plants, the approval process is quite informal and might be done in an office or sitting around a small conference table. For others, the alarm management team will make formal presentations to the interested audience, followed by questions, positioning (yes, people are people), and discussion. There may even be a few follow-up items to attend to before sign-offs. You know how things like this will go in your plant. Eventually, you will get the final approval. Remember, the real approval came when the project was first approved. If you did the work well, their buy-in will have been reinforced during the rationalization and enhanced alarm design steps and at every checkpoint and review meeting they participated in.

Configuration

The configuration is fully specified. This means a complete copy of the "build" file including all alarm settings, priorities, and the like. This is the data that a PCS technician, not knowing what changes you made or why, can unequivocally install in the PCS. This file (or files or other format information) is now ready to load into the PCS.

Enhanced Alarm Features

We know that the set of all approved alarms, with activation points and priorities is not enough to ensure that every alarm activation will always require operator action. There are some situations when the alarm is needed and others when it might not be. The enhanced alarm features extend alarm activation control and management to the point of ensuring proper alarm design. In addition, it provides for the timely availability of operator assistance and guidance. The enhanced alarming is now explicitly coded and loaded so as to be ready to act on call.

Process Graphics and Other Displays

All the changes to process graphics and other user configurable displays and reports are now coded and loaded. The operator graphics, and the situation awareness that it is intended to provide, forms a pillar for safe and proper plant operations. See chapter 12, "Situation Awareness." Almost all major incidents have the loss of situation awareness as a key contributor to the damage. You will *not* want to miss this opportunity. If situation awareness is not significantly enhanced, none of the alarm improvements you can make will be enough. No operator will feel secure. No plant will be safe enough.

Procedures

One of the very important and synergistic benefits of rationalization is the creation of the alarm response documentation. It was done for every alarm that is configured in the new design. The documentation provides detailed support to the operator as each alarm activates. Naturally, you will want to plan to incorporate the availability of this information into the operating practices and procedures.

In addition, the procedures themselves should be modified to include relevant aspects of the alarm system. For example, during a particular plant operation (e.g., start-up, grade change, operating load change, etc.), you will include a list of alarms that might activate and the reasons why. Moreover, in the general operating instructions, you might want to include discussion for how to access the alarm response details and what to do if alarms do not seem to be working appropriately.

Training

Your alarm improvement project will have touched much of what the operator does and works with. Most existing training materials should be updated. If proper training materials do not exist or are significantly deficient, then they will need to be redone. Seriously deficient procedures and training may have been brought to light by the alarm improvement work, but it was not caused by it. Paying for it is a plant business policy. Blaming the alarm work for the costs is a management error.

So get the accountants out of the picture, and ensure that the procedures are updated, documented, and trained for.

Documentation

Modern plants live or die by the quality of their documentation. Let us not go into this easy and very cost-effective way to improve production efficiency and safety. That is the job of plant management. Here, it will be sufficient to make a list, and it is going to be a very long list, of all of the items that alarm improvement has modified or caused to be changed. Take that list and ensure that the documentation for each piece is flagged for updating. Have the appropriate entities update their part. Do your part. Ensure that all the rest is done, and you are done.

Infrastructure

Infrastructure is a tidy little way to point to the need to fill in all the bits and pieces of everything that matters in your plant to get the new alarm design to work. You are well aware that a key skill of instrumentation work is the ability to note and keep track of all the little things that matter. You also know that close is not close enough. More than any other plant practice, instrumentation has much esoterica that appear in nonintuitive ways. Consequently, it is not always apparent that a misplaced decimal point is there, a parameter is missing, or a logical test is wrong or missing. And to make matters even more vexing, there is so much detail and minutia. All must be found. All must fit into the plant schedule, the plant maintenance programs, the plant management style, and all the rest. Here is where it is checklisted and done.

Operability Review

Now the real test of the design is upon us. Most engineering systems are so broadly complex and so extensive in their reach that it is not fully possible to anticipate everything until it has been actually implemented and tested. This is the "test" for the new alarm design and the other infrastructure changes that go along with that design. The philosophy specified what the tests will be and how they will be evaluated. Those tests are done here and evaluated here.

It is not going to work perfectly. You know that. So make the punch lists, fix things, and repeat the tests until everyone who needs to be is satisfied.

Final Approval

A note is entered here, just to keep you in the know. An important critical success factor for obtaining approvals is the extent that you go through to keep the approvers in the loop. Savvy project leaders know that when they establish strong communication lines with both the approvers and the users during the working process, everyone gets a chance to learn what is going on. Just as important as learning, the rest of the plant gets a chance to participate in the decisions that are made and see any real difficulties as they arise. Rather than be presented with a final completed design, they, in effect, become informal partners in the whole design process.

Now is the time to get those sign-offs. Once everyone who needs to be satisfied is satisfied, you will need to make sure it is official. Who needs to sign off? Well, that list comes directly from the plant's MOC for the project. It would have been spelled out in the alarm philosophy. Generally, it should include operations, the new owner, as well as other senior plant officials. Get the sign-offs.

9.4 IMPLEMENTATION

It is already on paper. Now we make it real. The schedule and actual steps are left up to the plant's normal practices for projects of this magnitude. There is nothing special here. You have done it all before. Whether you call it commissioning, start-up, or cutover, it is just that. So figure out what you need to do. Draft the plans. Enlist the help. Schedule everything. And do it.

Simulators and Training

If you have a reasonably good simulation of the plant, this is an excellent opportunity to do a lot of win-win stuff. Use the simulator to verify the alarm design and the configuration downloads. You get to check the new alarm design and try it out to see if everything works. The things that need fixing can be repaired without actually affecting the plant. This would also be a great time to validate the procedures and train operators. Finally, at a time when plants might have thought that their simulator might be an unnecessary digression, here it can come out shining.

Cutover and Testing

Changing the entire configuration for a complete operator's area can be a scary thing. Some plants wait for a shutdown. Others cannot afford the delay. Those that do it on the fly, and it is done that way often enough, carefully prepare and carefully implement. It is usually "rolled in" over several shifts. Each portion is loaded and extensively tested. If problems arise, it is rolled out. Problems are fixed, the repaired configuration is rolled back in, and things move on. It goes without saying that there will be enough personnel on hand to handle the unexpected. Steps are also taken to ensure that the plant is operated at a forgiving operational mode that can provide as much movement as possible to accommodate the unusual and provide enough time for the experienced hands to operate safely.

Moving On

Once implementation is complete and you have validated it, go home and celebrate a bit. Then come back and prepare to follow your plan to keep it working this way for the future.

9.5 CONCLUSION

The chapter has reviewed the steps to move a good design into a good reality. Again, as has been pointed out often, you have seen that alarm improvement is the focused application of good engineering practices, albeit those engineering practices augmented by a body of specific and very useful fundamental knowledge. You have been introduced to all the bits and pieces. You are aware of what you started with and what you would like to end up with. You know what to do and how to do it. But there is more.

Please wait to do the actual work. Let's not rush into that implementation just yet. In the next two chapters, you will learn the methodology of moving the technology into the plant in a way that respects the magnitude of the work to be done and places the working process within the style of your enterprise.

CHAPTER 10

Life Cycle Management

> Concern for man and his fate must always form the chief interest of all technical endeavors . . . Never forget this in the midst of your diagrams and equations.
>
> —Albert Einstein

After implementation, that new alarm system will be humming! Everything is fresh. The problem to be solved was met full on. The prize for doing the job well was clear. Everyone was watching, helping, and wishing success. But later, when the "new" drifts into the "commonplace" and other pressing matters gain the upper hand, that is another situation. This chapter reviews the design and execution of a plan to continually assess the new alarm system's performance (your plant performance) to assure continued benefits. See Figure 10.2.1 for the roadmap location. You will use a program to periodically audit the alarm performance and make recommendations and modifications where indicated. All this will have been fully covered in your alarm philosophy.

10.1 KEY CONCEPTS

Every alarm activation is diagnostic	Once the alarm system is fully functional, every new alarm activation means that something in the plant did not perform as expected: investigate each for cause.
Periodic assessments keep the alarm system performing	Alarm performance assessments (APAs) done after alarm system improvement can be quite beneficial. Some "slippage" always occurs. This is the way to identify it and make repairs.

Interpreted results from periodic assessments are *leading indicators* for plant problems	Regular and careful analysis of alarm system and operator performance can be a powerful tool to uncover operational problems. The list of operational problems includes plant design issues, maintenance issues, operating procedure problems, operator training problems, operator overload, and problems with the alarm system itself.
Day-to-day temporary modifications must be managed	Most plants enforce the approved alarm system configuration by using a formal process to revert unneeded temporary changes back to the approved design.

10.2 ASSESS ALARM PERFORMANCE

Even a newly implemented alarm system will perform somewhat differently from what might be expected. The first task will be to redo the APA, described in chapter 4.

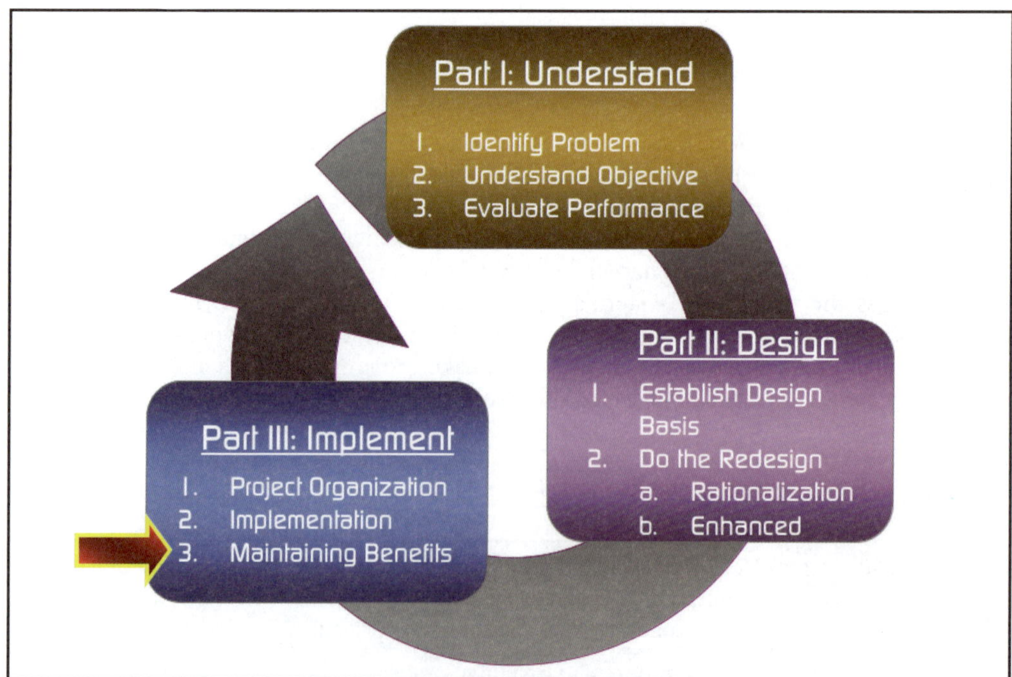

Figure 10.2.1. Life-cycle roadmap location

Initial Assessment

An initial evaluation is performed after the new design has settled in and most of the implementation issues resolved. This is normally done about 3 to 4 months in. A comparison with the earlier APAs as well as against the design metrics will reveal weak spots and other areas for attention. It also illuminates enterprise production improvement opportunities. Alarm floods and showers must be investigated and remediation efforts put in place. All significant process upsets and incidents are reviewed and compared with the alarm management philosophy and the alarm system design basis. These reviews should be conducted soon after the events to maximize information accuracy as well as facilitate early remediation. Most sites find it useful to include this review within their standard incident investigation procedure.

Periodic Assessment

Once the system has been in place for a while, and depending on the plant needs and operational level, most sites find a yearly review comprised of a small APA and a review of the past year's incidents and events to be more than sufficient. Particular attention is paid to continued training, ensuring all documentation is adequate and up to date, and ensuring that all MOCs have been completed.

With a well-working system, the emphasis now shifts to improving operator guidance as well as identifying and implementing enhanced alarm techniques. Particular attention is directed toward improving the operator's feel and use of the alarm system. This way the alarm system can continue to provide a tool for plant operations that enhances operability, improves safety, and facilitates user satisfaction.

Timing of Assessments

You will want to wait until the improved alarm system is in operation for a while before any real assessments are done. You want to ensure that enough time has elapsed for the rough edges to become worked smooth and everything is working as well as it can. For some plants, it may take a few months; for others, nearly a year. It all depends on how well things are working, how much time is available for this work, and who will do it. Certainly it is well beyond the break-in time. All those early problems would have been found and remediated as part of the commissioning or soon after. At this point, you are way beyond that. The exact timing and protocol was specified in your alarm philosophy. You will repeat the activities in this section often during the course of operation, for sure, at one or more planned times during the operating year. Additional assessments should be done after any serious mishap or as part of safety certification programs, whenever they occur.

Collection of Data

The data here are generally the same data categories that you collected in the very beginning of your work on alarms. Using the same lists, we can obtain the data for the plant after rationalization. It is not necessary to remind everyone that we are talking about data from the same operator area(s) that were rationalized earlier, is it? Included in the items we will want to examine are the following:

- Operator journal and log data
- Plant production data, including rates and quality assessments
- Incident reports
- New equipment or replacement equipment placed into service
- Other plant modifications
- All MOCs filed since the last evaluation
- New procedures or revisions made since the last periodic alarm review
- Alarm key performance indicators (KPIs)
- Alarm activation data and other data from the real-time event journals
- Changes in the configuration

Every Alarm Activation Points to Opportunity

Notwithstanding a program to periodically assess alarm performance, once the alarm system has been fully implemented and is working well, each alarm activation means that something went amiss.[1] Yes, there is no reason an alarm should ever have to activate if everything is performing. Before we go off on a tangent here about blame, let's make it clear that this point has nothing to do with anything except making things better. With that said, let's get to the point. Each alarm activation is now an opportunity to find out what went wrong and fix it. What might have gone wrong? Well, you already know the list, but here it is again:

- Equipment in the plant is either broken or degraded.
- Raw materials coming into the plant may have changed.
- The process might have inherent weaknesses that were just exposed.
- A procedure might not be as correct as needed.
- Some particular situation was not easy enough for the operator to see coming.
- This situation may not have been appropriately trained for.
- The process might have been operating in a regime that was not fully understood.

The list continues, but you get the idea. Many plants make it a daily activity for the process engineer or senior operator to look over all alarms that have activated in the past 24 hours and examine them for cause. Many will just be the "noise" of normal operation. While the alarm is important and needed, that "once in a while" misstep happened and set it off. No change might be needed. But others will definitely point toward improvement opportunity—a built-in indication that is always reliable and will almost always tell the truth.

10.3 INTERPRETATION OF PERIODIC ASSESSMENTS

Now that you have obtained the data, it is time to see what it means.

Evaluate

You can expect the alarm and event data to be obtained using the same assessment tools and procedures that you earlier used for your APA and any other alarm performance evaluations that were done before the project was initiated. You perform similar sets of analyses to the ones you did then. Also plan on adding the new ones that were called for in the philosophy for follow-up reviews. Consider doing other additional evaluations that might be useful. Identify any new nuisance alarms. Find any alarms that are standing, that are stale, that take a long time to clear, that the operator seems to take a long time acknowledging, and so on.

With this assessment information at hand, identify alarm problems, process problems, training problems, and any other alarm-related issues or concerns that suggest modification or improvement of the alarm system configuration. Pay particular attention to the alarm system performance during significant upsets. Did the alarm system work as designed? Did the alarm system help or hinder the operator? Did the incident reports, incident investigation reports, and all HazOps appropriately identify alarm concerns and performance issues?

Look for Added Benefits

Poor alarm performance before alarm redesign was what you needed to justify to everyone that work was needed. After redesign, everything works much better and the plant should be significantly less vulnerable. And that is a fine reason for doing the work. But please take note that there is a bit of magic waiting for us. Now, after the alarm redesign has been implemented, each and every alarm activation has meaning. It is this new window for observation that we use to gain more insight into plant operations.

Most plants have someone (often the process engineer) looking after each alarm activation. Sometimes, reports are e-mailed. They are discussed in daily or weekly operations meetings. They might be just informally investigated on a casual basis. But, in one way or another, they should be examined for cause and purpose. You will ask why the alarm activated. Did it suggest a procedural or operating protocol problem? Did it suggest a maintenance matter? Did it point to a raw material or utility problem? Did it reflect a specific process design issue? Was it caused by an alarm design problem, missing issue, or other configuration problem? Each item is examined and changes noted or made.

Modify and Repair

Take the results of these evaluations, update the philosophy where needed, and make all other alarm improvements and modifications (consistent with the philosophy). It is done exactly the same way this information was used during the initial alarm improvement project, except now, the alarm system has been working in its improved design, so everyone is much more alarm-savvy.

Monitor and Enforce

So far, all this chapter has done is to open up the floodgate for changes and modifications to what had been a rigid, fixed, dependable, and fully documented alarm system. Now we move away from tracking and understanding the real-time operational data into the configuration part of things. This is the alarm design. This is what the alarm system had been configured to do. And, supposedly, this is what you want it to continue to do. We now examine the configuration to find out everything that is different from what would be considered the norm. Monitoring is the way you find any changes. These changes might have been a welcome addition at the time or even unauthorized or worse. They must not persist beyond the time of need. Consequently, your plans will ensure that all the changes keep their relevance. Enforcement is the act of restoring alarm parameters to an accepted norm or baseline state. Section 10.6, "Enforcement," will cover this topic in detail.

Nuisance Alarms

You remember the phrase "the best laid plans." Trite, yes. True, also yes. So even for the best designed and implemented alarm system, things get missed and things change. Do the alarm activation analyses on a periodic basis. Identify the bad actors and fix them. However, this is not as straightforward as it might appear. Be sure that you identify the root cause of each candidate nuisance alarm. Root causes were discussed in some detail in chapter 4. There is a real problem that is often seen by plants who thought they were doing a good job at fixing their nuisance alarms. What they found out was that those same nuisance alarms that were fixed a while back reappear. No, the "fixes" were not coming undone; it was that the fix was too tailored to the specific plant and alarm conditions.

For example, they might have noticed that the deadband needed to be enlarged a bit to eliminate the effects of process noise on alarms. So it was enlarged. But later, at another plant operating condition, the process noise was greater and hence the deadband was insufficient. But merely enlarging the deadband might not be the wise choice either.

If it were to become too wide, then there might not be sufficient "room" between the alarm clear and normal operating conditions. This would in effect keep an alarm active when the plant was actually well away from any abnormal operation. Again, the important message here is that all alarm system tune-ups should be carefully done. Refer to the philosophy for guidance; keep ever mindful of the sage advice of "avoiding a permanent solution to a temporary problem." Nuisance alarms often can be in the way for only a very narrow portion of the operational spectrum. If they become annoying, it might be better practice to have the operator just shelve it for a while.

Alarm Creep

Alarms themselves appear to be very useful to the operator. No matter how much thought might go into the selection of an alarm or the full understanding of an abnormal situation, when some small part of the operation bothers an operator and he thinks that having an alarm would be very convenient, adding another alarm might happen. Multiply this event a dozen times over a month, and it is not much of a stretch of the imagination to see more and more alarms added to the configuration. After all, it is just one alarm here and one alarm there. We have just described how alarm creep works. Figure 10.3.1[2] illustrates how such situations usually unfold.

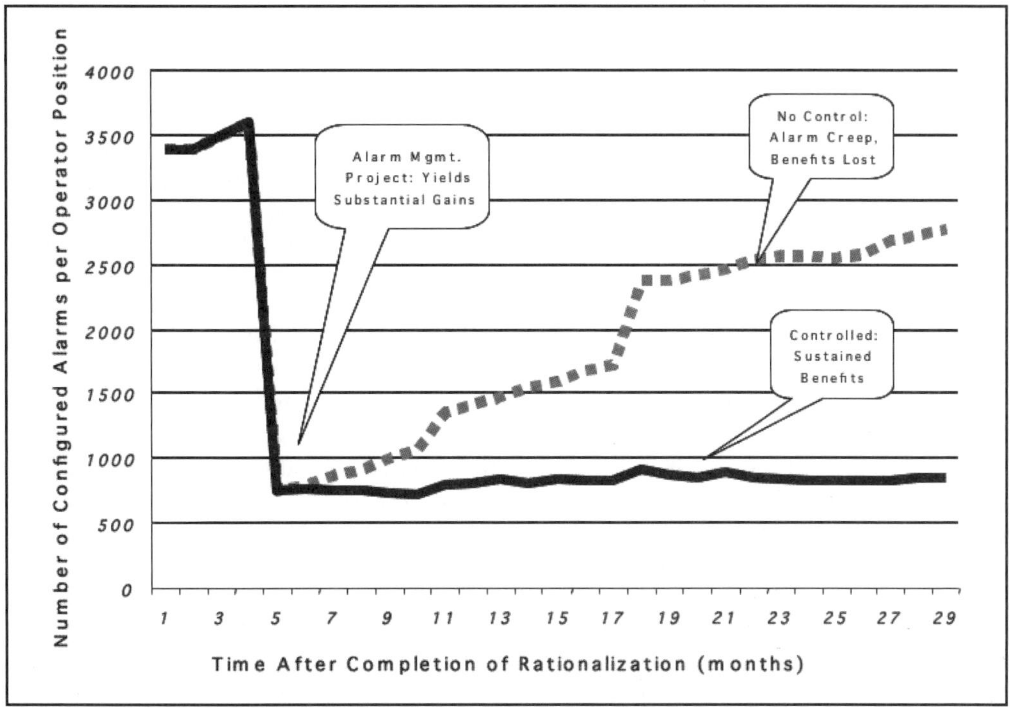

Figure 10.3.1. Alarm creep without proper controls

Keep in mind that the "bad guy" here is nothing nefarious; it is just human nature. Follow the plant requirements and MOCs to keep things working the way the plant has asked them to be.

Adding and Removing Alarms

Not to spoil the moment, but let me say right now that solely getting down to a low number of alarms is not my idea of a good alarm system. Alarms have a purpose. We need them. We just want to be sure that they are proper and do their job. So getting rid of an alarm is a good idea only if it fails to meet the need and standards you set forth. "Yes, Virginia, it is okay to add alarms." If they are needed, follow the process and add them. Just be careful it does not go to your head.

10.4 ADVANCED INTERPRETATION OF PERIODIC ASSESSMENTS

You have done the periodic alarm performance assessments. You have made the normal repairs and modifications. It is time to find out what more these results might bring to light. Earlier you examined individual alarm activations. Now we turn to a synergistic view of alarm performance interpretation.

Nomenclature and Design

There are three main players in this drama: the plant, the alarm system, and the operator. Figure 10.4.1 illustrates our stage and explains the basic key for the graphical depictions. We focus on the action between the players. The action can be either heavy or light. Basically, heavy interaction suggests a heavy load, which is undesirable. Light interaction suggests a light load, which is beneficial.

Interaction "A": Plant with Alarm System

This aspect tracks the number of abnormal situations. This is not the actual number of alarms (see Interaction "B," next). An abnormality might be a distillation tower that is flooding. A flooding tower will likely have abnormal temperatures on several trays; it might have too much recycle, and it might have too little heat input into the reboiler or too low a temperature in the inlet feed—all at the same time. However, it counts as only one abnormal situation.

Unfortunately, this is a bit difficult to actually track using most of the conventional alarm analysis tools. It might be done by examining how the enhanced alarming is controlling alarm activations in the areas of related alarms, consequential alarms, alarms that are dynamically suppressed (or inhibited or disabled), and any other similar dynamic filtering. Since analyses of this kind are done after alarm redesign, it is often done manually by examination of specific actual cases as they occur.

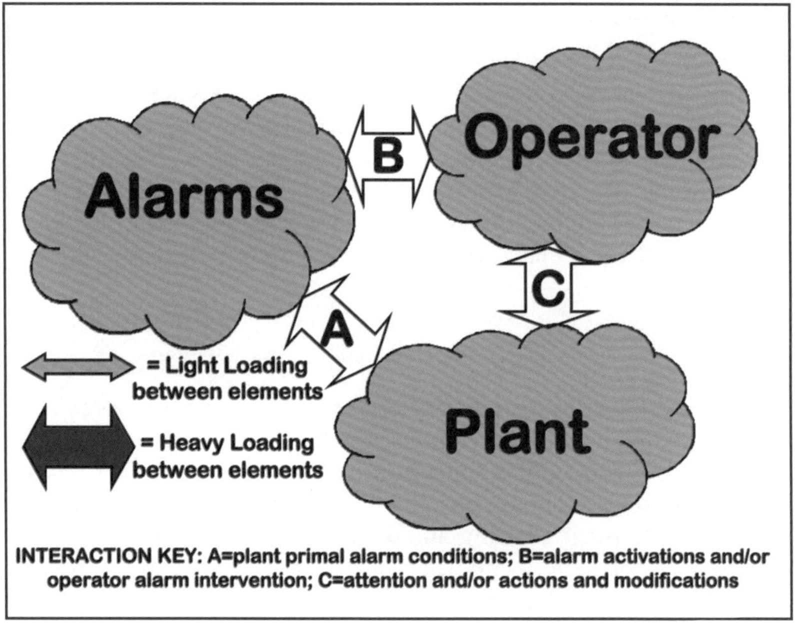

Figure 10.4.1. Key to figures for alarm design interactions

Interaction "B": Alarm System with Operator

Here we are looking for a measure of how much stuff the alarm system is throwing at the operator and how often he might duck the load. This is generally straightforward to determine. It is simply measured by any useful combination of running averages of alarm activations, alarm flood periods, and the like. It includes a penalty weighting for long running average time-to-acknowledge and low average acknowledgment ratios. Simply put, we want to know how hard the alarm system is pummeling the operator and whether he is playing the game or ignoring most of it. A modest number of alarm activations and floods that are all totally ignored will count as a high interaction, while the same modest number of alarm activations and floods that are all worked consciously would be a low interaction.

This is also a good place to recall how the implications of alarm loading metrics affect available operator resources. Recall that if an operator is subjected to more than five or six alarms an hour, then all the operator's available time is used to deal with those alarms. Nothing will be left over for anything else. For plants designed for a lot of operator activity to keep things running, those extra duties will directly compete with managing abnormal situations. This is a "zero-sum" competition—both cannot be handled. Either needed plant changes for normal operations get neglected or alarms get neglected. Either way, the plant will not be able to maintain good operation.

Interaction "C": Operator with Plant

Here we track both the number of actions the operator takes to manage his plant and the ratio of actions to alarms. A low ratio of actions to alarms will increase the measure of operator interactions inasmuch as the operator should be working but is somehow off the job. Remember, our alarm system has been redesigned so that every alarm requires action. In this case, modest rates of alarm activations that mostly are not actioned by the operator also count as a high rate of interaction. Why? Well, we really should not let the fact that our operator is not doing the job masquerade as a well-running plant, should we?

Value

Specific operational difficulties or deficiencies are called out in the cases reviewed. You will find that poor maintenance is a recurring theme. You will see operating procedures and operator training are also often contributors. There is another very important contributor: situation awareness. Simply put, if an operator is situationally aware, the operator is aware of what the plant is doing and is likely to do. Alarms are a small part—we actually want the operator to be aware before alarms activate. Poor situation awareness is not explicitly included in the cases, even though it is a vital aspect of good plant operation. Refer to chapter 12 for full, complete coverage of this topic. Please plan to spend time there. It is not included here because it is actually an important contributor to every case (except Case 1, of course).

Poor plant design is another recurring theme. This is not just a place to do finger pointing at the plant designer. This category is as broad as it is important. Included in the poor plant design list are the following:

1. Plants that were designed for one product and are now used to make others
2. Plants that are poorly integrated or too tightly integrated
3. Plants lacking appropriate automation (too few control loops, poor sensors and transmitters, obsolete instrumentation)

Remember the iceberg effect (chapter 1)? Alarms and alarm systems cannot be asked to patch over missing and inadequate plant infrastructure. The items listed above are important. The only job of alarms is to bring the operator to a place that enables the management of what happens that cannot be managed any other way than through the operator's efforts. Poor plant design is not one of them.

Cases

There are eight cases discussed. Each case has a unique pattern of interaction loading. For each, the results are summarized in a comparison table. For each case under the heading "Problem or Defect" in the tables, you will see discussions for both a rationalized alarm system and one that is not. Only the rationalization system implications will be discussed. Each case brings to view opportunities for steady improvements to the operational theater. Those who utilize Six Sigma improvement practices take note: this methodology provides an unusually direct and focused opportunity to pin down areas in an operation to examine and remediate.

Keep in mind that these analyses supplement the surgical analyses of individual alarm activations discussed in section 10.3, "Interpretation of Periodic Assessments." Those earlier recommendations were very highly focused and directed specific consideration of an individual alarm and what might it say when it activates regarding other related infrastructure items. The cases in this section can often provide evidence for a broader look at the operations improvement opportunities.

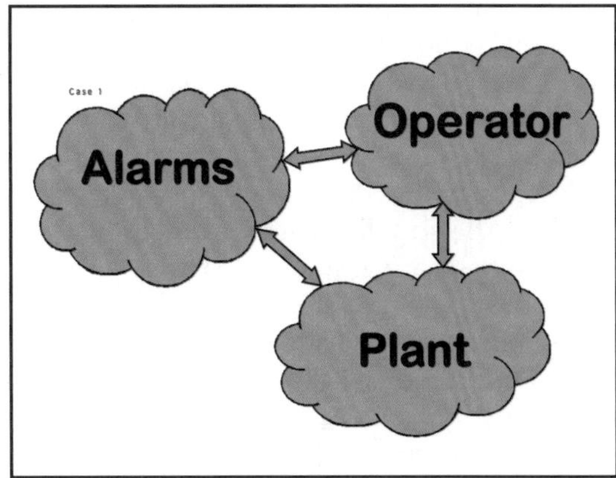

Figure 10.4.2. Case 1 diagram

Case 1: Low Abnormal Situations, Low Operator Load, and Low Plant Management Requirements

This is what we hope for when plants are designed and operated. You will want to plan for continual, periodic assessments to ensure that this performance does not slip.

	Problem or defect		**Interactions between elements**		
CASE 1	**(Supposedly a) well-designed alarm system**	**No design or poorly designed alarm system**	**"A" plant operations — Alarm system**	**"B" alarm system — Operator**	**"C" operator — Plant operations**
	Well-designed alarm system; well-designed and maintained plant with well-trained operator; [or] small and simple-to-manage plant with competent operator	Small and simple-to-manage plant with competent operator	Low	Low	Low

Table 10.4.1. Case 1 interpretations

Case 2: Low Abnormal Situations, High Operator Load, and Low Plant Management Requirements

In this case, the plant has relatively few abnormal situations and the plant requires very little direct attention by the operator to run well. However, the alarm system is initiating a very large number of alarms.

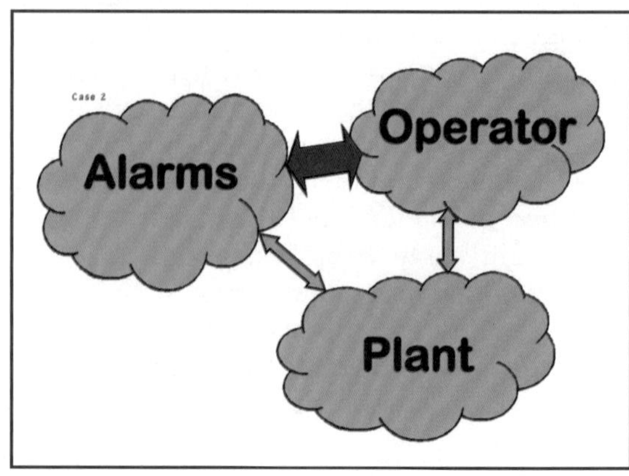

Figure 10.4.3. Case 2 diagram

This situation suggests that even though the number of abnormal situations is not that high, there are large numbers of individual alarm activations. We normally manage this situation by reassessing the alarm philosophy to see if it properly specified how the specific causes for each important abnormal situation should be handled. It might also be that the philosophy may be inadequate for those situations or might not have addressed them at all. On the other hand, the additional alarms can be important in their own right in some situations but be redundant for other unrelated situations. In this case, advanced alarm processing should be considered to ensure that each alarm is permitted only for the situation for which it is needed. You get the idea: the enhanced alarming needs to be improved.

	Problem or defect		**Interactions between elements**		
CASE 2	**(Supposedly a) well-designed alarm system**	**No design or poorly designed alarm system**	**"A" plant operations — Alarm system**	**"B" alarm system — Operator**	**"C" operator — Plant operations**
	Overly redundant alarm system with insufficient enhanced alarm design; well-designed and maintained plant; competent operator	Overly redundant alarm system and/or one with insufficient enhanced alarm design; well-designed and maintained plant; competent operator	Low	High	Low

Table 10.4.2. Case 2 interpretations

Case 3: Low Abnormal Situations, Low Operator Load, and High Plant Management Requirements

In this case, the number of abnormal situations is low and the number of alarms is also low, but the operator is working quite hard to keep the process operating. Before you ask, let me answer: the operator, having to work hard to keep things running, does not require the plant to move to abnormal situations that cause alarms to activate. After all, that was the purpose of alarm redesign. Plants must be able to move about in their own natural way for at least some levels of movement. Alarms are not being used for wake-up calls. The situation must become abnormal enough to require the operator to intervene if he has not already done so. A good example of this situation is plants that require operators to perform a large number of "manual" activities. Operators might have to manually introduce additives or watch a reaction for a color change and then switch on

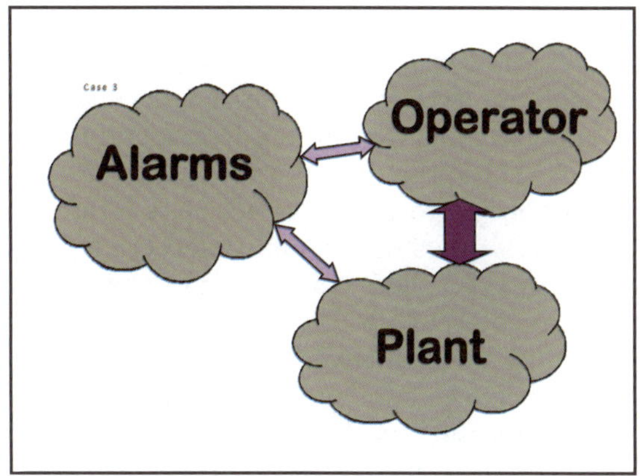

Figure 10.4.4. Case 3 diagram

	Problem or defect		Interactions between elements		
	(Supposedly a) well-designed alarm system	**No design or poorly designed alarm system**	**"A" plant operations — Alarm system**	**"B" alarm system — Operator**	**"C" operator — Plant operations**
CASE 3	Important and useful alarms are missing, resulting in excessive operator attention to the plant, [and/or] poorly maintained plant, [and/or] simple, but hard-to-manage plant indicative of weak or poor design, [or] too large a plant for one operator to manage	Small, but hard to manage plant, [and/or] important, alarms are missing, resulting in excessive operator attention to the plant, [and/or] too large a plant for one operator to manage	Low	Low	High

Table 10.4.3. Case 3 interpretations

a dump valve to send the product on its way. This situation is typical of some batch manufacturing operations found in food and pharmaceuticals.

As long as we are on the matter of increased operator action just to keep the plant operating reasonably normally, let us recognize that this work takes operator resources. There will be less left over for managing the really abnormal situations that really require operator intervention. Plant managers must recognize that saving money on automation might cost money by having less operator resource to manage upsets—upsets that can cost the enterprise dearly.

Returning to this case, there are a number of things we should carefully examine. The first is operator loading. Often the territory (amount of process, number of loops, etc.) that is assigned to any given operator is determined by how specific plant units can be naturally split up (divided between different operators). Splitting a highly related individual process between two separate operators should never be considered. The way to split between units should be guided by how the units interact with each other. Refer to the discussion on decomposition from chapter 7. The way things sometimes end up, some operators can be lightly loaded (OK for them), but others might be overloaded (not okay for them or the plant).

Check to see if a significant number of important alarms were missed during the initial rationalization work. If so, this suggests that the alarm system may not have been properly designed or appropriately implemented during the original project work.

Finally, our performance problems might just be a case where the current plant design is insufficient to promote good operation. It might not have been properly designed or properly built, or the procedures for operation are inadequate for the job. A careful review by experts in design should prove very beneficial.

Case 4: High Abnormal Situations, Low Operator Load, and Low Plant Management Requirements

This case represents a situation where the plant encounters in a large number of abnormal situations, yet the alarm system produces very few alarms and the operator appears to be required to do little to the plant.

This is a very unusual situation. What is happening is that the plant is going awry but the operator is uninformed, and hence is unaware of problems. This situation will continue until something much more serious occurs. When that happens, it will be too late for the operator to do anything effective.

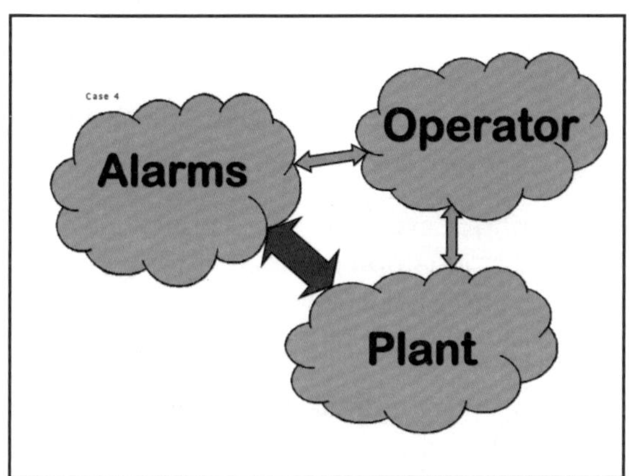

Figure 10.4.5. Case 4 diagram

	Problem or defect		Interactions between elements		
	(Supposedly a) well-designed alarm system	**No design or poorly designed alarm system**	**"A" plant operations — Alarm system**	**"B" alarm system — Operator**	**"C" operator — Plant operations**
CASE 4	Well-designed plant, but with an excessive number of configured primary alarms and an alarm management process that relies too heavily on enhanced alarm filtering and management logic	Well-designed plant, but with an excessive number of configured primary alarms that are suppressed with crude logic that often makes the operation vulnerable to true problems	High	Low	Low

Table 10.4.4. Case 4 interpretations

First, check to see if the numbers of alarms might be abnormally low. A low number of alarm activations is good. A very low number of configured alarms is not necessarily good. Alarms serve a useful purpose. Too few configured alarms place a very heavy load on the operator to keep on top of things. Check also to see that the configuration details are appropriate. If the number of alarms seems appropriate, then the enhanced alarm functionality is not working. Enhanced alarming is used to reduce nuisance alarms and ensure that all alarms that activate have a purpose. In this case, that functionality is actually preventing alarms that need to activate.

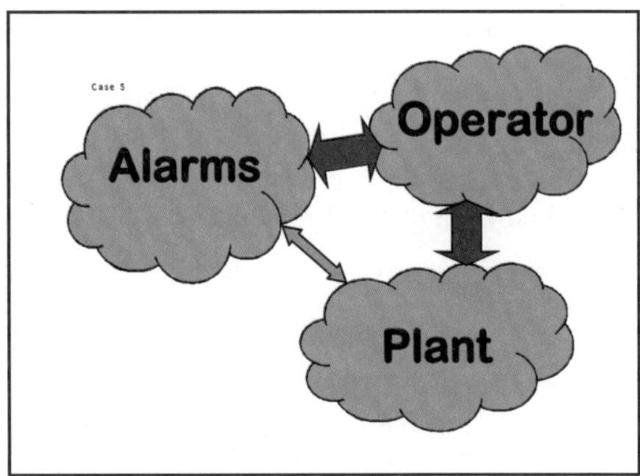

Figure 10.4.6. Case 5 diagram

Case 5: Low Abnormal Situations, High Operator Load, and High Plant Management Requirements

For this case, the plant experiences very few abnormal situations; however, the alarm system produces a

CASE 5	Problem or defect		Interactions between elements		
	(Supposedly a) well-designed alarm system	No design or poorly designed alarm system	"A" plant operations — Alarm system	"B" alarm system — Operator	"C" operator — Plant operations
	Important, useful, and/or critical alarms are missing and insufficient enhanced alarm functionality, resulting in excessive operator attention required to operate the plant	Small but hard-to-manage plant, [and] insufficient enhanced alarm functionality and/or important, useful, and/or critical alarms are missing, resulting in excessive operator attention required to operate the plant	Low	High	High

Table 10.4.5. Case 5 interpretations

high level of alarms and the plant requires a significant level of operator interaction and attention.

As the table suggests, examine the configuration and design to see if important alarms have been missed. If not, then there is a good possibility that the enhanced alarm functionality might be a bit too thin to do the proper job.

Case 6: High Abnormal Situations, Low Operator Load, and High Plant Management Requirements

This case represents a plant with a large number of abnormal situations that produce a small number of alarms, yet the operator is heavily engaged in managing the plant.

This case points rather directly to enterprise issues:

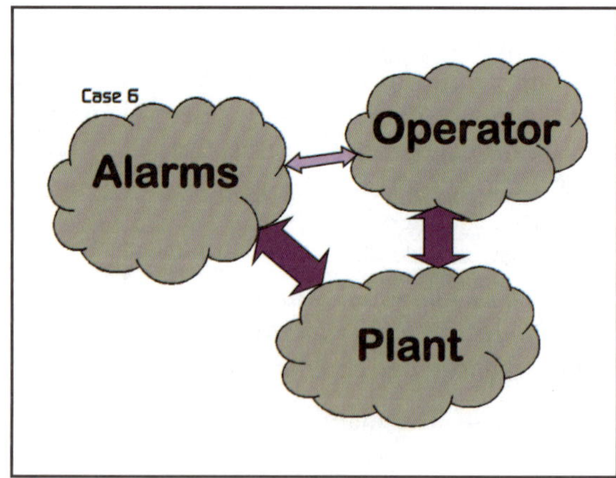

Figure 10.4.7. Case 6 diagram

	Problem or defect		Interactions between elements		
	(Supposedly a) well-designed alarm system	No design or poorly designed alarm system	"A" plant operations — Alarm system	"B" alarm system — Operator	"C" operator — Plant operations
CASE 6	Very poorly designed primary alarm system, but with effective enhanced alarm management capabilities, [and/or] Poorly designed and/or maintained plant, [and/or] poorly trained operators, [and/or] too large an operating area for one operator	Hard-to-manage plant with poor alarm configuration so that the operator virtually ignores most alarms. This plant is vulnerable to an eventual disaster, should the process be of a dangerous nature.	High	Low	High

Table 10.4.6. Case 6 interpretations

1. The original plant design is poor and an excessive number of alarms were used to try to accommodate for that deficiency.
2. The original plant design is good, but the plant is poorly maintained, or the alarm response information is either weak or incorrect.
3. The original plant design is good, but the level of instrumentation and/or automation is poor.
4. The plant is too large or difficult for one operator to manage.

Work down the list and locate the problems.

Case 7: High Abnormal Situations, High Operator Load, and Low Plant Management Requirements

This case covers the situation where the number of abnormal situations is high and the number of alarms is high, but the plant requires a low level of operator attention.

This plant is in a dangerous place. Even though the operator is not interacting with the plant very much, he needs to. Unfortunately, the alarm system is either poorly designed or the plant is poorly designed and/or poorly maintained. The bottom line is this plant is

CASE 7	Problem or defect		Interactions between elements		
	(Supposedly a) well-designed alarm system	No design or poorly designed alarm system	"A" plant operations — Alarm system	"B" alarm system — Operator	"C" operator — Plant operations
	Well-trained operator, [and] very poorly designed alarm system, [and/ or] poorly maintained plant, [and/ or] too large an operating area for one operator	Small or simple plant with poor alarm configuration so that the operator virtually ignores most alarms. This plant is vulnerable to an eventual disaster, should the process be of a dangerous nature.	High	High	Low

Table 10.4.7. Case 7 interpretations

experiencing regular, serious problems and is headed for even worse problems down the road. The only real difference between this case and the next is here the operator has given up, where the operator in Case 8 is overworked.

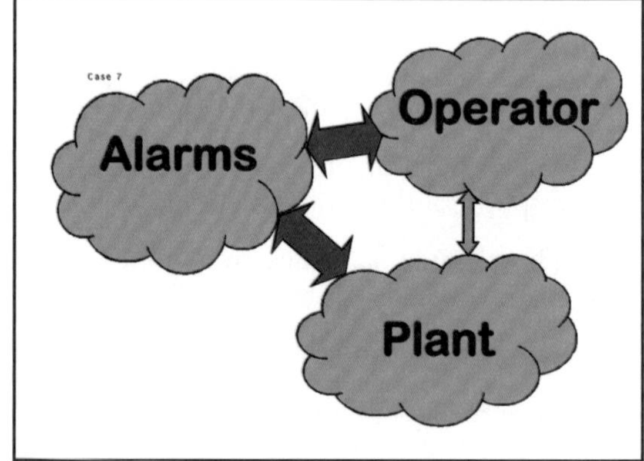

Figure 10.4.8. Case 7 diagram

Case 8: High Abnormal Situations, High Operator Load, and High Plant Management Requirements

For this case, everything that could go wrong does go wrong. The plant is experiencing a large number of abnormal situations, the alarm system is producing a large number of alarms, and the operator is overloaded by the operational demands of the plant. The first alarm redesign has failed. The design and principles that should have been used are a best practice. EEMUA and the rest is really good stuff. If the plant alarm system was redesigned, I am afraid that something went horribly wrong.

What you have on your hands is a serious design and/or implementation failure. It is unlikely that minor repairs can make things right. Please take the time to do a careful reexamination of the entire work. If you used outside firms to assist with the design or implementation, consider that their work would need independent evaluation. If the work

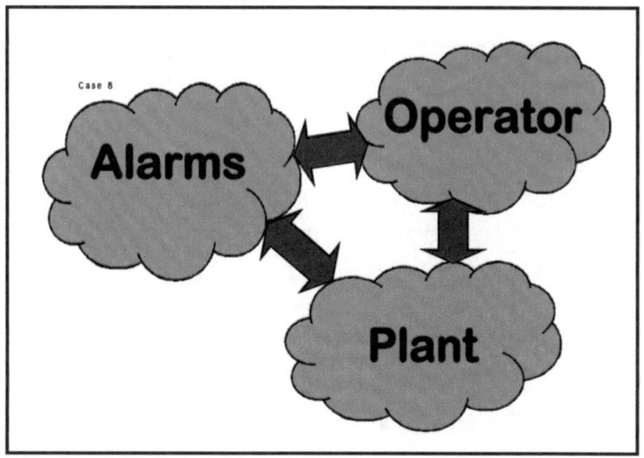

Figure 10.4.9. Case 8 diagram

	Problem or defect		Interactions between elements		
	(Supposedly a) well-designed alarm system	No design or poorly designed alarm system	"A" plant operations — Alarm system	"B" alarm system — Operator	"C" operator — Plant operations
CASE 8	Very poorly designed alarm system, [and/or] poorly designed and/or maintained plant, [and/or] poorly trained operators, [and/or] too large an operating area for one operator	Hard-to-manage plant with poor alarm configuration and high alarm load on the operator without providing any assistance for plant operations. This plant is vulnerable and likely has a recurring history of accidents or worse, should the process be of a dangerous nature.	High	High	High

Table 10.4.8. Case 8 interpretations

was done almost entirely in house, it is likely that there was insufficient experience to design or undertake a task of this complexity and magnitude. Consider bringing in an outside expert to advise your team.

Alarm improvement is not "rocket science." It is just good engineering, competently designed and carefully implemented. Some big things must have fallen through the cracks and been missed until it was too late. However, please take heart. Alarm improvement can be done at any site. The resulting design is doable and will work. Understand the methodology and process and take courage and begin again.

10.5 STATISTICAL PROCESS CONTROL AND ALARM MANAGEMENT

Statistical Process Control (SPC) is a well-developed methodology for quality management. In this section, we explore its relevance to alarm management and alarm improvement monitoring. From the outset, as of this publication, the author is unaware of any serious work or, actually, any accessible published work on this topic. The discussion in this section is a beginning.

Background

SPC is a methodology for monitoring repetitive events (processes) to ascertain whether these events are usual or unusual. It was pioneered in the 1920s by Walter A. Shewhart (Shewhart Charts) and later by W. Edwards Deming, one of the early giants of modern quality control. SPC attempts to identify two types of changes. The normal changes, called common causes, are attributable to the very nature of the events under study. While it is possible to manage the common causes of variation, doing so is normally outside the realm of practicality and need. Such causes include the subtle effect of weather on manufacturing production and the normal variations of raw materials, even carefully screened and managed raw materials. They are an expected part of life. These are all usually small and vary in a normal distribution pattern around their nominal values.

On the other hand, the unusual variations, called special variation, are attributable to growing defects in the process itself. Examples include drifting sensors, wearing of mechanical components, raw materials that are being contaminated, and similar. They are unexpected. This does not mean that they are unreasonable or that they are not a part of the reality of things; it is just that they come when we are not necessarily expecting them. These variations, while small initially, usually vary in a way that, if uncorrected, means the process is producing product that is or will soon be undesirable. Identifying the existence of these variations that are outside of acceptable tolerances is the reason for SPC. Once it is known that a given variation is too much, remediation efforts are initiated. The usefulness of this method is its ability to identify an unacceptable variation before its effect on the process causes direct defects in its quality.

Relevance to Alarm Management

Suppose one were to chart alarm performance using the methodology of SPC. Would these charts be able to identify deviations that suggest special cause alarm variations? Would the suggested special cause variations be linkable to alarm problems? And finally, if all this is possible, does SPC charting provide any real benefit in a way that alarm system improvement might be specifically identified? The answers to these three questions are yes, yes, and maybe. Clearly, the story is in the "maybe."

We continue this discussion with the understanding that we are doing SPC charting only on alarm performance parameters. For a full description of the possible alarm performance parameters we might consider for charting, refer back to chapter 3 on alarm performance. Before we get too far into this, we must ask whether these existing alarm performance indicators are good enough to be capable of identifying the "out of the ordinary" to be relied on as reliable triggers for action. If so, might they be better at that job than a more general tool like SPC? The current thinking is that existing alarm performance indicators appear to be effective. They provide a clean distinction between the normal alarm activities versus alarms that appear to be of little to no value. Therefore what we really might want to know is whether SPC can bring more to the table.

Before Alarm Improvement

Conventional alarm performance monitoring before alarm improvement is of little value except for identifying bad actors. And that is an easy task. Therefore, by analogy, any alarm performance SPC charting done before alarm improvement will probably not be productive beyond identifying the same bad actors either. If you recall the discussion in chapter 1, "Meet Alarm Management," in the section on Six Sigma, you understand that with the limited exception of bad actors (and improving them is a by-product, not a direct goal anyway), alarm performance problems do not point to any useful alarm improvement action. Without a comprehensive alarm design in use, you should already be convinced that the current alarm design is inadequate. All charting will do is to confirm that fact. The real work of alarm redesign will be based on going back to the fundamentals and building the appropriate infrastructure to do the job.

The bottom line here is that SPC charting is not effective for a process that is not in control. An alarm system before redesign is generally considered to be very much "not in control."

After Alarm Improvement

Here we have an improved alarm system. In this case, all alarm activations should be useful and acted on by the operator. If alarm activation SPC charting can be designed to show only the unusual alarms (from the alarm system point of view), then we are on to something. The issue is whether we can effectively separate process and production problems from alarm performance problems. For the former, we can use SPC charts on process and production parameters. For the latter, we are considering charting alarm performance parameters. This is the heart of it all.

Changes in alarm "charts" will probably be root caused by the following:

1. *Process is not especially variable, but alarm system is inadequacy designed and therefore alarms chart as problems.* Basically, this situation represents alarm charting where the appropriate alarm improvements have not been made. We are back to the "Before Alarm Improvement" situation earlier. Same conclusions hold now as then.
2. *Process is more variable, and alarms are fully functional and effective.* In this case, charting alarms will only serve to suggest that there are process management issues. However, unless you already know this, you are likely to become confused as to the true meaning of the alarm charting results. It will not be possible to separate process problems from alarm problems. SPC charting would appear to be of little use and actually might confuse matters.
3. *Process is more variable, and alarm accommodation problems appear as a result.* In this case, we have a number of possibilities. Our alarm system is well designed but is not adequately tracking plant state, so it fails to adjust properly to plant operational changes. Either (a) the plant has states that are not possible to track, and hence it is not possible to advise the alarm system so it may accommodate, or (b) the plant has trackable states, but the alarm system does not adequately incorporate that ability. Now we are left with an uncovered problem but no rational remediation pathway since it remains unclear how to construct an SPC chart of alarm issues that is capable of separating the alarm part from the process part.

Again, the bottom line here is to find a way that SPC charting can identify problems with the alarm system that will alert us to the issue that there is something wrong with the alarm system, regardless of what the process might be doing. So far, this is an open question.

Guidance

Now

One of the potentially useful benefits of using SPC might be to employ it to identify the level of interactions between the plant, the alarm system, and the operator. Recall from section 10.4 that we have a requirement for identifying the cases of periodic assessments. Here we must categorize each interaction as "light" or "heavy." Think of light as being usual and heavy as being unusual. Now we might just have a way to do that identification.

Down the Road

If you carefully charted the important process parameters and the useful alarm performance parameters, you might be able to identify situations where the process was in control but the alarm system was not and where the process was not in control but the alarm system was. For the former, you would have indication that the alarm system either needed a tune-up or was inadequate. How would you tell? For the latter, you might have an indication that the alarm system was either really robust or just unresponsive. How would you tell?

You might also identify cases where both charts were not in control. Here the alarm system could be just fine and simply responding to the process—more process problems, more alarms to assist the operator. Or the alarm system might not be of much value and simply showing its true colors. How could you tell? Since answering these questions poses more difficulty than following a more conventional alarm performance monitoring approach and comparing the results to the alarm system design, it is suggested that SPC be dedicated to charting the process and alarm performance be used for just that: alarm performance monitoring.

Before we leave this topic, let's observe that not far down the road, there may well be important and useful ways to use SPC charts for alarm performance monitoring. That we are not sure how to do it now should only serve to encourage those who would try, to do so. Keeping our alarm system functioning properly will be a lasting need. Useful methodology and tools to help will be important.

10.6 ENFORCEMENT

No plant has a lock on the configuration. Changes will be made. An emergency here might require an emergency change there. Plant personnel might find "due process" too burdensome, so someone makes changes without following procedures. And, just as impactful, some approved changes somehow get missed and are not implemented. The primary purpose of enforcement is to manage the balance between the operator's need to make alarm modifications with the plant's need to ensure that the alarm system remains intact as designed and documented. It is a key safeguard in the management of change process.

Monitoring is the activity by which the current configuration (alarm system that is being used by the PCS right now) is compared against the approved configuration. Any discrepancies are examined and resolved according to the alarm philosophy. A new, approved configuration is the usual result. This new, approved configuration becomes the required one, and therefore the current actual configuration is modified (to match) the current approved configuration. The process of "modifying to match" is called enforcement.

Another way of looking at this is that enforcement is like rebuilding the foundation wall back to its intended design. This is not to say that it cannot be modified or altered, but those changes need to be made according to the approved process. When done so, it becomes the current design. The manner and timing of enforcement is specified in the alarm philosophy. There are several methods commonly used for the enforcement.

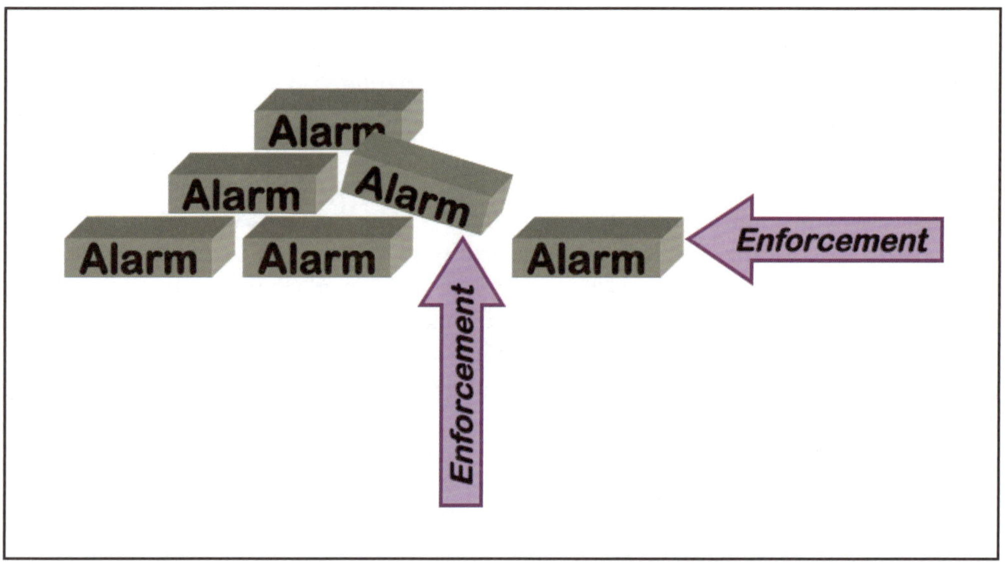

Figure 10.6.1. Alarm enforcement—rebuilding the foundation wall

Enforcement by Shift

This is the usual form. At the end of the shift, restore all alarm system modifications to their approved values. It is timed so that the operator scheduled to go off shift will observe all the reset values and deal with all the alarm issues raised by resetting the alarms before the next operator starts his shift. This way, the alarm system is stabilized before the next operator arrives. It facilitates the new shift operator's ability to better understand the condition of the plant, as well as to ensure that all modified alarms consider plant conditions.

Shift enforcement is implemented in one of two general methods.

Shift Changes Only

At the end of the shift, only those changes made by the current operator during the current shift are reverted. In order to properly manage this capability, it will be necessary to identify each alarm change during a shift as being initiated by the operator during the current shift. This form of enforcement does not address any alarm changes that were done by others at different times or changes done as a result of logic or other forms of configuration change.

Changes outside of shift changes are not reverted. These changes were allowed to persist longer than a shift since the conditions they reflect have a longer time horizon. Examples might include malfunctioning process elements that require repair that cannot be scheduled immediately due to lack of parts or other resource limitations. Those alarm changes were conditionally approved and so do not revert with the normal shift reversion.

All Changes

At the end of the shift, the entire alarm configuration is reverted to the master base configuration.

While this might seem draconian at first, it might not be so when we consider what will not be reverted here. Only explicit alarm changes to the configuration will be reverted. This includes restoration of alarm enable, alarm activation point, and alarm priority. It does not include any alarm configuration changes specified as a result of being linked or set by any of the preconfigured logical infrastructure built into the alarm management process. Only manually induced changes are reset.

Periodic Enforcement

Periodic enforcement looks much like enforcement by shift, except the time periods are longer. Some plants do this every 24 hours. Others do it weekly. They are done at a regular, predictable interval. Similar to the enforcement by shift, the process of reverting should occur with either the outgoing operator managing it fully or both the outgoing operator and the incoming operator together doing it during a special shift overlap period. This level of attention is important, since the longer the time period between enforcement events, the more changes are likely to be affected and the more important communications between operators becomes.

Aperiodic Enforcement

Unlike the enforcement done each shift, this one must be initiated manually. Moreover, since it is done much less often than each shift, it is usually done only after all changes have been reviewed and either approved to remain or reverted. This review may occur on a daily basis and be done by the respective process engineer as part of the daily watchful performance of duties, or be done weekly or even monthly. It can be used in addition to any form of shift enforcement. The rules and timing of this will be specified by the alarm philosophy.

10.7 NOTES

1. Mark McTavish, "Manage the Process, Not the Alarm," *ProcessingTalk*, October 18, 2007.
2. B. Hollifield, D. Oliver, I. Nimmo, and E. Habibi, *The High Performance HMI Handbook* (Houston, TX: PAS, 2008). http://www.pas.com.

CHAPTER 11

Project Development

Do one thing every day that scares you.

—Eleanor Roosevelt

Many of us have had the experience of considering a project, getting it started, getting somewhere along the way, and by some stroke of insight, magic, or fate, we paused for a "look-see." What we saw at that moment sometimes did not seem to make a lot of sense. Not at all. Not for a moment. So what we did was to rethink and redesign the work to get done what was needed. The lucky few get to see this early on. This chapter is all about sharing the insight from those who have looked ahead for us and wanted to share their lessons.

Alarm redesign is straightforward, uncomplicated technology. It asks for many judgments that are also being asked as part of the safe operation of industrial production plants. This chapter talks about how the actual working construction of alarm management projects can be decided. It comes here for a very pragmatic reason: we needed to know what is the full scope and breadth of a good design before we could appreciate how a successful project might actually be set up and worked. With understanding comes the insight so necessary to organize the work to fit into your plant's ways of doing things.

The large scope and broad effects of alarm system design and performance must be apparent by now. For traditional manufacturing enterprises composed of 1000 to 1500 control loops (counting control valves usually provides a good loop count), a project might take from 1 to 3 years to complete. The work may cost (total cost, including outside tools and expertise and most internal expenses) somewhere in the neighborhood of $600,000 to $1,400,000. It pays out quickly—after that it is profit. There are ways to efficiently do work of this size. A single project can be done, of course. Or depending on approval authority of local managers, other available resources, and the work practices of purchasing departments, it might be desirable to structure work into a number of smaller projects. The enterprise can be divided into individual operator areas and a separate project done for each area, or the work of alarm redesign can be divided into its logical staged parts and each stage done for the enterprise for each operator area in turn.

11.1 KEY CONCEPTS

Alarm improvement projects can be justified	Improving the alarm system improves the operability of the plant. This improvement is tangible (reduced maintenance costs, reduced insurance costs, and fewer costly abnormal incidents) and intangible (reduced variability, less environmental challenges, and more time for the operator to improve production).
Alarm improvement projects pay back, and quickly	Typical payback time is less than 18 months but the improvements last much longer.
Costs are manageable	General cost for a complete plant (four to six operator areas) is $600,000 to $1,400,000.

11.2 THE FIT OF ALARM IMPROVEMENT

There are a number of circumstances when the timing of an alarm improvement project would be a good, natural fit back into your plant, for example, when the plant needs to replace an outdated control system. You have been notified for the umpteenth time that all support for your favorite, loved PCS will be dropped, for good, on the first of July of this year. And somehow, this time, you believe it. Since it will neither be possible nor advisable just to map the entire old existing controls configuration into the new one, your plant will spend the effort to engineer the new system. It is the perfect opportunity. So off you go to the project lead and offer to make everyone look like a hero. Your mantra is, if we do not do it now, we will just have to do it later—so why pay twice and lose the potential benefits while waiting?

Well, project leads often think very differently from many of us. They just want to get the job done, on time and on budget with no injuries or other "history" lessons. Redoing the alarm system sounds an awful like "more to do" and that means more opportunities for things to go wrong. We are talking about a PCS replacement project that has not even been fully scoped. Notwithstanding this, it really makes sense to do it here.

This is the era of harsh cost justifications and equally harsh consequences for disasters. Many enterprises have adopted strategic initiatives as a way through these pressures. Most such initiatives have economic viability, safe operation, and market responsiveness at their core. Alarm improvement is a perfect fit to many of those initiatives. It should be an integral part of the initiative. Here are some useful steps for doing it:

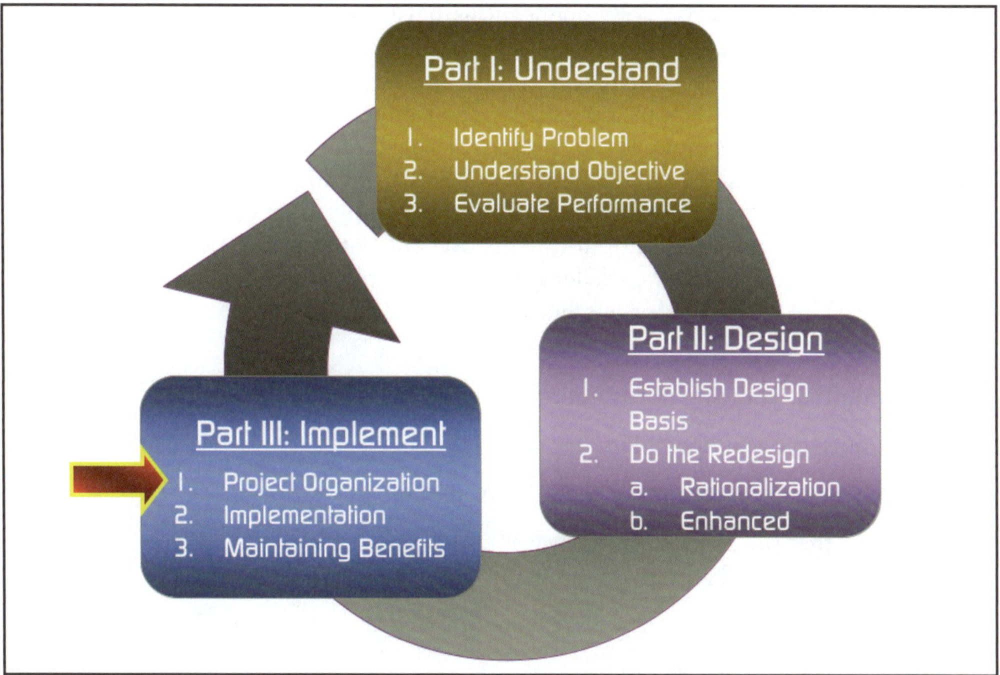

Figure 11.2.1. Project implementation roadmap location

- Link alarm improvement directly to strategic enterprise initiatives.
- Provide real, believable performance evaluations of the "old" and performance targets for the "new." Show how meeting the targets will result in the improvements everyone feels would be clearly beneficial.
- Make the business case.
- Use the best technology for the work and find experienced guides to help you.

11.3 THE BUSINESS CASE

This section describes how to approach the task of producing a workable business case for alarm system redesign. It must be pointed out from the outset that "business case" and "cost justification" are probably two of the most closely related and badly distinguished terms ever. If every alarm management project had to be strictly cost justified based on that wonderful management practice, we would have few, and the ones we did have would have been delayed. It all comes down to a very basic aspect of technology: it is very easy to identify what it might cost to do some task or another; it is extremely difficult to determine benefit values for the work once it is complete. The reason is that most well-thought-out technology applications have direct benefits and indirect benefits. Surprisingly, the indirect ones are often the most beneficial down the road. They add infrastructure that subsequent technology can exploit; they open new avenues of doing things and thinking about things that are of value.

The business case is simple: if your present alarm system fails the test of being useful and relevant (all of those EEMUA metrics and such), we know that it is getting in the way of operators doing their job and is likely to contribute to more serious problems as well. It really needs work. It continuously exposes the enterprise to poorer production, unsafe operation, environmental challenges, and other consequences of weak operations management. This makes important business sense. Making the business case involves the following:

- Showing that your site needs important improvements in its alarm system
- Explaining the real operability improvement benefits of a proper alarm system
- Identifying the costs to make the improvements
- Showing that the cost of the improvements is well recovered by the improvements being implemented and used

Percentage of Daily Losses

Recall the discussion (repeated here as fig. 11.3.1) from chapter 2, "Abnormal Situations," where the daily production was scored against plant capacity. We saw that tracking production this way is very revealing of how operations were affected by both outside forces and the inability to sufficiently manage variability. Sites have found that their focused efforts can recover 3%–8% of capacity. Some of this results from alarm improvement efforts. For your site, estimate the portion of improvement due to improved alarming and ask for a percentage of that to do your project. A good ballpark number is 5%–10%. Most projects pay out in 18 months or less. Let's take a look at an example situation to see what the numbers might look like as an illustration of the magnitude of what is there.

Example

1. Total plant production capacity expressed in gross value per year = $1,000,000,000.
2. Accumulated daily losses due to production irregularities = 3%–8%. This translates into results of between $30,000,000 and $80,000,000 per year. Market issues, personnel issues, and legal issues are not included. All other losses due directly or causally indirectly are included. Take these losses and allocate (or negotiate, or otherwise credibly determine) a percentage to alarm related causes and issues = 5%. This amounts to a working, conservative figure of between $1,500,000 and

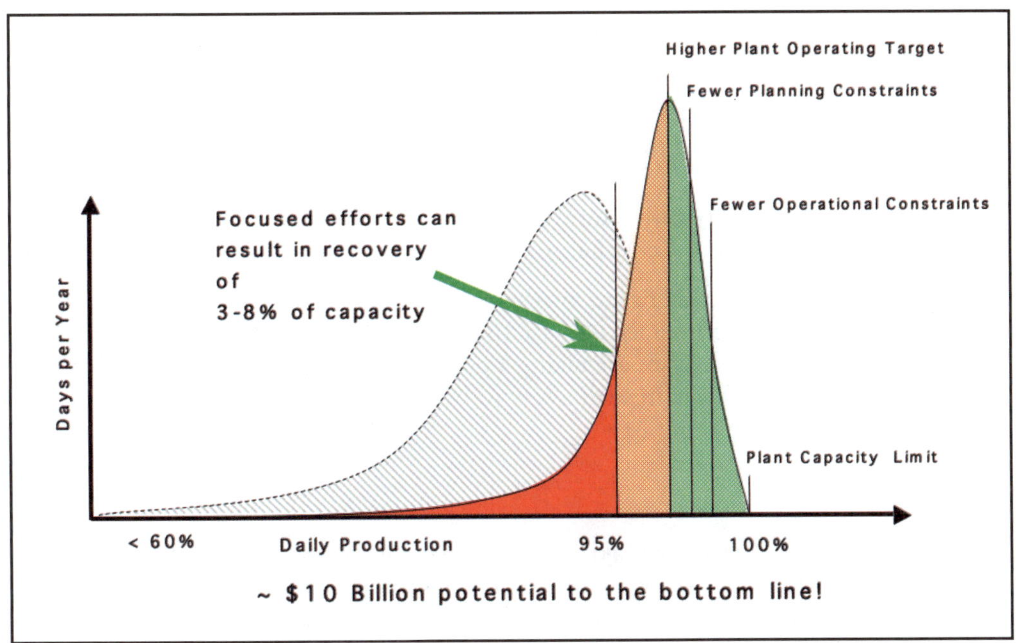

Figure 11.3.1. Daily production vs. capacity

$4,000,000. Interestingly enough, this figure is more than enough to fully fund most alarm improvement projects for an entire site.

Direct Calculation

At times, it might be possible to actually map out an alarm improvement project, do the costs and benefits, and obtain the numbers. However, all this is new to industry. Case experiences are either limited or closely held as proprietary. Really good numbers are hard to get. This can also make it more difficult to judge the appropriateness of costs for outside contracts to do alarm management work.

Negotiation

Negotiation has generally been an effective way to work through the benefits part of traditional industrial manufacturing projects. It would be equally applicable to alarm improvement projects. A typical run-through might look like this:

1. Document production upsets and problems that cost money (enumerate money costs for as many aspects of the losses as possible). This might be done by identifying

and cataloging the upsets or by appropriate estimates where all parties agree on their reasonableness.

2. Get the operations and engineering team together. The team must be capable of understanding operational losses and be able to link production problems and off-performance to these losses in a creditable way.
3. Work out, in a collaborative way, the "contribution" of the alarm system to the losses; reduce to either a cost figure or a percentage figure.
4. Negotiate a percentage figure of the contribution above to allocate to alarm improvements, usually 5%–10%, so this will be a very conservative working amount. Note here that we are taking a percentage of a percentage. The results of this effort must be done in such a way to be creditable to management.

Bottom Line

The objective of the fuss over making a business case is that at the end of the day, you will be able to take all your engineering, technology, and operating experiences and transform them into a way forward to improve your alarm system—a sufficiently funded project.

11.4 PROJECT DESIGN APPROACHES

Rationalization, we recall, is the activity by which a site ends up with an alarm system where every single alarm is needed, the proper activation level is determined, the proper priority is assigned, and the needed operator support information produced. It is the heart of alarm improvement. There are two basic methods: start from where you are or start from zero. The process is very different for each approach. The results are somewhat different as well. From earlier discussion in chapter 7, this might be an easy choice in the abstract, but when it comes down to actually deciding, many sites find it difficult to overcome their attachment to whatever "comfort" level their old alarm system might have provided. They are unwilling to cut the ties very readily.

Both approaches can be structured as a staged process so the parts can be more easily funded and managed. But, the bottom line, the real headline so to speak, the reason for phasing or staging is that at the end of each stage, the alarm system will be improved, demonstrably. Even if you have to slow down or stop, you have got real benefits from the work done so far. Before we go to the staging alternatives, be reminded that it is assumed that there is an approved alarm improvement project. For if not, one of the best ways to not get one down the road is to eliminate the bad actors up front, before the project, and then expect the plant to be prepared to provide ten to twenty times more cash to do the rest of the job. Bad actor elimination only rids the plant of the symptoms. But bad actor elimination is a very good way to show good faith and demonstrate competence *after a project is approved.*

Alarm Improvement by Starting from Where You Are

Starting from where you are means just that. Capture the existing alarm configuration database, alarm activation points, priorities, and all. Then using whatever strategy might appear to be useful (one-by-one, by classes, by physical entities, by what-have-you), decide YES, this alarm stays, and if so what is its new activation point and priority or NO, eliminate it completely. Do this entirely until the entire configured database has been reviewed.

There is, of course, only one practical problem with doing it this way. There is no way to take advantage of the work until it is completely done. There are no halfway benefits, no partial advantages. So we make a small modification in the procedure. We construct the process in stages. Here we have six stages. Figure 11.4.1 depicts their highlights.

- *Stage 1.* Identify the current performance of the alarm system, benchmark and document the findings to be used for later stages, and make all equipment repairs and modifications so that the alarm system is not called on to overcome or accommodate equipment deficiencies.
- *Stage 2.* Remove or otherwise eliminate all the nuisance or "bad actor" alarms. The complete list of these types of alarms can be found in chapter 6.
- *Stage 3.* Establish valid alarm activation points (alarm limits) and proper priorities for all alarms. If you are paying attention here, you realize that you are going to be doing a lot of work for many alarms that you will eliminate later. We do it now

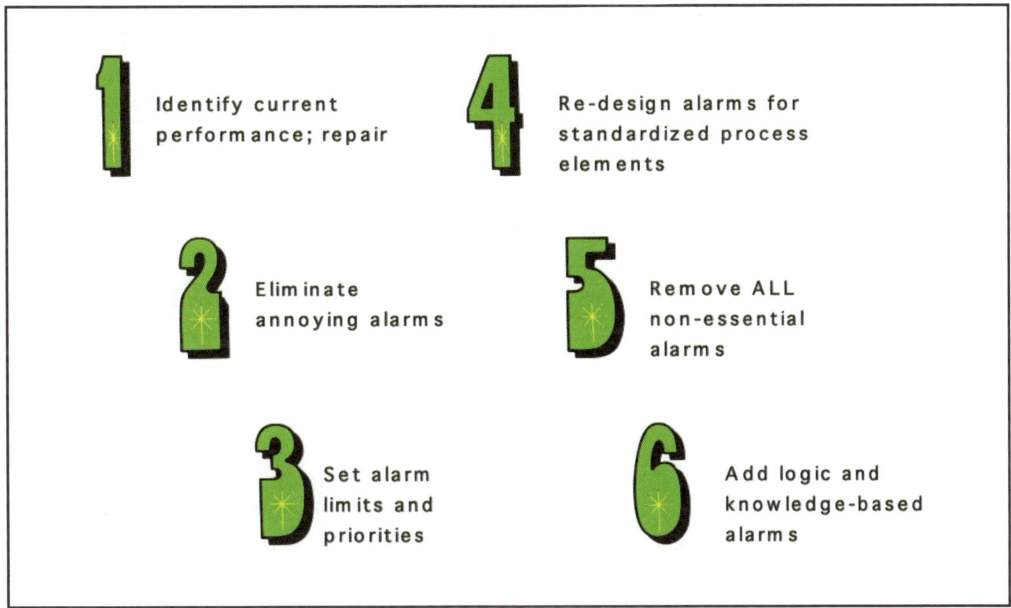

Figure 11.4.1. Alarm improvement project stages: starting from where you are

because it is very much easier to redo the limits and priorities using a consistent working procedure. They will not be optimal, but they will be a great improvement over the mess that is likely there in a legacy system. And if you can only go this far, you have something useful.

- *Stage 4.* Design basic alarm conventions for standard, conventional pieces of equipment and process; use these conventions as templates everywhere they are applicable. Basic pieces of equipment include pumps, heat exchangers, cooling fans, compressors, tanks, and such. Basic process elements include reactors, separation towers, boilers, and such.
- *Stage 5.* Remove all nonessential alarms. Well, we cannot put it off any longer. Here we go through the alarm database and get rid of all remaining nonessential alarms. But this is a much smaller list of alarms to look at than if we started without the earlier stages. The reason is that many of the unnecessary alarms were eliminated earlier in Stage 4.
- *Stage 6.* Add logic and other advanced and enhanced knowledge-based alarm features. Here we move away from looking at alarms point-by-point and start looking at combinations. Moreover, we also consider plant and process conditions as they reflect into operating states and situations.

Now that the individual stages are laid out, we can examine the benefits from and limitations of each stage. Table 11.4.1 summarizes this information.

Alarm improvement can pay off even if you do not actually get more than a good beginning. Completion of Stage 1 should completely eliminate all stress on the alarm system due to equipment not in good order. This distinction is important. The best that you can do here is to bring the plant into equipment conformance with its design. That in itself provides several important benefits. First, it means that all the design Procedures, practices, training, and documentation meet plant needs. Operators do not have to "make allowances" for things in the plant not being up to the (design-anticipated) need. Second, it helps to bring design and procedural deficiencies into clearer focus. If parts do not operate well now, the situation strongly suggests that external change should be considered. Contrast this with the before Stage 1 situation.

There are similar benefits to be derived by completion of each subsequent stage.

Alarm Improvement by Starting from Zero

Starting from zero means your alarm redesign starts with no configured alarms. The design team must add each alarm that will eventually end up in the configuration. Chapter 7 laid out this process in significant detail, so we will not go into it again here. Please refer back if you want a refresher. There are four stages, and they are summarized in Figure 11.4.2. Take careful note that since you start with no alarms, Stages 2 and 3 must be completed as a unit process (before implementation) in order to have a functioning alarm system.

Stage	Description	Benefits	Limitations
1	Assess and repair	Strengths known; good benchmark; data for philosophy; better operation	Alarm system not changed
2	Reduce nuisance alarms	Substantially less annoying alarms; less standing alarms	No flood control; no alarm consistency; no guidance to operator
3	Consistent alarm limits and priority	More time to respond; some "urgency" information; limited guidance	No flood control; modest consistency; minimal guidance; more work
4	Alarms for standardized parts of process	More meaningful alarms; improved guidance; modest consistency	Hint of flood control; care must be used before completion of Stage 5
5	Full rationalization	Alarms consistent throughout plant; basic flood control	Some flood control
6	Knowledge-based alarming	Effective operator guidance	Acceptable flood control (as good as best practice provides)

Table 11.4.1. Benefits and limitations of starting from where you are

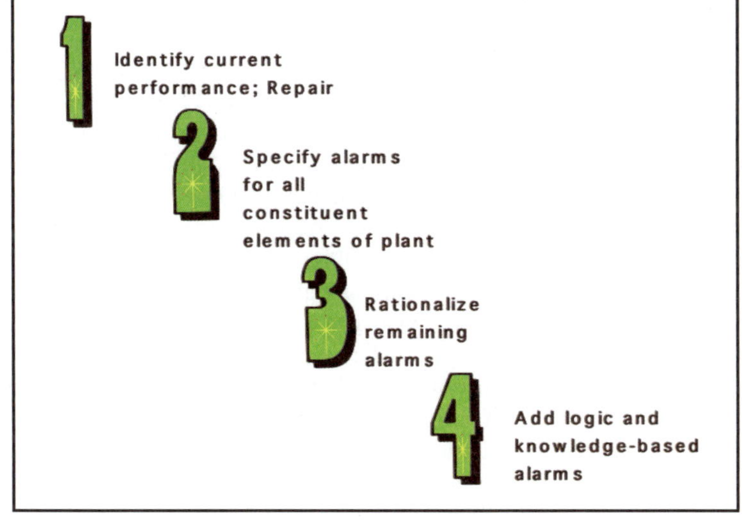

1
Identify current performance; Repair
2
Specify alarms for all constituent elements of plant
3
Rationalize remaining alarms
4
Add logic and knowledge-based alarms

Figure 11.4.2. Alarm improvement project stages: starting from zero

Here are the stages.

- *Stage 1.* Identify the current performance of the alarm system, benchmark, and document the findings to be used for later stages; make all equipment repairs and modifications so that the alarm system is not called on to overcome or accommodate equipment deficiencies. (This is essentially the same as Step 1 of starting from where you are.)
- *Stage 2.* Conceptually eliminate all existing configured alarms. Design comprehensive alarm conventions for each standard, conventional, and specialized piece of equipment and process; use these conventions as templates everywhere they are applicable, revise for special cases. Basic pieces of equipment include pumps, heat exchangers, cooling fans, compressors, tanks, and such. Basic process elements include reactors, separation towers, boilers, and such. (This is similar to Step 3 of starting from where you are, except that here much more attention is paid to understanding the fundamental aspects of each piece and capturing the abnormal operational situations in the alarms configured for it.) Continuation to Stage 3 is mandatory.
- *Stage 3.* Add any missing but essential alarms. These will be alarms for parts of the process that were not in the list for Stage 2. In general, they are found either in isolated variables or those parts of the process that connect the basic building blocks of the plant together to function as a unit. Their number is usually a small percentage of the total.
- *Stage 4.* Add logic and other advanced and enhanced knowledge-based alarm features. Here we move away from looking at alarms point-by-point and start looking at combinations. Moreover, we also consider plant and process conditions as they reflect into operating states and situations. (This is essentially the same as Stage 6 of starting from where you are.)

Now that the individual stages for this second approach are laid out, let's take a look into the benefits from and limitations of each stage. Table 11.4.2 summarizes this information. Again, alarm improvement will deliver benefit, even if you do not actually get more than a good beginning. As we discussed earlier, completion of Stage 1 should dramatically reduce or eliminate stress on the alarm system due to equipment inadequacies being out of design state. Stages 2 and 3 bring two very important benefits. First, in order to perform this, the site personnel will have visited the basic aspects and fundamental operating designs for all parts of the enterprise. Many times, this activity, if done at all way back in the plant design stage, was not well documented and largely forgotten when it came time to prepare operating instructions and policy. Second, all alarms configured here are going to be important, useful, and not distracting. And their number will be a lot smaller than the current configuration contains. Stage 4 ensures proper knowledge, minimal distractions, and adequate alarm flood control.

Stage	Description	Benefits	Limitations
1	Assess and repair	Strengths known; good benchmark; data for philosophy; better operation	Alarm system not changed
2	Design alarms for all sub-systems of process	More meaningful alarms; improved guidance; significant consistency	Hint of flood control; cannot implement without Stage 3
3	Finish rationalization	Alarms consistent throughout plant; basic flood control	Some flood control
4	Knowledge-based alarming	Effective operator guidance	Acceptable flood control (as good as best practice provides)

Table 11.4.2. Benefits and limitations of starting from where you are

Usefulness of Stages

At this point, we have broken down the work of alarm redesign into the technology steps. Next we consider the enterprise geography, the equipment on the ground, and work the redesign and implementation part together with the parts of the enterprise that need the work done.

11.5 PROJECT CONSTRUCTION ALTERNATIVES

Alarm redesign projects can be a significant undertaking if done for an entire site all at once. Some enterprises do not have the management foresight and financial freedom to do the work as one start-to-finish project. For this reason, alarm improvement work can be structured so that, in addition to single large projects, smaller ones can be considered for the work.

There is a note on costs, and this applies to all three ways of structuring projects. Costs here refer to the total assignable costs of the project. It will include training, tools, man power (site teams and operators), and other direct and indirect customary and usual project costs. It does not include lost production due to equipment being out of service for repair and modifications or during upgrades and configuration changeovers and the like. It does not include the costs for repairing broken items and revising worn-out procedures and training. It does not include preparation and delivery of training beyond that for the revised alarm system. And finally, the last stage (Stage 6 or Stage 4, depending) is not estimated for either costs or time to complete. This work required for this phase is so highly variable, site supporting infrastructure so unique, and the extent to which any individual site may chose to go is so open that estimates just are not very effective. However, their costs are not out of line with other parts of the project or for other projects using similar technology and implementation. It is important to do it.

Before we go into the details, let's be sure we are all thinking about the same terminology for sites and units and such. Here is a quick review:

- *Comprehensive.* All work done complete, from start to finish for every aspect in the plan.
- *Site (Sitewide).* An entire physical plant (may be part of a larger corporate structure) and generally comprised of several units or operating areas. Total operating areas number about four to six with 1000 to 1500 control loops total.
- *Unit.* One operator's area; part of a larger plant; generally containing 200 or so control loops. This is the smallest portion of the plant that can have its redesigned alarm system implemented!

The three basic approaches are as follows:

- Sitewide, comprehensive
- Sitewide, staged
- Sitewide, unit-by-unit, comprehensive

Sitewide, Comprehensive

A sitewide, comprehensive alarm improvement project is the single project whose scope covers the entire enterprise site. It will be approved with an overall budget and time frame. At the successful conclusion, the site will have an installed and operating redesigned alarm system based on a single unified design. The cost is in the range of between $600,000 and $1,100,000 and it will usually take from 1 to 3 years to fully complete.

This approach provides the most benefits the earliest. Since the work will cross all unit boundaries and involve a cross section of the enterprise staff, it will have a good opportunity to effect a culture change—improved operations can be had and here is a good way to start. And since changes are going to be sitewide, the interunit infrastructure can be changed.

On the concern side:

- The purchase order for this job will be a big bite and last more than a year.
- There will be no previous successes to pave the way, both financially and energetically.
- Proving the benefits will take longer since you really cannot find out much until everything is done.

- Like most large, long projects, it might be hard to control and keep a consistent vision and all the personnel.

However, projects like this are the most efficient and most plants would do them this way if other unavoidable aspects would not prevent them.

Sitewide, Staged

Breaking down the alarm improvement project into a series of smaller projects according to the sitewide, staged approach is probably the simplest way to do the stages. It ends up being a bit less efficient for the site, as there are built-in inefficiencies due to the long times between teams working on a specific unit. Here is how it works. Decide which rationalization approach you will use. Based on the approach, you will have either six or four stages of work. Starting with Stage 1, work this stage for each of the units in rotation for the site. Then do the same for Stage 2 and so on until all planned stages are complete. Each stage can be one enterprise-funded project, or a few stages can be combined for a single project. Costs for this range from $800,000 to $1,400,000 and usually take from 1 to 3 years to fully complete. It is more than the single sitewide project, but there are built-in realities that make it so.

This approach provides early but limited benefits. Once the first stage is complete for the first unit, "before" and "after" comparisons can be made and improvements noted. However, early stage successes are generally less impressive for the bottom line than later ones. Nonetheless, these early successes can be used to help justify the next projects yet to be approved. Individual purchase orders and costs will be lower and time frames usually less than 1 year.

On the concern side, there is a limited ability to include infrastructure that crosses unit boundaries, since one unit will be the one worked on and the other will not have any work done at all, and therefore may be unprepared to utilize or mesh with any changes. Since only a small part of the entire alarm improvement work will be done at any one stage, it is harder to keep a consistent long-term vision of it all. Finally, a misstep or other visible mistake can sometimes be blown out of proportion and used to terminate the entire plan. Care is needed to ensure that this is not the case. Consider arranging for expert assistance to get things going the first few times.

Sitewide, Unit-by-Unit, Comprehensive

For this approach, we again divide the overall large project into a series of small projects. There is usually one project for each separate unit. For each unit, we do a complete alarm improvement project—start to finish. Use the same plan and move it to each of the other units on the site, one at a time. The costs for this approach will range from $800,000 to $1,400,000 and will take from 2 to 4 years to fully complete. Again, it is more than a single sitewide project, but again there are built-in realities that make it so. The added completion time recognizes the inherent difficulties in moving rationalization and design teams from one unit to another. Usually projects divided this way will find that a separate team will be used for each unit. Starting up and coordinating the teams to ensure adequate vision and capture the lessons learned from preceding units can take more time.

This approach also provides early benefits. In contrast to the previous approach, those benefits will be for an entire unit, as it will have been completely redesigned. Thus, these results are going to be more valid for the process and provide more useable data. Early successes can be used to help justify the next projects yet to be approved. Purchase orders and costs will be lower and time frame usually less than 1 year. And there is less chance for a misstep.

On the concern side here, as we found for the other staged process, there is a limited ability to include infrastructure that crosses unit boundaries, since one unit will be the one worked on and the other will not have any similar work done at all, and therefore be unprepared to utilize or mesh with any changes. Since the entire alarm improvement work will be done at only one unit, it is harder to keep a consistent long-term vision between units. And, as pointed out earlier, this organization of project sometimes makes it more difficult to use the same team(s) for each of the units. The practicality of retaining personnel dedicated to this work over several years means that inevitable changes and substitutions are common. Different players have different backgrounds, and later ones might not be as excited about the process or the project as those on the ground floor.

Review

Table 11.5.1 summarizes the three approaches. It is important to observe that the project approaches discussed represent three distinctly different but functionally related ways

Type	Costs	Time frame
Sitewide, comprehensive	\$600,000 to \$1,100,000	1 to 3 years
Sitewide, staged	\$800,000 to \$1,400,000	1 to 3 years (assuming no delay between stages)
Sitewide, unit-by-unit, comprehensive	\$800,000 to \$1,400,000	2 to 4 years (assuming minimal delays between units)

Table 11.5.1. Summary of approaches to alarm system redesign

to do the work. Often, it will not be possible to pick one of the three and decide that will be your approach. Even if an initial choice were to do a single all-encompassing project, any number of subsequent events might impact it. Approvals can be withdrawn—for cause or not. Business climates change. Essential personnel move about or otherwise become unavailable.

It is entirely possible that your alarm improvement project might involve a customized combination of the basic approaches we have reviewed. This is not an important issue so long as the project fits the enterprise needs. Take care, however, that your work is consistent with the alarm philosophy and in keeping with good alarm system design practices. Do be careful to be consistent with the stages outlined in section 11.4, "Project Design Approaches." These stages have been constructed to provide clear-cut divisions between the smallest consistent pieces of work.

11.6 WHY SOME PROJECTS FAIL

You are set to get started. You have done all the preliminary work, planned your approach, written the project proposal, and through some magical process got it all approved. But before you sound the bell at the starting gate, let's take a look at some of the important reasons that have been found to cause failure down the road. Knowing them early just might make the difference between success and ho-hum. Go for success!

- Weak or confusing philosophy
- Confusion about the purpose and benefits of the philosophy
- Weak or poorly defined goals
- Overly optimistic goals
- Overly pessimistic goals
- Bad alarm performance treated as a "crisis" rather than a culture change
- Failure to work project according to plan—acceptance of "on the fly" modifications
- Parts of the team lack vision or think they "know better"
- Weak or uneven accountability of team individuals
- Early enthusiasm replaced by reluctance to change
- Each team doing their own thing
- Within teams, members failing to share ideas and concerns
- Weak or missing attention to revision of operator graphics
- Weak or missing attention to procedures, training, and performance standards

By the way, if you think this list is "just for the other guy," please keep in mind that for everyone else but you, you are the other guy!

11.7 "LOW-HANGING" FRUIT

Low-hanging fruit refers to the ad hoc approach sometimes used by inexperienced but well-meaning individuals to improve their alarm system. The process is as simple as it is appealing. To do it, all one does is to take a short, serious look at the current performance of their alarm system. Find all the obvious things wrong and just fix them. So what is wrong with that? It sounds perfectly fine.

The list of low-hanging fruit includes the following:

- *Bad actors.* All of those nuisance alarms including chattering, repeating, duplicate, and shutdown or out-of-service equipment indicators
- *Alarms with no response.* Activations that occur but for which the operator either needs to do nothing or for which he has no idea what is needed and does nothing in response
- *Totally stale alarms.* On the screen for weeks, months, or even years
- *Functionally disabled and/or shelved alarms.* "Permanently" out of use because of operator action, yet remain in the configuration database

A number of things are wrong. All are important. Last, but perhaps most vital, let's say for argument's sake that you did pick the low-hanging fruit and did it very well—so well, in fact, that operators who were after you to fix the alarm system now find it working so much better that they now no longer want to spend any more money on it. Your part was quick and effective and that is all they thought it needed anyway. They are done, and now you are left knowing that the alarm system really needs an overhaul. While it seems to work better when little is going wrong, you are convinced that real process upsets will derive little benefit from the alarm system as it stands, and serious problems will likely be made worse by lack of proper alarm system support. Since the deeper deficiencies of the alarm system become exposed only after a serious alarm system examination (like the one you are doing by reading this book), you would know what they are and how important it is to resolve them. But now you have lost your support.

Sure, it is just fine to work them. You just might want to wait until your project is designed and the teams have started work. Do it as part of Stage 1. Fixing them will endear your team to the operations guys and not jeopardize the real job of serious and fundamental improvement of the alarm system.

11.8 CONCLUSION

This chapter has provided a number of alternatives for you to consider. None of the alternatives and options is any more important than any other. You are encouraged to be creative to find a way to structure alarm improvement work at your site so that it fits your local ways of doing projects.

CHAPTER 12

Situation Awareness

A mind stretched by a new idea can never go back to its original dimensions.

—Oliver Wendell Homes

The plant operator has an extremely valuable and important responsibility. For an entire shift, rotation after rotation, he is the force and energy managing a capital enterprise easily worth hundreds of millions of dollars. He operates an industrial plant that produces or impacts a daily revenue stream of around a million dollars, give or take. Within his area of responsibility and authority, he can view every control loop, most pieces of equipment, and much of the supporting utilities, and then adjust as appropriate. We ask him to be ever mindful of what the plant might be doing. We ask him to be capable of finding every little problem before it grows into a big one. We ask him to shoulder the burden of everything that goes wrong during his watch, all without any recognition when nothing does, and precious little (if not actual blame) when it goes amiss and he manages to manage.

Chapter 12 covers the emerging industrial practices that enhance operator awareness of the true nature and conditions of the plant in their charge. We call this situation awareness. This increased awareness can be enhanced through careful design of operator graphics and navigation. It will be a major contributor to the operator's ability to manage abnormal situations. It is a vital partner of effective alarm management.

12.1 KEY CONCEPTS

Situation awareness is more important than the alarm system	The plant operator's ability to know where the plant is operating and understand current threats to good operation is one of the most important benefits to successful plant management.
Excellent methodology available	Paying attention to human factors principles is extremely effective for plant systems designs including graphical interfaces.
Poor use of color in graphics is a critical impediment to situation awareness	Color is only used for information.

12.2 OPERATOR SUPPORT NEEDS

There is a surprising dearth of tools and technology presently in place to assist operators. As their primary information, operators are provided only simple values of variables, sometimes augmented by desired targets, and less usually, complex calculated or inferred parameters. Trend plots are available for most variables. Sketchy operational status is provided for many pieces of equipment. There is an alarm system. Using these simple tools, we ask operators to identify what must be going on deep inside the operating units, decide what is normal and what is not, and if not, figure out what to do to return it to normal and then take only the proper corrective actions to cause just that result. Few plants are so readily understood. Few operators are so favored by such an unusual talent.

The Hat

There is an interesting story to tell that drives the message of situation awareness home.[1]

> The story takes place a bit more than a few years ago, during the coat, tie, and hat era. There was a very good industrial equipment salesman who, as often as not, would be found at his customers' sites. If the sales call was planned for an office visit, the usual garb included a nice business hat. It was during such an office call on a nice sunny day with just the hint of a wind that our salesman was prevailed upon to accompany his contact out to the plant to offer some important advice on a vexing problem. As luck would have it, on the way back from

the site to the office, they passed a bit too close to a tower being repainted. Some over-spray dripped on the hat.

The next day our salesman submitted his usual expense reimbursement form to his boss. All went well until the boss noticed the entry for the cost of a new hat. "What's this? We don't buy hats for employees."

"Oh, the hat. Well, yesterday, when I was visiting our customer to drop off the revised estimate for a new compressor, I was asked to assist a technician to align that new inertial coupling we sold them a while back. They were having trouble with a back-tension adjustment. I didn't expect this and so I was in office dress. Unfortunately, the trip into the plant went too close to a repainting job and some excess paint dripped onto my hat and ruined it."

"Well, I just can't approve this. Take it off and resubmit the expense report to me."

The next day, a new expense form was in the boss's hands. He looked it over and saw that the hat was indeed not there. "Wait a minute, he said. I don't see the hat. But your final total is just the same as it was yesterday when the hat was there. Where's the hat?"

Our salesman replied, "If you think the hat is there, then you find it."

Interesting story? Perhaps. Where is the connection? The connection is the analogy: are we not doing the same thing to our plant operators? We provide lots of controls, instrumentation, sensor readings, and other interesting displays. Then we ask them to rummage around everywhere to find out where the process is not working right. We know that the needed information must be out there somewhere, but no one knows for sure where it is or what it might look like. Find the problem. *Find the hat!*

The Disaster Chain

Of course you remember the saying, "A chain is only as strong as its weakest link." This wisdom is often used to motivate us into doing the right thing: into doing our share of the work. If we are the one whose actions cause the chain to break, we are naturally the bad guy. Without the chain intact, the prize at the end cannot be pulled along the pathway to our success. But chains work both ways. Yes, we want the chain to pull what we would like to have. But if the chain is attached to something that we do not want to have, we now want it to break. We do not want what is at the end.

Figure 12.2.1. The disaster chain

We would rather not, thank you just the same, have the chain pull a disaster to us. Same chain, but now it has a vastly different connotation. A single broken or missing link will provide the chance to avoid the problem. A link might stand for improper training, another for faulty alarms, and still others for inadequate maintenance, a faulty HazOp policy, failed lockout/tag-out policy, and so on. If any one of these safeguards were to work, the corresponding link in the disaster chain would break leaving the disaster behind. We would be safe.

Sadly, most awful incidents have many more than one link in their disaster chain. Milford Haven[2] had seven links:

1. Plant shutdown not properly monitored
2. Faulty display design
3. Process operational fault not recognized
4. Process modification not properly assessed
5. Failure to maintain situation awareness ←
6. Faulty alarm system design
7. Faulty operational goals and policies

Piper Alpha[3] had nine links:

1. Faulty maintenance isolation procedures
2. Faulty permit-to-work system
3. Faulty/missing return to service procedures
4. Ineffectual managerial inspections
5. Defective fire control systems
6. Inadequate attention to explosion risks during the design stage
7. Inoperable fire control equipment
8. No emergency response policy to ensure safety and life
9. No plans or training for interplant emergencies

BP Texas City[4] had eight links:

1. Violating safe start-up policy
2. Faulty installation of a critical level transmitter
3. Alarm activation failures
4. Failure to follow procedures
5. Failure to maintain situation awareness ←
6. Faulty relief valve operation

7. Failure to understand basic process operations
8. Failure to manage exposure to personnel danger within hazardous process limits

Situation awareness provides a safeguard to help "break" the disaster chain.

Need for Situation Awareness

Start with the assumption that no one wants an accident and that no one would choose disaster over success. But accidents and disasters happen. We now know to a high degree of certainty that they happen because those in charge of ensuring that they do not happen are not aware that they are happening. They fail to see. No, they are not blind. They do see things, but what they see fails to inform. Or worse, misinforms. In short, they fail to know the situation. They are unaware of what is really going on, what is likely to happen, or what is not happening that they think is. The problem shows up with the operator. A solution is enhanced by the systems designer.

The failure to maintain situation awareness has been present in almost every disaster event that was not the result of spontaneous complete surprise. The 1994 Milford Haven incident, which led to the beginning of Europe's seriousness with faults in alarm

HSE Report Recommendations

- #1 and #2 ... Safety management systems ...
- #3 "Display systems should be configured to provide an overview of the condition of the process including, where appropriate, mass and volumetric balance summaries"
- #4 "Operators should know how to carry out simple volumetric and mass balance checks whenever level or flow problems are experienced within a unit"
- #5 "... clear guidance on when to initiate controlled or emergency shutdowns ... "

Figure 12.2.2. HSE Milford Haven recommendation requiring effective plant overview displays

system design, documents over 5 hours of operational uncertainty. The 2005 BP Texas City disaster lasted about 12 hours. Interestingly enough, both disasters had almost exactly the same tragic cause—sending excessive volatile liquids to a flare system unable to cope with them.

Both the HSE (in the United Kingdom, after Milford Haven) and the Chemical Safety Board (in the United States, after Texas City) cited significant alarm system faults and the inability to maintain situation awareness as causes for the incidents. The bottom line of effective operator performance is the ability to understand where the process is and what the likely threats to continued proper operation are. Situation awareness is the key. A proper overview is a vital component of situation awareness.

Visualizations

This is about what we see. It is about how to show things so we can see. And once seen, to know what it means and be able to act. What to do and how to do it has already been covered elsewhere in this book. This is about seeing. There are going to be lots of things to see. Please see on!

12.3 THE DEVIATION DIAGRAM

A very good example of the ability to observe and understand complex interrelated situations can be demonstrated by a simple display. This type of display, called a deviation diagram, was first developed by the Foxboro Company. It accompanied the launch of their entry into the DCS race with a product called VideoSpec. Refer to Figure 12.3.1. Please keep in mind that this figure is not about time. The order of the bars reflects the major production order or processing order of the plant. Recall the order of variables in the transformational analysis diagram of chapter 7. The deviation diagram order would be similar but would represent not process steps but key variable values within all steps.

A deviation diagram is a representational construct of the whole of a plant and its relationship within itself and to its operational goals. It is constructed by first listing the major plant processed entities from entry at battery limits (from outside or another portion of the plant) to exit from battery limits (to another portion of the plant or shipping). For most continuous plant manufacturing facilities, we track mass (raw materials, intermediates, and finished products). Each significant mass is listed as a single vertical bar for a single aspect (pressure, temperature, pH, etc.). The example contains thirty-six bars (from 1 to 36), each one an entity. An example of an individual bar would be a single precursor chemical with its most revealing attribute shown: a propane *flow*; later on, the polymeric reaction *temperature*; and still later, the *conversion rate* of reaction. Multiple precursor chemicals will each have their own separate bar. Significant processing variables would each have a bar: for example, a critical separator *temperature* and *pressure*. Adjacent bars mean that the entities they represent are functionally (if not physically) closely related in their processing steps. A bar to the left of another bar means that the one on the left comes before the one to the right in the normal "flow"

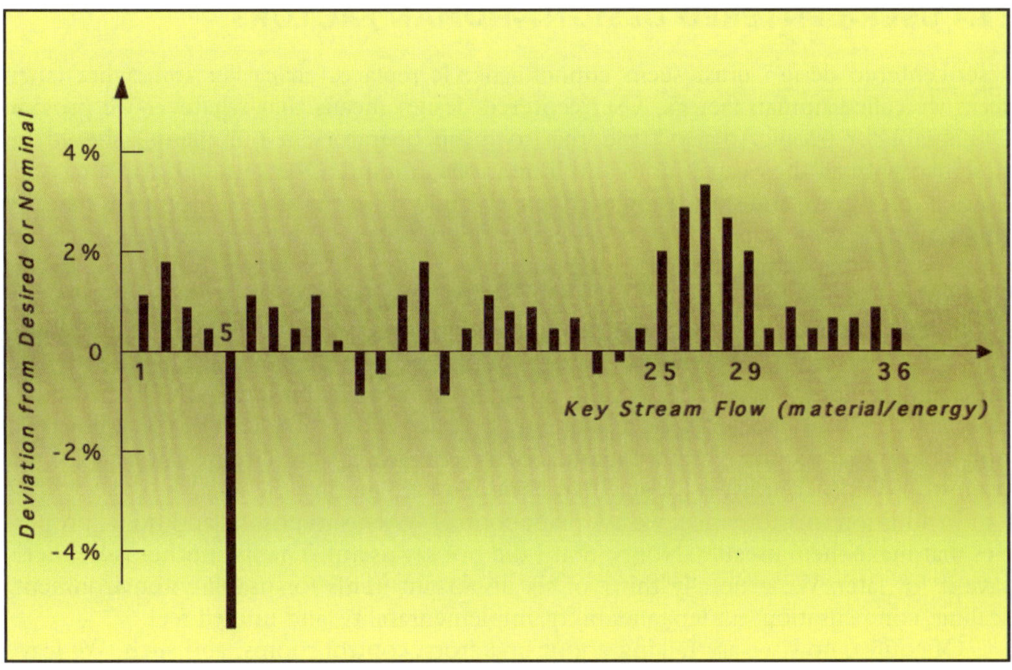

Figure 12.3.1. The deviation diagram

of production. For each bar, the height shows how far away it is from a proper target. A "target" is not normally the related controller setpoint. Rather, the target represents the expected value of the particular attribute needed for longer-term effective production. Right on target would be a bar of zero height. Values higher than target would be a bar height above zero; values below target would be a bar below zero.

Now we get to the part about what the diagram is intended to illustrate. First and foremost, everywhere there is abnormality in the plant is clearly visible. Noticeably high or low bar heights clearly point to a notable difference from normal. At those places, we see bars deviating very noticeably from the zero value. The higher (positive) or lower (negative) a bar is, the more the deviation—and hence a greater cause for concern. A single or isolated highly deviating bar (e.g., at position 5 in fig. 12.3.1) means that something is clearly wrong but its effect is strictly localized, for example, a failing transmitter that is not a part of a control loop. Groups of noticeably deviating bars (at positions 25 through 29 in our example) suggest that a broader area of the process is abnormal. Deviating bars at the left side of the diagram mean that our problems are in the earlier part of the plant. Deviating bars to the right mean problems are near the final part of the plant's production. With a glance, the operator has a broad overview of the entire process. If a good job was done selecting the variables to display, a good understanding of the overall process will ensue. It is a simple yet powerful visual agent! It is only one of many.

12.4 USER-CENTERED DESIGN—HUMAN FACTORS

User-centered design must seem euphemistic. It replaces an earlier somewhat fancy term we called human factors. User-centered design means that whatever we provide ("things") for people to use (read this to mean operators in our plants) should be designed with all their capabilities and limitations in mind. One should be able to walk up to it, pick it up, and start using it effectively right from the start. This means that the thing must, as part of its very fabric of being, instruct us as to how to use it. Easy things must be obvious. More difficult things must be readily discernable. Impossible things should be evident. That is simple enough. We call it "affordance."

Affordance is that property that a physical thing has as a result of purposeful design. It suggests, or teaches, the user how the thing is to be used. A pail with a handle sticking up would "suggest" to anyone walking up to it that it can be picked up by the handle. A doorway that has deliberately been made to look like an invisible part of the wall would not. Things that require reading a manual before anyone can even figure out how to turn it on would not. Your list is longer than mine. You are getting the idea.

In addition to affordance, we ask plant control systems to possess additional properties that make them useable. Notice that I did not say useful. That is another issue. Let's save it for later. We generally think of six important items for usability: environment, scaling, compensation, understandability, implementability, and unified feel.

Of course, now we are talking about operators, control rooms, and such. We leave the abstract behind.

Human Factors Details

Let's review the key components. The discussion will start at the 12 o'clock position of Figure 12.4.1 and progress clockwise around the diagram.

Environment

The control room must provide appropriate temperature and other environmental controls such that no operator feels a need to even think about them. Included here are noise controls, visual distraction elimination, management of seating, viewing distances and angles, and all the rest that make up the operator's work area. Assumed is the proper design and sufficient supply of storage for equipment and resources that might be needed and used.

Specifically, when applied to the operator's ability to gain situation awareness, this implies that the display, keyboard, and other operator view and entry equipment have been chosen and arranged for comfortable, easy, and convenient use. Operators coming back to shift after long weekends off, vacations, and even extensive jury duty or sick leave must feel welcomed by the equipment. Everything should be where it is expected to be and work the way it is expected to work.

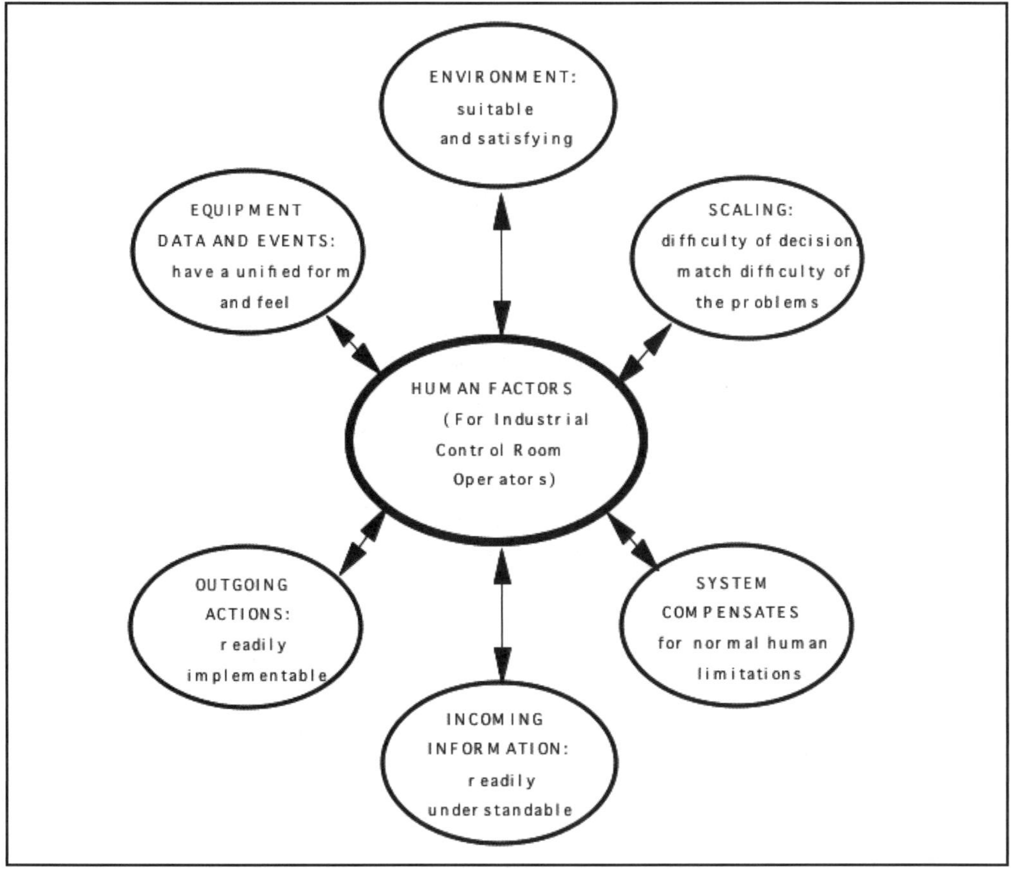

Figure 12.4.1. Human factors diagram

Scaling

Proper scaling means that the general difficulty for the operator to perform certain operations should be affected by the basic difficulty of the task itself. Easy things should not be hard to do; hard things should not be impossibly difficult to do. In order to be able to avoid scaling problems, it is rare that a "one size fits all" approach to design and implementation can be used. That approach usually is either biased toward the easy to do things, making it much more involved and difficult to do the harder things, or it is designed around the more difficult-to-do tasks, therefore placing an undue burden on the easy tasks. We see this commonly in decision-making aids. Often, they are designed to work their way down a set of sequential queries, with the appropriate answers chosen at each step. Since the problem may be a difficult one, each step along the process must be finely honed to elicit the proper amount of thinking to arrive at the correct choice. Then we move on to the next query. In this manner, the operator can work his way toward the needed support. Such a laborious process might be useful for a very involved and difficult situation, but it will be extremely vexing for simpler ones.

Compensation

There is an old story that describes the attendants at an "ideal" control room. The only two occupants are a man and a dog. The man is there only to feed the dog. The dog is there only to make sure that the man does not touch anything.

The message is clear enough. Giving in to the temptation to "tweak" things a bit, the operator often contributes to operational abnormalities—not the production improvement intended. Almost as often, the problem was due less to "tweaking" or "tinkering" than to the foibles of human error. We are not perfectly created engines of production operation support. None of us can shoulder the mantle of invincibility just by entering a control room to manage an important and perhaps dangerous production operation. We come as we are, to borrow from an interesting social party situation event.

It is the responsibility of the equipment and technology designers and providers to ensure that normal human errors are accommodated. Accommodation means that terrible things should not happen just because someone makes a small mistake. Two illustrative examples come to mind. The first example that has occurred more often than we would like to admit centers on operator consoles with several stacked video displays. Each pair of displays is managed by a shared data entry device (e.g., keyboard and mouse). Often enough, the operator's eyes were on the upper display, but the keyboard was on the lower display. The second example relates to the practice of directly entering controller setpoint or output values via a keyboard. Our operator enters "15" when he meant to enter "1.5" instead. But the controller cannot recognize intent. The new setpoint becomes 15.

Understandability

One of the best characteristics of modern PCSs is their ability to provide a great deal of data to the operator. In addition to the control variables and measurements surrounding control loops, it can provide elaborate temperature profiles, imbedded analyzer values, valve positions, and a host of other seemingly important and relevant data. And, as you suspect, the operator has too much data and perhaps too little information. User-centered design principles suggest that much less data are needed; much more context and interpretation is required. Moreover, the method of presentation itself can either obscure understanding or provide just the necessary context. A simple way to do this would be to provide graphs with goal marks located at appropriate levels on the same scale instead of tabular data with numerical targets. Recall the directness and clarity of the deviation diagram.

Implementability

You might suppose that this part of the user-centered design would be the least offended requirement. Operator workstations have been designed and built around providing all of the necessary "handles" for operators to manipulate the things operators do during their shift. Let me suggest that such is not always the case. All too frequently an operator has to "hunt around" for the right valve to check or observe one process value while he is required to manipulate another process value—but neither can be found on the same display. And there are the situations where the operator is required to shut down part of a process by gradually reducing the setpoints for a number of controllers in strict unison to avoid thermal or other operational stresses. Not only are the controllers on different displays, but also many of the variables needed to check the progress of the work are not colocated. The operator has to do a lot of shifting around and remembering.

The message for achieving good design is to ensure that routine operations and strategically important operations can be carried out in ways that ensure success and provide for a minimum of operational risk.

Unified Feel

When we ask for equipment to possess a unified feel, we are suggesting that once operators learn how to interact with some of the equipment, they will know how to interact with all the equipment. A very good example of this requirement can be found by approaching any conventional personal computer. Whether it be run under a Windows system, a UNIX system, an Apple Mac system, or any of a number of others, it would be obviously clear to the user how the keyboard is used and what should happen when a mouse or other pointing device is moved. We know what should happen when one "right clicks" the pointing device. This is also to say that the activities that are used by employing the keyboard and pointing devices are also going to be interacted with according to an expected feel.

12.5 OUR BIOLOGICAL CLOCK

We humans are designed to be generally compatible to our surroundings.

If we trust Darwin, we believe that those of us who adapted best to our environment have survived. Adapting means being in harmony. We get enough sleep, have the ability to reason quickly and efficiently, and expect the dependability of our bodily engine to process foods and rid of waste. One of the most observable adaptations is the synchrony of our bodies to the daily cycle of life. For all but the briefest of moments in human history, the sun has governed our activities. It is little wonder that our inner clocks are so linked to this powerful diurnal rhythm.[5]

Alas, the demands of a modern manufacturing society, facilitated by the invention of ways to produce artificial light, have made us necessary and able to shift our personal clocks different from the natural sun clock. Doing so has exacted a sometimes-tragic price for such rashness. Yes, we can move our inner clocks. Certain parts of the world have needed to do so because of the very large variability of this natural clock—witness what goes on in the highest latitudes. And take note of their accommodation difficulties. These natural variations, however, occur very slowly when compared to the artificial, forced shift-time changes of the modern workforce.

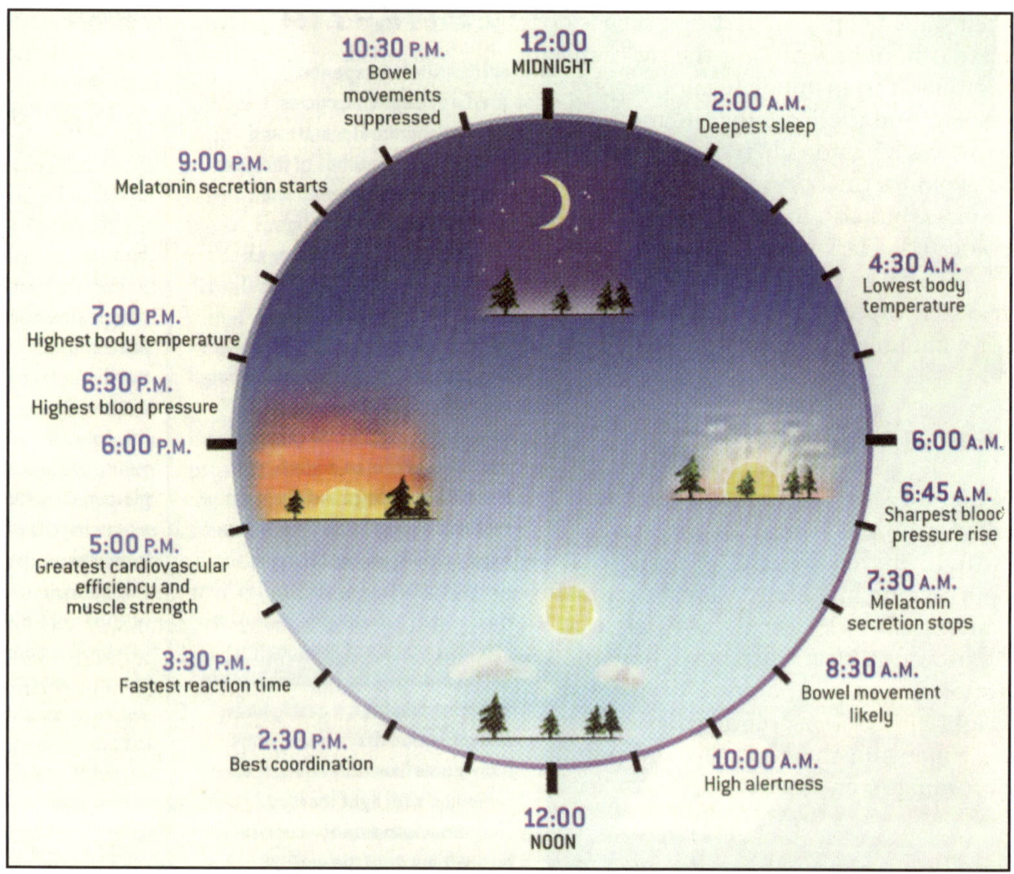

Figure 12.5.1. Circadian rhythms; our internal biological clock

It has been observed that a healthy adult person takes from 1 to 3 days to accommodate to each hour time shift of their inner clock. And this accommodation is often aided by the shifting of the sun clock at the same time—as in jet plane travel. On the other hand, a totally artificial shifting of the inner clock has no such lucky assistance. Little wonder that those shifting inner clocks have difficulty performing for days and days. And little wonder that such difficulty may affect their stress performance response. Accidents result. Situation awareness must be able to reach even those individuals.

12.6 OTHER OPERATOR SUPPORT ISSUES

Understanding the limitations of operator awareness is another vital resource in our kit bag of operator support. What is it that we see? What we sometimes see might not always be what is there.

Intent Recognition

Let's suppose for a moment that you were looking over the shoulder of an operator at his station. And let's suppose that the operator was working on an important task, one that you understood very well. And let's further suppose that the operator was less familiar than you and somehow faltered in making his way down an action path toward managing a situation. Assuming that your presence was for assisting and not for testing, you would likely ask the operator what he was trying to do. And once you assured yourself that it was the task you understood, you might suggest that he do "thus and so" to get the job done. We call this help.

Continuing with this line of thought, suppose that instead of you, there was an "agent" (software and such) watching the operator at his duties. This agent determined, by whatever means, that it strongly suspected the operator was trying to perform a specific task. It would be extremely helpful if the agent could query the operator to determine if the operator was, in fact, trying to perform the detected task. Having affirmative confirmation, the agent might also suggest ways for the operator to do the intended task or, in the alternative, and with permission, do the task for the operator. This is intent recognition.

Are you convinced that you know how it works but just as sure that such assistance is not very useful? Let me assure you that this is far from the case. Developing hundreds of intent recognition scenarios for routine tasks would be hardly beneficial. But developing a few scenarios for tasks or situations that the operator must do well but is often called on to do under high stress might just be the ticket to better and safer operations. Part of situation awareness technology would be to identify just those situations where intent recognition and assistance could be critical. The other part would be to provide this assistance when needed. Implementation of intent recognition and assistance normally follows the operator consent construction introduced in chapter 8.

Operator Vigilance

Operators have relatively long periods of time on duty. Eight hours is usual. Twelve hours is increasingly popular. To make matters worse, those 8 hours do not occur during the same clock time of the day for each day of the working month. We know from the study of the effects of attempts at changes in human circadian rhythms (section 12.5) that what our bodies are trying to do is strongly associated with where our inner clock is at the moment, regardless of the actual daily clock time. If they are mismatched, we are not at our best. However, clocks being out of match are not the only reasons for operators not being at their best. Personal health, home matters, and exciting plans for the future all play a part in occupying and distracting our minds. Distracted minds pay little attention to tasks, important or otherwise.

Operator vigilance is a process by which a specialized "agent" (software or otherwise) monitors operators to determine the degree of attention they are able to pay or appear to be paying to their jobs at hand. How this monitoring is done is not a subject for this book. Where diligent operator attention is required, the plant environment must ensure that it happens. This might be done without any specialized monitoring through the use of a program of careful attention to varying operator duties. It might be done by varying control room ambience or by construction of efficient reminders of routine tasks or by other comfortable and effective means. However it is done, it must be done. It is unfortunate that most control room protocols do this only as a by-product of coffee, operator jokes on one another, and other idle distractions attempting to break the heavy mantle of monotony.

To Push or to Pull

In information processing language, "push" means that a possessor of information actively provides that information to a party of expected need. In effect, they push the information to them. "Pull" means that the party that thinks itself in need must find the possessor of that information and grab it. Again, in effect, pull it (by sharing, not taking possession) from where it is and toward oneself. Both modes have their uses and problems. For example, alarms are always a "push." Operators are immediately notified—they never have to request. State changes for equipment do not push. Yes, certainly, if the operator was looking at information that normally includes the current state, then the current state would update to reflect that change. But otherwise, state changes are generally not "push" events. This section will not cover the push/pull subject in much more detail.

Before we leave the push/pull discussion, there is one misuse that has been suggested by responsible operator station designers. They design so that operators get a chance to see everything they should see by pushing it all to them. The normal way it is pushed is to design a series of displays that rotate on a time basis between several screens. That way, based on a clock timer, each screen appears according to schedule so the operator can view it. A common use comes to mind: a security guard stationed in front of several monitors; each monitor is split between perhaps four cameras. For high-security operations, no rotating screens are used or operators are changed very frequently—say every half hour. The normal approach, however, is to switch the four camera views every 5, 10, or 15 seconds in a continuous rotational loop. In short order, all the cameras are displayed for the guard to see. It is well known that in such situations, the security guard rapidly succumbs to the events of screen changes but does not pay much attention to their content. From a human factors or user-centered design point of view, what is wrong with this situation? In short, everything. In long, there are several interesting aspects to point out:

- Display movement distracts from all other tasks and generally reduces the attention span for everything.
- Forced rotation of screen images actively prevents operator focus on anything but the grossest of detail, or they blank entirely and see nothing at all.
- Display movement (the changing of camera images) reduces the ability to discern image movement within a given screen, save the grossest.

- Forced rotation requires actively interfering in order to return to a previous mode or even stay focused on a desired one. At the same time, other views might contain important information, but since viewing is the only way to know that, it will be missed. Nefarious individuals often take advantage of this phenomenon by creating a distraction to draw attention away while the bad stuff takes place somewhere else, off view.

12.7 OPERATOR DISPLAYS

The video display unit (VDU) would seem to represent a significant step forward in human-machine interaction. Actually, it was a step backward. But not in the way you might think. Only recently have humans been faced with interacting with "things" that were not real and seen with their own eyes. Herds of animals did not need a video display to be seen. Crops during a drought did not require them either; neither did sailing ships, managing armies, nor cooking soup in a pot over the fire. It is when the stuff that needed to be managed was composed of things, real or virtual, that cannot be seen with eyes directly that we are required to employ surrogate machines to visualize them.

Unfortunately, it was the evolution of displays, not their intrinsic faults or limitations, that led us down the wrong early paths. The earliest displays were able to show lots of (weakly formatted) text, and later on, limited graphics. These early video displays for process control started from where the predecessor hardware equipment left off. They first mimicked control stations as faceplates on the VDUs. They were arranged in rows with most of the display agents and handles that were found on their physical analog counterpart it so quickly replaced. What was gained, of course, was the ability to place many times more displays easier and cheaper on video units than on metal control walls.

Once the faceplate barrier was broken, so to speak, the world of graphic design opened up. As soon as color made the scene, all the primary PCS manufacturers started a race to see who could use the more appealing and flashy colors to preen in front of prospective buyers. I do not actually know who won that race, but I do know that the purchasers were the collective losers. What was lost was the objective of what the video display should do and how best to do it. Now, many decades later, we know a bit more. This chapter will help you to understand.

Physical Display Architecture

Ironically, it was the very flexibility of the early VDUs that contributed to our failure to appreciate the need for a physical architectural arrangement that incorporated multiple units. Add the high cost of the proprietary VDUs to the equation and rarely would one find more than two or three units per operator station. Early on, there was an inherent conflict between flexibility and navigational difficulty. Operators became overburdened with their primary task: keeping an eye on things. In those days, keeping an eye on things required constantly shifting displays to find all the relevant states and situations. Only now do we understand how difficult that burden was. Being difficult, it was often done incompletely or ineffectively. It should not be surprising that alarms were (over) used to be watchful servants for the operator. It was a natural way to automatically be able to keep an eye on his plant. Everything that might be abnormal would cause an appropriate alarm on operational displays.

We know that more screens are as necessary as more displays. Figure 12.7.1 shows a recommended architecture. This architecture, first proposed in the late 1980s, has very important structural aspects. First, we have an expected location for all necessary information needed for general operations. Second, there are enough screens so that most tasks, including monitoring, can be viewed at the same time without requiring switching displays on a screen. The locations are arranged so that related information is naturally located.

The choice of which resident displays to locate on which screens is made so that those requiring close interactions are center and lower. Note the two working screens (1 and 2) are at the bottom center and right. Here is where the operator would be monitoring specific control points and related variables. Or he may be intervening to manage an abnormal situation or other event, for example, by altering a controller setpoint or moving a valve. Close at hand (bottom row, to the left) would be other advisories to assist him. For example, the screen could provide assistance, as needed to augment the operator's current activities. This assistance would show procedures, provide relevant background analytical data, alarm diagnostic assistance data, and the like.

The displays that provide more global information are located above. These displays, as a package, complement the operator's role of observing and managing. To the top left

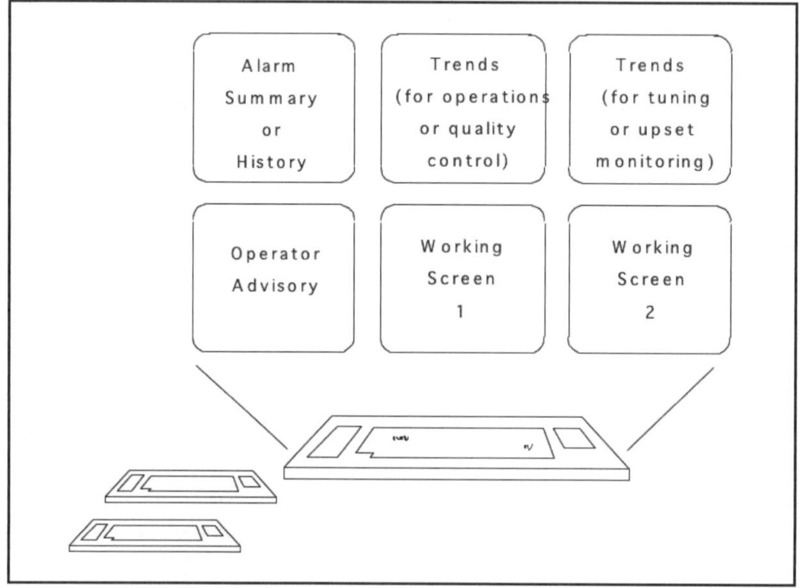

Figure 12.7.1. Suggested physical display arrangement

would be the screen dedicated to the alarm system. The screen above center provides overview information on how well the process is working. This aids the operator working to ensure the plant does not go astray. He may also be working on process improvements. The displays that support improvements are located on the upper-right screen. Again, we have a hierarchy of detailed displays:

- Task displays within easy reach in the front and sides
- Overview information above, within easy sight

Modern Displays

Six screens that were so radical and extravagant 20 years ago are now rather commonplace. Local control rooms, once located in satellite areas with each area separate from the other, have been replaced with a single centrally located one in a protected area remote from the actual production equipment. Figures 12.7.2 and 12.7.3 show photographs of a typical central control room at a modern petroleum refinery. There are six operator areas within view and a seventh (out of view) for engineering and maintenance.

Figure 12.7.2. Typical modern operator area

Figure 12.7.3. Typical operator areas within a control room

There is also a growing tendency to group closely linked plants with separate operators together into a structured super operator station. Figure 12.7.4 illustrates one. Notice that this arrangement also includes the increasingly common addition of large overview screens located high above the operator areas so as to be visible to others. This permits important information to be shared among operators, so vital for crisis management.

Also present are optional special-purpose hard-wired alarms and the ever-present collection of emergency shutdown switches and interlock management hardware.

A view of the expanded operator area is shown in Figure 12.7.5. It is easy to appreciate how the screen layouts and presence of large overhead screens can facilitate operators viewing and sharing information and collaboratively working problems.

We now turn our attention away from the number and arrangement of screens and toward the content and purpose of the displays placed on those screens. This is actually where the awareness of situation is to be found.

Hierarchical Display Architecture

A three-level display hierarchy (fig. 12.7.5) delivers a robust structure that encourages ready access to information while at the same time keeping important situation context and promoting efficient navigation to go deeper. The first two levels follow an expected progression from the general to the more detailed. But the third one is a departure. It does not provide a more detailed view. Rather, it is presumed that if the first and second levels are not capable of providing the needed detail, another paradigm is needed. That paradigm is support, not more detail. This structure is a very clear departure from

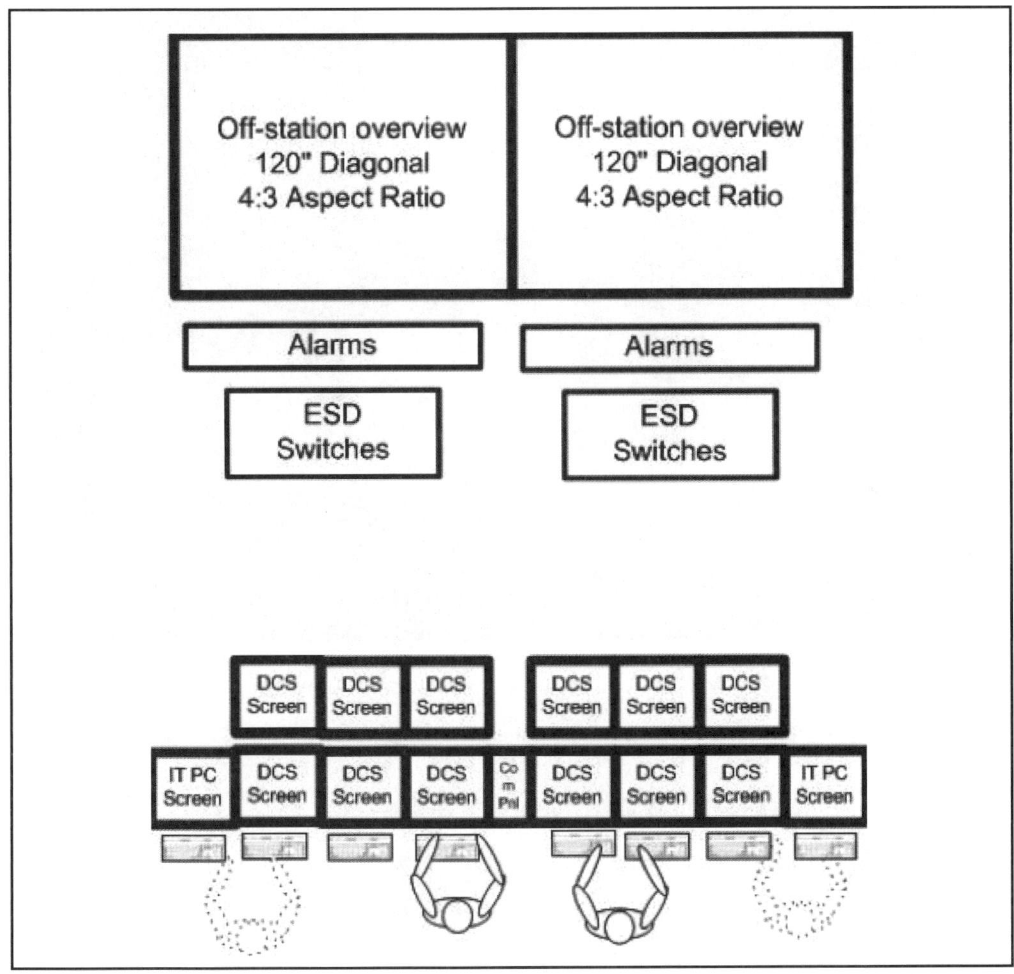

Figure 12.7.4. Expanded screen layout architecture

the widely accepted norm for navigation and structure: "clicking to oblivion" that characterizes most personal computer operator interface exchanges.

Clicking one's way down a "tree" of choices to locate information is a tedious, time-consuming activity. If the tree arrangement logic differs from the mental expectations of the user, it can be distracting. Even if there is a good match, it takes time and requires a constant stream of locating the click target area, using pull-down menus to provide the accepted choices, deciding which choice matches the current need (and often finding no good match, one must click down and back to attempt to find the closest or the least not-closest match), selecting the choice, and going down to the next level. As you might imagine, this approach is always frustrating and often becomes nonproductive during

Figure 12.7.5. Expanded screen control room

FIRST
Produce an overview
(provides a one-view glimpse of everything that might go wrong)

SECOND
Provide ready access to important details
(second line of defense—confirm or deny)

THIRD
Provide detailed guidance
(if problem requires support—here is where it is)

Figure 12.7.6. Display hierarchy levels

times of operator stress. At the very time it needs to be at its best, it fails. This is why operator-machine interaction uses as little clicking down as possible.

The Overview Level

An overview must provide a way to take in the entire operator's area of responsibility. It must show summarized alarm statistics, grouped status information, and important conditions of both upstream and downstream units and provide ready navigation to everywhere. It must communicate whether or not the production unit is operating well. If well, how well? If not, then how not and where? And it must do this without requiring the operator to search it out.

Let's take a look at an example of how individual displays might be organized to provide both the continuous ability to maintain an overview of the process as well as the necessary information and handles required to manage problems. We start with an unusual form of overview, shown by Figure 12.7.7. It is called a taxonomy view. This example permits a very clear illustration of the essential elements of an overview, without the distraction of the familiar. But please do not go out and build one just yet. The illustrated process is a refinery boiler unit. Colocated there are the plant steam and plant

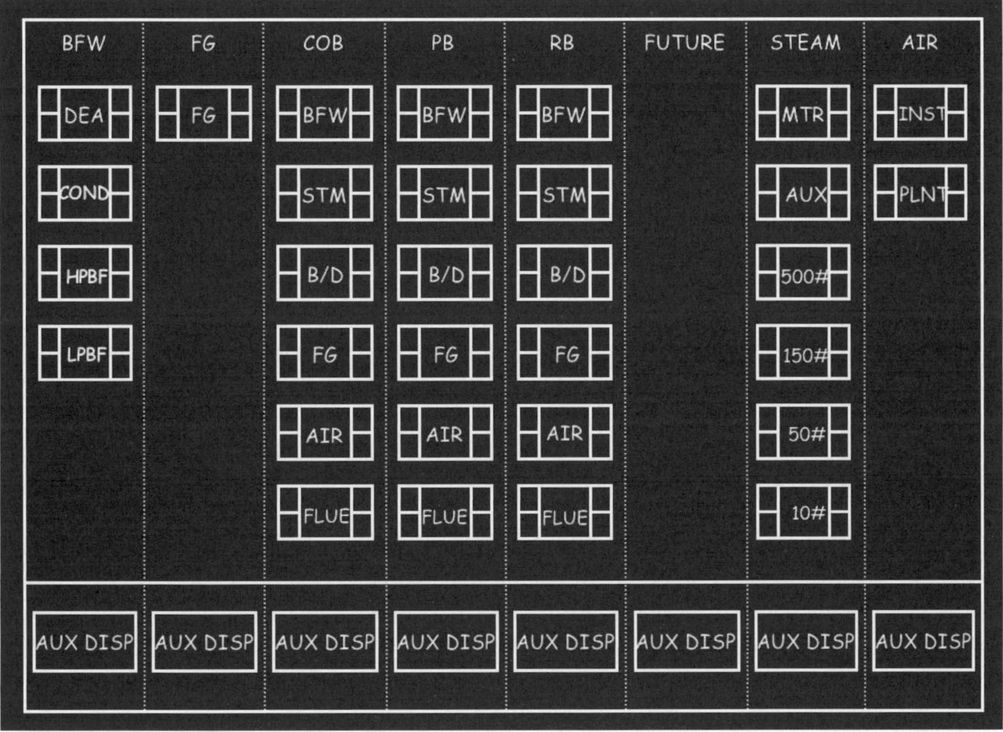

Figure 12.7.7. Taxonometric-graphic overview level

air systems. Taxonomy views most often are used when the particular plant has a number of very similarly functioning units, regardless of how different their physical appearances might be. This arrangement has the benefit of providing information in the same format for each element. Normal means the same thing for each, and so for the abnormal. Knowing what to do with one during an abnormal situation will serve as a close guide for what to do for the others. Shown in the figure are places for four boilers: COB stands for a CO boiler, PB is a Power Boiler, RB is a Riley Boiler, and FUTURE is for a future one. Each boiler has six constituent components: BFW is the boiler feedwater, STM is the produced steam, B/D is the blow-down and condensate, FG is the combustion fuel gas, AIR is the combustion air, and FLUE is the burned gas exhaust. The columns for BFW and FG are for the shared boiler feedwater system and fuel gas supply. The BFW component includes the following: DEA for deaerator, COND for condensate system, HPBF for the high-pressure boiler feed, and LPBF for the low-pressure boiler feed. STEAM and AIR are the colocated utilities.

Those curious rectangles with small square boxes in each corner represent the subsystems. The central large square within each rectangle is also a navigational target. The corner boxes are placeholders for alarms. In this plant there are four priorities of alarms. Each one has its own place. If there were to be three high-priority alarms active in the steam system of the CO boiler, then the upper-right-corner box would be flashing red with a white "3," and likewise for the other areas and alarms. At a glance, it would be clear to any operator what the alarm status of his plant would be. The boxes at the column bottom navigate to a preplanned auxiliary display.

At perhaps the other end of the design spectrum, we have a text-based overview display shown in Figure 12.7.8. Like the taxonomy view, our display is divided into its constituent areas with each subarea clearly defined. Alarms are shown in boxes also. However, this view contains extensive process data that have been deemed by the designers to be important for an overview.

It is likely that neither of the two overview types is familiar to you. Their use here is not a recommendation. They were employed to demonstrate certain concepts that would have been too subtle to show in a more conventional overview. The recommended overview is found later on in the chapter.

The Secondary Level

The secondary level is for details. It shows alarms, the process details, provision for monitoring and adjusting where needed, and again, navigation to other needed places. Figure 12.7.9 illustrates a graphic-based secondary-level display view. Control handles and specialized information displays will appear as "pop-ups" here, not as another level below. Pop-ups have the advantage of maintaining situation perspective, which we require. But they can be distracting if not done carefully. This particular secondary illustration was taken from the example that used a text-based overview level. A change in level does not mean that the design paradigm cannot be shifted.

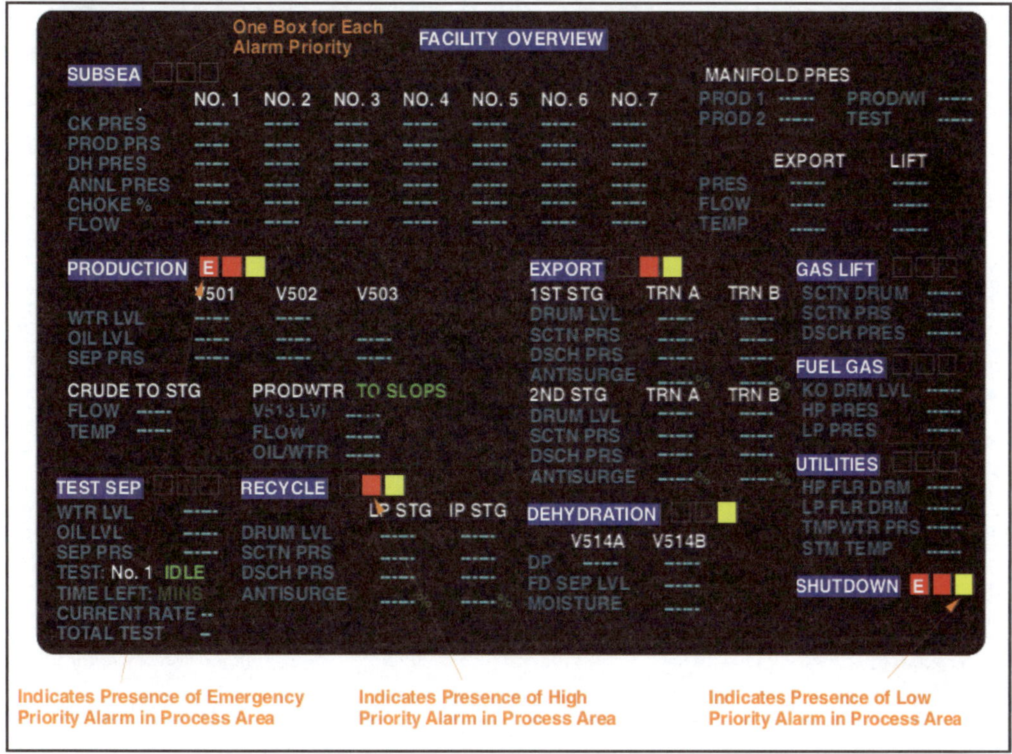

Figure 12.7.8. Text-based overview level

The Tertiary Level

By the time the operator needs to go to a tertiary level, more detail will not likely help. Level three is not designed to provide even more detail than the secondary level. We must presume at this point that the operator needs to better understand things. Usually better understanding is best supported by advice and amplification. This situation is not unlike initially reading a newspaper. From far away, all that can be seen is that there appears to be a paper. Moving closer, we can easily read the headlines and perhaps key lines. Moving closer still, we can read the text. Moving closer still just provides larger text; there is not much more to gain from it.

If our operator needs more than provided by the secondary level, we know that something different is required. Here is the place for highly specific assistance. Examples of more and different assistance include the following: access to procedures and design documents or historical incident documents, detailed laboratory reports, and such. And again, we need navigation to all other levels. Figure 12.7.10 illustrates the alarm response information relating to an active alarm. Recall the discussion on alarm response sheets in chapter 7 on rationalization.

There are other interesting ways to provide operator guidance. Figure 12.7.11 illustrates a causality inference diagram. This diagram is a quick graphical way to show what

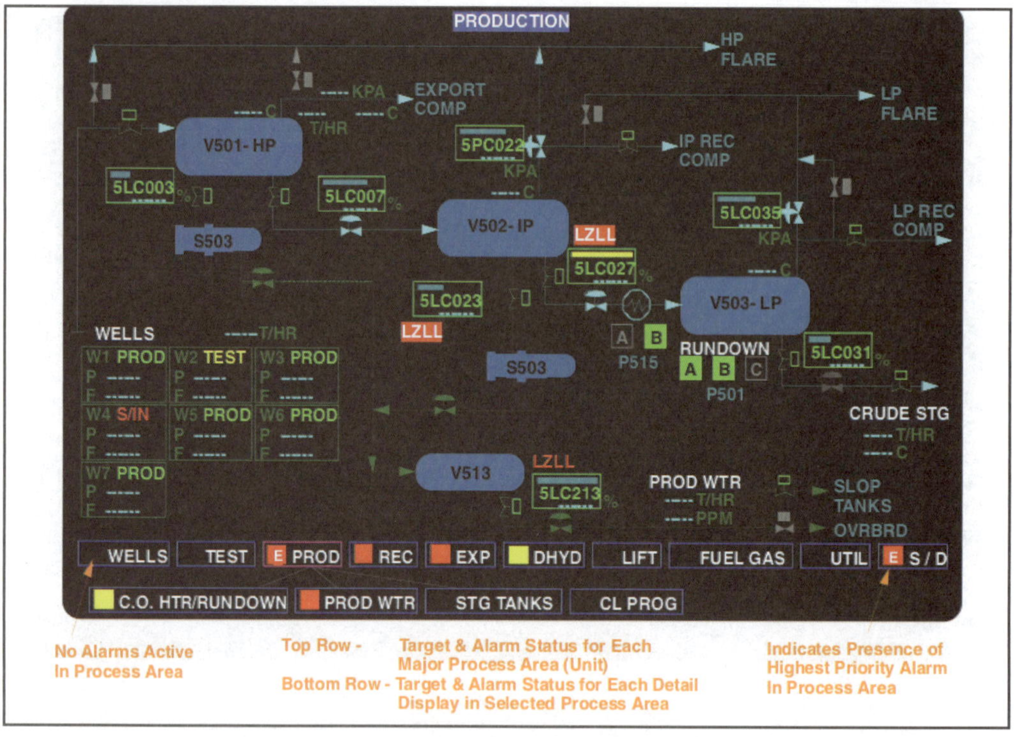

Figure 12.7.9. Graphic-based secondary level

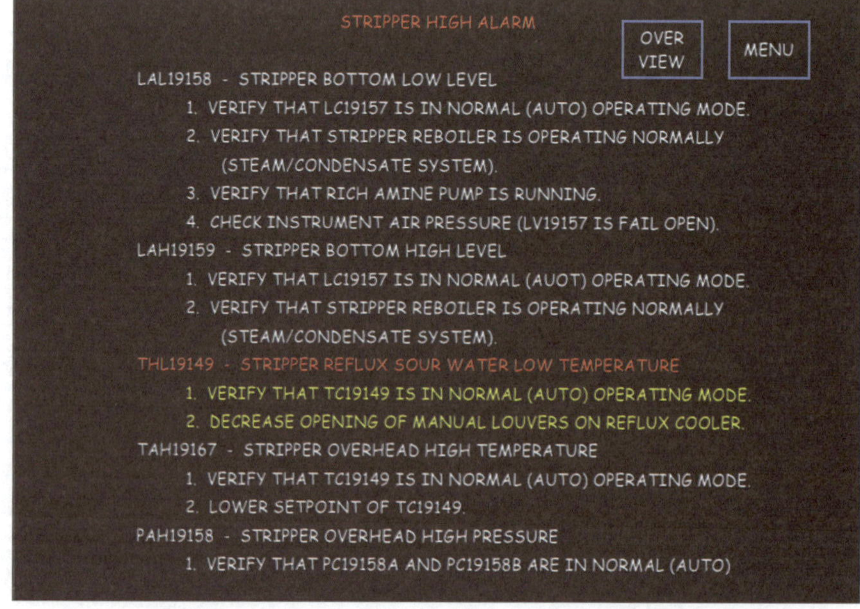

Figure 12.7.10. Alarm response at the tertiary level

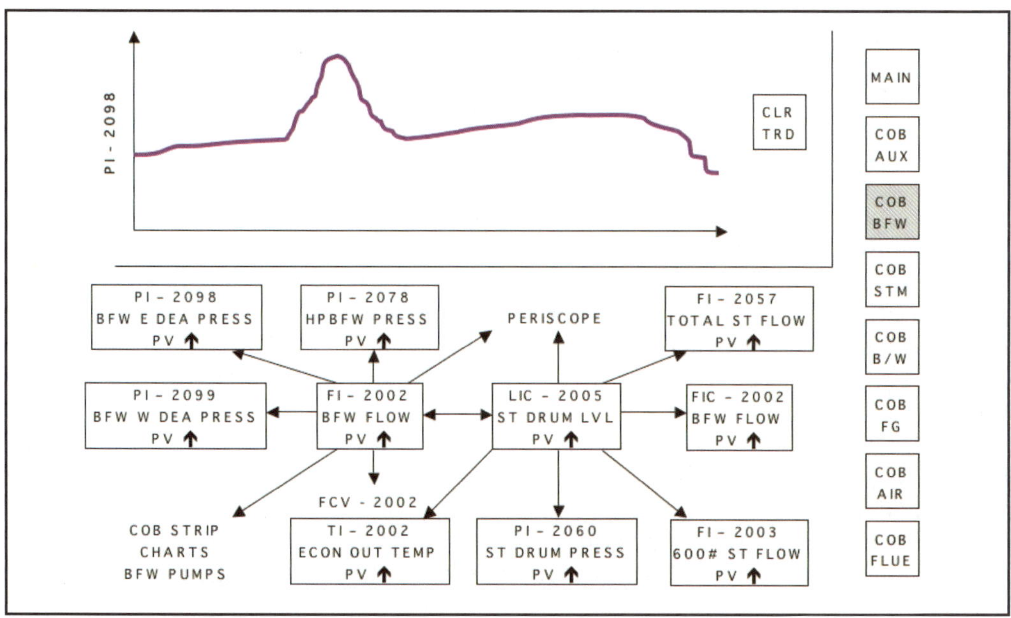

Figure 12.7.11. Causal-inference tertiary level

variables affect which process conditions. Figure 12.7.11 is an example diagram together with a graph of the offending process variable (which is either in alarm or abnormal, and is our operator's center of attention). Key variables are shown as boxes, and directional arrows show the causal links between those variables.

Here is how it works. Assume that there is a low-pressure alarm on the boiler feedwater deaerator pressure PI-2098 in the COB. The graph in the inset is for this pressure trend. The operator first looks for direct reasons why the pressure PI-2098 might be increasing (problems with a valve, controller malfunction, etc.). Locate the PI-2098 "box" in the upper-left corner of inference diagram below the graph. Within the box you can see a signal directionality arrow pointing up. Observe that the only other box linked to PI-2098 is FI-2002. Since the arrow points from FI-2002 toward PI-2098 and this is the only such box, except for a direct problem within PI-2098 (which is the first thing to be investigated), this is the only other basic process malfunction cause for an abnormal PI-2098. Note that the signal directionality for both is up. Signal directionality is depicted by a small arrow inside each of the "tag" boxes. Thus, increasing FI-2002 leads to increasing PI-2098, so we look for any reason why FI-2002 might be abnormally increasing.

12.8 NAVIGATION

There is no wind at our backs and no smell of salt in the air, but navigation is navigation, be it on the open sea or sitting at the consoles of a production control system. The job is to know where one is and how to get to where one needs to go. Some will say that the job of control room operator is even more difficult than that of sailor. With only recollection to fall back on, there are precious little tools available to find the lost. Clear and effective support for operator navigation is an important tool. Situation awareness requires locating and following up on clues, hunches, and problems. Operators must easily and quickly be able to locate and move to appropriate displays to view information and make control or other operational actions for the manufacturing process.

Ever wonder how vexed an operator can get when he has a good idea of the specific tag he is looking for but cannot seem to find it? His tools for "dead reckoning" include hunting around various screens thought to be close and, if that fails, keying in guess after guess for how the tag might be configured in the database, in an often lengthy trial and error process to locate the needed point. Who knows how FIC 20710 might be coded? It could be FIC-20710, or FC-20710, or FC 20710, or FIC-020710, or. . . . You have probably experienced this yourself.

Figure 12.8.1 illustrates an actual start-up for a plant. This one lasted nearly 6.5 hours. The process was significantly disturbed for over 2 hours during which there were a number of alarms. In searching around for what to see and where to go, our operator made 459 moves; 356 were made during the upset period. A portion of this searching time is shown in the figure. We can see 257 of the searches. The directional lines linking

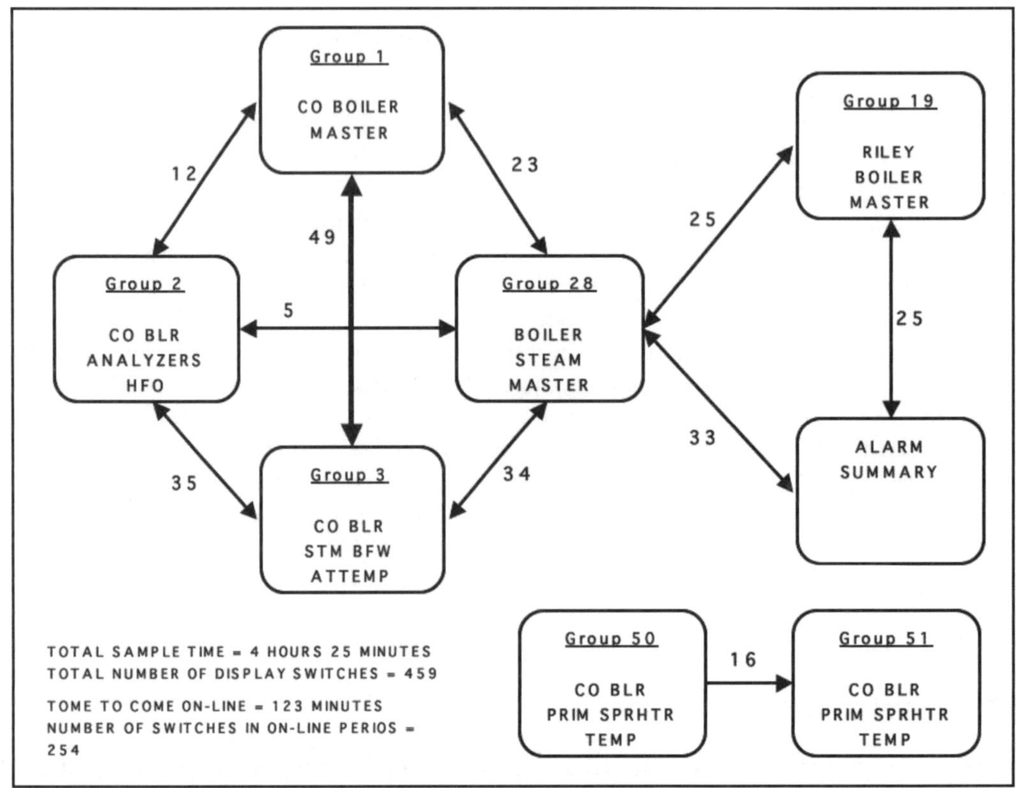

Figure 12.8.1. Example display linking during plant start-up

the boxes depict a screen change. Most are bidirectional, meaning that there was a lot of going back and forth between the linked displays. The numbers near the lines show the number of changes. The ultimate reason of the upset was a misbehaving flow measurement. Ultimate cause was a plugged impulse line. It was not discovered until the unit had lined out.

Effective navigation is facilitated by the following:

- Logically grouping information and controls so that the operator can see and change things a bit without writing down or remembering, and without having to change views or displays
- Placement of clear and efficient navigation targets on relevant displays that have the ability, when needed, to be "in context" with the situation
- Ability to bookmark and place mark displays so that operators can easily and quickly call and recall appropriate displays

12.9 NOTIFICATIONS INSTEAD OF ALARMS

Recall the point made earlier in this book that one does not solve a problem by subtraction alone. It is not enough just to keep our alarm system from being overused when it is used to provide "wake-up calls" for the unusual, or notices of things to remember, or any of the other excesses we have mistakenly placed on alarms. Yes, we do remove those from the alarm system, but these other needs remain legitimate. Most, if not all, needed to be communicated to the operator somehow.

In this light, alerts, also called notification systems, were born.[6] What is shown in Figure 12.9.1 is an early prototype design developed for the Honeywell TDC3000 system. It is nothing short of magic. What it represents is a fully integrated capability within the PCS. It is structured to be self-documenting, with a full complement of management and control tools. It is able to link easily into the plant tag database with the same integrity as any other controls function and incorporates a robust set of logical tools. Pictured is a particular notification designed by our imaginary operator Steve Waite. The notification monitors "High Steam Flow to the Exchangers" by comparing FIC-0023 against a target value of 1275 that will alert him whenever that flow is greater than the target for 10 consecutive minutes. The notification message will read, "FIC-0023 is greater than 1275 tph [tons per hour]."

The really important part is what it means. It means that although Steve Waite felt it very necessary to watch the heat exchanger flow to prevent a problem, he did not have to do that "watching" himself. He appointed an agent to do it for him. He could actually forget about it and go on to the many other tasks and distractions so much a part of operating a modern manufacturing plant. Whenever, if ever, the watched-for situation occurred, Steve would be notified appropriately.

One of the more interesting stories surrounding the early use of notifications is who found them the most useful. Operator notifications were created for the board operator to free up the alarm system from a large area of abuse. Interestingly, the earliest adopters were outside operators. For the first time, they were able to create "alarm clocks" for all

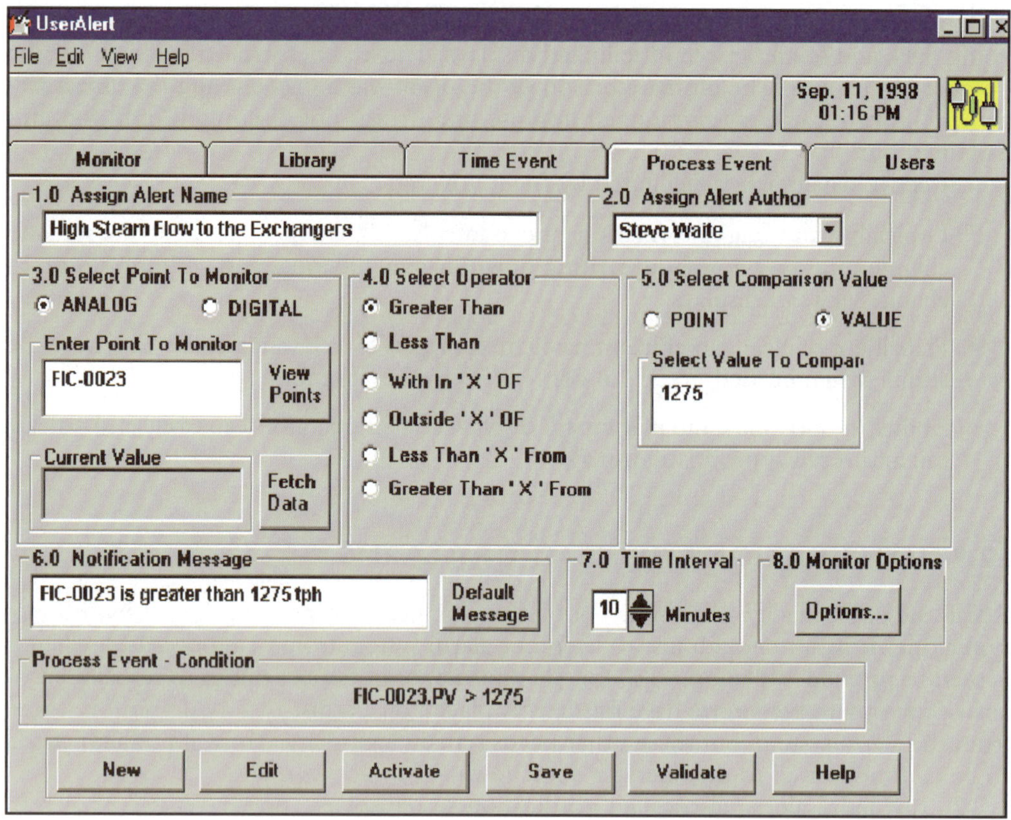

Figure 12.9.1. Notifications construction screen

of their important scheduled duties during shift. An outside operator's world is divided between routine walking around and generally checking things out, with specific duties to take samples and keep an eye on equipment that might be thinking about getting into trouble. It is easy to become absorbed in the task of the moment. And being absorbed, it is just too easy to miss other scheduled duties.

12.10 PERCEPTION PROBLEMS WITH VIDEO DISPLAYS

> **Truth**
>
> It is often said that the "truth shall set you free." And that might be so, but we really know that we are never guided by the truth! Never! We are guided by what we believe to be the truth or, as often as not, what we believe to "not be true."

It is now time for a bit of fun. In this section we will see how our eyes can trick us. And if we trick our eyes and do not know that they are being tricked, we trick our minds as well. And if we trick our minds, we perform false acts. False acts are seldom benign. We therefore hurt or get hurt. So all this is not simply parlor games.

Relationships and Size

The following illustrations and illusions are shown to sensitize us into becoming aware not to ask operators to make deductions, comparisons, or other judgments without providing an adequate frame of reference. Figure 12.10.1 asks one to judge whether lines imbedded within a distracting framework are straight and, if straight, whether they are parallel to each other. Even when what might appear to be convincing reference lines are added, the answer is not a foregone observation (see fig. 12.10.3). All of the horizontal lines in the figure are straight and parallel to each other.

Figure 12.10.1. Are these lines straight?

Another similar situation occurs often. We present material to the operator, either on different displays that have to be switched back and forth or on the same one, but presented it in a way to make a called-for judgment difficult. Such an example is shown in Figure 12.10.2.

Both box A and box B represent exactly the same virtual area.

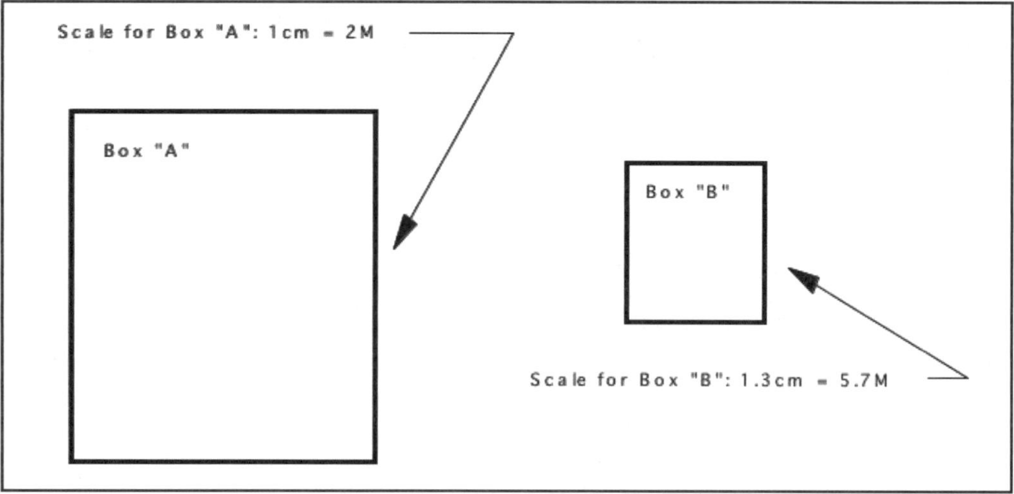

Figure 12.10.2. Areas and scale factors

Figure 12.10.3. Lines parallel or not (with red straight reference lines added)

Coding Conflicts

You have heard the old saying, "Do what I say, not what I do." This type of conflicting messages is also easily found in display designs. For example, work your way down the list of words in Figure 12.10.4. Speak out loud the *color* of the word—do not read the actual word! See how fast you can go.

Distracting, was it not? Observe that if these words were written in Chinese or Icelandic, you would have no problem at all. You would not be distracted by the actual

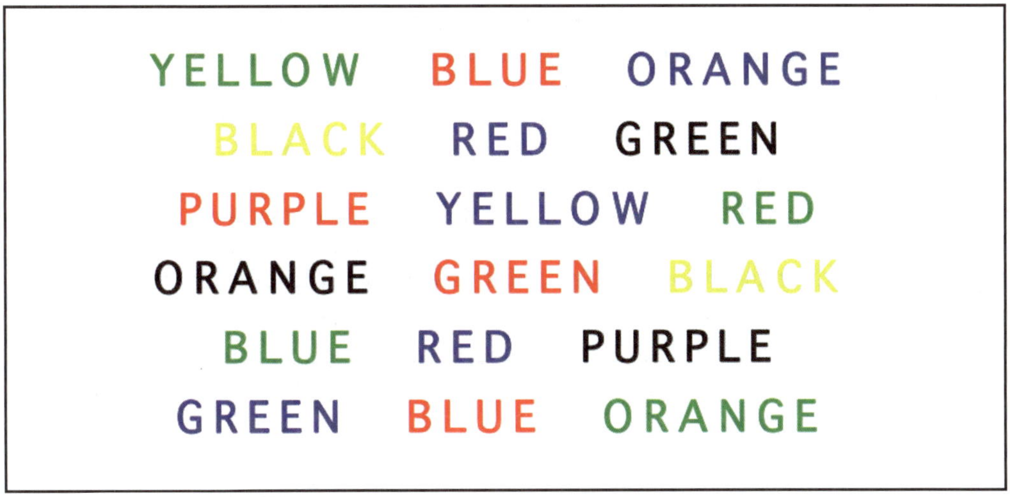

YELLOW BLUE ORANGE
BLACK RED GREEN
PURPLE YELLOW RED
ORANGE GREEN BLACK
BLUE RED PURPLE
GREEN BLUE ORANGE

Figure 12.10.4. Visualization and information conflicts

words, since they would be meaningless to you. The "conflict" between the color and the words with meanings would not exist.

Let's take another more subtle and much more common example. Figure 12.10.5 contains two identical situations. The physical parts being represented are exactly the same. Each half contains a pump connected to a vessel. The pump and vessel are connected at exactly the same place. Yet the way they are drawn to appear on the page (read that "shown on the operator's screen") suggests that they are different. They are not. But the down-leg of the left half gently suggests that something important might be indicated. The simple

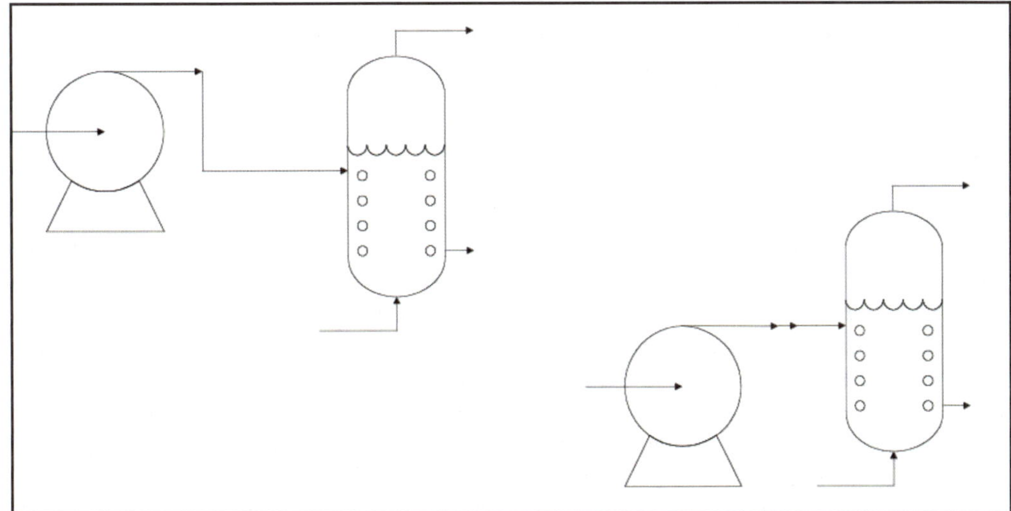

Figure 12.10.5. Visual positional coding conflicts

act of changing the direction of an otherwise uneventful connecting line can cause one to wonder, "What is it that I need to pay attention to but might be missing?"

If you think this might be conveniently contrived, think about the case where the plant installed two heat exchangers in parallel to achieve the needed heat transfer. The plant may have purchased them this way since two smaller ones may have been less expensive than one larger one. Or it could have been any number of other reasons quite unrelated to any specific need for duplicate ones. The recommended depiction is not two items. Since they are instrumented as one and operated as one, they should be depicted as one. Yet there are many graphics that depict both exchangers as shown in Figure 12.10.6.

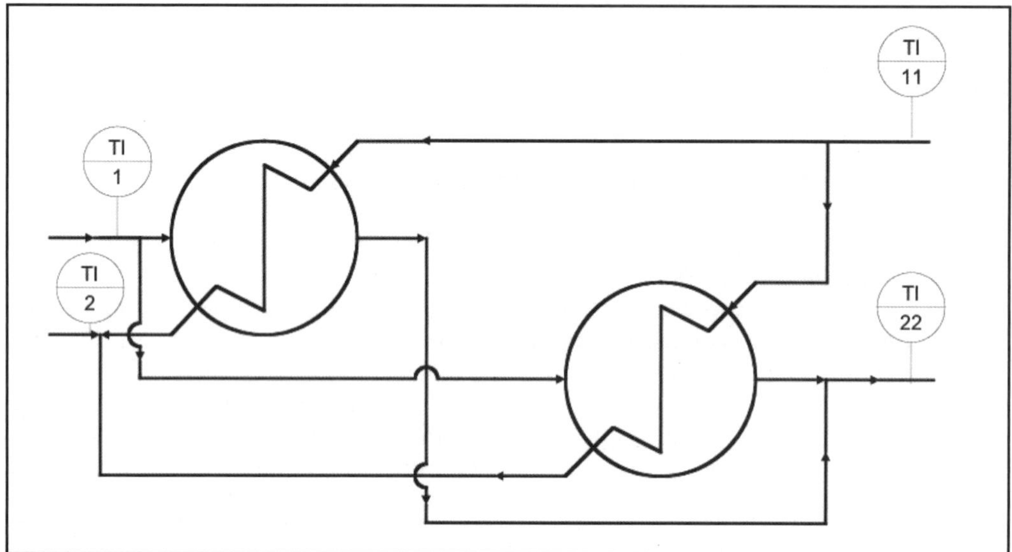

Figure 12.10.6. Two exchangers in parallel—a not recommended depiction

Color

We now come to a powerful lesson. Please follow along carefully if you want to feel its full impact. And please, please, refrain from reading ahead. For if you do, you might surely miss the feeling of discovery.

One of the first principles for the use of color in operator graphics is that color is used to delineate, not to decorate. Misused, color can obscure or even hide the most obvious of things. Consider an illustration for the concept. Take a close look at the color blocks shown by Figure 12.10.7.

What you see is a set of blatantly colored blocks. Please imagine this to represent a sophisticated, highly colored operator display. The color might be used to denote different processing aspects of the raw materials. It might be used to identify the difference between natural gas and propane, between hot and cold lines, or anything else

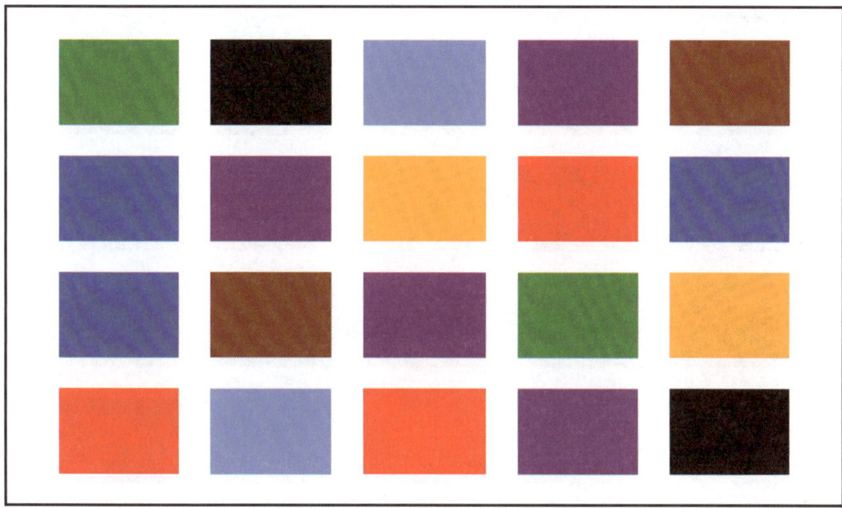

Figure 12.10.7. Example of a highly colored operator display

the designer thought was worth identifying as important to ensure that they were not confused with something else. The first thing we want to notice is that the red blocks stand out. And so do the black ones. And so does the diagonal pattern of the deep purple ones. What are the most important parts of the graph that its designer wanted to point out to you? Not sure, are you? In fact, the more you look at the collection of colors, the less it seems important or makes sense. And you are right.

But the fullness of the lesson of the colored displays is yet to be played out before you. Please do not turn the page yet. Turning too soon can deprive you of the discovery of this important lesson. Patience will be rewarded. Figure 12.10.8 (on the next page, but read this before turning) shows the same processing unit as the one above. You are going to be asked to find whether anything important changed.

Without flipping the page back, now turn the page forward, stay there, and have a look.

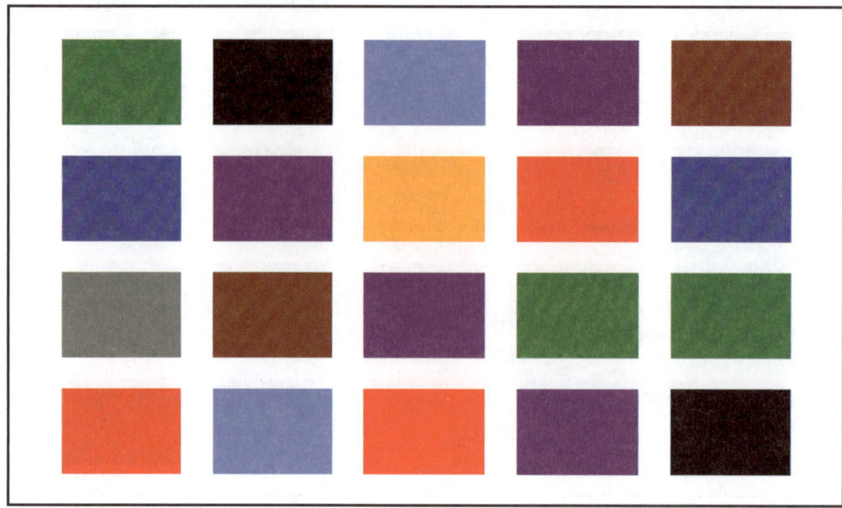

Figure 12.10.8. Changes to a highly colored operator display

OK, what do you see? Hard to figure out, is it not? All this color seems to get in the way of information. Do you see any changes? How many? Are you sure? Let's check. Now turn back quickly and see the first display again. Go ahead, turn back and take another look.

By now, you ought to be well on your way to becoming convinced about the power and the danger of color. Color has only one purpose: to identify and separate the important from the less important or unimportant. That is it. If we attempt to use it for more, you see what will happen.

See page 424 for the solution.

We return to our example. Here is what changed between Figure 12.10.7 and Figure 12.10.8. In the leftmost column, the third box from the top started out as blue and then changed to gray. In the rightmost column, the third box from the top started out as orange and changed to green (fig. 12.10.9). That is all. Nothing more changed.

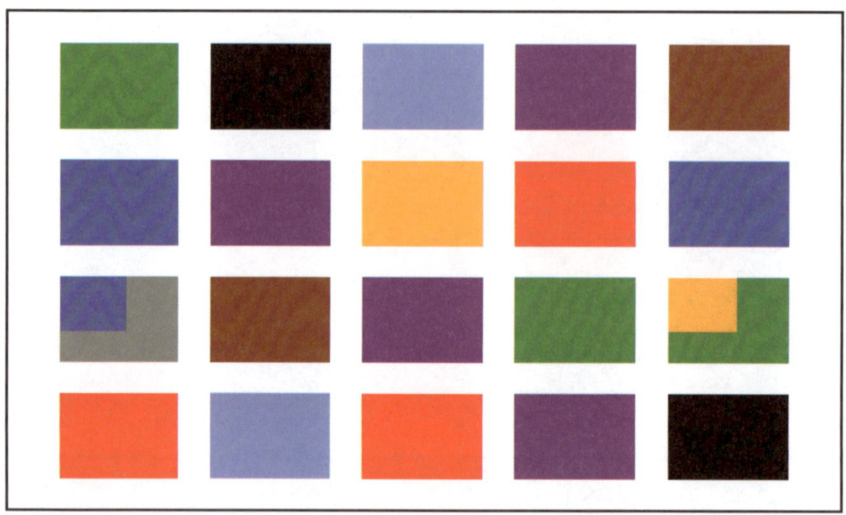

Figure 12.10.9. Changes to a highly colored display—what changed

Now, consider a neutral-based display with color used only for information (see page 425).

Figure 12.10.10 represents a slightly different scenario in exactly the same plant as the earlier highly colored one. There are the same plant components (twenty units all arranged exactly the same on the display). Color is now used to show one area in abnormal condition and another in upset. Without a thought, you can tell which ones they are. You will base your judgment on the colors red and orange and the accepted differentiation that red is considered to be more serious than orange. You will easily spot which one was the more seriously affected. Now, in exactly the same fashion as we did earlier, let's see what our view looks like when something changes. Figure 12.10.11 (on page 426) shows this situation.

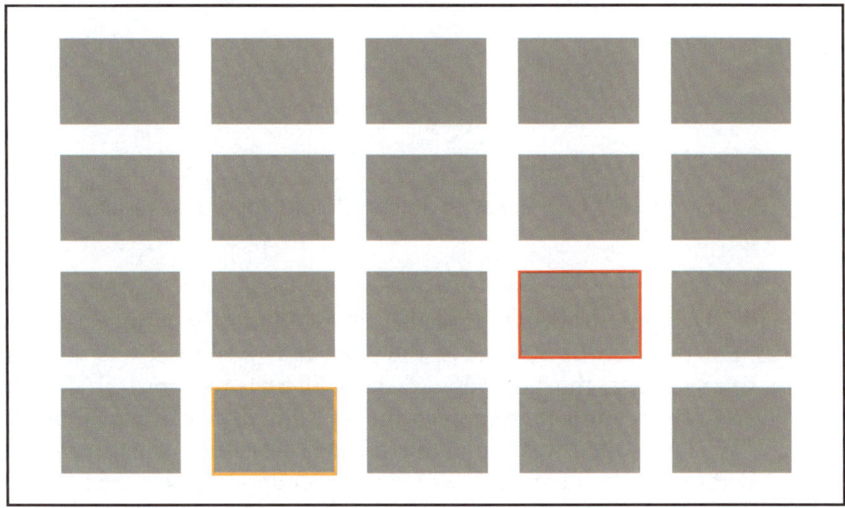

Figure 12.10.10. Example of a neutral-color operator display

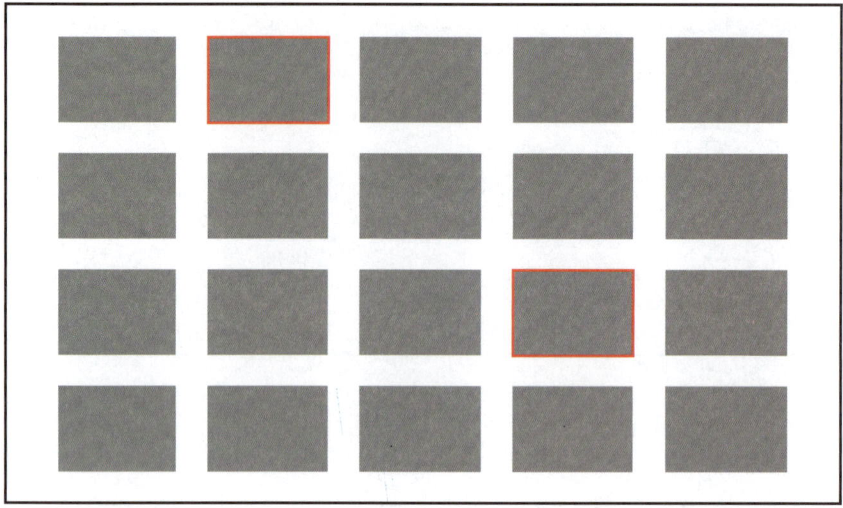

Figure 12.10.11. Changes to a neutral-color operator display

It is dramatically simple to decide what has changed and how out-of-the-ordinary the change was. For the record, the part of the plant that was upset was the box in the fourth column from the left, third one down (shown with a red border). The part that was abnormal was in the second column from the left, fourth one down (shown with an orange border). What changed was the abnormal area became normal (orange border disappeared) and another area became upset (red border appeared on the first box in the second column from the left).

You should be convinced. Now, please consider graphical standards and usage to match.

Comments

You have clearly seen the importance of color, object placement, and connection orientation on the viewer's ability to orient and focus. Moreover, you have seen how easily the eye can be confused or misled by our intrinsic human visual process. But please do not for a moment dismiss all this as fluff. You must not feel that way now that you know about those things; they will not confuse you in the future. Yes, these items were raised here somewhat in the context of "parlor talk." However, they are as real as anything else we perceive and as omnipotent as gravity. One relaxed moment and the eyes and brain are going to go back to what they do. If what they do is not what you would like to see, make sure you manage the situation. To reduce the problems, we design in the solutions.

12.11 NEW OPERATOR DISPLAY DESIGN

We are now ready to "cut to the chase," as they say in movies. The previous material in this chapter was preparation to understand and appreciate what the new operator display and navigation designs are all about. Now it is time to take that look! What you will see here depicts what is considered the best practices for graphical operator displays. They were pioneered by the groundbreaking work of the Abnormal Situation Management (ASM) Consortium.[7]

Coding Schemes and Icons

Icons are powerful codes. Icons can be selected to evoke a ready understanding. These codes are not hidden (as in spy and intrigue stories). They are meant to be in-your-face obvious. That is what affordance is all about. This is the stuff that speeds up understanding and minimizes confusion. Icons both evoke a meaning and confirm that what is evoked is correct. While it might seem to be redundant, such redundancy is important. Our minds are always testing truth. When two points come together and the message is the same, we tend to believe.

Examine a set of icons for operator communication of messages and warnings. Figure 12.11.1 illustrates that color, shape, and duplication are all used to communicate and reinforce. For example, four classes of communications are identified: malfunctions, goals, alarms, and collaboration. They represent the full spectrum of what could be flagged for operators. Each communication has a unique shape. The diamond is used for

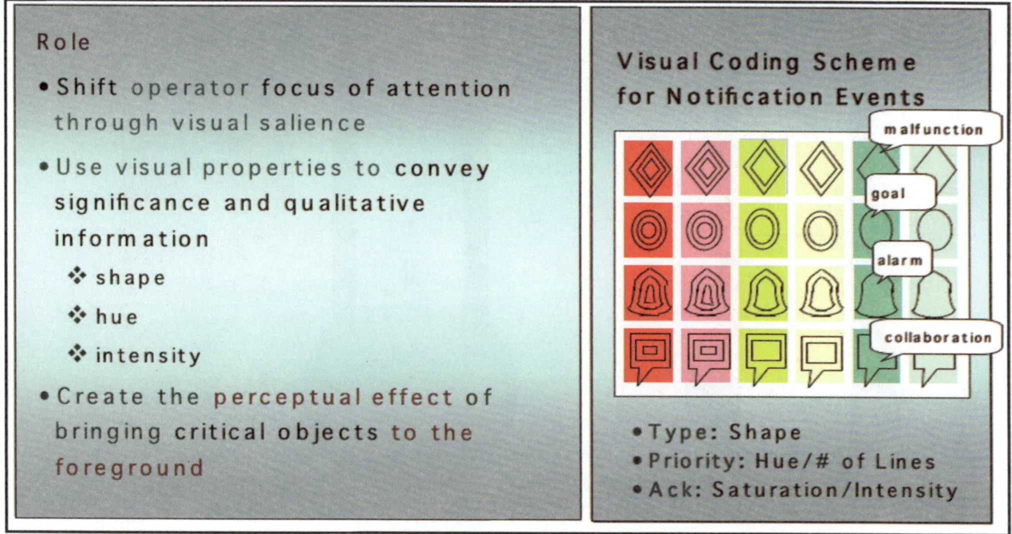

Figure 12.11.1. ASM-style visual coding scheme

malfunction, the oval for goals, the bell for alarms, and the call-out box for collaboration. Each can have three levels of urgency or violation. For the lowest level of urgency, we find that the shape is a single outline and the background color is a green. Two shades of color are used: the darker shade indicates that the operator has not yet acknowledged the information; the lighter shade indicates operator acknowledgment. This color-coding is repeated for the middle level of urgency, yellow, and the highest level, red. Moreover, for the middle level of urgency, the outline shape is duplicated. For the highest level, it is triplicated. Redundancy is used for reinforcement. Anywhere we need to inform the operator of an important aspect of his plant, we can use the appropriate icon and it will be immediately obvious what type of message is being proffered and what its importance is. This is simple and effective.

Icons are more than placeholders and announcers of warnings. They also communicate status. Take the thermometer-like icon shown in Figure 12.11.2. We call them context icons. First, the obvious. This icon is used to convey temperature. It is used on graphics to flag important temperatures that the operator will want to observe. They serve as indicator measurements of normalcy or pending abnormal operation. Looking at the left icon in the figure, note that all that is visible is the shape (a temperature-like thermometer), a range (the white vertical box), and the current measurement (the blue bar about halfway up the range). What you "conclude" from this is that the displayed temperature is within its normal range of expected values. We know it is normal because it does not look abnormal. We easily see in the right icon that our temperature is abnormal. It is clearly too high. First, the normal range box now has all its abnormal areas

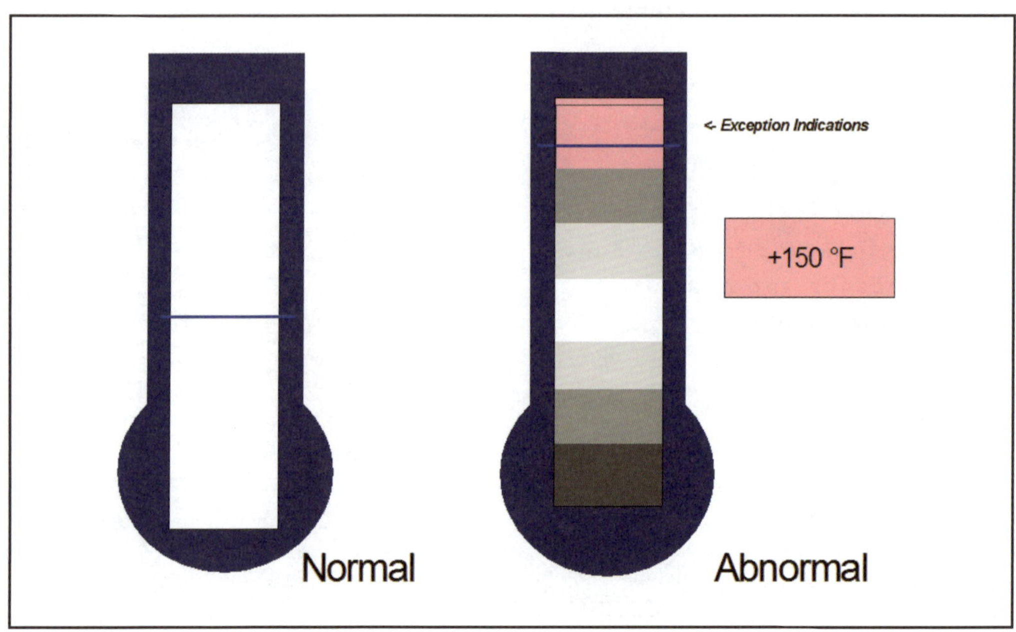

Figure 12.11.2. ASM-style context icons

identified, from slightly abnormal to grossly so. We see that the current value, again a simple blue bar, is in a red area. Red means trouble. So this temperature must be badly amiss. And to reinforce this, we have the current actual value displayed against a red background. It all looks abnormal.

Overview Level

Put it all together. Illustrated in Figure 12.11.3 is a best-practice overview level display. Notice that all the process elements are represented by neutral gray. This is what we expect. Process elements are placeholders. They are only there to show causal relationships and whether things that are OK or wrong. Arrangement conveys causality. Color is used for information only. There is the legacy video inset for the flare camera! It always needs to be in view.

Let's take a tour. The central area shows the five main components of the plant: feed preheater, combustion air, riser/regenerator, waste heat, and fractionator systems. In addition to the four critical riser/regenerator temperatures, we find five other flows

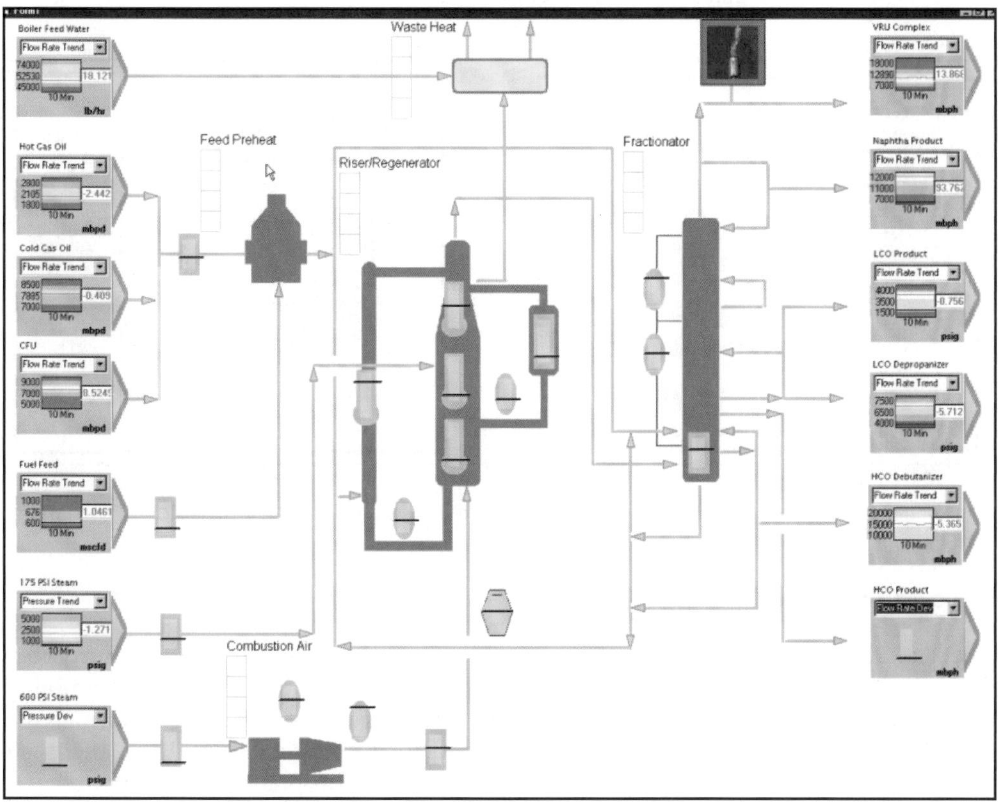

Figure 12.11.3. ASM-style primary (overview) level display

(vertical rectangles on flow lines), six pressures (ovals), and one analytical value (elongated hexagon). Each is located appropriately. We are assured that if the operator can see that these variables are normal, there is an exceedingly good chance that the entire process will be normal as well.

You are well aware that a process can be normal, yet outside influences may change and cause problems. Therefore, also shown in the overview are the seven most critical inflow variables to our plant and the six critical products our plant produces. These are shown as miniature graphs along the input side (left) and output side (right) of the overview. These graphs use the same paradigm as icons (fig. 12.11.4). Where normal, the range box would be white. If abnormal, the range box would have all regions identified (similarly to the temperature icon previously illustrated). Graphs will be used because they convey significantly more information than just current values.

Shown on the overview, but barely visible, are the areas for operator notifications and alarms. You should just be able to make out four-box stacks near each of the major plant elements. Any of our four types of event notifications (from fig. 12.11.1) can appear there in the appropriate color, shape, and flash or not.

Our overview is complete. Using it, the operator can observe at a glance what is okay and what is not. And that is what an overview is for.

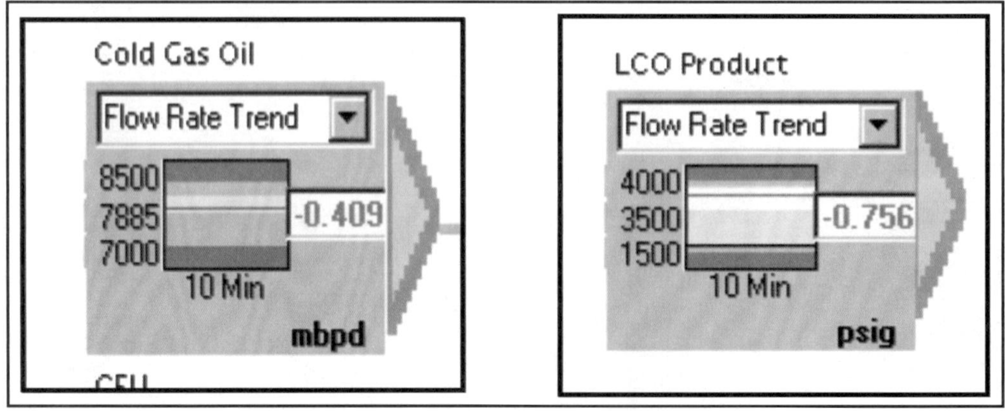

Figure 12.11.4. ASM overview—key input and output variables

Secondary Level

The secondary level provides more details to support what is already known or suspected from viewing the overview or primary level. If the overview looks normal, examining the secondary displays can be queried to verify this. If the overview looks abnormal, the offending secondary display view can be queried for more details. And yes, there are more than one secondary display views; each covers an appropriate area component.

The template of views for the secondary level is illustrated by Figure 12.11.5. In our example, it is divided into four information regions. The upper left shows an overview

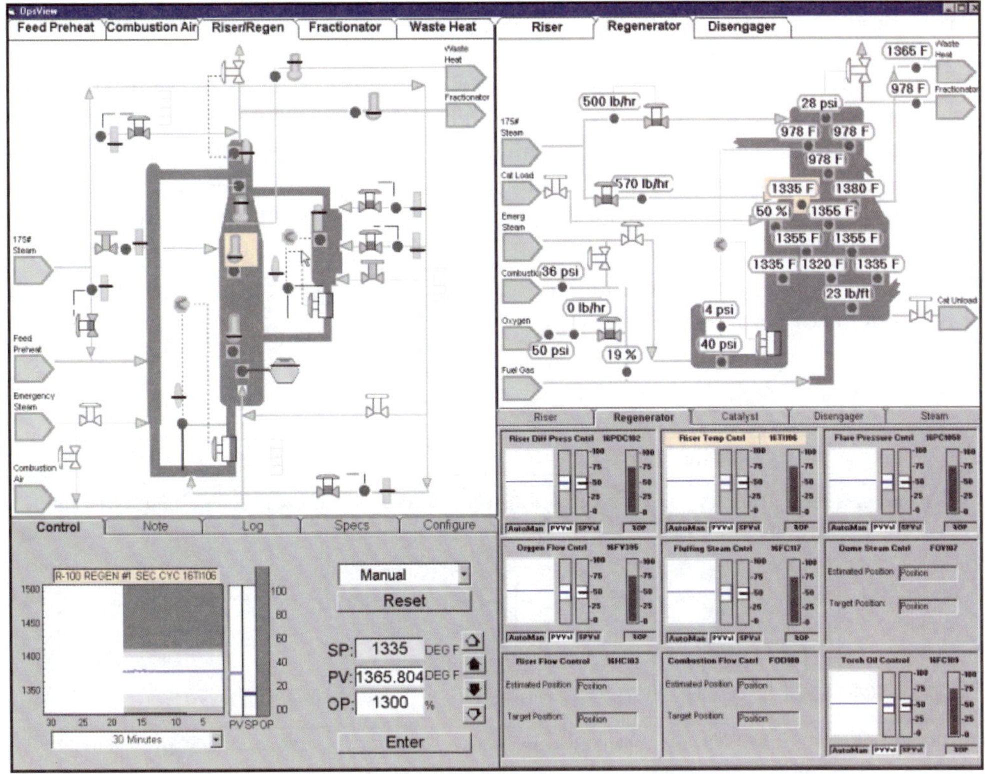

Figure 12.11.5. ASM-style secondary display

but at the unit level. For this example, there are five units or components. A selection tab arranged along the top of the subframe identifies each. The upper right shows a detail from one of the components of the selected unit at the left. It is the regenerator portion of the riser/regenerator. The other portions, the riser and the disengager, can be selected by one of the three tabs arranged along the top of this subframe. For both of these subviews, the same paradigm of showing critical variables, critical inflows, and outflows are continued.

Our example illustrates an alarm of a critical temperature in the regenerator section. Hence, clicking on the riser/regenerator in the overview would have linked this secondary display view on the appropriate screen designated for secondary displays. This particular linked set of subviews would be "yoked" together due to the presence of an alarm and the clicking of the alarmed unit in the overview.

Because our alarm is in the regenerator, the specific set of nine critical variables necessary to understand this alarm will be displayed in the lower-right subframe. These variables come directly from the information on the alarm response documentation produced at rationalization. Again, using earlier display paradigms, we see items in context. Note that there will be other appropriate variable sets for the riser, the catalyst system, the disengager, and the steam system. Refer to the tabs arranged along the top of Figure 12.11.6.

Finally, since there is an alarm, the lower-left subframe centers (a contextual yoke) at the controller block for that alarmed point. From here, it will be possible to do all the normal manual interventions operators might need to do to examine the abnormality and return things to normal. In addition to the standard faceplate view for control, there is access to a notes database, the appropriate log, the loop specifications, and the actual control configuration. The tabs illustrate this across the top of Figure 12.11.7.

The depictions use the same context display paradigms as other views. We are exploiting the benefits of a unified feel.

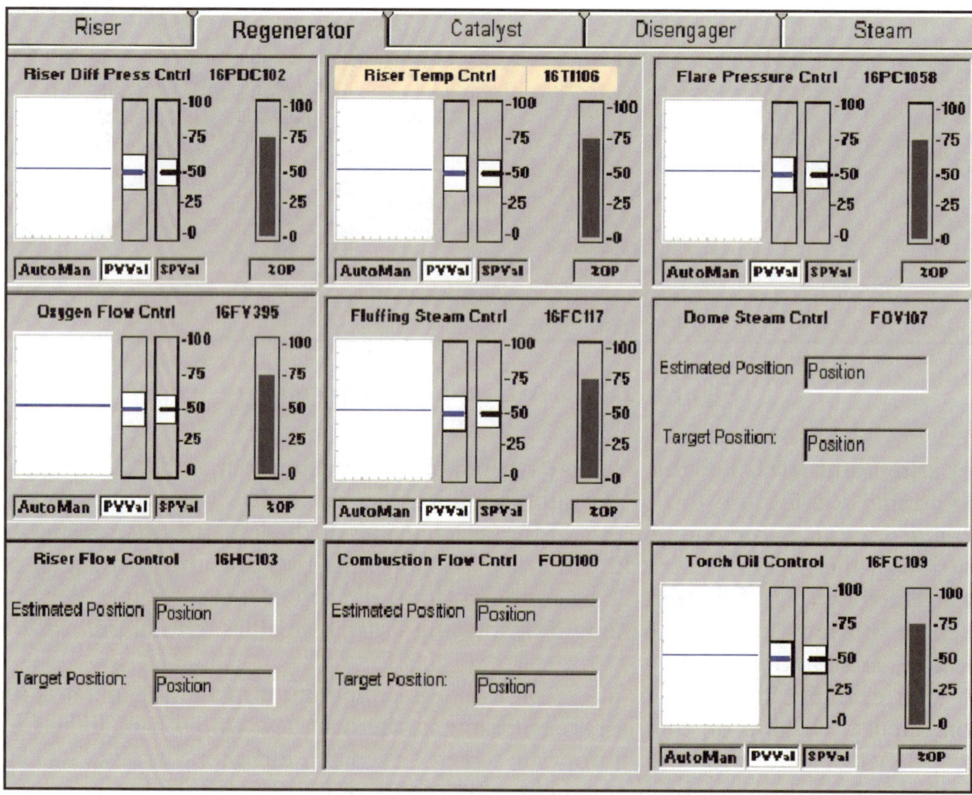

Figure 12.11.6. ASM-style secondary display—key variables

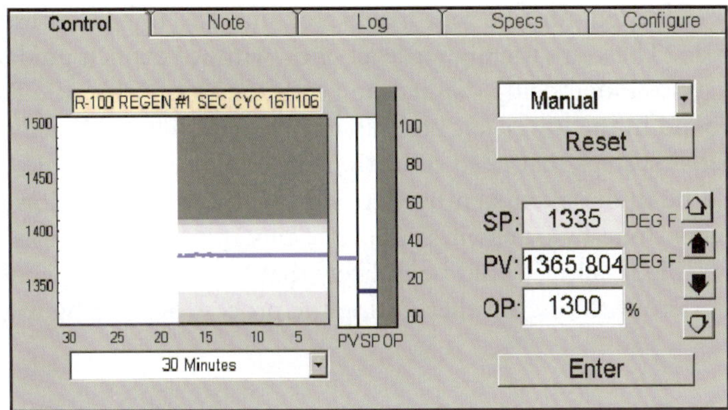

Figure 12.11.7. ASM-style secondary display—controller block

Tertiary Level

We already know that if our problem cannot be adequately addressed at the secondary display level, we would consult the tertiary one for guidance. In our case, a tertiary level shows the plant undergoing a significant process operation. It might just have well been an upset, but here a start-up is illustrated. It was during this start-up that our problem arose. Start-ups traditionally involve lots of coordinated activities by a number of individuals. Figure 12.11.8 shows such a tertiary display.

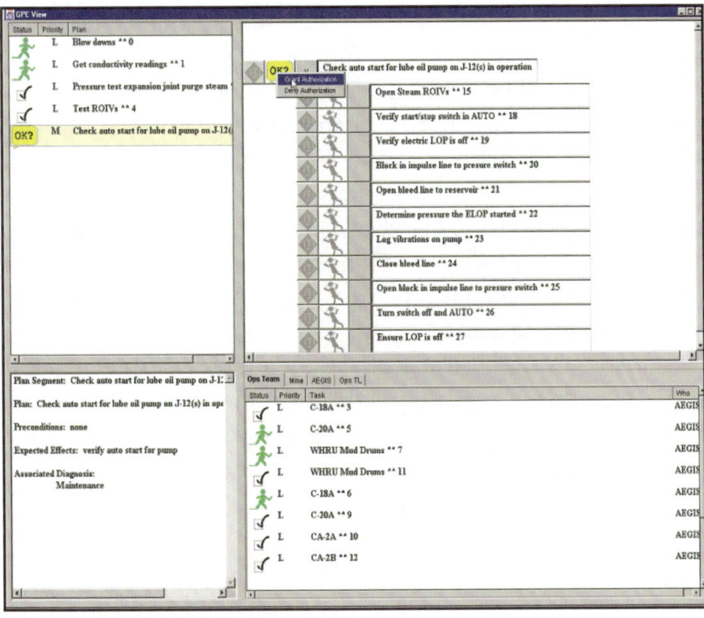

Figure 12.11.8. ASM-style tertiary display

The display is divided into the four major sections. The upper left lists the tasks for our operator to do. These are the higher-level ones, without explicit guidance or details. The bottom step is the one being worked on (with the "OK?" block). The description for this task is found in the lower-left box: a narrative describing this activity. The upper right lists the detailed steps that need to be done to complete that single OK step at the left. They would be undertaken step-by-step. Finally, the box at the lower right shows general tasks being done by others in support of this start-up. With this level of detail and support available to all, it would be unlikely that team members would skip important steps or interfere with one another during the process of working together.

Do ASM-Style Displays Work?

So, ASM-style displays sound like a good idea. And they really are! Here is a study that illustrates the benefits. Luckily, tragic incidents happen rarely, so no plant that has employed the new designs has reported having one. Not to mention that with the new design, they are even less likely to see one at all. So we do not have actual case histories to demonstrate this. But you already know that this all makes extremely good sense. Some of you already know that this is very likely a best practice. Let's see how it performs.

In place of actual incidents, a controlled set of carefully constructed operator simulations was conducted. Each test used a high-fidelity simulator to represent the plant. The "plants" were controlled by two sets of highly trained, experienced operators. Half of the operator subjects (results on the left) used traditional displays, and the other half used the new ASM displays (results on the right). Both groups worked exactly the same scenarios. Four scenarios are depicted in Figure 12.11.9. The vertical scale measures total time for the scenario. The colored vertical rectangular regions show the range of time for all operators to manage a given scenario. For scenario 1 (blue), using traditional displays, operators took from between 4 and 30 minutes to manage. Getting it in 4 minutes was due to a "lucky guess" by one operator. Bars represent the median time. Other operators using ASM displays took between 7 and 8 minutes. Yes, more time was taken at the bottom end, but this time was used more for a proper orientation (who wants to run a plant on luck) and to make sure the situation was understood. From what you may recall of human nature, this is a good thing.

The overall statistics from this test are very encouraging.[8] In general, those using the ASM-type displays were twice as fast in managing and almost 40% again more accurate. Even more revealing, those using the ASM displays were four times faster at early event detection. That is what we call better situation awareness!

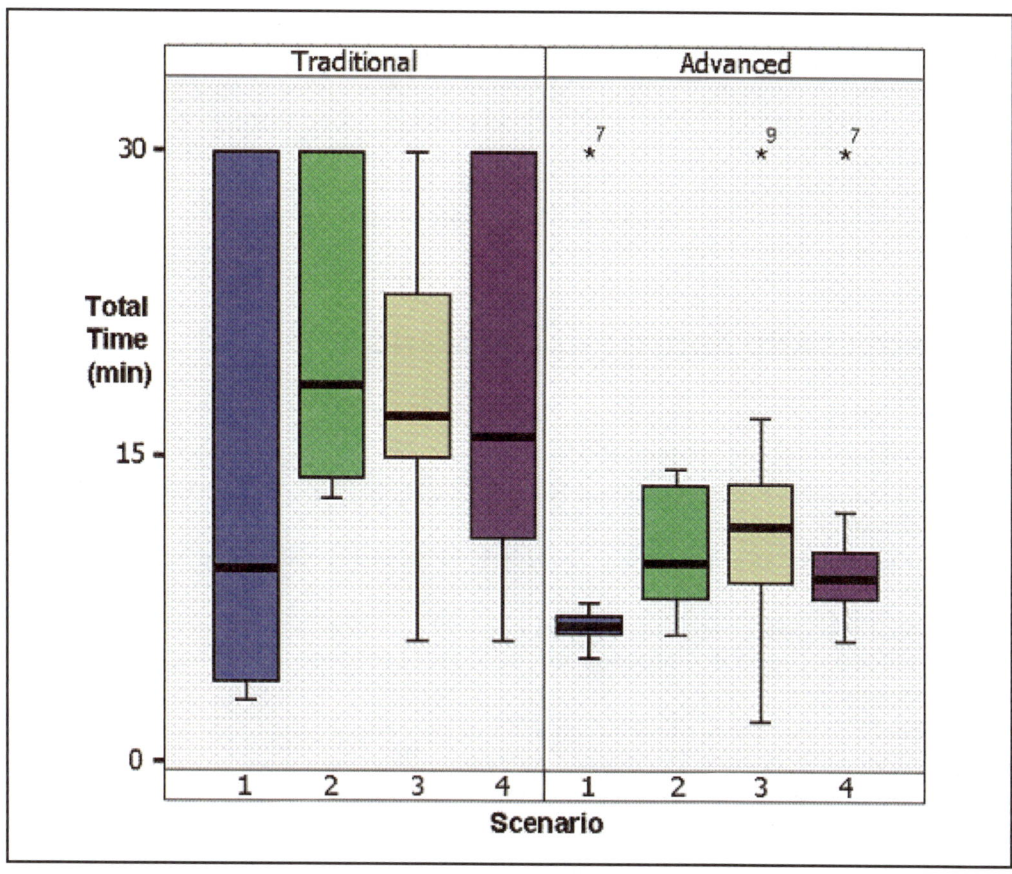

Figure 12.11.9. Effectiveness of ASM-style displays

12.12 WRAP-UP

Situation awareness is fundamental to operators being able to keep abreast of the plant and being ready to intervene and adjust as needed to steady the course or reduce the operating risk by curtailing production. The technological part has been covered in this chapter. The best practices have been illustrated, and the need for them has been motivated. The best that alarm management has to offer is contained within these pages. I trust that it was accessible. I hope it will be useful.

We end this chapter, and this book, full circle back where we started, hopefully much richer and more informed than at the start. You should be energized with hope that the job can be done, trust that this technology is well-enough developed and strong enough to do the job right, and confident that you can do the work and your resulting alarm system will deliver.

The operator's job is to run the plant as intended, as safely as required, and as economically as practical. The first line of success is a well-conceived plant, properly built, effectively planned for operation, well maintained, and well managed. The second line of success is a watchful, resourceful operator—an operator who is able to maintain an awareness of what the plant is doing. The third line of success is a proper warning tool of situations that require more than usual care to manage—your alarm system. The fourth line of success is the safety infrastructure and physical protection in existence for those rare instances when even the most diligent of care is not enough.

It is now time to rest this book and think about what to do with what you've learned here. I wish you every success.

12.13 NOTES AND ADDITIONAL READING

Notes

1. Curt Hill, personal conversation, 1981.
2. Health and Safety Executive, Milford Haven Incident, Sudbury, Suffolk, UK, http://www.hse.gov.uk.
3. Lord Cullen, *The Public Inquiry into the Piper Alpha Disaster* (London: HM Stationery Office, 1990).
4. Chemical Safety and Hazard Investigation Board, "U.S. Chemical Safety and Hazard Investigation Board—Urgent Recommendation [BP Texas City Explosion and Fire, March 2004]," news release, August 17, 2005.
5. William G. Sirois, "The Myths & Realities of Fatigue: Reducing the Costs, Risks, and Liabilities of Fatigue in 24-Hour Operations," *Circadian*, 2009.
6. Douglas Rothenberg and David Beach, *Notifications Management*, ASM White Paper, ASM Consortium, 1995. http://www.asmdashboard.com.
7. *Effective Operator Display Guidelines*, ASM Consortium, http://www.asmdashboard.com. 2002.
8. Jamie Errington and Peter Bullemer, *Advanced Operator Interface* (Human Centered Solutions, 2008), http://ApplyHCS.com.

Recommended Additional Reading

Hollifield, B., D. Oliver, I. Nimmo, and E. Habibi. *The High Performance HMI Handbook*. Houston, TX: PAS, 2008. http://www.pas.com.

Lett, Corbett D. *Man-Machine Interface*. Houston, TX: Litwin Process Automation, 1990.

NAMUR standard AK 2.9 Human Machine Interface (HMI). *Human Machine Interface*. NAMUR Standard AK 2.9. Potsdam, Germany. http://www.namur.de.

Nimmo, Ian. "Alarm Management & Graphics Projects." User Centered Design Services. http://www.mycontrolroom.com/sitedata/articles/archive/alarms%20%20Graphic%20Projects.pdf.

Noyes, Jan, ed., and Mathhew Bransby. "People in Control—Human Factors in Control Room Design." *IEE Control Engineering Series 60.* London: IEE, 2002.

Engineering Equipment Materials Users' Association. *Process Plant Control Desks Utilising Human-Computer Interfaces—A Guide to Design, Operational and Human Interface Issues.* EEMUA Publication No. 201. London: EEMUA, 2002. http://www.eemua.co.uk.

APPENDIX I

Definitions of Terms, Abbreviations, and Acronyms

What's in a name?

—William Shakespeare

Term	Description
	A
Abnormal	The state of not being *normal.* See *normal.*
Absolute alarm	Standard construction for most alarms. The present condition of a potential alarmed variable is compared against the preset *alarm activation point.* When exceeded, the alarm becomes active. Dynamic modification of the *alarm activation point* and/or *priority* is permitted. See *enhanced alarming.*
Ack	Acknowledge. See *acknowledge.*

Ack'd	An alarm that has been *acknowledged.* See *Ack.*
Acknowledge	The first *operator action* that indicates the recognition of an *alarm* to the alarm managing portion of the *control system.*
Activate	The *alarm* becomes "in alarm" by the monitored entity passing over the *alarm activation point* from the *normal* into the *abnormal* condition.
Activation point	The threshold value or discrete state of a process variable that triggers the alarm into the *active state.*
Active state	The status of a configured *alarm* actually being in alarm (process or condition value on the alarm side of normal and within deadband).
Advanced alarm **Advanced alarming**	See *enhanced alarm.*
Alarm	An audible or visible means of indicating to the operator an equipment or process malfunction or *abnormal* condition requiring a response.
Alarm acknowledge	See *acknowledge.*
Alarm activation point	See *activation point.*
Alarm class	Misleading term. Dividing alarms into classes is not useful for operators or for designing alarm systems in general. Priority should be used instead to capture this aspect.
Alarm deadband	The range through which an input must be varied from the *alarm activation point* necessary to *clear* the alarm.
Alarm flood	The situation when the number of alarm activations exceeds the operator's ability to process them.
Alarm group	The set of alarms associated with a specific plant area. See *plant area model.*
Alarm historian	The database that contains the long-term record of all alarm (and other) activities associated with alarms. See also *alarm summary.*

Alarm horn	Any type of audible sounds (including voice response commands and warnings) that are initiated when an *alarm* activates and designed to alert the operator by use of sounds to the presence of an active alarm. This will silence whenever the operator silences it or *acknowledges* the alarm. Some plants use different sounds and/or tones to distinguish between various operator positions and/or alarm priority. This practice must be *very* carefully designed to avoid confusion and unnecessary distraction during times of stress.
Alarm indication	All means of indicating to the operator that an *alarm* has been activated, *acknowledged*, suppressed, inhibited, bypassed, cleared.
Alarm limit	Archaic use. See *alarm activation point.*
Alarm log	The historical record of *alarm indications.*
Alarm management	The processes and practices for determining, documenting, designing, monitoring, and maintaining alarm systems.
Alarm message	The text message that is normally used to convey identification information about an *alarm* when it *activates.* This is an important part of the *alarm summary.*
Alarm philosophy	The guiding document for design or redesign of the entire *alarm* system.
Alarm points	See *alarm activation point*, *alarm.*
Alarm priority	The attribute of an *alarm* that defines a specific level of importance to the alarm to be used by the *operator* in deciding which alarms to work in which order.
Alarm rationalization	See *rationalization.*
Alarm readiness state	The physical state of an *alarm* with regard to whether it is expected to *activate* when the *alarm activation point* is passed.
Alarm response manual	The set of all *alarm response sheets.*
Alarm response sheet	All information about a given *alarm* that fully documents every operational aspect of the alarm. It is extremely useful to both the operator and the alarm system designer. The format may be printed forms, online printed forms, or dynamically assembled information relevant to the alarm and plant states.

Alarm setpoint	Misleading term. See *alarm activation point.*
Alarm summary	A graphic *display* that lists alarm indications over a period of time. List includes the status of all active alarms and all recently cleared alarms. In general, the display is preconfigured by the PCS manufacturer with only minor display customization allowed.
Alarm summary manual	See *alarm response manual.*
Alarm summary sheet	See *alarm response sheet.*
Alarm system	The collection of hardware and software that detects an *alarm* state, transmits the indication of that state to the *operator*, and records the transmission. See also *alarm summary, alarm historian.*
Alarm trip point	Archaic use. See *alarm activation point.*
Alert	A message or visible and/or audible means of indicating to the *operator* of plant situation or equipment condition that does not require a response. Useful to replace old *alarms* that were only for the purpose of notifying the *operator* of conditions or status of equipment.
Auto shelve	An *advanced alarm* or *enhanced alarm* function that detects the need for *shelving* an *alarm* and does so without *operator* intervention according to a preplanned procedure.
	B
Bad actors	Any alarm, usually thought of as a significant number of alarms, that activates too often to require operator actions for each. Taken as a class, these result in a significant distraction to the operator both to acknowledge them and because they can block or hide more useful alarms, even very important ones.
BPCS	Basic process control system. See *primary controls platform.*

Bypass	To manually modify a function to prevent its activation. (This term is not normally used to describe alarm readiness states.) This is used to imply that some mechanical method (jumper wires, etc.) has been employed to ensure that the affected alarm will not activate. See also *inhibit*, *disable*.
C	
Chattering alarm	An alarm that repeatedly and rapidly transitions between the alarm state and the return to normal state; generally indicating nothing wrong with the process. This type of alarm is generally associated with *discrete* or *digital alarms*. See *nuisance alarms*, *bad actors*. See also *cycling alarms*.
Classification	The process of separating rationalized alarms into categories based on the type of consequences.
Clear	See *cleared alarm*.
Cleared alarm	An *alarm* that has returned to *normal*.
Configure	To modify the *control system* parameters by entering appropriate values into a prestructured database provided by the equipment manufacturer.
Consequential alarms	An alarm that activates often enough after one or more specific other alarms activate that it may be considered unnecessary when the predecessor alarm(s) activates. See also *nuisance alarms*.
Console	The interface for a single *operator* to monitor the process, which may include multiple displays and other annunciators and communications equipment. See also *display*, *screen*, *operator area*.
Control system	See *BPCS*, *PCS*, *PLC*.
Cut out	To automatically prevent the transmission of the alarm indication to the operator through a designed function, usually on a temporary basis.

Cycling alarm	An alarm that repeatedly transitions between the alarm state and the return to normal state; generally indicating nothing wrong with the process. This type of alarm is generally associated with analog-type alarms. See *nuisance alarms, bad actors.* See also *chattering alarms.*
	D
DCS	Distributed control system.
Deviation alarm	An alarm that is activated based on the difference between two other measured entities. This alarm is configured to be either "high" when the deviation is positively signed or "low" when the deviation is negatively signed. See also *mismatch alarm.*
Digital alarm	An alarm that is activated based on a logic variable being either on or off, zero or one. Most often, this alarm arises from a process switch.
Disable	To manually or automatically remove the alarm function from a process indication See also *inhibit, bypass.*
Discrepancy alarm	See *mismatch alarm.*
Discrete alarm	See *digital alarm.*
Display	A given rendition of a graphical depiction for showing on a *screen.* See also *screen.*
	E
Eclipse	An alarm that becomes redundant or otherwise unnecessary once another alarm has activated. See also *nuisance alarms.*
Enforcement	The process of resetting the alarm system to the configuration state specified as the base or norm. It may be done in any manner consistent with the alarm philosophy.
Enhanced alarm	See *enhanced alarming.*

Enhanced alarming	Any method of modifying alarm parameters or information to ensure that the alarm maintains its relevance for the current operating conditions. This also includes the provision of advice and information to the operator regarding the interpretation and managing of alarms.
ESD	Emergency shutdown.
F	
Fault tolerance time	See *process safety time.*
G	
Grouping	The alarm process of identifying a number of different alarms, but whose alarm response would be the same.
H	
HMI	Human machine interface.
Horn	See *alarm horn.*
I	
Inhibit	To manually or automatically prevent the transmission of the alarm indication to the operator, usually on a temporary basis. See also *bypass, disable.*
Initiating event	A malfunction, failure, or other condition that can cause an alarm indication.
Intelligent fault detection	See *enhanced alarming.*
J	
K	
L	
Latching alarm	An alarm that remains in alarm state after the process has returned to normal and requires an operator action beyond acknowledgment before it will clear.
Logic-driven alarm	Any form of calculated or recipe-determined alarm designed to alert the operator that certain situations are present that require action to remediate. Included are all forms of calculations, including statistical and inferred.

M	
Major event	An abnormal situation in which the plant or production unit is disturbed to the point of seriously jeopardizing its operational integrity. See also *alarm flood*.
Mismatch alarm	Used for discrete values (e.g., switches) and mostly for valves that do not open or close, or motors that do not start or stop when they are supposed to. See also *deviation alarm*.
MOC	Management of change.
N	
Normal	An *alarm state* that indicates that the particular point is not in *alarm*; however, the alarm has not been inhibited or otherwise prevented from activation normally.
Nuisance alarm	An alarm that transitions to the alarm state but does not require action from the operator. See *bad actors*.
O	
Operating mode	The particular situation under which a plant is operating; generally, it is presumed that the various operating modes are sufficiently different to affect how it is managed and perhaps how the alarm system must work.
Operator	The primary person responsible for ensuring the process parameters are maintained within limits.
Operator area	The specific portion of the plant that a single *operator* has been assigned to operate.
Operator console	The equipment used by an *operator* to manage aspects of production that are facilitated by remotely operated controls and/or where operational parameters are observed where they are not directly observed on the equipment.
Operator response time	The time between the annunciation of the alarm and the time the *operator* takes the (presumed) correct action in response to the alarm. See also *SUDA*.
Operator, inside	An *operator* whose primary working location is at an operator console, usually in a sheltered area which may be within but separate from the physical plant.

Operator, outside	An *operator* whose primary working location is within the equipment of physical plant as opposed to the controls console.
Out of service **Out-of-service plant**	Used to designate a plant or part of a plant that is not actively being used for production. Usually it has been specially prepared to ensure that dangerous situations or conditions are not present, but such cannot and should not be presumed. Hence out-of-service plants may have active alarms and alarms not suppressed or otherwise inhibited or bypassed.
P	
P&I **P&ID**	Piping (or Process) and Instrumentation Diagram.
PCS	See *primary controls system.*
PFD	Process flow diagram or probability of failure on demand.
PHA	Process hazards analysis.
Philosophy	See *alarm philosophy.*
Plant	A coordinated or physically contiguous collection of areas.
Plant area	The basic part of a plant that is managed by one operator position (but which may be actually operated by more than one physical operator, so long as the equipment and controls interface contains all of the pieces for which the operator is responsible and none other).
Plant area model	A construction by which the plant is divided up into constituent parts in a structured way that enables the ability to identify and control alarms as a group with respect to *related alarms*, *consequential alarms*, and others.
Plant enterprise	The largest entity in a corporate structure; generally consisting of multiple plants located on one or more sites.
Plant equipment	Individual items in a plant module.
Plant module	Collections of plant equipment; the smallest general entity for which alarms and/or alarm readiness states are managed.
Plant response time	The time it nominally will take a plant to respond to an operator action that is designated to bring the plant to an acceptable state of operation.
Plant site	A single, physical geographical location for manufacturing.
Plant unit	See *unit.*

PLC	See *programmable logic controller.* One instantiation of a BPCS or PCS. See also *BPCS, PCS.*
PreOp safety review **Pre-operation Safety Review**	A formal process by which an enterprise reviews the design, construction, and procedures of manufacturing plant (or portion thereof) to ensure that it would be safe to operate. "Safe to operate" normally includes financial integrity and environmental integrity.
Pre-start-up safety review	See *PreOp safety review.*
Primary controls system **Process control system**	The hardware and software combined that manages the process. It usually contains the configuration of alarms and has the ability and authority to generate alarms.
Prioritization	The process of assigning a specific level of importance to the alarm. See also *alarm priority.*
Priority	See *alarm priority.*
Process response time **PRT**	The time that a process takes to become adjusted enough to avoid unwanted consequences due to an upset after an operator (or automated) change has been made.
Process safety time **PST**	The amount of time between an *abnormal* process event and the inception of serious consequences. Effective management of the *abnormal* event, if done in time, can usually prevent the inception of the serious consequences.
PSAT	Pre-start-up acceptance test. See also *PreOp safety review.*
PSSR	See *PreOp safety review, pre-start-up safety review.*
	Q
	R
Rate-of-change alarm	An alarm indicating that the monitored value is changing at either too high a positive rate or too low a negative rate. Positive and negative refer to direction of change. *Rate-of-change* is not applied to values that are NOT changing fast enough.

Rationalization	The process of rejecting or accepting a potential *alarm* using the *alarm philosophy*, determining and documenting the rationale, and specifying all design requirements for the alarm (including *activation point* and *priority*).
Redundancy logic	Use of more than one method of determining that an alarm condition exists and therefore activates only a single appropriate alarm. See also *enhanced alarming*.
Redundant alarm	Two or more alarms that, if activated, will indicate the same process condition. Therefore all might be removed except the best indicator of the abnormal condition. Where removal of redundant alarms is not possible, redundancy logic can be used to limit the activations to the first one. See also *enhanced alarming*.
Related alarms	An alarm that activates often enough either before or after one, or more specific other alarms activate. This differs from *consequential alarms* in that any *related alarm* may activate either before or after rather than after the other alarm. See also *nuisance alarms*, *enhanced alarming*.
Release	To manually prevent the transmission of the alarm indication to the operator through a managed list or shelf, usually on a temporary basis. However, an alarm on *release* is not reactivated until it has been *cleared*. See also *shelve*.
Remote alarm	An alarm from a remotely operated facility and often mirrors a local alarm at that facility.
Retriggering	The process of automatically causing an alarm, regardless of its acknowledge status, to realarm (and thereby activate the horn, add an entry into the alarm summary, and otherwise indicate the presence of an active, unacknowledged alarm). This is NOT GOOD PRACTICE and should not be used.
Return or return to normal	The alarm system indication that an alarm condition has cleared.
Risk assessment	Any form of determination that provides useful information about the probability and impact of any event.
RTN	Return to normal. See also *normal*.

S	
SCADA	Supervisory control and data acquisition (system).
Screen	A single piece of hardware that is capable of depicting a visual graphical rendition normally associated with operator consoles. Any number of displays may be viewed (serially or by combining views) on a given screen, up to the limit of the hardware and software controlling them. See also *displays*.
Shelve	To manually prevent the transmission of the alarm indication to the operator through a managed list or shelf, usually on a temporary basis.
SIF	Safety instrumented function.
SIL	Safety integrity level—provides a universally understood safety performance rating based on legal codes and such.
SIS	Safety instrumented system.
Stale alarm	See *standing alarm*.
Standing alarm	An alarm that remains in the alarm state for an extended period of time (well beyond any usefulness to the operator)
Station	A single human machine interface within the operator console. An operator console may have a number of stations. See *display*.
SUDA	See, understand, decide, and act. This is the activity of an *operator* in responding to an alarm activation. It begins with the operator seeing and recognizing an alarm activation and continues through all the activities needed for the operator to make what is anticipated as the proper corrective action to the process. See *SUDA time*.
SUDA time	The time it takes for the entire SUDA activity to take place. See *SUDA*.
Suppress	To automatically, based on designed logic, prevent the indication of the alarm to the operator when the base alarm condition is present and the alarm has actually activated. GOOD PRACTICE AVOIDS THIS USE.

T	
Tag	A unique label identifier that is used for BPCS data. Typical ones are temperatures, flows, and the like. Displays, alarms, logs, and other aspects are usually organized by their *tag*.
Time to manage fault	The amount of time required for the operator to become aware of an *abnormal* event, understand it, take action to manage it, and the process to respond in a way that prevents serious consequences. So long as the *time to manage fault* is less than the *process safety time*, serious consequences can be thought of as being amenable to operator action. See also *process safety time*.
U	
UNack'd	Unacknowledged. See also *unacknowledged*.
Unacknowledged	An alarm in the alarm state that has not been acknowledged by the *operator*.
User alert	See *alert*.
Unit	A general term usually used to identify a major portion of an operator area that performs one or more important manufacturing functions in a somewhat self-contained way.
User notification	See *alert*.
V	
W	
X	
Y	
Z	

APPENDIX 2

Twenty-Four Hours of Alarms

Date & Time	Unit	Alarm State	Tag	Alarm Type (or Descriptor)	
03/11/98 00:10:38	SC 31	ACK	AI0114	PVHI	
03/11/98 01:43:28	SY 19		NODE 46		
03/11/98 01:43:28	SY 19		NODE 46		
03/11/98 01:48:38	HM 26		TC0091	HGO RETURN TO BED 3	
03/11/98 02:19:29	SC 21	ALM	AI0114	PVHI	50.000
03/11/98 02:33:37	SC 31	ACK	AI0114	PVHI	
03/11/98 02:59:16	RX 21	ALM	PDI1203	PVLO	100.000
03/11/98 02:59:52	RX 21	RTN	PDI1203	PVLO	100.000
03/11/98 03:00:21	RX 31	ACK	PDI1203	PVLO	
03/11/98 04:36:54	SY 19		NODE 04 $P4		
03/11/98 04:36:58	SY 19		NODE 04 $P4		
03/11/98 05:22:39	SY 16		NIM 30	LCN DRIVER	000
03/11/98 06:43:43	DP 26		LC0053	V-33 DEPROP OH ACCUM LVL	
03/11/98 06:43:49	U1 21	ALM	TC5030	PVLO	100.000
03/11/98 06:52:59	HM 26		TC5277	FRAC BOTTOM TEMP	
03/11/98 06:58:27	HM 26		FC4095	E-47 SLURRY BYPASS	
03/11/98 06:58:32	HM 26		FC4095	E-47 SLURRY BYPASS	
03/11/98 07:00:27	HM 26		CA0110	LAB UPDATED GRAVITY -CLO	
03/11/98 07:00:27	HM 26		CA0110	LAB UPDATED GRAVITY -CLO	
03/11/98 07:05:11	HM 26		TC5277	FRAC BOTTOM TEMP	
03/11/98 07:12:55	SP 21	ALM	AC0117	BADPV	
03/11/98 07:13:34	HM 26		FC1102	SLURRY BYPASS WHB	
03/11/98 07:13:56	DP 26		LC0053	V-33 DEPROP OH ACCUM LVL	
03/11/98 07:18:26	SP 31	ACK	AC0117	BADPV	
03/11/98 07:18:50	SP 21	ALM	LC0207	PVHI	75.000
03/11/98 07:18:59	SP 31	ACK	LC0207	PVHI	
03/11/98 07:21:13	SP 26		FC1211	6 IN COLD SPLITTER BTMS	
03/11/98 07:24:13	SP 26		FC1211	6 IN COLD SPLITTER BTMS	
03/11/98 07:27:43	RX 26		LC0002B	RX TOP SLIDE VALVE	
03/11/98 07:27:55	SP 21	RTN	AC0117	BADPV	
03/11/98 07:28:03	RX 26		LC0002B	RX TOP SLIDE VALVE	

Alarm	Descriptor (or failure mode)	Mode (or value)	Units	Other		
HIGH	D-10 KO POT H2S			1		
	IF00 DEVICE FAILED					00
	IF00 DEVICE FAILED					00
MODE	MAN	AUTO	DEG F	CONS	1	
HIGH	D-10 KO POT H2S	50.860				
HIGH	D-10 KO POT H2S			1		
HIGH	RX SV DIFF PRESSURE	99.771				
HIGH	RX SV DIFF PRESSURE	112.077				
HIGH	RX SV DIFF PRESSURE			1		
	LP00 DEVICE AVAILABLE		OFFLINE			01
	LP00 DEVICE AVAILABLE		DEVICE READY			01
SLOT 000	CMD REG 0200 NODE	000031 SECD 019	001F	0000	0000	
SP	50.0000	55.0000	PERCENT	CONS	1	
LOW	PROP SCRUB BTM OVERRIDE	99.980				
SP	680.2010	682.0000	DEG F	CONS	1	
MODE	CAS	MAN	BPH	CONS	1	
OP	0.0000	---------	BPH	CONS	1	
OPRR	- 2.7000	---------	API	CONS	1	
OPRT	10 Mar 98 07:38:38	11 Mar 98 07:00:26		CONS	1	
SP	682.0000	684.0000	DEG F	CONS	1	
LOW	C-6 SLPLIT BTM 99% POINT					
OP	- 6.0000	100.0000	BPH	CONS	1	
SP	55.0000	50.0000	PERCENT	CONS	1	
LOW	C-6 SLPLIT BTM 99% POINT			1		
HIGH	SPLITTER REBOILER LEVEL	75.345				
HIGH	SPLITTER REBOILER LEVEL			1		
OP	11.1000	13.1000	B P H	CONS	1	
OP	13.1000	17.1000	B P H	CONS	1	
OP	95.0000	50.0000	PERCENT	CONS	1	
LOW	C-6 SLPLIT BTM 99% POINT					
OP	50.0000	95.0000	PERCENT	CONS	1	

```
03/11/98 07:31:43   SP 21 RTN      LC0207     PVHI                  75.000
03/11/98 07:31:53   SP 21 ALM      LC0207     PVHI                  75.000
03/11/98 07:31:57   SP 31 ACK      LC0207     PVHI
03/11/98 07:32:33   SP 21 RTN      LC0207     PVHI                  75.000
03/11/98 07:33:27   HM 26        FC1102      SLURRY BYPASS WHB
03/11/98 07:35:31   HM 26        FC1102      SLURRY BYPASS WHB
03/11/98 07:36:21   HM 26        FC1102      SLURRY BYPASS WHB
03/11/98 07:46:24   DP 26        PC0042      C-12 HOT VAPOR BYPASS
03/11/98 07:46:32   DP 26        PC0042      C-12 HOT VAPOR BYPASS
03/11/98 07:46:50   DP 26        PC0042      C-12 HOT VAPOR BYPASS
03/11/98 07:50:17   DP 26        HC4037      SOUTH E-45 OUTLET
03/11/98 07:50:28   DP 26        HC4039      NORTH E-45 OUTLET
03/11/98 07:50:36   DP 26        HC4038      MID E-45 OUTLET
03/11/98 07:52:16   DP 26        HC4039      NORTH E-45 OUTLET
03/11/98 07:53:14   HM 26        FC4095      E-47 SLURRY BYPASS
03/11/98 07:53:39   SP 26        FC1211      6 IN COLD SPLITTER BTMS
03/11/98 07:54:42   DP 26        HC4037      SOUTH E-45 OUTLET
03/11/98 07:56:13   DP 26        HC4039      NORTH E-45 OUTLET
03/11/98 07:56:19   DP 26        HC4038      MID E-45 OUTLET
03/11/98 07:56:44   HM 26        FC4095      E-47 SLURRY BYPASS
03/11/98 07:57:46   DP 26        FC0105      REFLUX TO C-12 TOWER
03/11/98 07:57:55   DP 26        FC0105      REFLUX TO C-12 TOWER
03/11/98 07:58:07   HM 26        FC4095      E-47 SLURRY BYPASS
03/11/98 07:59:49   DP 26        FC0105      REFLUX TO C-12 TOWER
03/11/98 08:00:54   HM 26        FC4095      E-47 SLURRY BYPASS
03/11/98 08:01:37   HM 26        FC4095      E-47 SLURRY BYPASS
03/11/98 08:02:41   HM 26        FC4095      E-47 SLURRY BYPASS
03/11/98 08:03:15   DP 26        HC4037      SOUTH E-45 OUTLET
03/11/98 08:03:24   DP 26        HC4039      NORTH E-45 OUTLET
03/11/98 08:03:27   DP 26        HC4038      MID E-45 OUTLET
03/11/98 08:03:59   DP 26        FC0105      REFLUX TO C-12 TOWER
03/11/98 08:04:06   HM 26        FC4095      E-47 SLURRY BYPASS
03/11/98 08:04:35   DP 26        FC0105      REFLUX TO C-12 TOWER
03/11/98 08:05:11   HM 26        FC4095      E-47 SLURRY BYPASS
03/11/98 08:05:35   DP 26        HC4037      SOUTH E-45 OUTLET
03/11/98 08:05:40   DP 26        HC4039      NORTH E-45 OUTLET
03/11/98 08:07:06   DP 26        HC4037      SOUTH E-45 OUTLET
03/11/98 08:07:10   DP 26        HC4039      NORTH E-45 OUTLET
03/11/98 08:07:15   DP 26        HC4038      MID E-45 OUTLET
03/11/98 08:12:31   DP 26        FC0105      REFLUX TO C-12 TOWER
03/11/98 08:17:05   HM 26        FC1102      SLURRY BYPASS WHB
03/11/98 08:17:56   HM 26        FC1102      SLURRY BYPASS WHB
03/11/98 08:18:31   HM 26        FC1102      SLURRY BYPASS WHB
03/11/98 08:19:05   SP 21 ALM      LC0207     PVHI                  75.000
03/11/98 08:19:15   SP 26        FC1211      6 IN COLD SPLITTER BTMS
03/11/98 08:19:18   SP 31 ACK      LC0207     PVHI
03/11/98 08:19:27   SP 21 RTN      LC0207     PVHI                  75.000
```

HIGH	SPLITTER REBOILER LEVEL	73.934			
HIGH	SPLITTER REBOILER LEVEL	75.162			
HIGH	SPLITTER REBOILER LEVEL			1	
HIGH	SPLITTER REBOILER LEVEL	73.936			
OP	100.0000	0.0000	BPH	CONS	1
OP	0.0000	100.0000	BPH	CONS	1
OP	100.0000	0.0000	BPH	CONS	1
MODE	CAS	MAN	PSIG	CONS	1
OP	60.0000	30.0000	PSIG	CONS	1
OP	30.0000	50.0000	PSIG	CONS	1
OP	50.5977	44.5977	PERCENT	CONS	1
OP	35.9468	33.9468	PERCENT	CONS	1
OP	25.3197	23.3197	PERCENT	CONS	1
OP	33.9468	31.9468	PERCENT	CONS	1
OP	- 6.9000	4.0000	BPH	CONS	1
OP	17.1000	13.1000	B P H	CONS	1
OP	44.5977	42.5977	PERCENT	CONS	1
OP	31.9468	29.9468	PERCENT	CONS	1
OP	23.3197	21.3197	PERCENT	CONS	1
OP	4.0000	8.0000	BPH	CONS	1
MODE	CAS	AUTO	BPH	CONS	1
SP	1573.470	1300.000	BPH	CONS	1
OP	8.0000	12.0000	BPH	CONS	1
SP	1300.000	1200.000	BPH	CONS	1
OP	12.0000	18.0000	BPH	CONS	1
OP	18.0000	22.0000	BPH	CONS	1
OP	22.0000	24.0000	BPH	CONS	1
OP	42.5977	38.5977	PERCENT	CONS	1
OP	29.9468	25.9468	PERCENT	CONS	1
OP	21.3197	19.3197	PERCENT	CONS	1
SP	1200.000	1100.000	BPH	CONS	1
OP	24.0000	28.0000	BPH	CONS	1
SP	1100.000	1300.000	BPH	CONS	1
OP	28.0000	32.0000	BPH	CONS	1
OP	38.5977	34.5977	PERCENT	CONS	1
OP	25.9468	23.9468	PERCENT	CONS	1
OP	34.5977	26.5977	PERCENT	CONS	1
OP	23.9468	19.9468	PERCENT	CONS	1
OP	19.3197	17.3197	PERCENT	CONS	1
MODE	AUTO	MAN	BPH	CONS	1
OP	0.0000	100.0000	BPH	CONS	1
OP	100.0000	50.0000	BPH	CONS	1
OP	50.0000	25.0000	BPH	CONS	1
HIGH	SPLITTER REBOILER LEVEL	75.006			
OP	13.1000	17.1000	B P H	CONS	1
HIGH	SPLITTER REBOILER LEVEL			1	
HIGH	SPLITTER REBOILER LEVEL	73.941			

```
03/11/98 08:20:32   DP 26       FC0105        REFLUX TO C-12 TOWER
03/11/98 08:20:53   DP 26       FC0105        REFLUX TO C-12 TOWER
03/11/98 08:24:21   DP 26       HC4038        MID E-45 OUTLET
03/11/98 08:26:31   DP 26       FC0105        REFLUX TO C-12 TOWER
03/11/98 08:26:43   HM 26       FC4095        E-47 SLURRY BYPASS
03/11/98 08:26:48   HM 21 ALM      FC4095      PVLO            1600.00
03/11/98 08:26:54   HM 31 ACK      FC4095      PVLO
03/11/98 08:29:14   DP 26       HC4039        NORTH E-45 OUTLET
03/11/98 08:29:18   DP 26       HC4038        MID E-45 OUTLET
03/11/98 08:32:24   HM 26       FC4095        E-47 SLURRY BYPASS
03/11/98 08:33:22   DP 26       FC0105        REFLUX TO C-12 TOWER
03/11/98 08:33:51   DP 26       FC0105        REFLUX TO C-12 TOWER
03/11/98 08:34:29   DP 26       HC4037        SOUTH E-45 OUTLET
03/11/98 08:36:27   DP 26       FC0006        C-12 BOTTOMS TO #2 DEBUT
03/11/98 08:37:03   DP 26       FC0006        C-12 BOTTOMS TO #2 DEBUT
03/11/98 08:37:18   DP 26       FC0006        C-12 BOTTOMS TO #2 DEBUT
03/11/98 08:37:48   HM 26       FC4095        E-47 SLURRY BYPASS
03/11/98 08:39:26   DP 26       FC0105        REFLUX TO C-12 TOWER
03/11/98 08:40:32   AB 26       TC1607        TRAY 4,5 C-10 STRIPPER
03/11/98 08:42:19   HM 26       FC4095        E-47 SLURRY BYPASS
03/11/98 08:42:25   RX 21 ALM      PDI1203     PVLO            100.000
03/11/98 08:42:33   RX 31 ACK      PDI1203     PVLO
03/11/98 08:42:42   HM 26       FC4095        E-47 SLURRY BYPASS
03/11/98 08:43:43   DP 26       FC0006        C-12 BOTTOMS TO #2 DEBUT
03/11/98 08:43:54   DP 26       FC0105        REFLUX TO C-12 TOWER
03/11/98 08:44:13   HM 26       FC4095        E-47 SLURRY BYPASS
03/11/98 08:44:33   RX 21 RTN      PDI1203     PVLO            100.000
03/11/98 08:44:45   DP 26       PC0042        C-12 HOT VAPOR BYPASS
03/11/98 08:46:10   HM 26       FC1102        SLURRY BYPASS WHB
03/11/98 08:46:23   HM 26       FC1102        SLURRY BYPASS WHB
03/11/98 08:46:34   U1 21 RTN      TC5030      PVLO            100.000
03/11/98 08:46:49   HM 26       FC1102        SLURRY BYPASS WHB
03/11/98 08:48:09   DP 26       HC4095        1032A SLURRY TO E-47
03/11/98 08:48:12   SP 21 ALM      LC0207      PVLO             30.000
03/11/98 08:48:23   SP 26       FC1211        6 IN COLD SPLITTER BTMS
03/11/98 08:48:24   SP 31 ACK      LC0207      PVLO
03/11/98 08:48:32   DP 26       HC4095        1032A SLURRY TO E-47
03/11/98 08:48:49   FD 26       TC5127        E-4 FEED BYPASS
03/11/98 08:48:55   SP 26       FC1211        6 IN COLD SPLITTER BTMS
03/11/98 08:49:41   DP 26       PC0042        C-12 HOT VAPOR BYPASS
03/11/98 08:50:33   DP 26       PC0042        C-12 HOT VAPOR BYPASS
03/11/98 08:50:51   SP 21 ALM      FC1210      PVLO             10.000
03/11/98 08:51:24   DP 26       PC0042        C-12 HOT VAPOR BYPASS
03/11/98 08:51:37   DP 26       PC0042        C-12 HOT VAPOR BYPASS
03/11/98 08:52:24   SP 21 ALM      PI4046      PVLO            100.000
03/11/98 08:52:24   SP 26       FC1210        3 IN SPLIT BTMS TO FIELD
03/11/98 08:52:26   SP 21 RTN      PI4046      PVLO            100.000
```

```
 OP          100.0000           80.0000      BPH      CONS  1
 OP           80.0000           90.0000      BPH      CONS  1
 OP           17.3197           15.3197      PERCENT  CONS  1
 OP           90.0000          100.0000      BPH      CONS  1
 OP           32.0000           38.0000      BPH      CONS  1
LOW      E-47 SLURRY BYPASS          1572.38
LOW      E-47 SLURRY BYPASS                              1
 OP           19.9468           17.9468      PERCENT  CONS  1
 OP           15.3197           13.3197      PERCENT  CONS  1
 OP           38.0000           40.0000      BPH      CONS  1
 OP          100.0000           90.0000      BPH      CONS  1
 OP           90.0000           80.0000      BPH      CONS  1
 OP           26.5977           24.5977      PERCENT  CONS  1
 SP          1293.920          1350.000      BPH      CONS  1
 SP          1350.000          1300.000      BPH      CONS  1
 SP          1300.000          1250.000      BPH      CONS  1
 OP           40.0000           44.0000      BPH      CONS  1
 OP           80.0000           90.0000      BPH      CONS  1
 SP          155.8670          158.0000      DEG F    CONS  1
 OP           44.0000           46.0000      BPH      CONS  1
HIGH     RX SV DIFF PRESSURE           99.643
HIGH     RX SV DIFF PRESSURE                             1
 OP           46.0000           48.0000      BPH      CONS  1
 SP          1250.000          1200.000      BPH      CONS  1
 OP           90.0000          100.0000      BPH      CONS  1
 OP           48.0000           52.0000      BPH      CONS  1
HIGH     RX SV DIFF PRESSURE          112.073
 OP           50.0000           60.0000      PSIG     CONS  1
 OP           25.0000          100.0000      BPH      CONS  1
 OP          100.0000            0.0000      BPH      CONS  1
LOW      PROP SCRUB BTM OVERRIDE      115.131
 OP            0.0000           15.0000      BPH      CONS  1
 OP          100.0030           70.0000      BPH      CONS  1
HIGH     SPLITTER REBOILER LEVEL       29.900
 OP           17.1000           11.1000      B P H    CONS  1
HIGH     SPLITTER REBOILER LEVEL                         1
 OP           70.0000           60.0000      BPH      CONS  1
 OP          100.0000           50.0000      DEG F    CONS  1
 OP           11.1000            3.1000      B P H    CONS  1
 OP           60.0000           10.0000      PSIG     CONS  1
 OP           10.0000            0.0000      PSIG     CONS  1
JOURNAL  3 IN SPLIT BTMS TO FIELD       9.250
 OP            0.0000            5.0000      PSIG     CONS  1
 OP            5.0000           10.0000      PSIG     CONS  1
EMERGNCY P-18 SPLTR BTM DISCH          92.365
 MODE         CAS              MAN           B P H    CONS  1
EMERGNCY P-18 SPLTR BTM DISCH         105.141
```

```
03/11/98 08:52:27    SP 21 ALM       PI4046       PVLO           100.000
03/11/98 08:52:27    SP 26       FC1210          3 IN SPLIT BTMS TO FIELD
03/11/98 08:52:29    SP 31 ACK       PI4046       PVLO
03/11/98 08:52:29    SP 31 ACK       PI4046       PVLO
03/11/98 08:52:30    SP 21 RTN       PI4046       PVLO           100.000
03/11/98 08:52:33    SP 26       FC1211          6 IN COLD SPLITTER BTMS
03/11/98 08:52:46    SP 26       FC1211          6 IN COLD SPLITTER BTMS
03/11/98 08:53:33    SP 21 ALM       PI4046       PVLO           100.000
03/11/98 08:53:38    SP 21 RTN       PI4046       PVLO           100.000
03/11/98 08:53:56    SP 26       FC1311          SPLITTER BTMS/CAT NAPTHA
03/11/98 08:54:06    SP 21 ALM       PI4046       PVLO           100.000
03/11/98 08:54:12    SP 21 RTN       PI4046       PVLO           100.000
03/11/98 08:54:38    SP 26       FC1311          SPLITTER BTMS/CAT NAPTHA
03/11/98 08:54:57    DB 21 ALM       AC0111       BADPV
03/11/98 08:54:58    SP 21 ALM       PI4046       PVLO           100.000
03/11/98 08:54:59    SP 21 RTN       PI4046       PVLO           100.000
03/11/98 08:55:22    DB 26       CX0203          E29 #2 DEB REB NORM DUTY
03/11/98 08:55:47    DB 31 ACK       AC0111       BADPV
03/11/98 08:55:47    SP 31 ACK       PI4046       PVLO
03/11/98 08:56:14    HM 26       TC5277          FRAC BOTTOM TEMP
03/11/98 08:56:16    DP 26       HC4037          SOUTH E-45 OUTLET
03/11/98 08:56:18    DP 26       HC4038          MID E-45 OUTLET
03/11/98 08:56:20    DP 26       HC4039          NORTH E-45 OUTLET
03/11/98 08:56:59    SP 26       FC1311          SPLITTER BTMS/CAT NAPTHA
03/11/98 08:58:30    SP 26       FC1311          SPLITTER BTMS/CAT NAPTHA
03/11/98 08:58:42    SP 26       FC1311          SPLITTER BTMS/CAT NAPTHA
03/11/98 08:59:01    SP 21 RTN       LC0207       PVLO            30.000
03/11/98 08:59:20    DP 26       PC0042          C-12 HOT VAPOR BYPASS
03/11/98 09:00:35    SP 26       FC1210          3 IN SPLIT BTMS TO FIELD
03/11/98 09:00:46    SP 26       LC0207          SPLITTER REBOILER LEVEL
03/11/98 09:01:59    SP 21 RTN       FC1210       PVLO            10.000
03/11/98 09:02:07    DP 26       PC0042          C-12 HOT VAPOR BYPASS
03/11/98 09:02:11    DP 26       HC4039          NORTH E-45 OUTLET
03/11/98 09:02:51    SP 26       FC1311          SPLITTER BTMS/CAT NAPTHA
03/11/98 09:10:07    SP 21 ALM       LC0207       PVHI            75.000
03/11/98 09:10:12    SP 31 ACK       LC0207       PVHI
03/11/98 09:10:19    SP 21 RTN       LC0207       PVHI            75.000
03/11/98 09:10:39    SP 26       FC1211          6 IN COLD SPLITTER BTMS
03/11/98 09:10:53    DP 26       PC0042          C-12 HOT VAPOR BYPASS
03/11/98 09:11:13    SP 21 ALM       LC0207       PVHI            75.000
03/11/98 09:11:21    SP 31 ACK       LC0207       PVHI
03/11/98 09:11:50    SP 21 RTN       LC0207       PVHI            75.000
03/11/98 09:12:24    SP 21 ALM       LC0207       PVHI            75.000
03/11/98 09:12:27    SP 31 ACK       LC0207       PVHI
03/11/98 09:12:35    SP 26       FC1211          6 IN COLD SPLITTER BTMS
03/11/98 09:12:55    SP 21 RTN       LC0207       PVHI            75.000
03/11/98 09:14:52    RX 21 ALM       PDI1203      PVLO           100.000
```

```
EMERGNCY P-18 SPLTR BTM DISCH              99.451
 OP           22.9739          -  6.9000         B P H    CONS  1
EMERGNCY P-18 SPLTR BTM DISCH                                   1
EMERGNCY P-18 SPLTR BTM DISCH                                   1
EMERGNCY P-18 SPLTR BTM DISCH             111.605
 OP            3.1000          -  0.9000         B P H    CONS  1
 OP          -  0.9000         ---------         B P H    CONS  1
EMERGNCY P-18 SPLTR BTM DISCH              92.728
EMERGNCY P-18 SPLTR BTM DISCH             106.274
 MODE         AUTO             MAN               B P H    CONS  1
EMERGNCY P-18 SPLTR BTM DISCH              94.852
EMERGNCY P-18 SPLTR BTM DISCH             109.928
 OP           80.2287            64.2287         B P H    CONS  1
LOW      AVG C3 AT C5&7 OH
EMERGNCY P-18 SPLTR BTM DISCH              99.640
EMERGNCY P-18 SPLTR BTM DISCH             108.747
 SP          675.0000           630.0000         MMBTUF/H CONS  1
LOW      AVG C3 AT C5&7 OH                                      1
EMERGNCY P-18 SPLTR BTM DISCH                                   1
 SP          684.0000           680.0000         DEG F    CONS  1
 OP           24.5977            26.5977         PERCENT  CONS  1
 OP           13.3197            15.3197         PERCENT  CONS  1
 OP           17.9468            19.9468         PERCENT  CONS  1
 OP           64.2287            53.4287         B P H    CONS  1
 OP           53.4287            55.7287         B P H    CONS  1
 MODE         MAN              AUTO              B P H    CONS  1
HIGH     SPLITTER REBOILER LEVEL           31.275
 OP           10.0000            20.0000         PSIG     CONS  1
 MODE         MAN              NORMAL            B P H    CONS  1
 SP           48.6791            50.0000         PERCENT  CONS  1
JOURNAL  3 IN SPLIT BTMS TO FIELD          13.636
 OP           20.0000            25.0000         PSIG     CONS  1
 OP           19.9468            17.9468         PERCENT  CONS  1
 SP          905.0410           950.0000         B P H    CONS  1
HIGH     SPLITTER REBOILER LEVEL           75.087
HIGH     SPLITTER REBOILER LEVEL                                1
HIGH     SPLITTER REBOILER LEVEL           73.980
 OP          -  6.9000            3.1000         B P H    CONS  1
 OP           25.0000            30.0000         PSIG     CONS  1
HIGH     SPLITTER REBOILER LEVEL           75.168
HIGH     SPLITTER REBOILER LEVEL                                1
HIGH     SPLITTER REBOILER LEVEL           73.888
HIGH     SPLITTER REBOILER LEVEL           75.212
HIGH     SPLITTER REBOILER LEVEL                                1
 OP            3.1000             5.1000         B P H    CONS  1
HIGH     SPLITTER REBOILER LEVEL           73.940
HIGH     RX SV DIFF PRESSURE               99.977
```

```
03/11/98 09:14:59    RX 31 ACK        PDI1203       PVLO
03/11/98 09:17:07    DP 26         HC4095         1032A SLURRY TO E-47
03/11/98 09:17:13    RX 21 RTN        PDI1203       PVLO            100.000
03/11/98 09:18:13    DP 26         HC4095         1032A SLURRY TO E-47
03/11/98 09:20:40    DP 26         HC4039         NORTH E-45 OUTLET
03/11/98 09:20:45    DP 26         HC4038         MID E-45 OUTLET
03/11/98 09:20:53    DP 26         HC4037         SOUTH E-45 OUTLET
03/11/98 09:21:28    HM 26         FC4095         E-47 SLURRY BYPASS
03/11/98 09:24:20    DB 26         TC0203         1 DEBUT TOWER TOP TEMP
03/11/98 09:24:45    DB 26         CX0205         E20 #1 DEB REB NORM DUTY
03/11/98 09:26:14    DP 26         PC0042         C-12 HOT VAPOR BYPASS
03/11/98 09:26:28    DB 26         FC0222         #1 DEBUT REFLUX
03/11/98 09:26:44    DB 26         FC0222         #1 DEBUT REFLUX
03/11/98 09:27:10    HM 26         FC4095         E-47 SLURRY BYPASS
03/11/98 09:27:50    SP 21 ALM        LC0207        PVHI             75.000
03/11/98 09:27:52    DB 26         FC0222         #1 DEBUT REFLUX
03/11/98 09:28:01    SP 31 ACK        LC0207        PVHI
03/11/98 09:28:08    DB 26         FC0221         2 DEBUT REFLUX FLOW
03/11/98 09:28:11    DB 26         FC0221         2 DEBUT REFLUX FLOW
03/11/98 09:28:22    SP 26         FC1211         6 IN COLD SPLITTER BTMS
03/11/98 09:28:30    DP 26         HC4095         1032A SLURRY TO E-47
03/11/98 09:28:45    DB 26         FC0222         #1 DEBUT REFLUX
03/11/98 09:28:52    DB 26         FC0221         2 DEBUT REFLUX FLOW
03/11/98 09:29:26    SP 26         FC1211         6 IN COLD SPLITTER BTMS
03/11/98 09:31:17    DB 26         FC0205         1 DEBUT REBOILER HT MED
03/11/98 09:31:20    DB 26         FC0205         1 DEBUT REBOILER HT MED
03/11/98 09:32:15    HM 26         FC4095         E-47 SLURRY BYPASS
03/11/98 09:32:17    SP 21 RTN        LC0207        PVHI             75.000
03/11/98 09:34:07    SP 26         FC1211         6 IN COLD SPLITTER BTMS
03/11/98 09:34:58    DB 26         FC0222         #1 DEBUT REFLUX
03/11/98 09:35:12    DP 26         HC4095         1032A SLURRY TO E-47
03/11/98 09:36:55    DB 21 ALM        LC0204        PVHI             70.000
03/11/98 09:37:02    DB 31 ACK        LC0204        PVHI
03/11/98 09:37:33    DB 26         FC0221         2 DEBUT REFLUX FLOW
03/11/98 09:38:20    DB 26         FC0205         1 DEBUT REBOILER HT MED
03/11/98 09:38:46    SP 21 ALM        LC0207        PVLO             30.000
03/11/98 09:38:47    SP 31 ACK        LC0207        PVLO
03/11/98 09:38:55    SP 26         FC1211         6 IN COLD SPLITTER BTMS
03/11/98 09:39:05    SP 26         FC1211         6 IN COLD SPLITTER BTMS
03/11/98 09:39:28    SP 26         FC1210         3 IN SPLIT BTMS TO FIELD
03/11/98 09:39:29    SP 26         FC1210         3 IN SPLIT BTMS TO FIELD
03/11/98 09:39:35    SP 26         FC1210         3 IN SPLIT BTMS TO FIELD
03/11/98 09:39:50    SP 21 ALM        FC1210        PVLO             10.000
03/11/98 09:40:36    DB 21 RTN        LC0204        PVHI             70.000
03/11/98 09:40:58    SP 21 RTN        LC0207        PVLO             30.000
03/11/98 09:41:13    SP 26         FC1210         3 IN SPLIT BTMS TO FIELD
03/11/98 09:41:19    SP 26         LC0207         SPLITTER REBOILER LEVEL
```

HIGH	RX SV DIFF PRESSURE			1	
OP	60.0000	20.0000	BPH	CONS	1
HIGH	RX SV DIFF PRESSURE	112.646			
OP	20.0000	10.0000	BPH	CONS	1
OP	17.9468	15.9468	PERCENT	CONS	1
OP	15.3197	13.3197	PERCENT	CONS	1
OP	26.5977	24.5977	PERCENT	CONS	1
OP	52.0000	60.0000	BPH	CONS	1
SP	143.6830	145.0000	DEG F	CONS	1
SP	823.1320	850.0000	MMBTUF/H	CONS	1
OP	30.0000	15.0000	PSIG	CONS	1
MODE	CAS	MAN	B P H	CONS	1
OP	61.6062	43.6062	B P H	CONS	1
OP	60.0000	70.0000	BPH	CONS	1
HIGH	SPLITTER REBOILER LEVEL	75.074			
OP	43.6062	41.6062	B P H	CONS	1
HIGH	SPLITTER REBOILER LEVEL			1	
MODE	CAS	MAN	B P H	CONS	1
OP	44.6417	40.6417	B P H	CONS	1
OP	5.1000	9.1000	B P H	CONS	1
OP	10.0000	50.0000	BPH	CONS	1
OP	41.6062	37.6062	B P H	CONS	1
OP	40.6417	36.6417	B P H	CONS	1
OP	9.1000	23.1000	B P H	CONS	1
MODE	CAS	MAN	B P H	CONS	1
OP	.8933	76.8933	B P H	CONS	1
OP	70.0000	100.0000	BPH	CONS	1
HIGH	SPLITTER REBOILER LEVEL	73.889			
OP	23.1000	19.1000	B P H	CONS	1
OP	37.6062	39.6062	B P H	CONS	1
OP	50.0000	20.0000	BPH	CONS	1
LOW	#2 DEBUT. ACCUM. LEVEL	70.050			
LOW	#2 DEBUT. ACCUM. LEVEL			1	
OP	36.6417	40.6417	B P H	CONS	1
OP	76.8933	74.8933	B P H	CONS	1
HIGH	SPLITTER REBOILER LEVEL	29.909			
HIGH	SPLITTER REBOILER LEVEL			1	
OP	19.1000	0.0000	B P H	CONS	1
OP	0.0000	---------	B P H	CONS	1
MODE	CAS	MAN	B P H	CONS	1
OP	54.7600	0.0000	B P H	CONS	1
OP	0.0000	- 6.0000	B P H	CONS	1
JOURNAL	3 IN SPLIT BTMS TO FIELD	9.998			
LOW	#2 DEBUT. ACCUM. LEVEL	68.916			
HIGH	SPLITTER REBOILER LEVEL	31.230			
MODE	MAN	NORMAL	B P H	CONS	1
SP	31.5606	50.0000	PERCENT	CONS	1

```
03/11/98 09:41:25   SP 21 ALM      LC0207      PVLO              30.000
03/11/98 09:41:50   DB 26        FC0205       1 DEBUT REBOILER HT MED
03/11/98 09:42:14   SP 21 RTN      LC0207      PVLO              30.000
03/11/98 09:42:30   SP 31 ACK      LC0207      PVLO
03/11/98 09:42:35   SP 21 ALM      LC0207      PVLO              30.000
03/11/98 09:42:42   SP 31 ACK      LC0207      PVLO
03/11/98 09:43:24   SP 21 RTN      LC0207      PVLO              30.000
03/11/98 09:43:28   DP 26        PC0042       C-12 HOT VAPOR BYPASS
03/11/98 09:44:13   DP 26        HC4095       1032A SLURRY TO E-47
03/11/98 09:44:17   HM 21 ALM      FI4095      BADPV
03/11/98 09:44:19   HM 21 ALM      FC4095      BADPV
03/11/98 09:44:22   HM 31 ACK      FC4095      BADPV
03/11/98 09:44:22   HM 31 ACK      FI4095      BADPV
03/11/98 09:45:40   SP 21 RTN      FC1210      PVLO              10.000
03/11/98 09:45:44   DB 26        CX0203       E29 #2 DEB REB NORM DUTY
03/11/98 09:45:48   DP 26        FC0006       C-12 BOTTOMS TO #2 DEBUT
03/11/98 09:48:45   DB 21 ALM      AI0212      PVHI               3.000
03/11/98 09:48:56   DB 31 ACK      AI0212      PVHI
03/11/98 09:48:57   DB 21 ALM      AC0212      BADPV
03/11/98 09:49:03   DB 31 ACK      AC0212      BADPV
03/11/98 09:49:28   DB 26        FC0205       1 DEBUT REBOILER HT MED
03/11/98 09:50:43   DB 26        FC0222       #1 DEBUT REFLUX
03/11/98 09:51:26   DP 26        FC0105       REFLUX TO C-12 TOWER
03/11/98 09:51:43   DP 26        FC0105       REFLUX TO C-12 TOWER
03/11/98 09:51:50   DB 26        FC0222       #1 DEBUT REFLUX
03/11/98 09:52:04   SP 21 ALM      LC0207      PVHI              75.000
03/11/98 09:52:19   SP 26        FC1211       6 IN COLD SPLITTER BTMS
03/11/98 09:53:04   DP 26        PC0042       C-12 HOT VAPOR BYPASS
03/11/98 09:53:38   SP 26        FC1211       6 IN COLD SPLITTER BTMS
03/11/98 09:54:14   DB 26        FC0222       #1 DEBUT REFLUX
03/11/98 09:54:31   SP 26        FC1211       6 IN COLD SPLITTER BTMS
03/11/98 09:55:07   SP 31 ACK      LC0207      PVHI
03/11/98 09:55:08   RX 21 ALM      PDI1203     PVLO             100.000
03/11/98 09:55:13   RX 31 ACK      PDI1203     PVLO
03/11/98 09:56:13   SP 26        FC1211       6 IN COLD SPLITTER BTMS
03/11/98 09:56:45   SP 21 RTN      LC0207      PVHI              75.000
03/11/98 09:56:52   RX 21 RTN      PDI1203     PVLO             100.000
03/11/98 09:57:21   SP 21 ALM      AI0211      PVHI               3.000
03/11/98 09:57:31   SP 26        FC1211       6 IN COLD SPLITTER BTMS
03/11/98 09:57:49   SP 31 ACK      AI0211      PVHI
03/11/98 09:57:58   DB 21 ALM      AC0211      BADPV
03/11/98 09:57:59   DB 21 ALM      AC0213      BADPV
03/11/98 09:58:05   DB 31 ACK      AC0213      BADPV
03/11/98 09:58:05   DB 31 ACK      AC0211      BADPV
03/11/98 09:58:26   DB 26        FC0221       2 DEBUT REFLUX FLOW
03/11/98 09:58:38   DB 26        TC1202       #2 DEBUT TOWER TOP TEMP
03/11/98 09:59:07   SP 26        FC1211       6 IN COLD SPLITTER BTMS
```

```
HIGH     SPLITTER REBOILER LEVEL          29.822
 OP           74.8933                 .8933         B P H     CONS   1
HIGH     SPLITTER REBOILER LEVEL          31.186
HIGH     SPLITTER REBOILER LEVEL                              1
HIGH     SPLITTER REBOILER LEVEL          29.960
HIGH     SPLITTER REBOILER LEVEL                              1
HIGH     SPLITTER REBOILER LEVEL          31.243
 OP           15.0000              10.0000          PSIG      CONS   1
 OP           20.0000              10.0000          BPH       CONS   1
LOW      E-47 SLURRY IN MED
LOW      E-47 SLURRY BYPASS
LOW      E-47 SLURRY BYPASS                                   1
LOW      E-47 SLURRY IN MED                                   1
JOURNAL  3 IN SPLIT BTMS TO FIELD         13.192
 SP           630.0000            600.0000          MMBTUF/H  CONS   1
 SP           1200.000            1300.000          BPH       CONS   1
LOW      2-DEBUT OVHD C-5S                 3.924
LOW      2-DEBUT OVHD C-5S                                    1
LOW      PV VALIDATION FOR AI0212
LOW      PV VALIDATION FOR AI0212                             1
 OP              .8933             76.8933          B P H     CONS   1
 OP           39.6062              35.6062          B P H     CONS   1
 OP           100.0000             50.0000          BPH       CONS   1
 OP           50.0000              60.0000          BPH       CONS   1
 OP           35.6062              31.6062          B P H     CONS   1
HIGH     SPLITTER REBOILER LEVEL          75.102
 OP         -  6.9000              15.1000          B P H     CONS   1
 OP           10.0000              25.0000          PSIG      CONS   1
 OP           15.1000              19.1000          B P H     CONS   1
 OP           31.6062              27.6062          B P H     CONS   1
 OP           19.1000              33.1000          B P H     CONS   1
HIGH     SPLITTER REBOILER LEVEL                              1
HIGH     RX SV DIFF PRESSURE              99.863
HIGH     RX SV DIFF PRESSURE                                  1
 OP           33.1000              35.1000          B P H     CONS   1
HIGH     SPLITTER REBOILER LEVEL          73.843
HIGH     RX SV DIFF PRESSURE             112.279
LOW      1-DEBUT OVHD C-5S                 5.546
 OP           35.1000              31.1000          B P H     CONS   1
LOW      1-DEBUT OVHD C-5S                                    1
LOW      #1 DEBUT OVERHEAD C5
LOW      AVG DEBUT OH C5
LOW      AVG DEBUT OH C5                                      1
LOW      #1 DEBUT OVERHEAD C5                                 1
 MODE         MAN                 NORMAL            B P H     CONS   1
 SP           150.4490            148.0000          DEG F     CONS   1
 OP           31.1000              29.1000          B P H     CONS   1
```

03/11/98	09:59:14	DP 31 ACK	FC0022	PVLL	
03/11/98	09:59:18	DP 21 ALM	FC0022	PVLO	100.000
03/11/98	09:59:20	DP 21 ALM	FC0022	PVLL	75.000
03/11/98	09:59:30	DP 21 RTN	FC0022	PVLL	75.000
03/11/98	09:59:31	DP 21 RTN	FC0022	PVLO	100.000
03/11/98	09:59:51	DP 21 ALM	FC0022	PVLO	100.000
03/11/98	09:59:52	DP 21 ALM	FC0022	PVLL	75.000
03/11/98	09:59:53	DP 26	FC0105	REFLUX TO C-12 TOWER	
03/11/98	09:59:55	DP 21 RTN	FC0022	PVLL	75.000
03/11/98	09:59:56	DP 21 RTN	FC0022	PVLO	100.000
03/11/98	10:00:06	DP 26	HC4095	1032A SLURRY TO E-47	
03/11/98	10:00:09	HM 21 RTN	FI4095	BADPV	
03/11/98	10:00:10	DP 21 RTN	CX4095	BADPV	
03/11/98	10:00:10	HM 21 RTN	FC4095	BADPV	
03/11/98	10:00:15	DP 26	PC0042	C-12 HOT VAPOR BYPASS	
03/11/98	10:00:32	DP 26	HC4095	1032A SLURRY TO E-47	
03/11/98	10:00:37	HM 21 ALM	FI4095	BADPV	
03/11/98	10:00:38	HM 21 ALM	FC4095	BADPV	
03/11/98	10:00:40	DP 21 ALM	CX4095	BADPV	
03/11/98	10:00:43	DP 26	FC0105	REFLUX TO C-12 TOWER	
03/11/98	10:01:00	DP 31 ACK	CX4095	BADPV	
03/11/98	10:01:00	DP 31 ACK	FC0022	PVLL	
03/11/98	10:01:00	DP 31 ACK	FC0022	PVLO	
03/11/98	10:01:00	HM 31 ACK	FI4095	BADPV	
03/11/98	10:01:00	HM 31 ACK	FC4095	BADPV	
03/11/98	10:02:42	CP 21 ALM	PAL0295	OFFNORM	
03/11/98	10:02:55	CP 31 ACK	PAL0295	OFFNORM	
03/11/98	10:03:01	DB 21 ALM	CA0212	BADPV	
03/11/98	10:03:02	SP 26	FC1211	6 IN COLD SPLITTER BTMS	
03/11/98	10:03:25	DB 31 ACK	CA0212	BADPV	
03/11/98	10:04:00	DB 21 ALM	CA0213	BADPV	
03/11/98	10:04:39	DP 26	HC4095	1032A SLURRY TO E-47	
03/11/98	10:04:42	HM 21 RTN	FI4095	BADPV	
03/11/98	10:04:42	HM 21 RTN	FC4095	BADPV	
03/11/98	10:04:45	DP 21 RTN	CX4095	BADPV	
03/11/98	10:05:38	DP 26	FC0105	REFLUX TO C-12 TOWER	
03/11/98	10:06:05	DP 21 ALM	FC0022	PVLO	100.000
03/11/98	10:06:07	DP 21 ALM	FC0022	PVLL	75.000
03/11/98	10:06:08	DB 31 ACK	CA0213	BADPV	
03/11/98	10:06:08	DP 31 ACK	FC0022	PVLL	
03/11/98	10:06:08	DP 31 ACK	FC0022	PVLO	
03/11/98	10:07:00	DP 21 RTN	FC0022	PVLL	75.000
03/11/98	10:07:01	DP 21 RTN	FC0022	PVLO	100.000
03/11/98	10:07:03	DP 21 ALM	FC0022	PVLO	100.000
03/11/98	10:07:04	DP 21 ALM	FC0022	PVLL	75.000
03/11/98	10:07:08	DP 31 ACK	FC0022	PVLL	
03/11/98	10:07:08	DP 31 ACK	FC0022	PVLO	

HIGH	V-33 ACCUM RELEASE			1	
LOW	V-33 ACCUM RELEASE	97.818			
HIGH	V-33 ACCUM RELEASE	70.191			
HIGH	V-33 ACCUM RELEASE	93.919			
LOW	V-33 ACCUM RELEASE	138.839			
LOW	V-33 ACCUM RELEASE	86.591			
HIGH	V-33 ACCUM RELEASE	73.358			
OP	60.0000	50.0000	BPH	CONS	1
HIGH	V-33 ACCUM RELEASE	80.530			
LOW	V-33 ACCUM RELEASE	124.155			
OP	10.0000	30.0000	BPH	CONS	1
LOW	E-47 SLURRY IN MED				
LOW	E47 DEPROP REB NORM DUTY				
LOW	E-47 SLURRY BYPASS				
OP	25.0000	35.0000	PSIG	CONS	1
OP	30.0000	10.0000	BPH	CONS	1
LOW	E-47 SLURRY IN MED				
LOW	E-47 SLURRY BYPASS				
LOW	E47 DEPROP REB NORM DUTY				
OP	50.0000	0.0000	BPH	CONS	1
LOW	E47 DEPROP REB NORM DUTY			1	
HIGH	V-33 ACCUM RELEASE			1	
LOW	V-33 ACCUM RELEASE			1	
LOW	E-47 SLURRY IN MED			1	
LOW	E-47 SLURRY BYPASS			1	
LOW	VIB CAB PURGE SW [2-A/B]	LO_PRESS			
LOW	VIB CAB PURGE SW [2-A/B]			1	
LOW	#2 DEBUT OH C5				
OP	29.1000	13.1000	B P H	CONS	1
LOW	#2 DEBUT OH C5			1	
LOW	AVG DEBUT OVHD C5				
OP	10.0000	40.0000	BPH	CONS	1
LOW	E-47 SLURRY IN MED				
LOW	E-47 SLURRY BYPASS				
LOW	E47 DEPROP REB NORM DUTY				
OP	0.0000	40.0000	BPH	CONS	1
LOW	V-33 ACCUM RELEASE	93.051			
HIGH	V-33 ACCUM RELEASE	66.428			
LOW	AVG DEBUT OVHD C5			1	
HIGH	V-33 ACCUM RELEASE			1	
LOW	V-33 ACCUM RELEASE			1	
HIGH	V-33 ACCUM RELEASE	80.556			
LOW	V-33 ACCUM RELEASE	114.439			
LOW	V-33 ACCUM RELEASE	85.501			
HIGH	V-33 ACCUM RELEASE	.247			
HIGH	V-33 ACCUM RELEASE			1	
LOW	V-33 ACCUM RELEASE			1	

03/11/98	10:07:15	DP 26	HC4095	1032A SLURRY TO E-47	
03/11/98	10:07:17	DP 21 RTN	FC0022	PVLL	75.000
03/11/98	10:07:17	DP 21 RTN	FC0022	PVLO	100.000
03/11/98	10:07:21	DP 21 ALM	FC0022	PVLO	100.000
03/11/98	10:07:22	DP 21 ALM	FC0022	PVLL	75.000
03/11/98	10:07:34	DP 21 RTN	FC0022	PVLO	100.000
03/11/98	10:07:34	DP 21 RTN	FC0022	PVLL	75.000
03/11/98	10:07:39	DP 21 ALM	FC0022	PVLO	100.000
03/11/98	10:07:41	DP 21 ALM	FC0022	PVLL	75.000
03/11/98	10:07:43	DP 31 ACK	FC0022	PVLO	
03/11/98	10:07:43	DP 31 ACK	FC0022	PVLL	
03/11/98	10:07:51	DP 21 RTN	FC0022	PVLO	100.000
03/11/98	10:07:51	DP 21 RTN	FC0022	PVLL	75.000
03/11/98	10:07:58	DP 21 ALM	FC0022	PVLO	100.000
03/11/98	10:07:59	DP 21 ALM	FC0022	PVLL	75.000
03/11/98	10:08:02	DP 31 ACK	FC0022	PVLL	
03/11/98	10:08:02	DP 31 ACK	FC0022	PVLO	
03/11/98	10:08:06	DP 21 RTN	FC0022	PVLO	100.000
03/11/98	10:08:06	DP 21 RTN	FC0022	PVLL	75.000
03/11/98	10:08:13	DP 21 ALM	FC0022	PVLO	100.000
03/11/98	10:08:15	DP 21 ALM	FC0022	PVLL	75.000
03/11/98	10:08:22	DP 21 RTN	FC0022	PVLO	100.000
03/11/98	10:08:22	DP 21 RTN	FC0022	PVLL	75.000
03/11/98	10:08:30	DP 21 ALM	FC0022	PVLO	100.000
03/11/98	10:08:31	DP 21 ALM	FC0022	PVLL	75.000
03/11/98	10:08:37	DP 21 RTN	FC0022	PVLO	100.000
03/11/98	10:08:37	DP 21 RTN	FC0022	PVLL	75.000
03/11/98	10:08:42	DP 21 ALM	TC0005	PVHI	175.000
03/11/98	10:08:45	DP 21 ALM	FC0022	PVLO	100.000
03/11/98	10:08:47	DP 21 ALM	FC0022	PVLL	75.000
03/11/98	10:08:50	DB 21 ALM	AI0112	UNREASBL	
03/11/98	10:08:51	DP 21 RTN	FC0022	PVLL	75.000
03/11/98	10:08:52	DP 21 RTN	FC0022	PVLO	100.000
03/11/98	10:09:00	DP 21 ALM	FC0022	PVLO	100.000
03/11/98	10:09:02	DP 21 ALM	FC0022	PVLL	75.000
03/11/98	10:09:07	DP 31 ACK	FC0022	PVLL	
03/11/98	10:09:07	DP 31 ACK	FC0022	PVLO	
03/11/98	10:09:07	DP 31 ACK	TC0005	PVHI	
03/11/98	10:09:09	DP 21 RTN	FC0022	PVLO	100.000
03/11/98	10:09:09	DP 21 RTN	FC0022	PVLL	75.000
03/11/98	10:09:22	DP 21 ALM	FC0022	PVLO	100.000
03/11/98	10:09:24	DP 21 ALM	FC0022	PVLL	75.000
03/11/98	10:09:29	DP 31 ACK	FC0022	PVLL	
03/11/98	10:09:29	DP 31 ACK	FC0022	PVLO	
03/11/98	10:09:31	DP 21 RTN	FC0022	PVLO	100.000
03/11/98	10:09:31	DP 21 RTN	FC0022	PVLL	75.000
03/11/98	10:09:44	DP 21 ALM	FC0022	PVLO	100.000

OP	40.0000	20.0000	BPH	CONS 1
HIGH	V-33 ACCUM RELEASE	109.955		
LOW	V-33 ACCUM RELEASE	109.955		
LOW	V-33 ACCUM RELEASE	85.691		
HIGH	V-33 ACCUM RELEASE	.401		
LOW	V-33 ACCUM RELEASE	115.671		
HIGH	V-33 ACCUM RELEASE	115.671		
LOW	V-33 ACCUM RELEASE	99.187		
HIGH	V-33 ACCUM RELEASE	70.853		
LOW	V-33 ACCUM RELEASE			1
HIGH	V-33 ACCUM RELEASE			1
LOW	V-33 ACCUM RELEASE	112.563		
HIGH	V-33 ACCUM RELEASE	112.563		
LOW	V-33 ACCUM RELEASE	88.577		
HIGH	V-33 ACCUM RELEASE	74.851		
HIGH	V-33 ACCUM RELEASE			1
LOW	V-33 ACCUM RELEASE			1
LOW	V-33 ACCUM RELEASE	109.217		
HIGH	V-33 ACCUM RELEASE	109.217		
LOW	V-33 ACCUM RELEASE	97.788		
HIGH	V-33 ACCUM RELEASE	69.842		
LOW	V-33 ACCUM RELEASE	120.738		
HIGH	V-33 ACCUM RELEASE	120.738		
LOW	V-33 ACCUM RELEASE	85.421		
HIGH	V-33 ACCUM RELEASE	.396		
LOW	V-33 ACCUM RELEASE	110.911		
HIGH	V-33 ACCUM RELEASE	110.911		
LOW	C-12 TRAY 10 TEMP	175.009		
LOW	V-33 ACCUM RELEASE	93.991		
HIGH	V-33 ACCUM RELEASE	67.457		
JOURNAL	2-DEBUT OVHD C-3S	18.300		
HIGH	V-33 ACCUM RELEASE	79.772		
LOW	V-33 ACCUM RELEASE	119.395		
LOW	V-33 ACCUM RELEASE	96.826		
HIGH	V-33 ACCUM RELEASE	69.178		
HIGH	V-33 ACCUM RELEASE			1
LOW	V-33 ACCUM RELEASE			1
LOW	C-12 TRAY 10 TEMP			1
LOW	V-33 ACCUM RELEASE	122.997		
HIGH	V-33 ACCUM RELEASE	122.997		
LOW	V-33 ACCUM RELEASE	90.411		
HIGH	V-33 ACCUM RELEASE	64.512		
HIGH	V-33 ACCUM RELEASE			1
LOW	V-33 ACCUM RELEASE			1
LOW	V-33 ACCUM RELEASE	120.966		
HIGH	V-33 ACCUM RELEASE	120.966		
LOW	V-33 ACCUM RELEASE	94.742		

```
03/11/98 10:09:46   DP 21 ALM      FC0022       PVLL              75.000
03/11/98 10:09:48   DP 31 ACK      FC0022       PVLL
03/11/98 10:09:48   DP 31 ACK      FC0022       PVLO
03/11/98 10:09:53   DP 21 RTN      FC0022       PVLO             100.000
03/11/98 10:09:53   DP 21 RTN      FC0022       PVLL              75.000
03/11/98 10:10:06   DP 21 ALM      FC0022       PVLO             100.000
03/11/98 10:10:07   DP 26        HC4095       1032A SLURRY TO E-47
03/11/98 10:10:08   DP 21 ALM      FC0022       PVLL              75.000
03/11/98 10:10:11   HM 21 ALM      FI4095       BADPV
03/11/98 10:10:12   HM 21 ALM      FC4095       BADPV
03/11/98 10:10:14   DP 21 RTN      FC0022       PVLL              75.000
03/11/98 10:10:15   DP 21 RTN      FC0022       PVLO             100.000
03/11/98 10:10:15   DP 21 ALM      CX4095       BADPV
03/11/98 10:10:22   DP 31 ACK      CX4095       BADPV
03/11/98 10:10:22   DP 31 ACK      FC0022       PVLL
03/11/98 10:10:22   DP 31 ACK      FC0022       PVLO
03/11/98 10:10:22   HM 31 ACK      FI4095       BADPV
03/11/98 10:10:22   HM 31 ACK      FC4095       BADPV
03/11/98 10:10:28   DP 21 ALM      FC0022       PVLO             100.000
03/11/98 10:10:30   DP 21 ALM      FC0022       PVLL              75.000
03/11/98 10:10:33   DP 26        FC0022       V-33 ACCUM RELEASE
03/11/98 10:10:34   DP 21 RTN      FC0022       PVLL              75.000
03/11/98 10:10:35   DP 21 RTN      FC0022       PVLO             100.000
03/11/98 10:10:36   DP 26        FC0022       V-33 ACCUM RELEASE
03/11/98 10:10:49   DP 21 ALM      FC0022       PVLO             100.000
03/11/98 10:10:51   DP 21 ALM      FC0022       PVLL              75.000
03/11/98 10:10:53   DP 26        FC0022       V-33 ACCUM RELEASE
03/11/98 10:10:55   DP 21 RTN      FC0022       PVLL              75.000
03/11/98 10:10:55   DP 31 ACK      FC0022       PVLL
03/11/98 10:10:55   DP 31 ACK      FC0022       PVLO
03/11/98 10:10:56   DP 21 RTN      FC0022       PVLO             100.000
03/11/98 10:11:10   DP 21 ALM      FC0022       PVLO             100.000
03/11/98 10:11:12   DP 21 ALM      FC0022       PVLL              75.000
03/11/98 10:11:17   DP 21 RTN      FC0022       PVLL              75.000
03/11/98 10:11:18   DP 21 RTN      FC0022       PVLO             100.000
03/11/98 10:11:20   DP 26        FC0022       V-33 ACCUM RELEASE
03/11/98 10:12:05   DP 21 ALM      FC0022       PVLO             100.000
03/11/98 10:12:08   DP 21 RTN      FC0022       PVLO             100.000
03/11/98 10:12:10   DP 31 ACK      FC0022       PVLO
03/11/98 10:12:10   DP 31 ACK      FC0022       PVLL
03/11/98 10:12:34   CP 21 RTN      PAL0295      OFFNORM
03/11/98 10:12:57   DB 21 RTN      AC0212       BADPV
03/11/98 10:12:57   DP 21 ALM      FC0022       PVLO             100.000
03/11/98 10:13:00   DP 31 ACK      FC0022       PVLO
03/11/98 10:13:01   DP 21 RTN      FC0022       PVLO             100.000
03/11/98 10:13:23   SP 21 ALM      LC0207       PVHI              75.000
03/11/98 10:13:37   SP 26        FC1211       6 IN COLD SPLITTER BTMS
```

HIGH	V-33 ACCUM RELEASE	67.724			
HIGH	V-33 ACCUM RELEASE			1	
LOW	V-33 ACCUM RELEASE			1	
LOW	V-33 ACCUM RELEASE	109.935			
HIGH	V-33 ACCUM RELEASE	109.935			
LOW	V-33 ACCUM RELEASE	94.657			
OP	20.0000	10.0000	BPH	CONS	1
HIGH	V-33 ACCUM RELEASE	67.841			
LOW	E-47 SLURRY IN MED				
LOW	E-47 SLURRY BYPASS				
HIGH	V-33 ACCUM RELEASE	87.706			
LOW	V-33 ACCUM RELEASE	132.515			
LOW	E47 DEPROP REB NORM DUTY				
LOW	E47 DEPROP REB NORM DUTY			1	
HIGH	V-33 ACCUM RELEASE			1	
LOW	V-33 ACCUM RELEASE			1	
LOW	E-47 SLURRY IN MED			1	
LOW	E-47 SLURRY BYPASS			1	
LOW	V-33 ACCUM RELEASE	90.677			
HIGH	V-33 ACCUM RELEASE	65.426			
MODE	CAS	MAN	BPH	CONS	1
HIGH	V-33 ACCUM RELEASE	83.229			
LOW	V-33 ACCUM RELEASE	123.385			
OP	99.8767	75.0000	BPH	CONS	1
LOW	V-33 ACCUM RELEASE	91.066			
HIGH	V-33 ACCUM RELEASE	65.495			
OP	75.0000	70.0000	BPH	CONS	1
HIGH	V-33 ACCUM RELEASE	95.873			
HIGH	V-33 ACCUM RELEASE			1	
LOW	V-33 ACCUM RELEASE			1	
LOW	V-33 ACCUM RELEASE	134.397			
LOW	V-33 ACCUM RELEASE	90.812			
HIGH	V-33 ACCUM RELEASE	65.516			
HIGH	V-33 ACCUM RELEASE	80.732			
LOW	V-33 ACCUM RELEASE	117.212			
OP	70.0000	50.0000	BPH	CONS	1
LOW	V-33 ACCUM RELEASE	98.322			
LOW	V-33 ACCUM RELEASE	110.652			
LOW	V-33 ACCUM RELEASE			1	
HIGH	V-33 ACCUM RELEASE			1	
LOW	VIB CAB PURGE SW [2-A/B]	NORMAL			
LOW	PV VALIDATION FOR AI0212				
LOW	V-33 ACCUM RELEASE	92.580			
LOW	V-33 ACCUM RELEASE			1	
LOW	V-33 ACCUM RELEASE	108.905			
HIGH	SPLITTER REBOILER LEVEL	75.064			
OP	13.1000	17.1000	B P H	CONS	1

```
03/11/98 10:13:49   SP 31 ACK      LC0207       PVHI
03/11/98 10:13:50   SP 21 RTN      LC0207       PVHI             75.000
03/11/98 10:14:53   SP 21 ALM      LC0207       PVHI             75.000
03/11/98 10:15:03   DP 26        PC0042       C-12 HOT VAPOR BYPASS
03/11/98 10:15:27   SP 21 RTN      LC0207       PVHI             75.000
03/11/98 10:15:53   CP 21 ALM      PAL0296      OFFNORM
03/11/98 10:16:01   CP 31 ACK      PAL0296      OFFNORM
03/11/98 10:16:01   SP 31 ACK      LC0207       PVHI
03/11/98 10:17:59   DP 26        PC0042       C-12 HOT VAPOR BYPASS
03/11/98 10:18:34   DB 26        FC0205       1 DEBUT REBOILER HT MED
03/11/98 10:19:25   CP 21 RTN      PAL0296      OFFNORM
03/11/98 10:20:07   DP 21 RTN      TC0005       PVHI            175.000
03/11/98 10:21:58   DB 21 RTN      AC0211       BADPV
03/11/98 10:21:59   DB 21 RTN      AC0213       BADPV
03/11/98 10:23:56   DB 21 RTN      AI0212       PVHI              3.000
03/11/98 10:29:26   DP 26        FC0022       V-33 ACCUM RELEASE
03/11/98 10:31:25   DP 26        FC0105       REFLUX TO C-12 TOWER
03/11/98 10:34:01   DP 26        FC0105       REFLUX TO C-12 TOWER
03/11/98 10:34:19   DP 21 ALM      FC0022       PVLO            100.000
03/11/98 10:34:22   DP 31 ACK      FC0022       PVLO
03/11/98 10:35:02   DP 26        HC4095       1032A SLURRY TO E-47
03/11/98 10:36:48   DP 26        FC0105       REFLUX TO C-12 TOWER
03/11/98 10:37:59   DP 26        HC4095       1032A SLURRY TO E-47
03/11/98 10:38:03   HM 21 RTN      FI4095       BADPV
03/11/98 10:38:04   HM 21 RTN      FC4095       BADPV
03/11/98 10:38:10   DP 21 RTN      CX4095       BADPV
03/11/98 10:39:52   DP 21 RTN      FC0022       PVLO            100.000
03/11/98 10:39:57   DP 26        HC4095       1032A SLURRY TO E-47
03/11/98 10:40:10   DP 26        FC0105       REFLUX TO C-12 TOWER
03/11/98 10:40:58   DP 21 ALM      AC0108       BADPV
03/11/98 10:41:00   DP 31 ACK      AC0108       BADPV
03/11/98 10:42:15   DP 26        FC0105       REFLUX TO C-12 TOWER
03/11/98 10:42:21   DP 26        HC4095       1032A SLURRY TO E-47
03/11/98 10:42:37   DP 26        FC0105       REFLUX TO C-12 TOWER
03/11/98 10:42:45   RX 21 ALM      PDI1203      PVLO            100.000
03/11/98 10:42:49   RX 31 ACK      PDI1203      PVLO
03/11/98 10:42:51   DP 21 ALM      FC0022       PVLO            100.000
03/11/98 10:42:53   DP 31 ACK      FC0022       PVLO
03/11/98 10:43:46   DP 26        FC0105       REFLUX TO C-12 TOWER
03/11/98 10:44:59   RX 21 RTN      PDI1203      PVLO            100.000
03/11/98 10:47:07   DP 26        HC4095       1032A SLURRY TO E-47
03/11/98 10:49:02   DP 26        FC0105       REFLUX TO C-12 TOWER
03/11/98 10:50:34   HM 26        FC1102       SLURRY BYPASS WHB
03/11/98 10:50:50   DP 26        FC0105       REFLUX TO C-12 TOWER
03/11/98 10:51:38   SP 21 ALM      LC0207       PVLO             30.000
03/11/98 10:51:43   SP 31 ACK      LC0207       PVLO
03/11/98 10:51:57   SP 26        FC1211       6 IN COLD SPLITTER BTMS
```

HIGH	SPLITTER REBOILER LEVEL			1	
HIGH	SPLITTER REBOILER LEVEL	73.942			
HIGH	SPLITTER REBOILER LEVEL	75.352			
OP	35.0000	25.0000	PSIG	CONS	1
HIGH	SPLITTER REBOILER LEVEL	73.775			
LOW	VIB CAB PURGE SW [CP-5]	LO_PRESS			
LOW	VIB CAB PURGE SW [CP-5]			1	
HIGH	SPLITTER REBOILER LEVEL			1	
OP	25.0000	15.0000	PSIG	CONS	1
OP	76.8933	74.8933	B P H	CONS	1
LOW	VIB CAB PURGE SW [CP-5]	NORMAL			
LOW	C-12 TRAY 10 TEMP	170.493			
LOW	#1 DEBUT OVERHEAD C5				
LOW	AVG DEBUT OH C5				
LOW	2-DEBUT OVHD C-5S	2.673			
OP	50.0000	44.0000	BPH	CONS	1
OP	40.0000	44.0000	BPH	CONS	1
OP	44.0000	58.0000	BPH	CONS	1
LOW	V-33 ACCUM RELEASE	99.997			
LOW	V-33 ACCUM RELEASE			1	
OP	10.0000	14.0000	BPH	CONS	1
OP	58.0000	54.0000	BPH	CONS	1
OP	14.0000	20.0000	BPH	CONS	1
LOW	E-47 SLURRY IN MED				
LOW	E-47 SLURRY BYPASS				
LOW	E47 DEPROP REB NORM DUTY				
LOW	V-33 ACCUM RELEASE	103.755			
OP	20.0000	24.0000	BPH	CONS	1
OP	54.0000	56.0000	BPH	CONS	1
LOW	C-12 DEPROP OH C2 VALID				
LOW	C-12 DEPROP OH C2 VALID			1	
OP	56.0000	64.0000	BPH	CONS	1
OP	24.0000	28.0000	BPH	CONS	1
OP	64.0000	66.0000	BPH	CONS	1
HIGH	RX SV DIFF PRESSURE	99.991			
HIGH	RX SV DIFF PRESSURE			1	
LOW	V-33 ACCUM RELEASE	99.951			
LOW	V-33 ACCUM RELEASE			1	
OP	66.0000	62.0000	BPH	CONS	1
HIGH	RX SV DIFF PRESSURE	112.192			
OP	28.0000	32.0000	BPH	CONS	1
OP	62.0000	66.0000	BPH	CONS	1
OP	15.0000	20.0000	BPH	CONS	1
OP	66.0000	70.0000	BPH	CONS	1
HIGH	SPLITTER REBOILER LEVEL	29.876			
HIGH	SPLITTER REBOILER LEVEL			1	
OP	17.1000	13.1000	B P H	CONS	1

03/11/98	10:52:04	SP 21 RTN	LC0207	PVLO	30.000
03/11/98	10:52:05	DP 21 ALM	FC0022	PVLL	75.000
03/11/98	10:52:10	DP 31 ACK	FC0022	PVLL	
03/11/98	10:52:34	DB 21 ALM	AI0111	UNREASBL	
03/11/98	10:52:37	SP 21 ALM	LC0207	PVLO	30.000
03/11/98	10:52:45	SP 31 ACK	LC0207	PVLO	
03/11/98	10:52:59	SP 26	FC1211	6 IN COLD SPLITTER BTMS	
03/11/98	10:53:15	DP 21 RTN	FC0022	PVLL	75.000
03/11/98	10:53:19	DP 21 RTN	FC0022	PVLO	100.000
03/11/98	10:53:27	SP 21 RTN	LC0207	PVLO	30.000
03/11/98	10:53:32	DP 26	FC0105	REFLUX TO C-12 TOWER	
03/11/98	10:53:40	DP 26	FC0022	V-33 ACCUM RELEASE	
03/11/98	10:53:45	HM 26	TC5277	FRAC BOTTOM TEMP	
03/11/98	10:53:46	SP 21 ALM	LC0207	PVLO	30.000
03/11/98	10:53:50	SP 31 ACK	LC0207	PVLO	
03/11/98	10:53:54	DP 21 ALM	FC0022	PVLO	100.000
03/11/98	10:53:57	DP 31 ACK	FC0022	PVLO	
03/11/98	10:54:25	DP 21 ALM	FC0022	PVLL	75.000
03/11/98	10:54:28	DP 31 ACK	FC0022	PVLL	
03/11/98	10:54:39	SP 21 RTN	LC0207	PVLO	30.000
03/11/98	10:55:07	DP 26	FC0022	V-33 ACCUM RELEASE	
03/11/98	10:55:08	SP 21 ALM	LC0207	PVLO	30.000
03/11/98	10:55:12	SP 31 ACK	LC0207	PVLO	
03/11/98	10:55:50	SP 21 RTN	LC0207	PVLO	30.000
03/11/98	10:56:34	DP 26	PC0042	C-12 HOT VAPOR BYPASS	
03/11/98	10:56:40	SP 21 ALM	LC0207	PVLO	30.000
03/11/98	10:56:49	SP 26	FC1211	6 IN COLD SPLITTER BTMS	
03/11/98	10:57:02	SP 21 RTN	LC0207	PVLO	30.000
03/11/98	10:57:49	DP 26	HC4037	SOUTH E-45 OUTLET	
03/11/98	10:58:12	DP 26	HC4039	NORTH E-45 OUTLET	
03/11/98	10:58:18	DP 26	HC4038	MID E-45 OUTLET	
03/11/98	10:58:44	SP 31 ACK	LC0207	PVLO	
03/11/98	10:59:23	DB 26	FC0205	1 DEBUT REBOILER HT MED	
03/11/98	11:01:55	DB 26	FC0205	1 DEBUT REBOILER HT MED	
03/11/98	11:01:59	DP 26	FC0105	REFLUX TO C-12 TOWER	
03/11/98	11:02:21	DB 26	FC0222	#1 DEBUT REFLUX	
03/11/98	11:03:09	DP 26	PC0042	C-12 HOT VAPOR BYPASS	
03/11/98	11:04:26	DP 26	PC0042	C-12 HOT VAPOR BYPASS	
03/11/98	11:04:27	SP 26	FC1211	6 IN COLD SPLITTER BTMS	
03/11/98	11:04:32	DP 26	HC4038	MID E-45 OUTLET	
03/11/98	11:04:37	DP 26	HC4037	SOUTH E-45 OUTLET	
03/11/98	11:04:43	DP 26	HC4039	NORTH E-45 OUTLET	
03/11/98	11:05:43	DP 26	FC0105	REFLUX TO C-12 TOWER	
03/11/98	11:06:51	DP 21 RTN	FC0022	PVLL	75.000
03/11/98	11:08:22	RX 21 ALM	PDI1203	PVLO	100.000
03/11/98	11:08:26	RX 31 ACK	PDI1203	PVLO	
03/11/98	11:08:39	DP 26	PC0042	C-12 HOT VAPOR BYPASS	

```
HIGH      SPLITTER REBOILER LEVEL          31.128
HIGH      V-33 ACCUM RELEASE               74.272
HIGH      V-33 ACCUM RELEASE                                     1
JOURNAL   1-DEBUT OVHD C-3S                19.321
HIGH      SPLITTER REBOILER LEVEL          29.795
HIGH      SPLITTER REBOILER LEVEL                                1
 OP             13.1000          7.1000          B P H    CONS  1
HIGH      V-33 ACCUM RELEASE               81.390
LOW       V-33 ACCUM RELEASE              104.185
HIGH      SPLITTER REBOILER LEVEL          31.076
 OP             70.0000         74.0000          BPH      CONS  1
 OP             44.0000         38.0000          BPH      CONS  1
 SP            680.0000        682.0000          DEG F    CONS  1
HIGH      SPLITTER REBOILER LEVEL          29.741
HIGH      SPLITTER REBOILER LEVEL                                1
LOW       V-33 ACCUM RELEASE               99.626
LOW       V-33 ACCUM RELEASE                                     1
HIGH      V-33 ACCUM RELEASE                 .609
HIGH      V-33 ACCUM RELEASE                                     1
HIGH      SPLITTER REBOILER LEVEL          31.100
 OP             38.0000         32.0000          BPH      CONS  1
HIGH      SPLITTER REBOILER LEVEL          29.914
HIGH      SPLITTER REBOILER LEVEL                                1
HIGH      SPLITTER REBOILER LEVEL          31.252
 OP             15.0000         19.0000          PSIG     CONS  1
HIGH      SPLITTER REBOILER LEVEL          29.891
 OP              7.1000       -  0.9000          B P H    CONS  1
HIGH      SPLITTER REBOILER LEVEL          31.153
 OP             24.5977         24.9977          PERCENT  CONS  1
 OP             15.9468         16.0000          PERCENT  CONS  1
 OP             13.3197         14.0000          PERCENT  CONS  1
HIGH      SPLITTER REBOILER LEVEL                                1
 OP             74.8933         74.2933          B P H    CONS  1
 OP             74.2933           .2933          B P H    CONS  1
 OP             74.0000         76.0000          BPH      CONS  1
 OP             27.6062         29.6062          B P H    CONS  1
 OP             19.0000         21.0000          PSIG     CONS  1
 OP             21.0000         23.0000          PSIG     CONS  1
 OP           -  0.9000          1.1000          B P H    CONS  1
 OP             14.0000         15.0000          PERCENT  CONS  1
 OP             24.9977         26.9977          PERCENT  CONS  1
 OP             16.0000         17.0000          PERCENT  CONS  1
 OP             76.0000         80.0000          BPH      CONS  1
HIGH      V-33 ACCUM RELEASE               78.847
HIGH      RX SV DIFF PRESSURE              99.708
HIGH      RX SV DIFF PRESSURE                                    1
 OP             23.0000         25.0000          PSIG     CONS  1
```

```
03/11/98 11:08:46    RX 21 RTN      PDI1203        PVLO           100.000
03/11/98 11:10:29    DP 26       PC0042         C-12 HOT VAPOR BYPASS
03/11/98 11:11:16    DP 21 RTN      AI0208         UNREASBL
03/11/98 11:11:34    DP 26       HC4095         1032A SLURRY TO E-47
03/11/98 11:13:50    DB 21 ALM      FC0208         PVLO           100.000
03/11/98 11:13:53    DB 31 ACK      FC0208         PVLO
03/11/98 11:13:57    DB 21 RTN      FC0208         PVLO           100.000
03/11/98 11:14:11    DB 21 ALM      FC0208         PVLO           100.000
03/11/98 11:14:13    DB 31 ACK      FC0208         PVLO
03/11/98 11:14:21    DB 21 RTN      FC0208         PVLO           100.000
03/11/98 11:14:24    DP 26       PC0042         C-12 HOT VAPOR BYPASS
03/11/98 11:14:30    DP 26       HC4095         1032A SLURRY TO E-47
03/11/98 11:14:33    DB 21 ALM      FC0208         PVLO           100.000
03/11/98 11:14:37    DB 31 ACK      FC0208         PVLO
03/11/98 11:14:48    DB 21 RTN      FC0208         PVLO           100.000
03/11/98 11:14:52    DB 26       TC1202         #2 DEBUT TOWER TOP TEMP
03/11/98 11:14:55    DB 21 ALM      FC0208         PVLO           100.000
03/11/98 11:14:57    DB 31 ACK      FC0208         PVLO
03/11/98 11:15:18    DB 26       CX0203         E29 #2 DEB REB NORM DUTY
03/11/98 11:16:31    DP 26       PC0042         C-12 HOT VAPOR BYPASS
03/11/98 11:17:40    DB 21 RTN      AI0111         UNREASBL
03/11/98 11:17:58    DP 26       PC0042         C-12 HOT VAPOR BYPASS
03/11/98 11:18:10    DP 26       HC4095         1032A SLURRY TO E-47
03/11/98 11:19:27    DP 26       FC0022         V-33 ACCUM RELEASE
03/11/98 11:19:32    DP 21 ALM      FC0022         PVLL            75.000
03/11/98 11:19:35    DP 31 ACK      FC0022         PVLL
03/11/98 11:20:10    DP 26       HC4038         MID E-45 OUTLET
03/11/98 11:20:14    DP 26       HC4037         SOUTH E-45 OUTLET
03/11/98 11:20:18    DP 26       HC4039         NORTH E-45 OUTLET
03/11/98 11:20:40    DP 21 RTN      PC0042         PVLO           210.000
03/11/98 11:21:38    BL 26       KC5929         HORN 3
03/11/98 11:21:42    DP 21 ALM      LC0059         PVHI            80.000
03/11/98 11:21:49    DP 31 ACK      LC0059         PVHI
03/11/98 11:22:17    DP 26       PC0042         C-12 HOT VAPOR BYPASS
03/11/98 11:23:12    BL 26       KC5929         HORN 3
03/11/98 11:23:16    DP 26       HC4095         1032A SLURRY TO E-47
03/11/98 11:24:21    DP 21 RTN      LC0059         PVHI            80.000
03/11/98 11:26:05    DB 21 RTN      FC0208         PVLO           100.000
03/11/98 11:26:16    DB 21 ALM      FC0208         PVLO           100.000
03/11/98 11:26:23    DB 31 ACK      FC0208         PVLO
03/11/98 11:26:28    DB 21 RTN      FC0208         PVLO           100.000
03/11/98 11:26:38    DP 26       HC4095         1032A SLURRY TO E-47
03/11/98 11:26:42    DP 26       HC4095         1032A SLURRY TO E-47
03/11/98 11:26:48    DB 21 ALM      FC0208         PVLO           100.000
03/11/98 11:26:52    DB 31 ACK      FC0208         PVLO
03/11/98 11:26:55    DB 21 RTN      FC0208         PVLO           100.000
03/11/98 11:27:23    DP 26       FC0022         V-33 ACCUM RELEASE
```

```
HIGH      RX SV DIFF PRESSURE           112.062
 OP             25.0000           27.0000          PSIG      CONS  1
JOURNAL   C-12 DEPROP  OH-C4              9.851
 OP             32.0000           36.0000          BPH       CONS  1
LOW       2 OVERHEAD RELEASE             99.937
LOW       2 OVERHEAD RELEASE                                  1
LOW       2 OVERHEAD RELEASE            103.805
LOW       2 OVERHEAD RELEASE             99.759
LOW       2 OVERHEAD RELEASE                                  1
LOW       2 OVERHEAD RELEASE            103.358
 OP             27.0000           30.0000          PSIG      CONS  1
 OP             36.0000           40.0000          BPH       CONS  1
LOW       2 OVERHEAD RELEASE             99.148
LOW       2 OVERHEAD RELEASE                                  1
LOW       2 OVERHEAD RELEASE            103.148
 SP            148.0000          152.0000          DEG F     CONS  1
LOW       2 OVERHEAD RELEASE             98.865
LOW       2 OVERHEAD RELEASE                                  1
 SP            600.0000          590.0000          MMBTUF/H CONS  1
 OP             30.0000           34.0000          PSIG      CONS  1
JOURNAL   1-DEBUT OVHD C-3S              19.500
 OP             34.0000           36.0000          PSIG      CONS  1
 OP             40.0000           50.0000          BPH       CONS  1
 OP             32.0000           20.0000          BPH       CONS  1
HIGH      V-33 ACCUM RELEASE             74.209
HIGH      V-33 ACCUM RELEASE                                  1
 OP             15.0000           16.0000          PERCENT   CONS  1
 OP             26.9977           28.0000          PERCENT   CONS  1
 OP             17.0000           18.0000          PERCENT   CONS  1
EMERGNCY  C-12 HOT VAPOR BYPASS         211.032
 OP             STOP            START                        CONS  1
HIGH      V-33 FOUL H2O BOOT             80.063
HIGH      V-33 FOUL H2O BOOT                                  1
 OP             36.0000           40.0000          PSIG      CONS  1
 OP             START           STOP                         CONS  1
 OP             50.0000           60.0000          BPH       CONS  1
HIGH      V-33 FOUL H2O BOOT             78.865
LOW       2 OVERHEAD RELEASE            103.012
LOW       2 OVERHEAD RELEASE             99.613
LOW       2 OVERHEAD RELEASE                                  1
LOW       2 OVERHEAD RELEASE            103.276
 OP             60.0000           86.0000          BPH       CONS  1
 OP             86.0000          105.0000          BPH       CONS  1
LOW       2 OVERHEAD RELEASE             99.988
LOW       2 OVERHEAD RELEASE                                  1
LOW       2 OVERHEAD RELEASE            103.614
 OP             20.0000           26.0000          BPH       CONS  1
```

```
03/11/98 11:28:36    DP 26        HC4037            SOUTH E-45 OUTLET
03/11/98 11:28:39    DP 26        HC4038            MID E-45 OUTLET
03/11/98 11:28:45    DP 26        HC4039            NORTH E-45 OUTLET
03/11/98 11:29:03    HM 26        FC4095            E-47 SLURRY BYPASS
03/11/98 11:29:04    DP 21 ALM        LC0059         PVHI            80.000
03/11/98 11:29:07    DP 31 ACK        LC0059         PVHI
03/11/98 11:29:18    DP 21 RTN        FC0105         PVLO           850.000
03/11/98 11:29:20    HM 21 RTN        FC4095         PVLO           1600.00
03/11/98 11:30:49    DP 21 RTN        LC0059         PVHI            80.000
03/11/98 11:30:56    DP 21 ALM        LC0059         PVLO            20.000
03/11/98 11:30:59    DP 31 ACK        LC0059         PVLO
03/11/98 11:31:19    HM 26        FC4095            E-47 SLURRY BYPASS
03/11/98 11:31:29    HM 21 ALM        FC4095         PVLO           1600.00
03/11/98 11:31:47    HM 26        FC4095            E-47 SLURRY BYPASS
03/11/98 11:32:01    DP 26        PC0042            C-12 HOT VAPOR BYPASS
03/11/98 11:32:06    DP 26        PC0042            C-12 HOT VAPOR BYPASS
03/11/98 11:32:13    DP 26        PC0042            C-12 HOT VAPOR BYPASS
03/11/98 11:32:17    DP 21 RTN        LC0059         PVLO            20.000
03/11/98 11:32:24    HM 26        FC4095            E-47 SLURRY BYPASS
03/11/98 11:32:26    HM 31 ACK        FC4095         PVLO
03/11/98 11:33:02    HM 26        FC4095            E-47 SLURRY BYPASS
03/11/98 11:33:15    AB 26        TC1607            TRAY 4,5 C-10 STRIPPER
03/11/98 11:33:40    HM 26        FC4095            E-47 SLURRY BYPASS
03/11/98 11:33:41    DB 26        FC0205            1 DEBUT REBOILER HT MED
03/11/98 11:34:46    DP 26        HC4037            SOUTH E-45 OUTLET
03/11/98 11:34:54    DP 26        HC4039            NORTH E-45 OUTLET
03/11/98 11:36:23    DP 21 RTN        FC0022         PVLL            75.000
03/11/98 11:36:23    DP 26        FC0022            V-33 ACCUM RELEASE
03/11/98 11:36:30    HM 26        FC4095            E-47 SLURRY BYPASS
03/11/98 11:37:41    SP 21 ALM        LC0207         PVLO            30.000
03/11/98 11:37:43    SP 31 ACK        LC0207         PVLO
03/11/98 11:37:52    SP 26        FC1211            6 IN COLD SPLITTER BTMS
03/11/98 11:37:53    SP 21 RTN        LC0207         PVLO            30.000
03/11/98 11:37:58    HM 26        FC4095            E-47 SLURRY BYPASS
03/11/98 11:38:07    HM 21 RTN        FC4095         PVLO           1600.00
03/11/98 11:38:30    DB 26        FC0205            1 DEBUT REBOILER HT MED
03/11/98 11:38:38    DB 26        FC0222            #1 DEBUT REFLUX
03/11/98 11:38:41    DP 26        PC0042            C-12 HOT VAPOR BYPASS
03/11/98 11:38:42    SP 21 ALM        LC0207         PVLO            30.000
03/11/98 11:39:08    SP 21 RTN        LC0207         PVLO            30.000
03/11/98 11:39:51    HM 26        FC4095            E-47 SLURRY BYPASS
03/11/98 11:39:53    SP 21 ALM        LC0207         PVLO            30.000
03/11/98 11:39:59    SP 31 ACK        LC0207         PVLO
03/11/98 11:40:17    SP 21 RTN        LC0207         PVLO            30.000
03/11/98 11:40:20    DB 26        FC0222            #1 DEBUT REFLUX
03/11/98 11:41:38    DP 26        FC0006            C-12 BOTTOMS TO #2 DEBUT
03/11/98 11:42:56    DB 26        FC0205            1 DEBUT REBOILER HT MED
```

```
 OP           28.0000          30.0000        PERCENT  CONS  1
 OP           16.0000          18.0000        PERCENT  CONS  1
 OP           18.0000          20.0000        PERCENT  CONS  1
 OP          100.0000          10.0000        BPH      CONS  1
HIGH   V-33 FOUL H2O BOOT            87.064
HIGH   V-33 FOUL H2O BOOT                               1
LOW    REFLUX TO C-12 TOWER         902.338
LOW    E-47 SLURRY BYPASS           1795.21
HIGH   V-33 FOUL H2O BOOT            69.938
LOW    V-33 FOUL H2O BOOT            18.696
LOW    V-33 FOUL H2O BOOT                               1
 OP           10.0000          95.0000        BPH      CONS  1
LOW    E-47 SLURRY BYPASS           1481.93
 OP           95.0000          85.0000        BPH      CONS  1
 MODE        MAN             AUTO             PSIG     CONS  1
 MODE        AUTO            NORMAL           PSIG     CONS  1
 SP          229.2470         230.0000        PSIG     CONS  1
LOW    V-33 FOUL H2O BOOT            21.308
 OP           85.0000          75.0000        BPH      CONS  1
LOW    E-47 SLURRY BYPASS                               1
 OP           75.0000          65.0000        BPH      CONS  1
 SP          158.0000         160.0000        DEG F    CONS  1
 OP           65.0000          55.0000        BPH      CONS  1
 OP             .2933          68.2933        B P H    CONS  1
 OP           30.0000          34.0000        PERCENT  CONS  1
 OP           20.0000          22.0000        PERCENT  CONS  1
HIGH   V-33 ACCUM RELEASE            79.885
 OP           26.0000          32.0000        BPH      CONS  1
 OP           55.0000          45.0000        BPH      CONS  1
HIGH   SPLITTER REBOILER LEVEL       29.982
HIGH   SPLITTER REBOILER LEVEL                          1
 OP            1.1000         ---------       B P H    CONS  1
HIGH   SPLITTER REBOILER LEVEL       31.119
 OP           45.0000          30.0000        BPH      CONS  1
LOW    E-47 SLURRY BYPASS           1649.60
 OP           68.2933          64.2933        B P H    CONS  1
 OP           29.6062          33.6062        B P H    CONS  1
 SP          230.0000         235.0000        PSIG     CONS  1
HIGH   SPLITTER REBOILER LEVEL       29.980
HIGH   SPLITTER REBOILER LEVEL       31.274
 OP           30.0000          20.0000        BPH      CONS  1
HIGH   SPLITTER REBOILER LEVEL       29.972
HIGH   SPLITTER REBOILER LEVEL                          1
HIGH   SPLITTER REBOILER LEVEL       31.022
 OP           33.6062          37.6062        B P H    CONS  1
 SP          1300.000         1250.000        BPH      CONS  1
 OP           64.2933          52.2933        B P H    CONS  1
```

```
03/11/98 11:43:25   HM 26       FC4095          E-47 SLURRY BYPASS
03/11/98 11:44:03   AB 26       FC1515          HEAT MED TO C-10 REBOILR
03/11/98 11:44:20   RX 21 ALM       PDI1203      PVLO          100.000
03/11/98 11:44:22   RX 31 ACK       PDI1203      PVLO
03/11/98 11:45:02   DB 26       FC0222          #1 DEBUT REFLUX
03/11/98 11:45:07   DP 26       FC0022          V-33 ACCUM RELEASE
03/11/98 11:45:12   DP 21 RTN       FC0022       PVLO          100.000
03/11/98 11:45:42   RX 21 RTN       PDI1203      PVLO          100.000
03/11/98 11:45:42   HM 26       FC1102          SLURRY BYPASS WHB
03/11/98 11:46:16   HM 26       FC1102          SLURRY BYPASS WHB
03/11/98 11:47:01   DP 26       FC0006          C-12 BOTTOMS TO #2 DEBUT
03/11/98 11:47:12   DB 26       FC0222          #1 DEBUT REFLUX
03/11/98 11:47:31   SP 21 ALM       LC0207       PVLO           30.000
03/11/98 11:47:34   SP 31 ACK       LC0207       PVLO
03/11/98 11:47:36   SP 31 ACK       LC0207       PVLO
03/11/98 11:47:57   DP 21 ALM       LC6024       PVHI           80.000
03/11/98 11:48:02   SP 21 RTN       LC0207       PVLO           30.000
03/11/98 11:48:08   DP 31 ACK       LC6024       PVHI
03/11/98 11:48:23   DP 21 RTN       LC6024       PVHI           80.000
03/11/98 11:48:25   SP 21 ALM       LC0207       PVLO           30.000
03/11/98 11:48:28   SP 31 ACK       LC0207       PVLO
03/11/98 11:48:30   DB 26       FC0222          #1 DEBUT REFLUX
03/11/98 11:49:24   HM 26       FC1102          SLURRY BYPASS WHB
03/11/98 11:50:04   DB 26       FC0205          1 DEBUT REBOILER HT MED
03/11/98 11:50:27   SP 26       FC1210          3 IN SPLIT BTMS TO FIELD
03/11/98 11:50:51   DB 26       FC0205          1 DEBUT REBOILER HT MED
03/11/98 11:50:55   SP 26       FC1210          3 IN SPLIT BTMS TO FIELD
03/11/98 11:50:58   DP 21 RTN       AC0108       BADPV
03/11/98 11:51:04   SP 26       LC0207          SPLITTER REBOILER LEVEL
03/11/98 11:51:17   DB 26       FC0205          1 DEBUT REBOILER HT MED
03/11/98 11:51:41   HM 26       FC1102          SLURRY BYPASS WHB
03/11/98 11:52:26   DP 26       FC0105          REFLUX TO C-12 TOWER
03/11/98 11:52:32   SP 21 RTN       LC0207       PVLO           30.000
03/11/98 11:53:22   DP 26       FC0022          V-33 ACCUM RELEASE
03/11/98 11:53:31   DB 26       FC0222          #1 DEBUT REFLUX
03/11/98 11:53:36   DB 26       FC0222          #1 DEBUT REFLUX
03/11/98 11:53:42   HM 26       FC1102          SLURRY BYPASS WHB
03/11/98 11:53:52   DB 21 ALM       FC0209       PVLO          100.000
03/11/98 11:55:34   HM 26       FC1102          SLURRY BYPASS WHB
03/11/98 11:55:43   FD 26       TC5127          E-4 FEED BYPASS
03/11/98 11:57:25   SP 21 ALM       LC0207       PVHI           75.000
03/11/98 12:01:59   DB 31 ACK       FC0209       PVLO
03/11/98 12:01:59   SP 31 ACK       LC0207       PVHI
03/11/98 12:02:07   SP 26       FC1211          6 IN COLD SPLITTER BTMS
03/11/98 12:02:57   DP 26       FC0022          V-33 ACCUM RELEASE
03/11/98 12:05:59   RX 21 ALM       PDI1203      PVLO          100.000
03/11/98 12:06:01   DB 21 RTN       CA0212       BADPV
```

OP	20.0000	15.0000		BPH	CONS	1
SP	353.4860	375.0000		BPH	CONS	1
HIGH	RX SV DIFF PRESSURE		99.973			
HIGH	RX SV DIFF PRESSURE				1	
OP	37.6062	39.6062		B P H	CONS	1
OP	32.0000	50.0000		BPH	CONS	1
LOW	V-33 ACCUM RELEASE		114.468			
HIGH	RX SV DIFF PRESSURE		112.120			
OP	20.0000	30.0000		BPH	CONS	1
OP	30.0000	20.0000		BPH	CONS	1
SP	1250.000	1200.000		BPH	CONS	1
OP	39.6062	51.6062		B P H	CONS	1
HIGH	SPLITTER REBOILER LEVEL		29.998			
HIGH	SPLITTER REBOILER LEVEL				1	
HIGH	SPLITTER REBOILER LEVEL				1	
LOW	PROP SCRUB FAT AMINE REL		80.043			
HIGH	SPLITTER REBOILER LEVEL		31.109			
LOW	PROP SCRUB FAT AMINE REL				1	
LOW	PROP SCRUB FAT AMINE REL		78.827			
HIGH	SPLITTER REBOILER LEVEL		29.952			
HIGH	SPLITTER REBOILER LEVEL				1	
OP	51.6062	57.6062		B P H	CONS	1
OP	20.0000	30.0000		BPH	CONS	1
OP	52.2933	50.2933		B P H	CONS	1
MODE	CAS	AUTO		B P H	CONS	1
MODE	MAN	AUTO		B P H	CONS	1
MODE	AUTO	NORMAL		B P H	CONS	1
LOW	C-12 DEPROP OH C2 VALID					
SP	23.4711	50.0000		PERCENT	CONS	1
SP	1147.120	1130.000		B P H	CONS	1
OP	30.0000	50.0000		BPH	CONS	1
OP	80.0000	76.0000		BPH	CONS	1
HIGH	SPLITTER REBOILER LEVEL		31.328			
OP	50.0000	40.0000		BPH	CONS	1
MODE	MAN	AUTO		B P H	CONS	1
SP	784.1910	830.0000		B P H	CONS	1
OP	50.0000	75.0000		BPH	CONS	1
LOW	#1 DEBUT OH RELEASE		99.433			
OP	75.0000	100.0000		BPH	CONS	1
OP	50.0000	75.0000		DEG F	CONS	1
HIGH	SPLITTER REBOILER LEVEL		75.154			
LOW	#1 DEBUT OH RELEASE				1	
HIGH	SPLITTER REBOILER LEVEL				1	
OP	- 6.9000	9.1000		B P H	CONS	1
OP	40.0000	35.0000		BPH	CONS	1
HIGH	RX SV DIFF PRESSURE		99.773			
LOW	#2 DEBUT OH C5					

```
03/11/98 12:06:25    FD 26        TC5127           E-4 FEED BYPASS
03/11/98 12:06:47    RX 21 RTN        PDI1203       PVLO            100.000
03/11/98 12:06:55    RX 31 ACK        PDI1203       PVLO
03/11/98 12:10:07    DB 21 RTN        FC0209        PVLO            100.000
03/11/98 12:14:35    SP 21 RTN        LC0207        PVHI             75.000
03/11/98 12:14:43    DB 26        FC0205           1 DEBUT REBOILER HT MED
03/11/98 12:14:56    DB 26        FC0222           #1 DEBUT REFLUX
03/11/98 12:14:57    HM 26        FC4095           E-47 SLURRY BYPASS
03/11/98 12:15:07    DB 26        TC0203           1 DEBUT TOWER TOP TEMP
03/11/98 12:15:26    RX 21 ALM        PDI1203       PVLO            100.000
03/11/98 12:15:48    RX 31 ACK        PDI1203       PVLO
03/11/98 12:16:06    RX 21 RTN        PDI1203       PVLO            100.000
03/11/98 12:16:50    DP 26        FC0105           REFLUX TO C-12 TOWER
03/11/98 12:20:04    HM 26        FC4095           E-47 SLURRY BYPASS
03/11/98 12:22:58    DP 21 RTN        AC0208        BADPV
03/11/98 12:28:00    DB 21 RTN        CA0213        BADPV
03/11/98 12:30:45    SY 19       NODE 04  $P4
03/11/98 12:30:46    SY 19       NODE 04  $P4
03/11/98 12:32:06    DB 26        TC0203           1 DEBUT TOWER TOP TEMP
03/11/98 12:32:30    SP 26        FC1211           6 IN COLD SPLITTER BTMS
03/11/98 12:33:37    HM 26        FC4095           E-47 SLURRY BYPASS
03/11/98 12:34:28    DB 21 RTN        AI0112        UNREASBL
03/11/98 12:34:36    DP 26        FC0105           REFLUX TO C-12 TOWER
03/11/98 12:34:42    DP 26        TC0005           C-12 TRAY 10 TEMP
03/11/98 12:39:10    HM 26        CA0110           LAB UPDATED GRAVITY -CLO
03/11/98 12:39:10    HM 26        CA0110           LAB UPDATED GRAVITY -CLO
03/11/98 12:39:59    HM 26        FC4095           E-47 SLURRY BYPASS
03/11/98 12:42:52    BL 26        KC5928           HORN 2
03/11/98 12:44:22    DP 26        TC0005           C-12 TRAY 10 TEMP
03/11/98 12:48:44    HM 26        FC4095           E-47 SLURRY BYPASS
03/11/98 12:52:32    HM 26        FC4095           E-47 SLURRY BYPASS
03/11/98 12:55:05    SC 21 ALM        AI0114        PVHH            200.000
03/11/98 12:55:09    SC 31 ACK        AI0114        PVHH
03/11/98 12:57:52    SP 21 ALM        LC0207        PVHI             75.000
03/11/98 12:58:09    SP 26        FC1311           SPLITTER BTMS/CAT NAPTHA
03/11/98 12:58:29    SP 21 RTN        LC0207        PVHI             75.000
03/11/98 12:58:37    SP 31 ACK        LC0207        PVHI
03/11/98 12:59:01    SP 21 ALM        LC0207        PVHI             75.000
03/11/98 12:59:04    SP 31 ACK        LC0207        PVHI
03/11/98 12:59:19    SP 21 RTN        LC0207        PVHI             75.000
03/11/98 13:00:02    DP 26        FC0022           V-33 ACCUM RELEASE
03/11/98 13:01:59    SC 21 RTN        AI0114        PVHH            200.000
03/11/98 13:05:52    DP 26        FC0022           V-33 ACCUM RELEASE
03/11/98 13:06:04    HM 26        FC4095           E-47 SLURRY BYPASS
03/11/98 13:09:45    SP 21 ALM        LC0207        PVHI             75.000
03/11/98 13:10:07    SP 26        FC1211           6 IN COLD SPLITTER BTMS
03/11/98 13:10:08    SP 21 RTN        LC0207        PVHI             75.000
```

OP	75.0000	100.0000	DEG F	CONS	1
HIGH	RX SV DIFF PRESSURE	112.037			
HIGH	RX SV DIFF PRESSURE			1	
LOW	#1 DEBUT OH RELEASE	105.684			
HIGH	SPLITTER REBOILER LEVEL	73.997			
MODE	AUTO	NORMAL	B P H	CONS	1
MODE	AUTO	NORMAL	B P H	CONS	1
OP	15.0000	13.0000	BPH	CONS	1
SP	159.1390	155.0000	DEG F	CONS	1
HIGH	RX SV DIFF PRESSURE	99.598			
HIGH	RX SV DIFF PRESSURE			1	
HIGH	RX SV DIFF PRESSURE	112.107			
OP	76.0000	74.0000	BPH	CONS	1
OP	13.0000	11.0000	BPH	CONS	1
LOW	C-12 DEPROP OH C4 VALID				
LOW	AVG DEBUT OVHD C5				
	LP00 DEVICE AVAILABLE		OFFLINE		01
	LP00 DEVICE AVAILABLE		DEVICE READY		01
SP	155.0000	150.0000	DEG F	CONS	1
OP	9.1000	5.1000	B P H	CONS	1
OP	11.0000	13.0000	BPH	CONS	1
JOURNAL	2-DEBUT OVHD C-3S	19.741			
MODE	MAN	NORMAL	BPH	CONS	1
SP	161.3310	162.0000	DEG F	CONS	1
OPRR	---------	- 1.8000	API	CONS	1
PPS	OFF	ON	API	CONS	1
OP	13.0000	15.0000	BPH	CONS	1
OP	STOP	START		CONS	1
SP	162.0000	165.0000	DEG F	CONS	1
OP	15.0000	14.0000	BPH	CONS	1
OP	14.0000	13.0000	BPH	CONS	1
EMERGNCY	D-10 KO POT H2S	200.703			
EMERGNCY	D-10 KO POT H2S			1	
HIGH	SPLITTER REBOILER LEVEL	75.268			
SP	950.0000	1000.000	B P H	CONS	1
HIGH	SPLITTER REBOILER LEVEL	73.816			
HIGH	SPLITTER REBOILER LEVEL			1	
HIGH	SPLITTER REBOILER LEVEL	75.219			
HIGH	SPLITTER REBOILER LEVEL			1	
HIGH	SPLITTER REBOILER LEVEL	73.897			
OP	35.0000	37.0000	BPH	CONS	1
EMERGNCY	D-10 KO POT H2S	195.339			
OP	37.0000	39.0000	BPH	CONS	1
OP	13.0000	11.0000	BPH	CONS	1
HIGH	SPLITTER REBOILER LEVEL	75.171			
OP	5.1000	7.1000	B P H	CONS	1
HIGH	SPLITTER REBOILER LEVEL	73.989			

03/11/98 13:10:57	SP 21 ALM	LC0207	PVHI	75.000
03/11/98 13:11:05	SP 31 ACK	LC0207	PVHI	
03/11/98 13:11:16	SP 26	FC1211	6 IN COLD SPLITTER BTMS	
03/11/98 13:11:19	SP 21 RTN	LC0207	PVHI	75.000
03/11/98 13:24:01	SP 21 ALM	LC0207	PVLO	30.000
03/11/98 13:24:24	SP 26	FC1211	6 IN COLD SPLITTER BTMS	
03/11/98 13:24:43	SP 21 RTN	LC0207	PVLO	30.000
03/11/98 13:25:05	SP 21 ALM	LC0207	PVLO	30.000
03/11/98 13:25:12	SP 31 ACK	LC0207	PVLO	
03/11/98 13:25:15	SP 26	FC1211	6 IN COLD SPLITTER BTMS	
03/11/98 13:25:46	SP 21 RTN	LC0207	PVLO	30.000
03/11/98 13:30:10	HM 26	FC4095	E-47 SLURRY BYPASS	
03/11/98 13:32:57	DB 21 RTN	AC0111	BADPV	
03/11/98 13:44:40	DP 26	FC0022	V-33 ACCUM RELEASE	
03/11/98 13:45:34	DP 21 ALM	LC0053	PVHI	70.000
03/11/98 13:45:42	DP 31 ACK	LC0053	PVHI	
03/11/98 13:47:09	DP 26	FC0022	V-33 ACCUM RELEASE	
03/11/98 13:48:27	DP 26	FC0022	V-33 ACCUM RELEASE	
03/11/98 13:52:11	HM 26	FC4095	E-47 SLURRY BYPASS	
03/11/98 13:53:50	DP 21 RTN	LC0053	PVHI	70.000
03/11/98 13:55:04	BL 26	KC5928	HORN 2	
03/11/98 13:55:21	BL 26	KC5929	HORN 3	
03/11/98 14:05:36	DP 26	FC0022	V-33 ACCUM RELEASE	
03/11/98 14:12:27	HM 26	FC4095	E-47 SLURRY BYPASS	
03/11/98 14:14:01	DP 26	FC0022	V-33 ACCUM RELEASE	
03/11/98 14:23:49	DP 26	FC0022	V-33 ACCUM RELEASE	
03/11/98 14:36:48	DP 26	FC0022	V-33 ACCUM RELEASE	
03/11/98 14:42:12	WG 26	FC1698	SCRUBBER WATER FLOW	
03/11/98 14:42:35	DP 26	FC0022	V-33 ACCUM RELEASE	
03/11/98 14:49:10	FD 26	CF0000	TOTAL FEED	
03/11/98 14:49:29	FD 26	CF0011	EAST HEATER FEED CONTROL	
03/11/98 14:49:34	FD 26	CF0010	WEST HEATER FEED CONTROL	
03/11/98 14:49:44	FD 26	CF0000	TOTAL FEED	
03/11/98 14:49:52	FD 26	CF0011	EAST HEATER FEED CONTROL	
03/11/98 14:49:57	FD 26	CF0010	WEST HEATER FEED CONTROL	
03/11/98 14:55:18	SC 21 RTN	AI0114	PVHI	50.000
03/11/98 14:58:44	SC 21 ALM	AI0114	PVHI	50.000
03/11/98 15:03:25	FD 26	FDMONITOR	COLD FEED LLIMIT MONITOR	
03/11/98 15:04:02	FD 26	FDMONITOR	COLD FEED LLIMIT MONITOR	
03/11/98 15:11:09	SC 31 ACK	AI0114	PVHI	
03/11/98 15:18:38	SP 21 ALM	LC0207	PVHI	75.000
03/11/98 15:19:06	SP 21 RTN	LC0207	PVHI	75.000
03/11/98 15:19:10	SP 31 ACK	LC0207	PVHI	
03/11/98 15:19:11	SC 21 RTN	AI0114	PVHI	50.000
03/11/98 15:19:17	SP 26	FC1211	6 IN COLD SPLITTER BTMS	
03/11/98 15:19:20	FD 26	FDMONITOR	COLD FEED LLIMIT MONITOR	
03/11/98 15:19:30	FD 26	FDMONITOR	COLD FEED LLIMIT MONITOR	

```
HIGH      SPLITTER REBOILER LEVEL         75.175
HIGH      SPLITTER REBOILER LEVEL                              1
 OP             7.1000               9.1000        B P H    CONS  1
HIGH      SPLITTER REBOILER LEVEL         73.930
HIGH      SPLITTER REBOILER LEVEL         29.912
 OP             9.1000               3.1000        B P H    CONS  1
HIGH      SPLITTER REBOILER LEVEL         31.080
HIGH      SPLITTER REBOILER LEVEL         29.956
HIGH      SPLITTER REBOILER LEVEL                              1
 OP             3.1000               1.1000        B P H    CONS  1
HIGH      SPLITTER REBOILER LEVEL         31.037
 OP            11.0000               9.0000        BPH      CONS  1
LOW       AVG C3 AT C5&7 OH
 OP            39.0000              45.0000        BPH      CONS  1
HIGH      V-33 DEPROP OH ACCUM LVL        70.012
HIGH      V-33 DEPROP OH ACCUM LVL                             1
 OP            45.0000              55.0000        BPH      CONS  1
 OP            55.0000              65.0000        BPH      CONS  1
 OP             9.0000               7.0000        BPH      CONS  1
HIGH      V-33 DEPROP OH ACCUM LVL        68.901
 OP          STOP                START                      CONS  1
 OP          STOP                START                      CONS  1
 OP            65.0000              64.0000        BPH      CONS  1
 OP             7.0000               3.0000        BPH      CONS  1
 OP            64.0000              62.0000        BPH      CONS  1
 OP            62.0000              58.0000        BPH      CONS  1
 OP            58.0000              54.0000        BPH      CONS  1
 SP           5750.000             5850.000        GPM      CONS  1
 OP            54.0000              52.0000        BPH      CONS  1
 PVTV          91.5000              93.0000        M B P D  CONS  1
 MODE        AUTO                CAS               BPH      CONS  1
 MODE        AUTO                CAS               BPH      CONS  1
 PVTV          93.0000              93.0000        M B P D  CONS  1
 MODE        CAS                 AUTO              BPH      CONS  1
 MODE        CAS                 AUTO              BPH      CONS  1
HIGH      D-10 KO POT H2S                 43.092
HIGH      D-10 KO POT H2S                 51.119
 PTEXECST    ACTIVE              INACTIVE          BPH      CONS  1
 $CLCMPST(01) ACTIVE             INACTIVE          BPH      CONS  1
HIGH      D-10 KO POT H2S                                      1
HIGH      SPLITTER REBOILER LEVEL         75.089
HIGH      SPLITTER REBOILER LEVEL         73.982
HIGH      SPLITTER REBOILER LEVEL                              1
HIGH      D-10 KO POT H2S                 45.682
 OP             1.1000               3.1000        B P H    CONS  1
 PTEXECST    INACTIVE            ACTIVE            BPH      CONS  1
 PPS         OFF                 ON                BPH      CONS  1
```

```
03/11/98 15:20:02   SP 21 ALM      LC0207      PVHI            75.000
03/11/98 15:20:11   SP 26       FC1211        6 IN COLD SPLITTER BTMS
03/11/98 15:20:19   SP 21 RTN      LC0207      PVHI            75.000
03/11/98 15:20:21   HM 26       FC4095        E-47 SLURRY BYPASS
03/11/98 15:21:26   SP 26       FC1211        6 IN COLD SPLITTER BTMS
03/11/98 15:23:45   SP 21 RTN      AI0211      PVHI             3.000
03/11/98 15:28:04   SP 21 ALM      LC0207      PVHI            75.000
03/11/98 15:28:22   SP 21 RTN      LC0207      PVHI            75.000
03/11/98 15:29:27   SP 21 ALM      LC0207      PVHI            75.000
03/11/98 15:29:42   SP 21 RTN      LC0207      PVHI            75.000
03/11/98 15:29:48   SP 31 ACK      LC0207      PVHI
03/11/98 15:30:09   SP 26       FC1211        6 IN COLD SPLITTER BTMS
03/11/98 15:30:45   SP 21 ALM      LC0207      PVHI            75.000
03/11/98 15:31:05   SP 21 RTN      LC0207      PVHI            75.000
03/11/98 15:32:09   SP 21 ALM      LC0207      PVHI            75.000
03/11/98 15:32:24   SP 21 RTN      LC0207      PVHI            75.000
03/11/98 15:32:25   SP 26       FC1211        6 IN COLD SPLITTER BTMS
03/11/98 15:33:30   SP 21 ALM      LC0207      PVHI            75.000
03/11/98 15:33:41   SP 21 RTN      LC0207      PVHI            75.000
03/11/98 15:34:36   SP 26       FC1211        6 IN COLD SPLITTER BTMS
03/11/98 15:34:43   SP 21 ALM      LC0207      PVHI            75.000
03/11/98 15:35:06   SP 21 RTN      LC0207      PVHI            75.000
03/11/98 15:35:35   SP 31 ACK      LC0207      PVHI
03/11/98 15:36:40   SP 26       FC1211        6 IN COLD SPLITTER BTMS
03/11/98 15:50:59   HM 26       FC1102        SLURRY BYPASS WHB
03/11/98 15:51:10   HM 26       FC1102        SLURRY BYPASS WHB
03/11/98 15:52:15   HM 21 ALM      FI4065      PVLO            40.000
03/11/98 15:52:16   HM 21 ALM      FI4064      PVLO            40.000
03/11/98 15:52:17   HM 21 RTN      FI4053      PVLO            40.000
03/11/98 15:52:17   HM 21 ALM      FI4053      PVLO            40.000
03/11/98 15:52:18   CR 21 ALM      FC1104      PVLO           350.000
03/11/98 15:52:18   CR 21 ALM      LC0101      DEVHI           10.000
03/11/98 15:52:19   CR 21 ALM      FC1104      BADPV
03/11/98 15:52:19   CR 21 ALM      FCAL1104    BADPV
03/11/98 15:52:19   HM 21 RTN      FI4064      PVLO            40.000
03/11/98 15:52:19   RX 21 ALM      ULVG0007B   BADPV
03/11/98 15:52:20   CR 21 ALM      CF1104      BADPV
03/11/98 15:52:20   HM 21 RTN      FI4065      PVLO            40.000
03/11/98 15:52:21   CR 21 RTN      FC1104      BADPV
03/11/98 15:52:21   RX 21 RTN      ULVG0007B   BADPV
03/11/98 15:52:22   CR 21 RTN      FCAL1104    BADPV
03/11/98 15:52:23   CR 21 RTN      CF1104      BADPV
03/11/98 15:52:23   HM 21 ALM      FI4065      PVLO            40.000
03/11/98 15:52:24   HM 21 RTN      FI4065      PVLO            40.000
03/11/98 15:52:26   CR 21 ALM      FCAL1104    BADPV
03/11/98 15:52:27   CR 21 ALM      CF1104      BADPV
03/11/98 15:52:28   CR 21 RTN      FCAL1104    BADPV
```

```
HIGH     SPLITTER REBOILER LEVEL          75.119
 OP            3.1000              7.1000          B P H    CONS  1
HIGH     SPLITTER REBOILER LEVEL          73.936
 OP            3.0000              1.0000          BPH      CONS  1
 OP            7.1000              9.1000          B P H    CONS  1
LOW      1-DEBUT OVHD C-5S                 2.793
HIGH     SPLITTER REBOILER LEVEL          75.072
HIGH     SPLITTER REBOILER LEVEL          73.985
HIGH     SPLITTER REBOILER LEVEL          75.109
HIGH     SPLITTER REBOILER LEVEL          73.910
HIGH     SPLITTER REBOILER LEVEL                                  1
 OP            9.1000             13.0000          B P H    CONS  1
HIGH     SPLITTER REBOILER LEVEL          75.293
HIGH     SPLITTER REBOILER LEVEL          73.926
HIGH     SPLITTER REBOILER LEVEL          75.099
HIGH     SPLITTER REBOILER LEVEL          73.950
 OP           13.0000             15.0000          B P H    CONS  1
HIGH     SPLITTER REBOILER LEVEL          75.040
HIGH     SPLITTER REBOILER LEVEL          73.844
 OP           15.0000             21.0000          B P H    CONS  1
HIGH     SPLITTER REBOILER LEVEL          75.062
HIGH     SPLITTER REBOILER LEVEL          73.921
HIGH     SPLITTER REBOILER LEVEL                                  1
 OP           21.0000             25.0000          B P H    CONS  1
 OP          100.0000             50.0000          BPH      CONS  1
 OP           50.0000            105.0000          BPH      CONS  1
LOW      P-2 GLAND FLUSH OIL              36.010
LOW      P-2B GLAND FLUSH OIL             37.005
LOW      FLUSH OIL TO P-5A                41.193
LOW      FLUSH OIL TO P-5A                39.287
HIGH     LCO RELEASE TO HDS              331.629
LOW      LCO STRIPPER LEVEL               59.232
LOW      LCO RELEASE TO HDS
JOURNAL  TOTAL LCO RELEASE CALC
LOW      P-2B GLAND FLUSH OIL             43.035
JOURNAL  FCCU LIQUID VOL. GAIN
JOURNAL  TOTAL LCO RELEASE
LOW      P-2 GLAND FLUSH OIL              42.712
LOW      LCO RELEASE TO HDS
JOURNAL  FCCU LIQUID VOL. GAIN
JOURNAL  TOTAL LCO RELEASE CALC
JOURNAL  TOTAL LCO RELEASE
LOW      P-2 GLAND FLUSH OIL              39.077
LOW      P-2 GLAND FLUSH OIL              41.586
JOURNAL  TOTAL LCO RELEASE CALC
JOURNAL  TOTAL LCO RELEASE
JOURNAL  TOTAL LCO RELEASE CALC
```

```
03/11/98 15:52:29    CR 21 RTN       CF1104        BADPV
03/11/98 15:52:43    CR 21 RTN       FC1104        PVLO             350.000
03/11/98 15:52:54    CR 31 ACK       FC1104        BADPV
03/11/98 15:52:54    CR 31 ACK       LC0101        DEVHI
03/11/98 15:52:54    CR 31 ACK       FC1104        PVLO
03/11/98 15:52:54    HM 31 ACK       FI4065        PVLO
03/11/98 15:52:54    HM 31 ACK       FI4053        PVLO
03/11/98 15:52:54    HM 31 ACK       FI4064        PVLO
03/11/98 15:53:13    SC 21 ALM       AI0114        PVHI              50.000
03/11/98 15:53:46    CR 21 RTN       LC0101        DEVHI             10.000
03/11/98 15:53:54    SC 31 ACK       AI0114        PVHI
03/11/98 15:54:58    CR 21 ALM       LC0101        DEVLO             10.000
03/11/98 15:55:28    SP 26         FC1211         6 IN COLD SPLITTER BTMS
03/11/98 15:56:43    CR 21 RTN       LC0101        DEVLO             10.000
03/11/98 15:58:52    CR 31 ACK       LC0101        DEVLO
03/11/98 16:14:34    DP 26         PC0042         C-12 HOT VAPOR BYPASS
03/11/98 16:21:48    SP 21 ALM       LC0207        PVHI              75.000
03/11/98 16:22:12    SP 31 ACK       LC0207        PVHI
03/11/98 16:28:51    SP 26         FC1211         6 IN COLD SPLITTER BTMS
03/11/98 16:29:57    SP 26         FC1211         6 IN COLD SPLITTER BTMS
03/11/98 16:30:45    HM 26         TC0091         HGO RETURN TO BED 3
03/11/98 16:31:59    SP 26         FC1211         6 IN COLD SPLITTER BTMS
03/11/98 16:32:55    FD 26         FC0027         HOT FEED TO HEATERS
03/11/98 16:33:11    FD 26         LC0008         V30 HOT FEED DRUM LEVEL
03/11/98 16:33:50    HM 26         FC4095         E-47 SLURRY BYPASS
03/11/98 16:35:54    SP 26         FC1211         6 IN COLD SPLITTER BTMS
03/11/98 16:41:44    RX 21 ALM       PDI1203       PVLO             100.000
03/11/98 16:42:44    RX 21 RTN       PDI1203       PVLO             100.000
03/11/98 16:44:16    DP 26         FC1202         LEAN AMINE TO PROP SCRUB
03/11/98 16:45:01    FD 26         CF0000         TOTAL FEED
03/11/98 16:45:36    RX 31 ACK       PDI1203       PVLO
03/11/98 16:46:06    SY 01            US 06   K$MMIK_CP      SOFTWARE
03/11/98 16:46:33    FD 26         CF0000         TOTAL FEED
03/11/98 16:46:37    SP 21 RTN       LC0207        PVHI              75.000
03/11/98 16:46:52    SP 21 ALM       LC0207        PVHI              75.000
03/11/98 16:47:26    SP 21 RTN       LC0207        PVHI              75.000
03/11/98 16:48:06    SP 31 ACK       LC0207        PVHI
03/11/98 16:50:21    FD 26         CF0011         EAST HEATER FEED CONTROL
03/11/98 16:50:37    FD 26         CF0010         WEST HEATER FEED CONTROL
03/11/98 16:53:34    FD 26         CF0010         WEST HEATER FEED CONTROL
03/11/98 16:55:14    FD 26         CF0011         EAST HEATER FEED CONTROL
03/11/98 16:57:05    DP 26         FC1202         LEAN AMINE TO PROP SCRUB
03/11/98 16:58:48    FD 26         CF0011         EAST HEATER FEED CONTROL
03/11/98 17:03:56    FD 26         FC0013         WEST HTR FUEL GAS FLOW
03/11/98 17:04:07    SP 26         FC1211         6 IN COLD SPLITTER BTMS
03/11/98 17:06:12    DP 26         FC0022         V-33 ACCUM RELEASE
03/11/98 17:06:23    FD 26         FC0013         WEST HTR FUEL GAS FLOW
```

JOURNAL	TOTAL LCO RELEASE				
HIGH	LCO RELEASE TO HDS	450.208			
LOW	LCO RELEASE TO HDS			1	
LOW	LCO STRIPPER LEVEL			1	
HIGH	LCO RELEASE TO HDS			1	
LOW	P-2 GLAND FLUSH OIL			1	
LOW	FLUSH OIL TO P-5A			1	
LOW	P-2B GLAND FLUSH OIL			1	
HIGH	D-10 KO POT H2S	50.379			
LOW	LCO STRIPPER LEVEL	57.802			
HIGH	D-10 KO POT H2S			1	
LOW	LCO STRIPPER LEVEL	38.900			
OP	25.0000	19.0000	B P H	CONS	1
LOW	LCO STRIPPER LEVEL	40.181			
LOW	LCO STRIPPER LEVEL			1	
SP	235.0000	233.0000	PSIG	CONS	1
HIGH	SPLITTER REBOILER LEVEL	75.182			
HIGH	SPLITTER REBOILER LEVEL			1	
OP	19.0000	31.0000	B P H	CONS	1
OP	31.0000	35.0000	B P H	CONS	1
MODE	MAN	AUTO	DEG F	CONS	1
OP	35.0000	41.0000	B P H	CONS	1
MODE	AUTO	BCAS	BPH	CONS	1
SP	40.1165	42.0000	PERCENT	CONS	1
OP	1.0000	---------	BPH	CONS	1
OP	41.0000	37.0000	B P H	CONS	1
HIGH	RX SV DIFF PRESSURE	99.878			
HIGH	RX SV DIFF PRESSURE	112.240			
SP	110.0000	105.0000	BPH	CONS	1
PVTV	93.0000	91.0000	M B P D	CONS	1
HIGH	RX SV DIFF PRESSURE			1	
	001A51E2 001A5382 0019E868	5	902	6	0
PVTV	91.0000	91.0000	M B P D	CONS	1
HIGH	SPLITTER REBOILER LEVEL	73.856			
HIGH	SPLITTER REBOILER LEVEL	75.075			
HIGH	SPLITTER REBOILER LEVEL	73.989			
HIGH	SPLITTER REBOILER LEVEL			1	
SP	1930.770	1915.000	BPH	CONS	1
SP	1968.210	1950.000	BPH	CONS	1
SP	1950.000	1850.000	BPH	CONS	1
SP	1915.000	1900.000	BPH	CONS	1
SP	105.0000	110.0000	BPH	CONS	1
SP	1900.000	1890.000	BPH	CONS	1
OP	7.5000	7.0000	MSCFH	CONS	1
OP	37.0000	31.0000	B P H	CONS	1
OP	52.0000	54.0000	BPH	CONS	1
OP	7.0000	6.5000	MSCFH	CONS	1

```
03/11/98 17:07:51   DP 21 ALM      LC0053     PVHI              70.000
03/11/98 17:07:59   DP 21 RTN      LC0053     PVHI              70.000
03/11/98 17:08:03   DP 21 ALM      LC0053     PVHI              70.000
03/11/98 17:08:13   DP 21 RTN      LC0053     PVHI              70.000
03/11/98 17:08:17   DP 21 ALM      LC0053     PVHI              70.000
03/11/98 17:08:18   DP 21 RTN      LC0053     PVHI              70.000
03/11/98 17:09:03   DP 31 ACK      LC0053     PVHI
03/11/98 17:09:32   DP 21 ALM      LC0053     PVLO              30.000
03/11/98 17:09:40   DP 21 RTN      LC0053     PVLO              30.000
03/11/98 17:09:42   DP 21 ALM      LC0053     PVHI              70.000
03/11/98 17:09:48   DP 31 ACK      LC0053     PVHI
03/11/98 17:09:48   DP 31 ACK      LC0053     PVLO
03/11/98 17:10:39   FD 26        FC0013     WEST HTR FUEL GAS FLOW
03/11/98 17:11:05   FD 26        FC0013     WEST HTR FUEL GAS FLOW
03/11/98 17:11:34   DP 21 RTN      LC0053     PVHI              70.000
03/11/98 17:11:44   DP 21 ALM      LC0053     PVHI              70.000
03/11/98 17:11:48   FD 26        FC0014     EAST HTR FUEL GAS FLOW
03/11/98 17:12:04   DP 21 RTN      LC0053     PVHI              70.000
03/11/98 17:12:06   DP 21 ALM      LC0053     PVHI              70.000
03/11/98 17:12:08   DP 21 RTN      LC0053     PVHI              70.000
03/11/98 17:12:11   DP 21 ALM      LC0053     PVHI              70.000
03/11/98 17:12:30   DP 21 RTN      LC0053     PVHI              70.000
03/11/98 17:12:31   DP 21 ALM      LC0053     PVHI              70.000
03/11/98 17:12:31   DP 21 RTN      LC0053     PVHI              70.000
03/11/98 17:12:40   DP 21 ALM      LC0053     PVHI              70.000
03/11/98 17:12:47   DP 21 RTN      LC0053     PVHI              70.000
03/11/98 17:12:50   DP 21 ALM      LC0053     PVHI              70.000
03/11/98 17:12:52   DP 21 RTN      LC0053     PVHI              70.000
03/11/98 17:13:42   DP 21 ALM      LC0053     PVLO              30.000
03/11/98 17:13:43   DP 31 ACK      LC0053     PVLO
03/11/98 17:13:43   DP 31 ACK      LC0053     PVHI
03/11/98 17:13:55   DP 21 RTN      LC0053     PVLO              30.000
03/11/98 17:13:56   DP 21 ALM      LC0053     PVHI              70.000
03/11/98 17:14:33   SC 21 ALM      AI0114     PVHH             200.000
03/11/98 17:15:42   DP 21 RTN      LC0053     PVHI              70.000
03/11/98 17:15:43   DP 21 ALM      LC0053     PVHI              70.000
03/11/98 17:15:46   DP 21 RTN      LC0053     PVHI              70.000
03/11/98 17:15:50   DP 21 ALM      LC0053     PVHI              70.000
03/11/98 17:16:02   DP 21 RTN      LC0053     PVHI              70.000
03/11/98 17:16:06   DP 21 ALM      LC0053     PVHI              70.000
03/11/98 17:16:20   DP 21 RTN      LC0053     PVHI              70.000
03/11/98 17:16:27   DP 21 ALM      LC0053     PVHI              70.000
03/11/98 17:16:39   DP 21 RTN      LC0053     PVHI              70.000
03/11/98 17:16:47   DP 21 ALM      LC0053     PVLO              30.000
03/11/98 17:17:23   DP 31 ACK      LC0053     PVLO
03/11/98 17:17:23   DP 31 ACK      LC0053     PVHI
03/11/98 17:17:23   SC 31 ACK      AI0114     PVHH
```

HIGH	V-33 DEPROP OH ACCUM LVL	.609			
HIGH	V-33 DEPROP OH ACCUM LVL	63.279			
HIGH	V-33 DEPROP OH ACCUM LVL	81.181			
HIGH	V-33 DEPROP OH ACCUM LVL	56.930			
HIGH	V-33 DEPROP OH ACCUM LVL	.073			
HIGH	V-33 DEPROP OH ACCUM LVL	61.029			
HIGH	V-33 DEPROP OH ACCUM LVL			1	
LOW	V-33 DEPROP OH ACCUM LVL	26.061			
LOW	V-33 DEPROP OH ACCUM LVL	45.211			
HIGH	V-33 DEPROP OH ACCUM LVL	89.176			
HIGH	V-33 DEPROP OH ACCUM LVL			1	
LOW	V-33 DEPROP OH ACCUM LVL			1	
OP	6.5000	6.0000	MSCFH	CONS	1
OP	6.0000	5.8000	MSCFH	CONS	1
HIGH	V-33 DEPROP OH ACCUM LVL	65.305			
HIGH	V-33 DEPROP OH ACCUM LVL	70.752			
OP	14.1000	13.6000	MSCFH	CONS	1
HIGH	V-33 DEPROP OH ACCUM LVL	67.530			
HIGH	V-33 DEPROP OH ACCUM LVL	70.989			
HIGH	V-33 DEPROP OH ACCUM LVL	68.337			
HIGH	V-33 DEPROP OH ACCUM LVL	71.670			
HIGH	V-33 DEPROP OH ACCUM LVL	67.979			
HIGH	V-33 DEPROP OH ACCUM LVL	70.954			
HIGH	V-33 DEPROP OH ACCUM LVL	68.222			
HIGH	V-33 DEPROP OH ACCUM LVL	70.471			
HIGH	V-33 DEPROP OH ACCUM LVL	68.799			
HIGH	V-33 DEPROP OH ACCUM LVL	71.449			
HIGH	V-33 DEPROP OH ACCUM LVL	68.228			
LOW	V-33 DEPROP OH ACCUM LVL	23.068			
LOW	V-33 DEPROP OH ACCUM LVL			1	
HIGH	V-33 DEPROP OH ACCUM LVL			1	
LOW	V-33 DEPROP OH ACCUM LVL	43.763			
HIGH	V-33 DEPROP OH ACCUM LVL	87.067			
EMERGNCY	D-10 KO POT H2S	202.367			
HIGH	V-33 DEPROP OH ACCUM LVL	68.823			
HIGH	V-33 DEPROP OH ACCUM LVL	.248			
HIGH	V-33 DEPROP OH ACCUM LVL	67.610			
HIGH	V-33 DEPROP OH ACCUM LVL	74.538			
HIGH	V-33 DEPROP OH ACCUM LVL	54.648			
HIGH	V-33 DEPROP OH ACCUM LVL	.409			
HIGH	V-33 DEPROP OH ACCUM LVL	67.323			
HIGH	V-33 DEPROP OH ACCUM LVL	71.988			
HIGH	V-33 DEPROP OH ACCUM LVL	67.404			
LOW	V-33 DEPROP OH ACCUM LVL	19.774			
LOW	V-33 DEPROP OH ACCUM LVL			1	
HIGH	V-33 DEPROP OH ACCUM LVL			1	
EMERGNCY	D-10 KO POT H2S			1	

```
03/11/98 17:29:52   SP 26       FC1211       6 IN COLD SPLITTER BTMS
03/11/98 17:30:13   DP 21 RTN      LC0053     PVLO              30.000
03/11/98 17:30:16   DP 21 ALM      LC0053     PVHI              70.000
03/11/98 17:31:38   HM 26       FC4095       E-47 SLURRY BYPASS
03/11/98 17:33:04   DP 21 RTN      LC0053     PVHI              70.000
03/11/98 17:33:07   DP 21 ALM      LC0053     PVHI              70.000
03/11/98 17:33:21   DP 21 RTN      LC0053     PVHI              70.000
03/11/98 17:33:24   DP 31 ACK      LC0053     PVHI
03/11/98 17:33:25   DP 21 ALM      LC0053     PVLO              30.000
03/11/98 17:33:28   DP 31 ACK      LC0053     PVLO
03/11/98 17:33:29   DP 21 RTN      LC0053     PVLO              30.000
03/11/98 17:33:32   DP 21 ALM      LC0053     PVHI              70.000
03/11/98 17:33:36   DP 31 ACK      LC0053     PVHI
03/11/98 17:34:03   DP 21 RTN      LC0053     PVHI              70.000
03/11/98 17:34:09   DP 21 ALM      LC0053     PVHI              70.000
03/11/98 17:34:22   DP 21 RTN      LC0053     PVHI              70.000
03/11/98 17:34:24   DP 21 ALM      LC0053     PVHI              70.000
03/11/98 17:34:43   DP 21 RTN      LC0053     PVHI              70.000
03/11/98 17:34:45   DP 21 ALM      LC0053     PVLO              30.000
03/11/98 17:34:46   SC 21 RTN      AI0114     PVHH             200.000
03/11/98 17:34:52   DP 31 ACK      LC0053     PVLO
03/11/98 17:34:52   DP 31 ACK      LC0053     PVHI
03/11/98 17:35:11   HM 26       FC4095       E-47 SLURRY BYPASS
03/11/98 17:38:12   HM 26       FC4095       E-47 SLURRY BYPASS
03/11/98 17:39:59   FD 26       CF0011       EAST HEATER FEED CONTROL
03/11/98 17:40:11   FD 26       CF0010       WEST HEATER FEED CONTROL
03/11/98 17:41:24   FD 26       FC0013       WEST HTR FUEL GAS FLOW
03/11/98 17:41:31   HM 26       FC4095       E-47 SLURRY BYPASS
03/11/98 17:41:36   FD 26       FC0014       EAST HTR FUEL GAS FLOW
03/11/98 17:42:17   FD 26       FC0014       EAST HTR FUEL GAS FLOW
03/11/98 17:43:22   SP 21 ALM      LC0207     PVLO              30.000
03/11/98 17:43:24   DP 26       FC0022       V-33 ACCUM RELEASE
03/11/98 17:44:01   SP 26       FC1211       6 IN COLD SPLITTER BTMS
03/11/98 17:45:13   SP 26       FC1211       6 IN COLD SPLITTER BTMS
03/11/98 17:45:28   SP 31 ACK      LC0207     PVLO
03/11/98 17:45:41   FD 26       CF0011       EAST HEATER FEED CONTROL
03/11/98 17:45:52   FD 26       CF0010       WEST HEATER FEED CONTROL
03/11/98 17:47:46   SP 21 RTN      LC0207     PVLO              30.000
03/11/98 17:48:35   SP 21 ALM      LC0207     PVLO              30.000
03/11/98 17:48:46   SP 31 ACK      LC0207     PVLO
03/11/98 17:48:57   SP 21 RTN      LC0207     PVLO              30.000
03/11/98 17:49:31   FD 26       CF0010       WEST HEATER FEED CONTROL
03/11/98 17:49:44   SP 21 ALM      LC0207     PVLO              30.000
03/11/98 17:49:47   FD 26       CF0011       EAST HEATER FEED CONTROL
03/11/98 17:50:12   SP 21 RTN      LC0207     PVLO              30.000
03/11/98 17:53:02   DP 21 RTN      LC0053     PVLO              30.000
03/11/98 17:53:43   DP 26       FC0022       V-33 ACCUM RELEASE
```

```
 OP              31.0000             29.0000       B P H    CONS   1
LOW       V-33 DEPROP OH ACCUM LVL          32.841
HIGH      V-33 DEPROP OH ACCUM LVL          73.013
 OP             - 6.9000              1.1000       BPH      CONS   1
HIGH      V-33 DEPROP OH ACCUM LVL          55.765
HIGH      V-33 DEPROP OH ACCUM LVL            .221
HIGH      V-33 DEPROP OH ACCUM LVL          61.026
HIGH      V-33 DEPROP OH ACCUM LVL                         1
LOW       V-33 DEPROP OH ACCUM LVL          28.808
LOW       V-33 DEPROP OH ACCUM LVL                         1
LOW       V-33 DEPROP OH ACCUM LVL          33.965
HIGH      V-33 DEPROP OH ACCUM LVL            .309
HIGH      V-33 DEPROP OH ACCUM LVL                         1
HIGH      V-33 DEPROP OH ACCUM LVL          68.522
HIGH      V-33 DEPROP OH ACCUM LVL          79.675
HIGH      V-33 DEPROP OH ACCUM LVL          57.592
HIGH      V-33 DEPROP OH ACCUM LVL            .030
HIGH      V-33 DEPROP OH ACCUM LVL          50.431
LOW       V-33 DEPROP OH ACCUM LVL          20.560
EMERGNCY D-10 KO POT H2S                   192.232
LOW       V-33 DEPROP OH ACCUM LVL                         1
HIGH      V-33 DEPROP OH ACCUM LVL                         1
 OP               1.1000              5.1000       BPH      CONS   1
 OP               5.1000              9.1000       BPH      CONS   1
 SP             1890.000            1880.000       BPH      CONS   1
 SP             1850.000            1840.000       BPH      CONS   1
 OP               5.8000              5.5000       MSCFH    CONS   1
 OP               9.1000             11.1000       BPH      CONS   1
 OP              13.6000             13.3000       MSCFH    CONS   1
 OP              13.3000             13.0000       MSCFH    CONS   1
HIGH      SPLITTER REBOILER LEVEL           29.868
 OP              54.0000             52.0000       BPH      CONS   1
 OP              29.0000             21.0000       B P H    CONS   1
 OP              21.0000             17.0000       B P H    CONS   1
HIGH      SPLITTER REBOILER LEVEL                          1
 SP             1880.000            1865.000       BPH      CONS   1
 SP             1840.000            1825.000       BPH      CONS   1
HIGH      SPLITTER REBOILER LEVEL           31.395
HIGH      SPLITTER REBOILER LEVEL           29.964
HIGH      SPLITTER REBOILER LEVEL                          1
HIGH      SPLITTER REBOILER LEVEL           31.400
 SP             1825.000            1810.000       BPH      CONS   1
HIGH      SPLITTER REBOILER LEVEL           29.945
 SP             1865.000            1850.000       BPH      CONS   1
HIGH      SPLITTER REBOILER LEVEL           31.186
LOW       V-33 DEPROP OH ACCUM LVL          32.718
 OP              52.0000             40.0000       BPH      CONS   1
```

```
03/11/98 17:54:43    DP 26        FC0022          V-33 ACCUM RELEASE
03/11/98 17:54:56    DP 21 ALM       FC0022        PVLO             100.000
03/11/98 17:55:22    DP 31 ACK       FC0022        PVLO
03/11/98 17:56:12    HM 26        FC4095          E-47 SLURRY BYPASS
03/11/98 17:57:32    DP 26        HC4037          SOUTH E-45 OUTLET
03/11/98 17:57:38    DP 26        HC4038          MID E-45 OUTLET
03/11/98 17:57:48    FD 26        CF0011          EAST HEATER FEED CONTROL
03/11/98 17:57:59    DP 26        HC4039          NORTH E-45 OUTLET
03/11/98 17:58:21    FD 26        CF0010          WEST HEATER FEED CONTROL
03/11/98 17:59:13    SP 31 ACK       LC0207        PVLO
03/11/98 17:59:16    DP 21 RTN       FC0022        PVLO             100.000
03/11/98 17:59:32    DP 26        FC0022          V-33 ACCUM RELEASE
03/11/98 18:00:59    DP 26        PC0042          C-12 HOT VAPOR BYPASS
03/11/98 18:02:04    DP 26        FC0022          V-33 ACCUM RELEASE
03/11/98 18:02:30    DP 26        FC0022          V-33 ACCUM RELEASE
03/11/98 18:04:17    DP 26        PC0042          C-12 HOT VAPOR BYPASS
03/11/98 18:16:04    HM 26        FC4095          E-47 SLURRY BYPASS
03/11/98 18:16:05    FD 26        CF0010          WEST HEATER FEED CONTROL
03/11/98 18:17:13    FD 26        CF0011          EAST HEATER FEED CONTROL
03/11/98 18:24:57    DB 21 ALM       AC0111        BADPV
03/11/98 18:25:06    DB 31 ACK       AC0111        BADPV
03/11/98 18:28:17    HM 26        FC4095          E-47 SLURRY BYPASS
03/11/98 18:37:02    FD 26        CF0011          EAST HEATER FEED CONTROL
03/11/98 18:38:20    FD 26        CF0011          EAST HEATER FEED CONTROL
03/11/98 18:45:22    FD 26        FC0013          WEST HTR FUEL GAS FLOW
03/11/98 18:45:41    FD 26        FC0014          EAST HTR FUEL GAS FLOW
03/11/98 18:47:56    FD 26        FC0014          EAST HTR FUEL GAS FLOW
03/11/98 18:48:14    FD 26        FC0013          WEST HTR FUEL GAS FLOW
03/11/98 18:53:44    FD 26        CF0011          EAST HEATER FEED CONTROL
03/11/98 18:56:19    FD 26        FC0013          WEST HTR FUEL GAS FLOW
03/11/98 18:59:23    SC 26        FC1201          LEAN AMINE TO D/G SCRUB
03/11/98 19:01:53    FD 26        FC0014          EAST HTR FUEL GAS FLOW
03/11/98 19:02:24    BW 26        FC0103A         OXYGEN ENRICHMENT
03/11/98 19:02:57    DB 21 RTN       AC0111        BADPV
03/11/98 19:03:18    BW 26        FC0103A         OXYGEN ENRICHMENT
03/11/98 19:03:41    BW 26        FC0103A         OXYGEN ENRICHMENT
03/11/98 19:03:47    BW 26        FC0103A         OXYGEN ENRICHMENT
03/11/98 19:04:08    BW 26        FC0103A         OXYGEN ENRICHMENT
03/11/98 19:08:25    FD 26        FC0014          EAST HTR FUEL GAS FLOW
03/11/98 19:09:43    FD 26        FC0014          EAST HTR FUEL GAS FLOW
03/11/98 19:10:58    FD 26        CF0010          WEST HEATER FEED CONTROL
03/11/98 19:11:08    FD 26        CF0011          EAST HEATER FEED CONTROL
03/11/98 19:12:17    FD 26        FC0013          WEST HTR FUEL GAS FLOW
03/11/98 19:36:26    DP 26        FC1202          LEAN AMINE TO PROP SCRUB
03/11/98 19:36:26    SC 26        FC1201          LEAN AMINE TO D/G SCRUB
03/11/98 19:45:18    OH 21 ALM       HPOS7002      PVHI              99.999
03/11/98 19:45:18    CT 21 ALM       ZA2791        LO DISCH
```

```
 OP            40.0000           34.0000       BPH       CONS  1
LOW      V-33 ACCUM RELEASE          99.299
LOW      V-33 ACCUM RELEASE                                 1
 OP            11.1000            7.1000       BPH       CONS  1
 OP            34.8705           38.8705       PERCENT   CONS  1
 OP            18.0000           20.0000       PERCENT   CONS  1
 SP           1850.000          1830.000       BPH       CONS  1
 OP            21.5243           24.0243       PERCENT   CONS  1
 SP           1810.000          1800.000       BPH       CONS  1
HIGH     SPLITTER REBOILER LEVEL                            1
LOW      V-33 ACCUM RELEASE         105.708
 OP            34.0000           40.0000       BPH       CONS  1
 SP           233.0000          240.0000       PSIG      CONS  1
 OP            40.0000           54.0000       BPH       CONS  1
 MODE        MAN              CAS              BPH       CONS  1
 SP           240.0000          237.0000       PSIG      CONS  1
 OP             7.1000            3.1000       BPH       CONS  1
 SP           1800.000          1775.000       BPH       CONS  1
 SP           1830.000          1825.000       BPH       CONS  1
LOW      AVG C3 AT C5&7 OH
LOW      AVG C3 AT C5&7 OH                                  1
 OP             3.1000        -   2.9000       BPH       CONS  1
 SP           1825.000          1815.000       BPH       CONS  1
 SP           1815.000          1800.000       BPH       CONS  1
 OP             5.5000            5.2000       MSCFH     CONS  1
 OP            13.0000           12.5000       MSCFH     CONS  1
 OP            12.5000           12.2000       MSCFH     CONS  1
 OP             5.2000            4.8000       MSCFH     CONS  1
 SP           1800.000          1775.000       BPH       CONS  1
 OP             4.8000            4.5000       MSCFH     CONS  1
 SP           260.0000          250.0000       B P H     CONS  1
 OP            12.2000           12.0000       MSCFH     CONS  1
 MODE        CAS              MAN              MSCFH     CONS  1
LOW      AVG C3 AT C5&7 OH
 OP            35.8795           25.8795       MSCFH     CONS  1
 MODE        MAN              CAS              MSCFH     CONS  1
 MODE        CAS              BCAS             MSCFH     CONS  1
 MODE        BCAS             CAS              MSCFH     CONS  1
 OP            12.0000            9.0000       MSCFH     CONS  1
 OP             9.0000            8.5000       MSCFH     CONS  1
 SP           1775.000          1765.000       BPH       CONS  1
 SP           1775.000          1765.000       BPH       CONS  1
 OP             4.5000            3.8000       MSCFH     CONS  1
 SP           110.0000          105.0000       BPH       CONS  1
 SP           250.0000          240.0000       B P H     CONS  1
LOW      #2 FRACT OH MOV POS        100.006
HIGH     S ELEC. PUMP DISCH         LO DISCH
```

```
03/11/98 19:45:18    CT 21 ALM        ZA2792        LO DISCH
03/11/98 19:45:18    CT 21 ALM        ZA0533        CLOSED
03/11/98 19:45:18    CT 21 ALM      37AI0102        PVLO               6.800
03/11/98 19:45:20    OH 21 ALM        HC6050        RUN
03/11/98 19:45:20    OH 21 ALM        HC6040        STOP
03/11/98 19:45:20    CT 21 ALM        ZA0530        TRIP
03/11/98 19:51:10    RX 21 ALM        PDI1203       PVLO             100.000
03/11/98 19:51:48    RX 21 RTN        PDI1203       PVLO             100.000
03/11/98 19:52:53    SC 26          FC1201         LEAN AMINE TO D/G SCRUB
03/11/98 20:02:04    RX 31 ACK        PDI1203       PVLO
03/11/98 20:12:52    RX 21 ALM        PDI1203       PVLO             100.000
03/11/98 20:13:24    RX 21 RTN        PDI1203       PVLO             100.000
03/11/98 20:33:50    SY 19         NODE 04  $P4
03/11/98 20:33:52    SY 19         NODE 04  $P4
03/11/98 20:37:01    RX 31 ACK        PDI1203       PVLO
03/11/98 21:00:05    DP 26          HC4037         SOUTH E-45 OUTLET
03/11/98 21:00:13    DP 26          HC4038         MID E-45 OUTLET
03/11/98 21:00:18    DP 26          HC4039         NORTH E-45 OUTLET
03/11/98 21:18:37    RX 21 ALM        PDI1203       PVLO             100.000
03/11/98 21:19:41    RX 21 RTN        PDI1203       PVLO             100.000
03/11/98 21:29:46    RX 31 ACK        PDI1203       PVLO
03/11/98 21:38:55    FD 26          HC4036A        FF PUMP SPEED CONTROL
03/11/98 21:40:10    FD 26          HC4036A        FF PUMP SPEED CONTROL
03/11/98 21:41:09    FD 26          LC0008         V30 HOT FEED DRUM LEVEL
03/11/98 21:42:24    FD 26          HC4036A        FF PUMP SPEED CONTROL
03/11/98 21:44:18    FD 26          HC4036A        FF PUMP SPEED CONTROL
03/11/98 21:47:39    HM 26          FC4095         E-47 SLURRY BYPASS
03/11/98 22:05:25    HM 26          FC4095         E-47 SLURRY BYPASS
03/11/98 22:07:34    DB 26          CX0203         E29 #2 DEB REB NORM DUTY
03/11/98 22:14:26    FD 26          FC0014         EAST HTR FUEL GAS FLOW
03/11/98 22:21:12    FD 26          CF0010         WEST HEATER FEED CONTROL
03/11/98 22:21:29    FD 26          CF0011         EAST HEATER FEED CONTROL
03/11/98 22:27:10    FD 26          HC4036A        FF PUMP SPEED CONTROL
03/11/98 22:28:21    RX 21 ALM        PDI1203       PVLO             100.000
03/11/98 22:30:50    RX 21 RTN        PDI1203       PVLO             100.000
03/11/98 22:32:22    FD 26          CF0011         EAST HEATER FEED CONTROL
03/11/98 22:32:32    FD 26          CF0010         WEST HEATER FEED CONTROL
03/11/98 22:33:08    RX 31 ACK        PDI1203       PVLO
03/11/98 22:36:21    FD 26          FC0013         WEST HTR FUEL GAS FLOW
03/11/98 22:37:10    FD 26          FC0013         WEST HTR FUEL GAS FLOW
03/11/98 22:38:50    FD 26          FC0013         WEST HTR FUEL GAS FLOW
03/11/98 22:41:09    FD 26          CF0011         EAST HEATER FEED CONTROL
03/11/98 22:41:18    FD 26          CF0010         WEST HEATER FEED CONTROL
03/11/98 22:46:04    RX 21 ALM        PDI1203       PVLO             100.000
03/11/98 22:47:30    OH 21 RTN        HPOS7002      PVHI              99.999
03/11/98 22:48:11    DB 26          CX0203         E29 #2 DEB REB NORM DUTY
03/11/98 22:49:00    DP 26          HC4037         SOUTH E-45 OUTLET
```

	HIGH	N ELEC. PUMP DISCH	LO DISCH			
	HIGH	CT STEAM GEN VLV	CLOSED			
	LOW	CT PH	0.000			
	HIGH	OVHD FAN #10W	RUN			
	HIGH	OVHD FAN #9W	STOP			
	HIGH	STM TURB CW TRIP	TRIP			
	HIGH	RX SV DIFF PRESSURE	99.863			
	HIGH	RX SV DIFF PRESSURE	112.222			
	SP	240.0000	230.0000	B P H	CONS	1
	HIGH	RX SV DIFF PRESSURE			1	
	HIGH	RX SV DIFF PRESSURE	99.906			
	HIGH	RX SV DIFF PRESSURE	112.057			
0	DEVICE AVAILABLE	OFFLINE		01		
0	DEVICE AVAILABLE	DEVICE READY		01		
	HIGH	RX SV DIFF PRESSURE			1	
	OP	38.8705	36.8705	PERCENT	CONS	1
	OP	20.0000	18.0000	PERCENT	CONS	1
	OP	24.0243	22.0243	PERCENT	CONS	1
	HIGH	RX SV DIFF PRESSURE	99.996			
	HIGH	RX SV DIFF PRESSURE	112.081			
	HIGH	RX SV DIFF PRESSURE			1	
	OP	97.0000	91.0000	PERCENT	CONS	1
	OP	91.0000	89.0000	PERCENT	CONS	1
	SP	42.0000	50.0000	PERCENT	CONS	1
	OP	89.0000	87.0000	PERCENT	CONS	1
	OP	87.0000	83.0000	PERCENT	CONS	1
	OP	- 2.9000	5.1000	BPH	CONS	1
	OP	5.1000	9.1000	BPH	CONS	1
	SP	669.2910	660.0000	MMBTUF/H	CONS	1
	OP	8.5000	8.0000	MSCFH	CONS	1
	SP	1765.000	1750.000	BPH	CONS	1
	SP	1765.000	1750.000	BPH	CONS	1
	OP	83.0000	81.0000	PERCENT	CONS	1
	HIGH	RX SV DIFF PRESSURE	99.580			
	HIGH	RX SV DIFF PRESSURE	112.196			
	SP	1750.000	1735.000	BPH	CONS	1
	SP	1750.000	1735.000	BPH	CONS	1
	HIGH	RX SV DIFF PRESSURE			1	
	OP	3.8000	3.3000	MSCFH	CONS	1
	OP	3.3000	3.1000	MSCFH	CONS	1
	OP	3.1000	1.8000	MSCFH	CONS	1
	SP	1735.000	1720.000	BPH	CONS	1
	SP	1735.000	1720.000	BPH	CONS	1
	HIGH	RX SV DIFF PRESSURE	99.597			
	LOW	#2 FRACT OH MOV POS	97.318			
	SP	660.0000	655.0000	MMBTUF/H	CONS	1
	OP	36.8705	37.9705	PERCENT	CONS	1

03/11/98	22:49:14	DP 26	HC4038	MID E-45 OUTLET		
03/11/98	22:49:31	RX 21 ALM		PAL6441	PRETRIP	
03/11/98	22:49:32	RX 21 RTN		PAL6441	PRETRIP	
03/11/98	22:49:33	DP 26	HC4039	NORTH E-45 OUTLET		
03/11/98	22:49:45	RX 21 ALM		PAL6441	PRETRIP	
03/11/98	22:49:46	RX 21 RTN		PAL6441	PRETRIP	
03/11/98	22:50:40	RX 21 RTN		PDI1203	PVLO	100.000
03/11/98	22:55:00	RX 31 ACK		PAL6441	PRETRIP	
03/11/98	22:55:00	RX 31 ACK		PDI1203	PVLO	
03/11/98	22:59:08	RX 21 ALM		PDI1203	PVLO	100.000
03/11/98	23:02:12	RX 21 RTN		PDI1203	PVLO	100.000
03/11/98	23:08:24	SP 21 ALM		FC1210	PVLO	10.000
03/11/98	23:08:58	SP 21 RTN		FC1210	PVLO	10.000
03/11/98	23:09:26	SP 21 ALM		FC1210	PVLO	10.000
03/11/98	23:10:18	SP 21 RTN		FC1210	PVLO	10.000
03/11/98	23:10:41	SP 21 ALM		FC1210	PVLO	10.000
03/11/98	23:10:43	RX 21 ALM		PDI1203	PVLO	100.000
03/11/98	23:11:10	RX 21 RTN		PDI1203	PVLO	100.000
03/11/98	23:11:30	SP 21 RTN		FC1210	PVLO	10.000
03/11/98	23:12:18	SP 21 ALM		FC1210	PVLO	10.000
03/11/98	23:12:42	SP 21 RTN		FC1210	PVLO	10.000
03/11/98	23:13:37	RX 31 ACK		PDI1203	PVLO	
03/11/98	23:13:44	SP 21 ALM		FC1210	PVLO	10.000
03/11/98	23:14:03	SP 21 RTN		FC1210	PVLO	10.000
03/11/98	23:14:47	SP 21 ALM		FC1210	PVLO	10.000
03/11/98	23:16:07	RX 21 ALM		PDI1203	PVLO	100.000
03/11/98	23:18:07	RX 21 RTN		PDI1203	PVLO	100.000
03/11/98	23:21:45	SP 21 ALM		LC0207	PVLO	30.000
03/11/98	23:21:56	SP 21 RTN		LC0207	PVLO	30.000
03/11/98	23:22:25	SP 26	FC1211	6 IN COLD SPLITTER BTMS		
03/11/98	23:22:44	SP 21 ALM		LC0207	PVLO	30.000
03/11/98	23:23:16	SP 21 RTN		LC0207	PVLO	30.000
03/11/98	23:24:02	RX 21 ALM		PDI1203	PVLO	100.000
03/11/98	23:24:34	SP 26	FC1211	6 IN COLD SPLITTER BTMS		
03/11/98	23:25:00	RX 21 RTN		PDI1203	PVLO	100.000
03/11/98	23:25:42	RX 31 ACK		PDI1203	PVLO	
03/11/98	23:25:42	SP 31 ACK		LC0207	PVLO	
03/11/98	23:30:25	SP 21 RTN		FC1210	PVLO	10.000
03/11/98	23:30:31	SP 21 ALM		FC1210	PVLO	10.000
03/11/98	23:30:39	SP 21 RTN		FC1210	PVLO	10.000
03/11/98	23:30:46	SP 21 ALM		FC1210	PVLO	10.000
03/11/98	23:30:52	SP 21 RTN		FC1210	PVLO	10.000
03/11/98	23:30:59	SP 21 ALM		FC1210	PVLO	10.000
03/11/98	23:31:08	SP 21 RTN		FC1210	PVLO	10.000
03/11/98	23:33:44	SP 21 ALM		FC1210	PVLO	10.000
03/11/98	23:33:53	SP 21 RTN		FC1210	PVLO	10.000
03/11/98	23:39:03	RX 21 ALM		PDI1203	PVLO	100.000

OP	18.0000	19.0000	PERCENT	CONS	1
LOW	RX S/V DIFF PRETRIP	PRETRIP			
LOW	RX S/V DIFF PRETRIP	NORMAL			
OP	22.0243	23.0000	PERCENT	CONS	1
LOW	RX S/V DIFF PRETRIP	PRETRIP			
LOW	RX S/V DIFF PRETRIP	NORMAL			
HIGH	RX SV DIFF PRESSURE	112.069			
LOW	RX S/V DIFF PRETRIP			1	
HIGH	RX SV DIFF PRESSURE			1	
HIGH	RX SV DIFF PRESSURE	99.788			
HIGH	RX SV DIFF PRESSURE	112.167			
JOURNAL	3 IN SPLIT BTMS TO FIELD	9.976			
JOURNAL	3 IN SPLIT BTMS TO FIELD	13.529			
JOURNAL	3 IN SPLIT BTMS TO FIELD	9.816			
JOURNAL	3 IN SPLIT BTMS TO FIELD	13.855			
JOURNAL	3 IN SPLIT BTMS TO FIELD	9.939			
HIGH	RX SV DIFF PRESSURE	99.912			
HIGH	RX SV DIFF PRESSURE	112.500			
JOURNAL	3 IN SPLIT BTMS TO FIELD	14.186			
JOURNAL	3 IN SPLIT BTMS TO FIELD	9.919			
JOURNAL	3 IN SPLIT BTMS TO FIELD	13.398			
HIGH	RX SV DIFF PRESSURE			1	
JOURNAL	3 IN SPLIT BTMS TO FIELD	9.723			
JOURNAL	3 IN SPLIT BTMS TO FIELD	14.038			
JOURNAL	3 IN SPLIT BTMS TO FIELD	9.486			
HIGH	RX SV DIFF PRESSURE	99.946			
HIGH	RX SV DIFF PRESSURE	112.157			
HIGH	SPLITTER REBOILER LEVEL	29.981			
HIGH	SPLITTER REBOILER LEVEL	31.059			
OP	17.0000	7.0000	B P H	CONS	1
HIGH	SPLITTER REBOILER LEVEL	29.921			
HIGH	SPLITTER REBOILER LEVEL	31.065			
HIGH	RX SV DIFF PRESSURE	99.988			
OP	7.0000	- 1.0000	B P H	CONS	1
HIGH	RX SV DIFF PRESSURE	112.070			
HIGH	RX SV DIFF PRESSURE			1	
HIGH	SPLITTER REBOILER LEVEL			1	
JOURNAL	3 IN SPLIT BTMS TO FIELD	13.305			
JOURNAL	3 IN SPLIT BTMS TO FIELD	9.933			
JOURNAL	3 IN SPLIT BTMS TO FIELD	13.626			
JOURNAL	3 IN SPLIT BTMS TO FIELD	9.697			
JOURNAL	3 IN SPLIT BTMS TO FIELD	13.207			
JOURNAL	3 IN SPLIT BTMS TO FIELD	9.837			
JOURNAL	3 IN SPLIT BTMS TO FIELD	13.230			
JOURNAL	3 IN SPLIT BTMS TO FIELD	9.515			
JOURNAL	3 IN SPLIT BTMS TO FIELD	13.499			
HIGH	RX SV DIFF PRESSURE	99.814			

```
03/11/98 23:41:50    RX 21 RTN      PDI1203        PVLO          100.000
03/11/98 23:42:55    BL 26       PC0313         DUAL CONTROL VALVE
03/11/98 23:44:13    BL 26       PC0313         DUAL CONTROL VALVE
03/11/98 23:46:30    RX 31 ACK      PDI1203        PVLO
03/11/98 23:56:58    HM 26       FC0101         CLARIFIED OIL RELEASE
03/11/98 23:57:34    BW 26       FC0103A        OXYGEN ENRICHMENT
03/11/98 23:58:20    BW 26       FC0103A        OXYGEN ENRICHMENT
03/11/98 23:58:34    BW 26       FC0103A        OXYGEN ENRICHMENT
```

HIGH	RX SV DIFF PRESSURE	112.065			
SP	145.4990	143.0000	PSI	CONS	1
SP	143.0000	140.0000	PSI	CONS	1
HIGH	RX SV DIFF PRESSURE			1	
SP	206.8980	210.0000	BPH	CONS	1
MODE	CAS	MAN	MSCFH	CONS	1
OP	11.7280	19.7280	MSCFH	CONS	1
MODE	MAN	CAS	MSCFH	CONS	1

APPENDIX 3

Operator Alarm Usefulness Questionnaire

Survey Contact Person: ____________________

Contact Telephone: ______________________

Date to Be Completed: ____________________

Return Location: ________________________

Date Survey Period Ends: __________________

NOTE:

While this survey is being worked on and when completed, it will be kept confidential with regard to the name of the person filling it out. Please wait to discuss any aspects of the survey with other operators until all surveys have been completed. Refer to the "Date Survey Period Ends" for that date. On that date, you are free to discuss any and all aspects of the survey, your answers, etc.

Tracking Number:	Date Out:

A3.1 OPERATOR ALARM USEFULNESS QUESTIONNAIRE

Explanation

We are evaluating the performance of your alarm system. We are doing this to find out what improvements might be necessary to enhance its usefulness for you. This questionnaire is part of a larger work. It is being given to a small sample of operators of varying experience and abilities. You are a key user of the alarm system. Consequently, your input and observations will form an important part of our evaluation.

Purpose

The results of this questionnaire, and others like it from other operators and even other company sites, will be reviewed and analyzed. The results will aid the plant alarm improvement team as they decide the nature and extent of any problems facing operators and work out a program for improvement, if needed.

General Instructions

You will notice that not all of the questions reach as far as the bottom of the page. That is done because the questions are grouped to make sure that the entire question being read and answered are on the same page.

Please keep all pages together. Do not detach any portion of this document.

There are two surveys in this document. The first is the Operator Alarm Usefulness Questionnaire. All operators will complete this one. The second is the Operator Alarm Quiet Period Questionnaire. Only those questionnaire forms that are marked "YES" in the beginning of that part should be completed. All other operators should leave that section blank.

Confidentiality

Your completed questionnaire will be combined with all others to provide a good picture of the overall operator concern and observations regarding the alarm system. No individual questionnaire will be identified as to who filled it in. In order to keep track of the questionnaires, a tracking number may be assigned to ensure that no one on the survey list was missed and that all questionnaires have been returned for evaluation.

Please refrain from discussing any of the details of this questionnaire with others until all requested personnel have had a chance to complete them. We'd like everyone's candid responses, please. Afterward, when the survey period is over, you are welcome to discuss whatever aspects you choose. The date the survey ends is shown on the cover page.

If you are unable to complete the questionnaire, please return what you've done. All questionnaires should be returned. If you find that you would prefer to have a fresh form for whatever reason, just ask.

Surveyors

The alarm performance evaluation team is conducting this survey. Members of the team will participate in various portions of the alarm performance assessment, alarm project formulation, and other various alarm improvement work.

Additional Information If You Have Questions

If you have any additional questions regarding how to fill out the questionnaire or the meaning or intent of any particular question or portion of this document, please contact the SURVEY CONTACT PERSON identified on the cover page.

Where Questionnaire Is to Be Returned

Please return the completed questionnaire by the requested date (see cover) to the SURVEY CONTACT PERSON identified on the cover page.

Please turn the page to start the questionnaire.

This questionnaire is based on EEMUA Publication 191 Appendix 12 and 13, and others.

OPERATOR ALARM USEFULNESS QUESTIONNAIRE

Plant Site Name: ______________________

Location: ____________________________

Operating Area: ______________________

Date Completed: ______________________

Please answer all questions carefully and thoughtfully. If the answer to any particular question is none or does not apply, please enter the word "none" or "n/a" as your answer. In some cases, you are to circle the correct answer. Other answers require inserting specific responses before the "unit of measure" (like years). Many cases just ask for a check mark. For "Remarks" please add any information you feel might help the survey to be more useful and/or better understood.

If you are unsure of your answer or have a problem with the meaning or wording of a question, please ask the survey contact person whose name and telephone number can be found on the cover. If you are still uncertain of your answer, please skip the question and leave it blank. Feel free to write in any comments or concerns you might have. They will be read and used to assist us.

Your completing this questionnaire will be appreciated!

1. How long have you worked for your current employer?	
Years	Months
Remarks:	

2. What other plants or other units have you worked on in the past (and for how long and how long ago)?
Others recently:
Others way in the past:

3. What is your job, and on what plant/unit?

4. Have you had experience/training on pre-DCS control systems?		
Yes	No	Too long ago to remember
Remarks:		

5. How long have you worked with the present alarm system?	
Years	Months

6. What percentage of time are you assigned to each of these operating roles?		
Inside operator:	Outside operator:	Supervisor:

7. What other plants or other units do you presently work in?

8. Have you had formal DCS training?		
Yes	No	Too long ago to remember
Training areas:		

9. How well does the alarm system support you during normal steady operations?			
Very good	OK	Poor	Very poor

10. How well does the alarm system support you during plant faults or trips?			
Very good	OK	Poor	Very poor

11. What about the number of alarms activating in your operating area? In your experience, are there:			
Too many alarms	Many but necessary	Few but adequate	Too few alarms

NORMAL STEADY OPERATION

12. How many alarms do you get in normal steady operation? (an estimate will do here)
____________________ per ______ (hour, etc.)

13. How often do you see an alarm that activates and appears to be a repeat of an alarm you have already seen within 5 minutes before?			
70–100% of alarms	40–70% of alarms	20–40% of alarms	Under 20% of alarms

14. Do your plant units suffer from the following nuisance alarms?			
	Often	Sometimes	Rarely
Alarms that are wrongly prioritized			
Alarms from plant that is shut down			
Two or more alarms occurring at about the same time that mean the same thing			
Alarms occurring after an equipment trip (or shutdown) which are only relevant in steady operation			
Remarks:			

15. What proportion of alarms is really useful to you in operating the plant?			
All essential	Most useful	Few useful	Very few useful

16. Do you fully understand each alarm message and know what to do about each?		
Almost always	Mostly	Sometimes

17. Consider a normal operating situation and 10 typical alarms. What percent (%) of the 10 alarms best fit into each of the indicated categories?	
a. Require you to take positive action; for example, operate a valve, speak to an assistant about something specific	(Insert % here. Should add to 100%)
b. Cause you to bring up a specific display or trend and monitor something closely	(Insert % here. Should add to 100%)
c. Are noted as useful information	(Insert % here. Should add to 100%)
d. Are read and quickly forgotten	(Insert % here. Should add to 100%)

18. Are the alarm messages displayed in a consistent manner (alarm name, text, etc.)?		
Almost always	Mostly	Sometimes

19. Do you keep an alarm list display on permanent view during normal steady state operation?	
Yes	No

20. How often do you look through the alarm list during normal steady state operation?			
Several times a minute	Once every couple of minutes	Once every 10 minutes	Less than once every 10 minutes

21. Is each alarm message relatively easy for you to understand?			
Almost always	Mostly	Sometimes	Almost never
22. Does the alarm system convey what action you should take?			
Almost always	Mostly	Sometimes	Almost never
23. Does the alarm system convey how much time you have to take action?			
Almost always	Mostly	Sometimes	Almost never

PLANT FAULTS AND TRIPS

24. How many alarms would you get during a large plant fault or trip, that is, power failure (best guess)?		
First minute	Next 10 minutes	Next hour

25. Do you keep an alarm list display on permanent display during a large plant fault or trip?	
Yes	No

26. How often do you look through the alarm list display during a large plant fault or trip?			
Several times a minute	Once every couple of minutes	Once every 10 minutes	Less than once every 10 minutes
27. How often in a large plant fault or trip are you forced to accept alarms without having time to read and understand them?			
Almost always	Quite often	Sometimes	Almost never

28. Does the alarm system help you to pick out key safety related events during a large plant fault or trip?			
Very well	Some help	Little help	No help

29. Does the alarm system help you to pick out key environmental related events during a large plant fault or trip?			
Very well	Some help	Little help	No help

30. Does the alarm system help you to pick out key economic related events during a large plant fault or trip?			
Very well	Some help	Little help	No help

31. Do the same alarms appear in multiple work areas (i.e., different operator positions)?	
Yes	No
Comment:	

32. When you acknowledge alarms, are they cleared from the alarm display?	
Yes	No

33. On the alarm display, do alarms appear in order of priority (highest at the top, lowest at the bottom)?	
Yes	No

34. When you acknowledge alarms, are they acknowledged on all screens in your area (all consoles that you use)?	
Yes	No

35. Are there information only alarms that really should appear on the alarm display?	
Yes	No

36. Do you use the following functions on the alarm display screens (circle your response)?		
Most recent alarms	Yes	No
Unacknowledged alarm summary	Yes	No
Acknowledged alarm summary	Yes	No
Alarm history	Yes	No
Other (describe)	Yes	No

37. Do you use the following buttons (targets) located on the alarm display (circle correct response for each line)?		
Acknowledge Page	Yes	No
Clear Page	Yes	No
Mute Horns	Yes	No
Block Detail	Yes	No
Other (describe)	Yes	No

38. When there is an audible alarm, where do you go first and why?
Describe where you go and why:

39. Do you normally acknowledge alarms from the process graphic or from the alarm display?	
Process graphic	Alarm display
If you don't like either or sometimes use process graphic and sometimes use the alarm display, please explain why here:	

40. Can you easily determine the alarm priorities that appear on your graphics (e.g., by color, or by shape, or . . .)?	
Yes	No

41. How many operating displays (a display is a physical display device) are present at your operating area?			
1 or 2	3 or 4	5 or 6	More than 6 (specify)

42. Considering only monitoring and control, how do you use combinations of your operating displays?
(Combinations mean for some task where the views on the displays are related to each other and you are working your issue using them together)

	Almost always	Mostly	Sometimes	Rarely
None together				
2 together				
3 together				
4 together				
5 or more together				

Remarks:

43. How many separate operating "windows" per display do you typically have open at the same time?	
When using 1 screen?	Insert an approximate number:
When using 2 screens?	Insert an approximate number:
When using 3 screens?	Insert an approximate number:
When using 4 or more screens?	Insert an approximate number:
Remarks:	

44. Please describe all of the alarm-specific documentation available to you in the control room.			
List items below	How do you use them?		
	Often	Sometimes	Almost never
	Often	Sometimes	Almost never
	Often	Sometimes	Almost never
	Often	Sometimes	Almost never
	Often	Sometimes	Almost never
	Often	Sometimes	Almost never
	Often	Sometimes	Almost never

45. How easy is it for you to find specific views or display windows?			
Very easy	Easy	Sometimes hard	More often than not, hard
Remarks:			

46. Is it possible for alarms in your area to be acknowledged by an operator in another area?	
Yes	No

47. How often are alarms that are useful to you acknowledged by another operator?			
Almost always	Quite often	Sometimes	Almost never

48. When you acknowledge any alarm from a graphic, do all alarms displayed in the graphic become acknowledged automatically?	
Yes	No

49. Is it easy to determine a new system alarm activates (e.g., a data highway glitch)?	
Yes	No

50. Does your operating area have a large overhead alarm display?	
Yes	No

51. If so, when do you use the large overhead alarm display in place of your normal workstation alarm display?
Please explain here:

52. Does your control room have more than one alarm horn sound for your area that is used to convey alarm priority?	
Yes	No

53. If you have them, are the different alarm horn sounds useful to determine alarm priorities?	
Yes	No
Remarks:	

54. Does your control room have alarm horn tones that are used to determine which operator area they are located in?	
Yes	No
Remarks:	

55. Are the tones useful for determining the operator area that the alarm is coming from?	
Yes	No
Remarks:	

56. How often do you place controllers in manual to help manage alarms?			
Almost always	Quite often	Sometimes	Almost never
Remarks:			

57. For loops where you suspect that the control loop is not properly tuned, do you find more alarms on poorly tuned ones?			
Almost always	Quite often	Sometimes	Almost never
Remarks:			

GENERAL

58. What do you think of your plant's procedures (actual, not necessarily the ones on paper) for getting changes made to alarm settings?			
Overly restrictive and cumbersome	Strict but safe	Easy to use—but you have to be careful what you do	Sloppy and uncontrolled
Remarks:			

59. Compared with the other things they do to improve your control system, does the plant put enough effort into improving and maintaining the alarm system?		
Too much	About right	Too little
Remarks:		

60. What features of the alarm system do you like best?
Please list:

61. What features of the alarm system do you like least?
Please list:

62. If you could change any part of the alarm system you use:
What features would you add to help you run the plant?
What features would you remove because they don't help or you don't like them?

63. Please tell us what you think might be useful for improving, managing, or understanding the alarm system.
Remarks:

64. Please tell us what you think might be useful for improving this survey.
Remarks:

STOP

Please do NOT complete the next questionnaire unless the box on the next page is marked "YES."

A3.2 QUIET PERIOD ALARM USEFULNESS QUESTIONNAIRE

Notice: Please do not complete this portion unless the "yes" box below is circled	
Yes	No

Explanation

This form is used to measure what proportion of alarms are found **useful** by operators when the plant is operating with much fewer problems than are encountered during abnormal situations. This does not mean that nothing is happening; just that upsets and alarms, if any, are coming in slowly.

Instructions

As each alarm occurs write its title (tag identifier) on a separate line in the table below. You may use abbreviations, but please be sure that others will understand what you write. For example "Boiler Press Hi."

We then want you to put a check mark (✓) in one of the five columns beside the alarm that best explains your thinking about how useful the alarm is. Please do this for ten consecutive different alarm activations that occur for you once you have started. If no alarms occur, please write "no alarms" somewhere in the table. If an alarm occurs more than once, use the row already labeled and just repeat the checks in the appropriate column. You may place more than one check in any given column for the repeated alarm row.

When all ten lines are complete, or the shift ends, the survey is over.

Please be sure to write down the starting time and ending time of the survey.

Column Definitions

- ACTION: Check this column if you took a positive action like operating a valve, changing a control setpoint, phoning your outside operator (or maintenance) to check something specific for you.
- CHECK: Check this column if you do not think you needed to make any positive actions (see above) but just wanted a check on the plant status; for example, checked a measurement already on display or called up a new display to have a quick look.
- NOTED: Check this column if you did not take any action or make any plant check, but you were still glad that the alarm was displayed to you.

- LITTLE USE: Check this column if the alarm was interesting but actually of little or no real use to you.
- NUISANCE: Check this column if the alarm was a complete waste of your time.

Start time:	*End time*:	*Date*:	*Plant/unit*:

Survey Data Table

Alarm title	Action	Check	Noted	Little use	Nuisance
(extra)					
(extra)					
1.					
2.					
3.					
4.					
5.					
6.					
7.					
8.					
9.					
10.					
(extra)					
(extra)					

PLEASE DON'T FORGET TO WRITE IN THE ENDING TIME OF YOUR SURVEY IN THE BOX ABOVE THE TABLE!

End of Survey

Summary

Survey Contact: ______________________________

Date Checked: ______________________________

Date Audited: ______________________________

Date Coded: ______________________________

General Remarks: ______________________________

____

____

____

Final Evaluator(s): ______________________________

____

____

APPENDIX 4

Alarm Philosophy from Honeywell European Users

Alarm Management Philosophy for Screen-based Control Systems.

Honeywell European User's Group Operator Interface Work-shop.

Revision 4: 22 July 1998

Preface

The Honeywell European User's Group Conference in Barcelona, September 1997, included a breakout session on "Alarm Handling". The session came out with a number of issues and requested the formation of a Work-shop to look at these and other alarm management matters. This request was discussed in the Users Group Steering Committee meeting and it was decided that alarm management should be handled by the already existing Operator Interface Work-shop.

At the first meeting, the Work-shop found that there was no internationally recognised standard for alarm handling in screen-based systems, but that a number of public enquiries into major incidents had raised concerns about the escalation of problems because of poor alarm management. Some public bodies are currently working on the subject, notably EEMUA, (Engineering Equipment and Materials Users Association, UK), ERGOS (Netherlands), CCH2K (Sweden). The Work-shop is co-operating fully with them and continuously looking for contacts with others working in this field.

We then considered that it would be in our best interest to develop our ideas on the subject into an "Alarm Management Philosophy for Screen-based Systems". This would be contributed as the views of informed and interested users to the bodies and other users working on this subject.

The document is structured on the basis of the operator's typical actions in a disturbance that involves his intervention. The requirements of alarm management are derived from first principles. The document is based on experience gained in a wide range of industries including: energy, oil and gas, bulk chemicals, fine chemicals, pulp and paper and steel. The Work-shop felt that other Honeywell users should have an opportunity to contribute to the development of this document. The document is, therefore, presented as a draft for discussion; comments, contributions and criticisms are invited.

Although I should thank all work-shop members for their continuous efforts and the enthusiasm they showed while producing this document, special thanks have to be addressed to John McCulloch of BP Chemicals Grangemouth UK. He has provided the work-shop with a splendid presentation and a first draft for the Alarm Philosophy document.

Although the document is seen as being produced and owned by the Honeywell Users Group, you can use and distribute it as long as the origin of the document is not removed or concealed. We hope that in this way, we can contribute a little bit to having less incidents.

Of course, producing this philosophy is only a start and not the final goal of the Work-shop. We will continue having discussions about alarm management philosophies (and refine the document), operating practices, dynamic alarm management, training, bench-marking, audible alarming, display issues, functional definitions, on-line tools, off-line tools, system speed and documentation. We will produce guidelines where appropriate, both for system developers and users.

The Work-shop will report back to you at the annual users group meeting and through its associated newsletter. Stay tuned. If you have questions, suggestions or remarks about the subject or the Work-shop, please don't hesitate to contact me.

Work-shop chairman,

Rudi Hörchert

Ineos

1. Introduction:

The purpose of this document is to define what an alarm is and what features are required in the DCS to satisfy these requirements. To do this, it is necessary to understand the interactions between the operator and the DCS in response to an alarm. The sequence of events can be split into four phases:

- Problem Detection
- Cause Identification
- Situation Assessment
- Corrective Action

The purpose of an alarm is to assist the operator in preventing or in dealing with upsets and disturbances on the plant. To achieve this, the DCS should offer:

- timely detection of an abnormal situation requiring operator action
- a fast route to the corrective actions required
- functionality to allow simultaneous events to be handled effectively
- the least possible opportunity for errors.

The typical steps that have to be taken in the process of responding to an alarm are summarised in the following table:

Phase	***Typical DCS Action***	***Operator***
Problem Detection	Sensor activation	
∨	Alarm audible warning	
Cause Identification		Operator silences alarm audible warning; acknowledges alarm #
∨		Operator requires to identify cause: calls alarm summary display
∨	Alarm summary or short-cut to process schematic on screen	
∨		Operator identifies cause of alarm
Situation Assessment		Operator requires to assess whole plant state: calls plant over-view schematic
∨	Plant over-view schematic on screen	*[An additional screen or window on which the plant over-view is already present would expedite this phase]*
∨		Operator situation assessment; makes corrective action decision
∨		Operator requires to perform corrective action: calls appropriate schematic
Corrective Action	Corrective action schematic on screen	*[If this is the same schematic as used for identification of cause, this would expedite this phase]*
∨		Operator performs corrective action through DCS and acknowledges alarm #

 # Traditional practice for many users is to acknowledge the alarm immediately it is identified. Leaving an alarm silenced but unacknowledged until the corrective action is complete can provide an important reminder of the alarm response actions that still have to be done. See appendix: "Alarm Acknowledgement Practice".

For clarity, the document is divided into the following subject areas:

- Alarm Definition
- Alarm Detection
- Alarm Prioritisation
- Audible Warning
- Identification of Cause
- Situation Assessment
- Corrective Action
- Operator Work-load
- System Management.

In each section, a sentence in bold italics is given as the statement of philosophy. There are then some accompanying paragraphs of explanation.

The document raises a number of associated issues that are handled in the appendices:

- Alarm acknowledgement practice
- First-up identification
- Display considerations
- Handling of non-alarm information

There is also a further appendix defining some of the terminology used.

2. Alarm Definition:

An alarm is advice to the operator that immediate action is required.

Some process disturbances will result in safety, environmental or economic problems unless the operator intervenes; here an alarm is required. Other process disturbances can occur that the control system will cope with; and in these cases no operator action is required, and nor is an alarm.

The risk of safety or environmental problems or the size of the economic penalty for failing to deal with the cause of an alarm increases with the passage of time. Thus, any delay is undesirable, and the consequences are minimised by taking immediate action. When an alarm occurs, the operator must make a decision on whether to react to it immediately or whether to deal with higher priority items first.

The types of action that the operator needs to take following an alarm are many and varied, and may include: shutting the plant down; checking the reading of an instrument; calling out the emergency services; starting a standby equipment; monitoring certain process parameters; logging a maintenance request. Most of them will involve some use of the DCS.

If an alarm exists on the system for which no immediate operator action is appropriate, it should be removed. Its presence on the system distracts the operator from his main task in an upset. The presence of many such alarms is a potential safety hazard, for the important alarms are thereby hidden from the operator.

An alarm is to be distinguished from other plant status indications, discrete value information, messages, and reports. Whilst the operator must have ready access to this information through his displays, he does not require to take any immediate action, and thus does not require it to be drawn to his attention.

3. Alarm Detection:

Alarm detection should provide timely advice that there is a problem requiring operator intervention whilst minimising unnecessary or nuisance alarms.

To achieve this the most appropriate alarm detection mechanism should be chosen for each parameter. Current DCSs typically provide:

- **Deviation alarms:** warn that the control system cannot deal with the disturbance.
- **Rate-of-change alarms:** These can often be the first indications of an upset, but should not be used on very noisy signals.
- **Recipe-driven alarms:** These are particularly useful in multi-mode or batch units where the alarm thresholds or priorities change with operating mode. Can also be used for dynamic prioritisation or intelligent alarm suppression.
- **Statistical alarms**: These are applicable where an SPC package is used, and can effectively filter out significant changes from process noise.
- **Absolute alarms:** These warn of the parameter approaching absolute limits such as trip or safety-valve settings.

Besides the above, some further DCS alarm detection mechanisms are proposed to address some needs that have been identified:

- **Combinational or sequential validation of alarms:** An alarm is only generated when a predefined combination or sequence of conditions indicate that a problem is present.
- **Adaptive alarms:** These use the 'rate-of-change' or 'deviation' principle in combination with absolute thresholds. Where the parameter is well clear of the absolute threshold, the allowable rate-of-change or deviation is relaxed; as the parameter approaches the absolute threshold, the limits on rate-of-change or deviation are tightened.
- **Automatic retriggering:** If an alarm has been active and silenced for more than a predefined time limit, a new audible warning is generated to remind the operator of its existence.
- **Dynamic Alarm Modification:** Such techniques can be used to reduce unnecessary and nuisance alarms by adjustment of priority, threshold, dead-band and other alarm settings according to the state or mode of operation of the plant.

4. Alarm Prioritisation:

Alarm prioritisation defines the degree of urgency associated with an alarm, and thus the order in which an operator should take corrective action when a number of alarms are present.

The degree of urgency of an alarm at any instant, and thus its priority, may be dependant on a number of factors, including:

- the severity of the consequences, (in safety, environmental and economic terms), of failing to take the corrective action associated with the alarm;
- the time available and required for the corrective action to be performed and to have the desired effect;
- the state of the rest of the plant, and, in particular, what other alarms are active;
- how long the alarm has been active.

None of these things are fixed; they can all vary in subtle and complex ways with the state of the plant. Techniques for dynamic prioritisation are available to assist with this process, and methodologies have been established for safety analysis of such techniques..

Many DCS manufacturers use the term "priority" to define the display and reporting functionality of alarms within their system. This may be entirely unrelated to the order in which the operator is required to take action. The definition of display and reporting functionality should be entirely independent of the identification to the operator of the order in which he has to deal with several alarms that are simultaneously present.

5. Audible Warning:

The audible warning should convey as much information as possible about the alarm priority and the plant area affected.

The reason for this is that the human brain responds far more quickly to audible information than to any other, and can very rapidly identify which of a large range of subtly different sounds is being heard. By appropriate codification of the information, the operator can already be much of the way towards identifying the problem before he has reached out his hand to silence the noise.

- **Audible Discrimination of Priority:** The volume and pulsation rate of the audible warning can be used to denote priority: a high volume sound with a rapid pulsation frequency denotes a high priority.
- **Audible Discrimination of Plant, Area or Unit:** For discriminating plant areas or units, it is necessary that the sounds have clearly distinguishable characteristics. It is not advisable to use different pitches; (few people have perfect pitch and in isolation they can be misinterpreted). Tones of the same fundamental pitch, though, can have quite different characteristics that are easily discriminated whether together or in isolation; even the musically unsophisticated can distinguish the same note played by: guitar, violin, flute, clarinet, trumpet or piano. The operator will rapidly learn to identify the particular sound that represents his area of responsibility, and, just as important, to ignore all others.

6. Identification of Cause

The operator should be able quickly to identify the cause of a problem that resulted in an alarm.

The way in which the alarms are presented to the operator is of utmost importance in determining how effectively he can take the appropriate corrective action in a process upset. Three main mechanisms are typically used in a DCS for identification of the alarm:

- **Current Alarm List:** This may be sorted by time, with the most recent alarm at the top, or by priority. It requires the operator to read text to be able to identify the event, and is the slowest and least informative of the mechanisms available. In many cases, alarms are attached to controllers or indicators that have other purposes. The descriptive text of the controller may not be the most appropriate for the alarm: a low flow alarm on a unit feed controller may indicate a pump failure.
- **Rectangular Grid:** This mimics the traditional panel-mounted alarm box. It is said that an experienced operator can glance at the panel and recognise a familiar event by the pattern of alarms illuminated. Studies have shown that pattern recognition is quicker to interpret than the reading of text. Some DCSs have many pages of such rectangular grids to display all the alarms; to perform such pattern recognition presupposes knowledge of which of several identical grids are on display.
- **Alarm Interpretation via Schematics:** Many DCSs employ process schematic diagrams as the primary route to interpret alarms. Each unit or area has a schematic diagram that is unique and clearly distinguishable from all others. Schematics should follow the "cool screen" concept where the normal situation appears in subdued shades of the cool colours:

blues greens and greys, and alarms are portrayed by hot, bright colours that highlight the element in alarm. (The "cool screen" display concept is further described in the appendix: "Display Considerations".) Alarm icon shapes may be used to indicate the type of alarm, (thermometer for temperature, gauge for pressure, &c). The location and function of the alarm are thus instantly visible and easily interpretable.

- **First-up Identification:** There are many situations where the consequences of an event include many alarms that are also possible causes of that event, (e.g. compressor trip). The whole sequence of events may last only a few tens of milliseconds. Correct identification of the first of the alarms is key to identifying the cause of the event. Traditional panel-based annunciators provided this feature by a different blink-rate. Some DCSs provide sequence-of-events journals which have to be searched to identify the cause, a tedious task. Other DCSs provide nothing. This is discussed further in the Appendix: "First-up Identification".

Whichever mechanism is used, the operator should be able to navigate to the display identifying the alarm as an immediate response to the audible warning. Ideally, this should be a single action.

7. Situation Assessment:

The operator should be able quickly to reach a sufficient understanding of the whole plant situation that he is able to make the correct alarm response decision.

The correct action following an alarm may well vary according to the state of the remainder of the plant. Besides providing the facilities for an operator to handle a single alarm efficiently and effectively, the system also has to respond in a sensible way when more than one alarm, (and perhaps many), is present.

Much human factors research has shown that a pattern recognition system allows an operator most quickly and accurately to assess a complex situation. As with the individual alarm identification, the DCS provides three main methods of assessing the whole plant situation:

- **Current Alarm List:** This may be sorted by time, with the most recent alarm at the top, or by priority. Whilst the size of the list may be an approximate indication of the state of the whole plant, it may require the reading of many lines of text to obtain a sufficient understanding of the whole plant situation. Not only is this a very slow mechanism for single alarm diagnosis, it is particularly unsatisfactory for appraisal of the overall situation.
- **Rectangular Grid:** A DCS often provides a single rectangular grid giving an overview of the whole plant, each box representing a unit with colours to indicate the alarm states. An operator quickly learns which box represents which process unit. The boxes can only provide alarm information, though, and situation assessment may require other considerations.
- **Alarm Over-view Schematic:** Many DCSs employ a process over-view schematic for situation assessment. Each process unit is represented by a clearly identifiable icon whose colour and/or shape is changed both according to the alarms present in it and by its state or operating mode. (For instance: the fill colour represents the alarm state and the outline colour is different for each of: off-line, stand-by, in-service and regeneration.) Process units are inter-linked by lines representing the main process streams. In addition, a small number of key process parameters are displayed to characterise performance, behaviour or state of each unit. This approach makes it possible to convey far more information than with simple boxes, yet with an effective pattern-recognition capability.

Some installations reserve a screen, or a window on a screen, for the plant over-view schematic, (see appendix: Display Considerations). The operator is thus constantly aware of the state of the whole

plant. Because he does not have to call the display and then interpret it, this can save an important amount of time in responding to an upset condition.

8. Corrective Action:

The operator must be able quickly to access the relevant schematic display and perform the appropriate corrective actions in response to an alarm.

When a new alarm occurs, having identified the cause and assessed the situation on the plant, the operator is now able to take the appropriate corrective actions. He needs to navigate quickly to the DCS schematic where those actions are performed, and to be able to perform them quickly and accurately.

There may optionally be an easily accessible reminder of the corrective action appropriate to that particular alarm. This can take the form of "Help" displays or windows. This is particularly useful for events that occur infrequently, and for inexperienced operators.

There should be a mechanism to remind the operator of which alarms are still waiting for attention. When several alarms are present, it may be difficult to identify which have already been dealt with and which are still waiting for attention. This is especially important at a shift change, but may also be important in a complex upset situation or when two or more operators are involved in the corrective action process. This is discussed further in the appendix: "Alarm Acknowledgement Practice".

There are a number of DCS features that have relevance to this issue:

- **Annunciator Buttons and Display Navigation:** Many DCSs provide annunciator buttons with LEDs that light up or flash when there is an alarm on their associated schematic and that call the schematic when pressed. There are rarely enough of these buttons to satisfy the requirements of larger plants and some systems have none; so some further display navigation facilities are required. The system should be configured to make display navigation as direct and efficient as possible. Some DCSs provide a single button to call the schematic associated with the highest priority unacknowledged alarm.
- **Task-Specific Displays:** It is clearly inefficient if some corrective action procedures require access to a number of schematics. Comprehensive use of task-specific displays will expedite the corrective action procedures following an alarm.
- **Alarm Acknowledgement Practice:** The list of unacknowledged alarms can provide a valuable check on what corrective actions have not yet been performed. To use this, the operator acknowledges each alarm from the corrective action schematic when he has completed the action. Further consideration of this subject is given in the appendix: "Alarm Acknowledgement Practice".
- **Links to Operating Procedures:** The use of 'help' displays or hypertext documents allow easy operator access to a check-list of alarm response actions.
- **DCS System Alarms:** Every DCS has a number of alarms associated with its own internal self-checking. They also require operator actions that may range from putting control valves onto by-pass to telephoning for call-out assistance or even shutting down the whole plant. These alarms need to be prioritised, disabled and associated with schematics in the same way as any other; they should be incorporated seamlessly into the overall alarm handling philosophy of the plant rather than having an entirely separate alarm handling mechanism.

The operator can clearly respond more quickly if the corrective action schematic is the same display that he initially called to identify the alarm.

9. Operator Work-Load:

The operator should not be presented with alarms at a faster rate than he can take appropriate action.

When an operator becomes overloaded, the effects are sudden, unpredictable, and plant-wide in scope, possibly resulting in multiple escalating process problems. A person in a state of panic makes irrational decisions. Traditional safety assessments take little account of the rate at which things happen, and assume no common mode failures, an assumption that cannot be made in this case.

It is common practice for safety studies and incident enquiries to deal with an identified problem by adding more alarms without fully considering the overall impact on the operator's task. A number of reports into major incidents have cited operator information overload as a key factor in the escalation of the event. If avoidance of disaster requires the operator to respond to alarms faster than he is capable of, disaster will occur and no amount of additional alarms will prevent it.

An acceptable rate of new alarms clearly depends on the response required from the operator, and must be considered for each case. Some studies have proposed the following guide-line for new alarm rates that the operator can properly deal with in a disturbance: a mean of 2 per minute with an occasional peak of 15 per minute.

The sequence of events that follow an alarm has been identified above. It is clear from this that each alarm requires a certain time for the operator's required response. It is, therefore, incumbent on those that perform process design, safety studies and incident enquiries to consider the implications of their decisions for new alarms on the totality of the operator work-load. In particular, the rate of occurrence of alarms should be quantified for each of the events considered. Where this exceeds the operator's capacity to deal with the alarms, design changes should be made until this criterion is satisfied. The design changes considered may include:

- process design;
- automation of some required responses;
- increase to the number of operators;
- DCS changes: additional screens, faster response;
- alarm reduction, automatic alarm suppression and dynamic prioritisation.

10. System Management:

The objectives of DCS system management are:

- ***safe and efficient operation of the plant;***
- ***prevention of upsets;***
- ***rapid recovery from disturbances;***
- ***facilitation of the operator's task.***

In particular, a significant part of the responsibility of the engineer in performing system management is to ensure that the issues discussed in the previous sections are fully satisfied. To do this effectively, a number of facilities are required:

- **Alarm Response Manual:** The alarm response manual is a record of everything in the alarm system. It contains, for each alarm:
  - ▹ Tag number and duty description: vessel location and sensor type.
  - ▹ Type of alarm: deviation, rate-of-change, high or low absolute.
  - ▹ Required operator response: immediate action on the alarm, and how this varies with the state of the plant.

- ▹ Justification: why it is there, with references to safety study reports, incident reports, &c., with a summary of the possible consequences of failing to respond correctly to the alarm.
- ▹ Configuration parameters: threshold, dead-band, priority, message text.
- ▹ Maintenance procedures: required testing frequencies, test methods and safety precautions.

- **Performance Monitoring:** The system engineer must regularly monitor the performance of the alarm system and of the operator's response to it. For this task, the alarm and discrete event history on the DCS should be treated as a database from which enquiries can be made and reports generated in a fully flexible manner. It is necessary to make use of this data for:
  - ▹ alarm auditing, a necessary accompaniment to any plant safety audit;
  - ▹ alarm rate monitoring, to ensure that information overload is not a risk;
  - ▹ post event analysis, for determining the course of events during a plant upset;
  - ▹ analysis of alarm frequency during normal operation to identify nuisance alarms, alarms with inappropriate thresholds, &c;
  - ▹ identification of standing alarms, so that steps can be taken to eliminate them, e.g. by automatic suppression mechanisms;
  - ▹ monitoring of alarms that are suppressed or inoperative, to check whether any have been suppressed which should not be.
- **Access Security and Audit Trail:** Many alarms have significant safety implications. It is required that access to change alarm settings or to suppress alarms be limited to authorised personnel, and that an audit trail be kept of all modifications to alarm parameters. The audit trail should identify the author of any change, and provide efficient enquiry mechanisms for searching and reporting the change history of any single alarm point or set of alarms.

Appendix 1. Typical DCS Alarm Handling:

The handling of alarms differ in detail between different DCSs. The following is a typical example.

- The DCS detects the parameter going into the alarm state.
  - Audible warning: A bleeper, horn or bell starts to sound and continues until the operator silences it.
  - Visual indication: A lamp or LED on the key-board blinks.
  - Journal: The event is recorded in the DCS historian system with the time and date, normally resolved to one second.
  - Print: The event generates a message that is spooled to a printer.
  - Graphic Display: The parameter changes colour and blinks where it appears on a graphical screen display.
  - Alarm List: The event appears in a special display called the current alarm list.
- The operator can press a button to silence the audible warning
- The display of the parameter continues to blink until the operator selects the item and presses a button to acknowledge it. Alarm acknowledgement may be recorded in the history.
  - If the parameter is still in the alarm state, it retains its alarm state colour but stops blinking.
  - If the parameter has returned to normal, the parameter stops blinking and returns to normal colour.
- If the parameter returns to the normal state before it has been acknowledged, the display of the parameter continues to blink in the alarm colour until acknowledgement. The time that the parameter returned to its normal state is recorded in the history.

Appendix 2. Alarm Acknowledgement Practice:

The current practice on alarm acknowledgement amongst many users is for the operator to acknowledge the alarm when he has identified it. There have been a number of incidents in which a large quantity of alarms has occurred in a complex situation; some of the less urgent alarms have been left to be dealt with later and have then been forgotten, resulting in an unnecessary escalation of the problem. This most often occurs at a shift change-over where the incoming shift makes different assumptions about the handling of the situation from the outgoing one.

Consideration should be given to the use of the alarm acknowledge status as a reminder of which alarms have been dealt with and which still require attention. For this, operators should be trained to leave each alarm unacknowledged until the associated corrective action has been performed. To assist with this, the only means of acknowledging an alarm is through a single point acceptance mechanism provided on the task-specific display where the alarm response is performed. This concept, that silencing the audible warning and acknowledging the alarm are fundamentally different actions, needs careful consideration.

If this is done, a prioritised list of unacknowledged alarms is a simple and immediate check on what actions are still outstanding. The highest priority unacknowledged alarm is thus the next item for the operator to deal with.

Some DCSs provide alarm acceptance mechanisms that cause all the active alarms on a schematic or on an alarm summary display to be acknowledged at a single key-stroke. This mechanism should be disabled if alarm acknowledgement is to be used in the way proposed here. For some purposes, however, (e.g. system testing), it may be desirable to have an additional group acknowledgement mechanism, perhaps under supervisor access privileges or key-lock.

Appendix 3. First-Up Identification:

There are many situations where the consequences of an event include many alarms that are also possible causes of that event, (e.g. compressor trip). The sequence of events between the first alarm occurring and the consequential alarms being active may take only a few tens of milliseconds. The alarms in question are all associated with events in a safety shut-down system. Identification of the alarm that occurred first is thus the identification of the cause of the problem and is, therefore, important to the operator in determining the correct course of action to take.

Traditional panel-based annunciator systems provided "first-up" indication as a standard option. DCSs, in general, do not provide this facility. The nearest approach is a "sequence of events" journal. This has the disadvantage that the journal may be very long with an admixture of events unrelated to the item of equipment of interest. To search the journal and identify the cause of the event may thus be tedious and time-consuming process.

Rapid identification of cause can often prevent problem escalation and allow rapid recovery. It would facilitate operator diagnosis of cause if the DCS were to provide first-up indication through the schematic displays. The requirements for the DCS to provide this are:

- millisecond discrimination of discrete event occurrence;
- automatic identification of the first occurrence within the group of alarms associated with the process unit;
- the ability temporarily to deselect faulty or other unwanted alarms from this group;
- discrimination of the first-up alarm on a graphical display by blink-rate, colour or icon.

Some of the required information on cause may be available by direct communication with the safety system.

Appendix 4. Display Considerations:

It should be clear from the above that DCS alarm handling has a fundamental impact on overall schematic display design. It is not the place of this document to provide a full definition of display design, but a number of implications derive directly from the above and are summarised here.

The "cool screen" concept requires that normal information is displayed in cool dull colours and that the hot bright colours are reserved for abnormal states. Thus, when a DCS schematic appears on the screen, the operator can see an abnormality at first glance; his eye is immediately drawn to the problem.

Blinking is used only for unacknowledged alarms; (the use of a blinking icon to indicate the perfectly normal and expected operation of an agitator is to be deplored). Systems that allow varying blink frequency may use this to denote priority, (a fast blink indicating a high priority), or to indicate the first-up alarm. Moving graphic objects should be reserved for situations that really justify their use, e.g. to indicate actual vehicle movements. Constant irrelevant movement on a screen is both tiring and distracting.

If text is made to blink, (to change from the invisible to the visible state), it takes the operator twice as long to read it. It is better to blink the back-ground of the text, or a box around it, or an object or icon beside it, and to leave the text steady. It should be noted that numbers or characters provide serial input to the operator and are slow to read; icons are faster sources of information and colours are fastest of all.

The use of colour for alarm states must be consistent across the plant, and should conform to people's expectations. Colour may be used to denote alarm priority, or for classification of alarm type, (for instance magenta may be reserved for instrument or system faults).

Alarms associated with parameters whose value is displayed can be denoted by a change to a brighter, hotter colour of the value text or of its back-ground or both. Where appropriate, one or two characters to distinguish the type of alarm, (high, low, deviation, rate, fault), can be displayed beside the value.

Alarms associated with parameters that are not available as analogue values in the DCS may be displayed either by short but unambiguous identifying text or by an icon or symbol, (e.g. thermometer for temperature, gauge for pressure). These may normally be invisible and become visible when the alarm is present, or they may change shape or colour or both. A key or touch-target should be provided to make invisible items temporarily visible for diagnostic and system management purposes.

Facilities should be provided for rapid access to the schematics needed by the operator in the alarm handling process. It should be possible for each display to be called by a single button or screen target. Mechanisms should be provided for attaching operating instructions or other context-sensitive help information to these displays.

The process of alarm response consists of: identification, situation assessment and corrective action. Clearly this process can be considerably expedited if one screen or screen window is reserved for a plant over-view display and another for the alarm interpretation and response schematic so that both can be visible to the operator at the same time.

An optional mechanism should be provided to prevent multiple alarm acknowledgement and to restrict single-point alarm acknowledgement to the display in which the corrective action is taken.

The display database of a plant should be interpretable using SQL-type commands so that the system manager can quickly and easily identify what alarm tags are used in which display. This will allow the identification and removal of errors and inconsistencies such as:

- an alarm that is not displayed,
- an alarm that can be acknowledged in an inappropriate place.

Appendix 5. Handling of Non-alarm Information

An alarm has been defined as: "advice to the operator that immediate action is required". This implies that there are other signal types, for which the alarm system has been employed in the past, where another means must be found for meeting the operational requirements of the plant and the operator's needs.

There are three such signal types that may be classified as follows:

- **Messages:** Messages are used in sequential operations or control systems to advise the operator that a particular state or step in the process has been reached.
- **Warnings or notifications:** Warnings advise that a parameter has crossed a threshold that requires operator awareness but no action.
- **Status indications:** Status indications relate to items of equipment that have a number of distinct operational states. Examples include valves open or closed; pumps running or stopped.

In these cases, no action is required of the operator, so no alarm is required. However, the information provided may affect his decision on what actions to take when other events occur that do require his intervention.

Some DCSs provide a separate message handling system. This consists of a light on the keyboard or a flag on the screen that indicates when a new message is present. It is steady, (not blinking), as no alarm is involved, and there is usually no sound associated with it. Pressing the button or selecting the screen target causes the message display or window to appear on the screen. Unread messages may be highlighted on this screen to facilitate the operator's task when several messages could appear simultaneously.

Many users use the alarm system for warnings or notifications, configuring each point to suppress the audible warning and flashing indication. The process displays then use the alarm status to highlight points where such warnings are present.

Most users display status information primarily through the process schematics. The "cool-screen" concept should be consistently observed. In particular, only when the condition is abnormal and could lead to exacerbation or propagation of a disturbance are hot colours used for status information. Otherwise change of icon shape is used to denote the state. In a typical instance, valves or pumps have hollow bodies when the pump is stopped or the valve is closed, and have a solid fill of the same colour when the pump is running or the valve open.

Some users may want one or more of the above to be accompanied by a sound. It is proposed that this sound should be: of short duration, (self-cancelling and not requiring the operator to silence it); non-intrusive; and easily distinguished from the sounds associated with alarm annunciation.

Appendix 6. Terminology:

Wherever possible, standard UK English terminology is used. However, to prevent the constant tedious repetition of long descriptive phrases, some terms are used in technically specific ways; these are defined below:

Alarm Configuration Parameters: The settings within the DCS that define an alarm. The parameters may include the value of the measurement that triggers the alarm, the dead-band, the priority, the display and reporting functionality, the message text associated with the alarm, the associated schematic display, &c.

Alarm Handling: The term Alarm Handling is used to denote the mechanisms within the DCS that respond to the change of an alarm state. They include: warning sounds; changes to the appearance of displayed data; logging of information to printers or journals; illumination of warning indicators.

Alarm Management: The term Alarm Management includes alarm handling and alarm response, but also covers the organisational aspects such as: the process of defining which parameters on a plant have alarms and how they are configured; alarm response definition; the use of sound; lamps and display colours for alarm identification.

Alarm Response: This term denotes the actions required of an operator in response to an alarm.

Batch Plant: A plant that performs a sequence of process operations on discrete quantities of material.

Continuous Plant: A plant that features continuous flows of feed and product.

Common-Mode Failures: This term is used to denote a situation where several different and usually unrelated problems can be caused by the same single initiating event.

Dead-Band: The gap between the value of the measurement that turns the alarm on and the value that turns it off. A wide gap is less susceptible to small random fluctuations in the process value; too wide a gap can mean that some alarms are missed.

DCS The term DCS is used here to denote any screen-based control or information system including distributed control systems and SCADA systems.

Dynamic Prioritisation: A mechanism for changing the priority of an alarm according to the mode of the plant or unit, or according to the state of other equipment on the plant.

Hypertext documents: Can be used for operator help displays. Typically the first screen accessed by the operator consists of brief and simple text descriptions of the procedure. Some of this text is distinguished by colour. Clicking the mouse pointer on coloured text brings up additional windows with more information about the selected item. Graphical information may be included.

Intelligent Alarm Suppression: Similar to dynamic prioritisation, but simply suppressing alarms when they are not needed.

Multi-mode unit: Many plant units have several modes of operation. Some are phases in a sequence of operations; typical states are: shut-down, start-up, standby, operating, regeneration. Some plants produce a range of products through different equipment line-up or different process conditions; each product represents a different mode of operation.

Plant: A plant is a set of process equipment producing a single main product or product group. There may be several plants on an operating site.

Plant Area: A plant area is a section of a plant that is operated by a single team of operators, each of whom has responsibility covering the whole of the area.

Plant Unit: A plant unit is an element of a plant area performing a single conversion or separation task. A distillation column with its reboiler, condenser, reflux drum and associated control systems is generally considered to be a single plant unit.

Operator Help displays: Displays containing operating procedures designed to assist the operator in performing his task. They can include simple fixed text displays; context sensitive text and graphic displays and hypertext documents.

Process noise: Process noise is the random small fluctuations in the values of measurements or process parameters that have no real significance in production terms.

Schematic Display: A graphical representation on the DCS screen of the process containing dynamically modified text or icons that represent the values or states of analogue or discrete measurements.

SPC Package: Statistical process control uses statistical techniques to discriminate significant changes from the expected random variations due to process noise.

S.Q.L.: Structured Query Language. A language used for making database enquiries.

Appendix 7. List of Operator Interface Workshop Contributors

The following table is a complete list of the Operator Interface Work-shop participants that have contributed to the development of this document. Their most valuable contribution is hereby acknowledged.

Name	Company	Tel./Fax	E-mail
Jan Andreasson	Scanraff Sweden	Fax: +46 523 660972	Jan-Jan-437.Andreasson@scanraff.msmail.telemax.no
Raf Broers	BASF Antwerp	Tel: +32 3 5613942 Fax: +32 3 561 4189	Rafael.broers@notes.basant.be
Luc Bryssinck	Fina Refinery Antwerp	Tel: +32 3 5455608 Fax: +32 3 545 5690	luc.bryssinck@fina.be
Donald Campbell Brown	BP Oil Sunbury	Tel.: +44 1 932 763316 Fax: +44 1 932 763449	Campbedc@bp.com
Willem D. Hazenberg	BPB De Eendracht Karton	Tel.: +31 596622432 Fax.: +31 596620570	Willem.d.hazenberg@wxs.nl
Rudi Hörchert	Inspec Belgium	Tel.: +32 3 250 9352 Fax: +32 3 250 9359	Rudi.horchert@ineos.com
Henk Jeursen	Hoogovens Ijmuiden	Tel.: +31 251494632 Fax: +31 251470743	Henk_jeursen@hoogovens.e-mail.com
Tom Koeken	DSM/EdeA	Tel.: +31 464766104 Fax: +31 464766532	Tom.koeken@dsm_group.com
Franck Nowak	Borealis Beringen	Tel.: +32 11 459402	F_nowak@iname.com
Geert Pauwels	Inspec Belgium		Geert.pauwels@ineos.com
Christer Mattiasson	Scanraff Sweden	Tel.: +46 523 669444 Fax: +46 523 660972	Christer.mattiasson@lysekil.mail.telia.com
John McCulloch	BP Chemicals Grangemouth	Tel: +44 1324 49 3142 Fax: +44 1324 493990	Mccullj2@bp.com
Jan Smets	Fina Refinery Antwerp	Tel: +32 3 5455502 Fax: +32 3 545 5501	Jan.smets@fina.be
Ralph Ullrich	DEA Mineralöl Wesseling	Tel: +49 2236792421 Fax: +49 2236792260	Ralph.Ullrich@dea-ag.de
Roger Van Dyck	BASF Antwerp	Fax: +32 3 561 3420	Roger.van-dyck@notes.basant.be
Hennie van Staden	DSM	Tel: +31 464763456 Fax: +31 464763271	Hennie.staden_van@dsm_group.com

Ignace Verhamme	Honeywell PACE	Tel.: +32 2 728 2804 Fax: +32 2 728 2696	Ignace.verhamme@belgium.honeywell.com
Bert Woestenburg	Hoogovens Ijmuiden	Tel.: +31 251 494842 Fax: +31 25170443	Bert_woestenburg@hoogovens.e-mail.com

Appendix 8. References:

A large proportion of the documents that were consulted in the preparation of this Alarm Management Philosophy are only available within the originating companies; for this reason, no reference is given to these items. Below is a short-list of those related documents that are in the public domain.

Bransby ML and Jenkinson J, The Management of Alarm Systems, *HSE Contract Research Report 166/1998 ISBN 07176 15154,* First published 1998

Campbell Brown DC, Alarm Management Experience in BP Oil, *Proceedings of the IEE Colloquium 'Best Practices in Alarm Management',* March 1998

Emigholz, KF, Improving the Operator's Capabilities; Observations from the Control House, *Proceedings of the AIChE Loss Symposium,* July 1995

Lyche NP, Alarm Management and System Design, *Proceedings of the 7th Annual Honeywell Australasian IAC Users Group Conference,* 1995

McCulloch, JG, 1996, Alarm Handling and Future DCS Developments, *Proceedings of the Honeywell European Users Group,* 1996.

Nimmo, I, Abnormal Situations Management: Giving your Control System the Ability to 'Cope', *Honeywell Journal for Industrial Automation & Control, Aug '95*

R.N. Pikaar e.a.: Mensgericht ontwerp van Alarm Beheersings-Systemen; aanpak en richtlijnen: (Human centred design of alarm management-systems Approach and guidelines.) *ERGOS Rapport B186-r3.alm.*

ERGOS : Signaleringen en alarmfilosofie; literatuurstudie. *Rapport B181-R2.alm.*

Bill R. Rodgers & Fred S. Petry: CCPS : Expert systems in process safety. *ISBN 0-8169-0680-7*

Directoraat-Generaal van de arbeid (the Dutch Health, Safety and Environment authority): Richtlijnen voor de ergonomie van werkplekken, (Guidelines for ergonomics in workplaces). *S59 mei 1989 ISBN 0921-9218/2.09.059/8904*

APPENDIX 5

Overview of Alarm Management for Process Control

It takes 20 years to build a reputation and 5 minutes to ruin it. If you think about that, you'll do things differently.

—Warren Buffett

This book embodies the current best practices of process control alarm systems for industrial manufacturing facilities. It is a comprehensive guide developed to help you understand, design, evaluate, and use alarm systems. The coverage is accurate and complete and, at the same time, easy to grasp. The book contains all the "what is" information about alarm management so that you will fully understand. There is an extensive "how to" that you can use to perform every aspect of alarm system redesign. The style is low-key. The technology is down-to-earth, solid, and based on strong design fundamentals.

The book is divided into three parts. Part I covers the alarm management problem. Part II lays out the solution. Part III provides the pathway to make it all real. There are twelve chapters. There is a progression of the development of the work from chapter 1 through chapter 12 with each chapter covering a specific area. It is suggested that the reader cover the material in that order; however, each chapter has value if read separately. This can be especially useful for those with a working knowledge of the technology who are looking for more detail or greater depth in selected topics. The book will work quite well as a working guide for project planning and execution.

Certain chapters work well as stand-alone treatments. Chapter 2 "Abnormal Situations," Chapter 5 "Permission to Operate," and Chapter 12 "Situation Awareness" are designed with this in mind. These topics bring out aspects of enterprise management that go beyond alarm systems in their overall importance. They elevate alarm improvement effectiveness to a level of value capable of delivering demonstrable enterprise improvement.

Upon completion, you will have a clear and solid foundation of the purpose of alarms, the rationale behind the state-of-the-practice design, and sufficient "how-to" knowledge to competently perform in the technology. This book is designed to provide a basis for the following competencies:

- Know the underlying defining attributes and purpose of an alarm
- Appreciate the importance of effective alarm management
- Understand the proper use of process control alarm systems
- Recognize alarm system performance problems including alarm flood
- Become knowledgeable in the best practices for alarm system design
- Learn the utility of third-party alarm data diagnostic tools
- Understand the entire process for designing and executing an alarm improvement project
- Understand the importance of good graphic interface design on plant operability and provide effective situation awareness
- Become a qualified participant in alarm improvement teams

A5.1 THE CHAPTERS

Part I: The Alarm Management Problem

Chapter 1: Meet Alarm Management

Chapter 1 lays out the objectives of the book and the approach used for the development of the work. It includes a historical review of alarm management concerns. Of special importance is the perspective it illuminates of why now, of all times, plants are experiencing more alarm activations than ever before and why they are leading to diminished ability to effectively mange production. The eye-opening Milford Haven Accident of 1994 serves to illustrate the increasingly direct contribution of alarm system performance to plant operational integrity. The strikingly similar incident at Texas City in 2005 reminds us of how much more we need to progress to achieve effective plant operations.

The chapter continues with the fundamental importance of plant process dynamics to the difficulty of detecting plant production problems as they are evolving. It brings out the challenging dichotomy between the implementation of corrective actions and the actual (or subsequent lack of) improvement in plant operation. Symptoms of alarm management problems are laid out. The discussion continues with the express acknowledgment that the current poor design and use of contemporary alarm systems was an unfortunate result of many well-intentioned, knowledgeable, and conscious practitioners. Unfortunately, their work was done before we understood how important alarms were and how carefully the alarm system needed to be designed. The chapter closes by assuring the reader that a well-designed and effectively performing alarm system is within reach and obtainable by the application of easily understandable technology and standard engineering practices.

Chapter 2: Abnormal Situations

Chapter 2 lays the foundation for understanding what constitutes an abnormal situation. It moves on to review the impact of abnormal situations on plant operations. This discussion provides a fresh awareness of how difficult it is to operate plants safely and productively. Key to the understanding of the importance of this topic is the recognition of the substantial magnitude of operational integrity risks and resulting losses experienced routinely by industry. For many enterprises, the ability to better manage production operations will determine the difference between subsistence and profitability, between danger and safety, between a society menace and public acceptance.

There is a dearth of operator decision support tools and process condition information supplied to plant operators. Modern plants can be highly interactive and significantly complex. There is some degree of regularity in their operation, but there are often times when their operation moves into unfamiliar and challenging territory. Past history and operational experience provide little guidance for real understanding and managing these unfamiliar situations. The ability to provide deeper understanding and guidance to operators by utilizing advanced state information and decision-support tools is a key goal of abnormal situation management approaches currently implemented as well as those under development.

The chapter calls into question the current thinking regarding the simplicity of root causes by providing a new protocol for examining and certifying safe practices. Several safe-operation evaluation methodologies are introduced that serve to promote the benefits and reinforce the utility of the methodology. A typical “upset” event is examined to illustrate the existence and vital importance of time delays affecting the plant control equipment’s ability to detect problems, the operator’s need to understand and react to problems, and the plant’s ability to respond to corrective actions—all in a time frame sufficient to manage substantial (and potentially catastrophic) losses. Out from the shadow of possible disaster comes a pivotal essential benefit: a fundamental way to determine appropriate settings for alarm activation points.

Chapter 3: Strategy for Alarm Improvement

With the stage set for understanding and appreciating the alarm management problem within the larger context of abnormal situations, the current best practices for alarm system design and implementation are laid out. The formal defining attribute of an alarm is established: an alarm is any important event that the operator might miss and for which the operator must act. The work advances to a broad discussion of the true purposes of alarms and alarm systems. Along the way, we learn the differences between poor alarm system performance symptoms and the underlying root causes. The eight lessons of alarm improvement are presented for guidance and humility.

The chapter moves on to supply details of a structure for approaching an alarm management study. The details include the following: effective alarm management process, attributes of "good" alarms, characteristics of "good" alarms, the make up of an alarm management strategy, and some of the important technology useful for successful alarm improvement projects. It then lays out the formal structure of the alarm improvement process—the "big picture" if you will. It reviews each of the four phases of the work. The phases are further broken down into twelve tasks. The importance of identifying the "owners" and "stakeholders" is stressed.

Chapter 4: Alarm Performance

Production facilities accumulate considerable information that is very useful in determining how well the alarm system is performing. There are two important databases: performance and configuration. The performance data consist of the minute-by-minute captures of alarm activations, acknowledgments, operator actions, and the like. These are the *dynamic data.* The control configuration data consist of all aspects of alarms that appear in the process control system's configuration and are termed *static data.* Chapter 4 reviews the basic aspects of obtaining the data, analyzing the data, and assigning conclusions to the results via an alarm performance assessment (APA) or an alarm performance review (APR).

Best-practice benchmarks (configuration and performance metrics) are introduced as a method of placing plant data into perspective. The usual measures include the number of tags, tags with alarms, and priority of alarmed tags. This is followed by metrics for the activation data including priority spreads, alarm activation rates, patterns of activations, standing alarms, time-to-acknowledge, time-to-clear, number of disabled alarms, chattering alarms, repeating alarms, and the class of correlated or otherwise related alarms. Sample results are presented derived from using some of the very effective and easy-to-use commercial "tools."

Beyond the numbers, the data contained in operator event logs, production interruption reports, incident reports, maintenance logs, and operator experience can provide a wealth of additional information. For example, if the operator logs reflect a shift without unusual problems or events but the alarm data contain records of significant activity, there is the suggestion of unnecessary nuisance alarms. These data provide the situational backdrop to compare the dynamic alarm data against actual plant operation. Within this context, the data analyses may provide the relevance so important for engineering evaluation. Yet, at the end of the day, the examination of current alarm performance is useful only for motivating the real work of alarm improvement. That work will progress along fundamental lines designed to produce a well-developed operator support tool.

Part II: The Alarm Management Solution

Chapter 5: Permission to Operate

The critical decision an operator faces when the plant is operating abnormally is when and if to cease attempts to maintain operation and shut down the plant. Sadly, the investigation of most serious industrial manufacturing incidents identifies "failure to stop" as one of the primary causes for the escalation of a manageable incident into an unmanageable disaster. Often, in the heat of the upset, the innate desire to "manage the problem" fails to make room for the more important decision of whether to continue to attempt to manage. This chapter reviews the process by which a decision to operate in the face of uncertainty is raised to the level of a strategic tool. When the supporting procedures and technology are in place to facilitate such a decision, a significant reduction in operating losses can be realized. The chapter concludes with an extension of these principles to the remaining players in the manufacturing enterprise operation.

Chapter 6: Alarm Philosophy

Effective design cannot be done on a haphazard framework. Robust, functional, and useful systems are based on a clear design that incorporates the essence of their function. Alarm systems are no exception. Chapter 6 explains how the alarm philosophy is developed and later used to provide the foundation for alarm system design. The use of the term "philosophy" is a bit of a historical misnomer. Think of "alarm philosophy," or alarm management philosophy, as a one-to-one synonym for a "complete design basis of an alarm system."

The manufacturing team under the leadership of senior plant management develops the alarm philosophy. It describes the intended alarm design/redesign process from beginning to end. It includes the following:

- Buy-in of site management through the critical success factors and the necessary detailed engineering specifications to do the job
- Adequate financing
- Realistic schedules
- Cooperation and participation of other key site players
- Working definition of alarms; their proper response
- Other details of the alarm system
- Integration with maintenance, training, and the remaining plant infrastructure
- Specific path to implementation to provide post-implementation robustness and relevance

Chapter 7: Rationalization

Alarm improvement typically requires a significant reduction in the number of alarm activations to a manageable number. Unfortunately, managing alarm activations directly is not possible. The only way to change things is by the careful selection of what to alarm and the parameters of these alarms. Rationalization is this activity. It is the procedure where each alarm is selected and all the configuration information and supporting information is decided. It is the heart of alarm management. In chapter 7, the reader will learn the objectives of alarm rationalization, how to select the working teams, and what the steps are for the rationalization activity. Rationalization is a structured process. It includes how to decide which control points to alarm (including calculated and imputed variables), the alarm activation point(s), the setting of priority, and the remaining alarm response information including potential causes, appropriate operator responses, and likely consequences of error.

The objective is to create a process control system configuration with the right number of configured alarms. The new configuration should result in significantly fewer alarm activations. Additionally, those alarms that do activate will be more important and provide more useful operator guidance. There are two basic approaches: (1) start with the existing configuration or (2) start from zero.

Starting with the existing configuration is really starting from where the plant is now. It involves a complete review of all existing alarms. During the review the decision will be made to keep the alarm as is, keep the alarm but modify it, or eliminate the alarm completely. This process also identifies any new alarms that are needed.

Starting from zero begins with a blank slate (or blank sheet of paper, the "white sheet," if you will) for a plant with no configured alarms. The plant is divided into its primal set of smaller subsystems, both for ease of understanding and to facilitate the design process. For each subsystem, an analysis is conducted to decide the minimum number and type of alarms needed to properly manage it.

For both approaches, the result is to make sure that each configured alarm will be understandable by the operator, prioritized to enable operators to select which ones to resolve first, relevant to the abnormal situation, unique, and timely enough to permit the operator to resolve the situation. The priority is determined according to the plant's assessment of the impact of each alarm, if not managed successfully. The alarm activation point is determined based on plant dynamics and the difficulty of managing the situation. The process is completed with a full compilation of all the important information operators need to understand and manage the abnormal situation announced by the alarm.

A successful rationalization yields a significant reduction in the number of configured alarms—on the order of five- to tenfold, with a concomitant reduction in eventual alarm activations during operation. It is not unusual to completely eliminate alarms during normal operations, to have very few alarms during abnormal operations, and to have a manageable number of alarms during upsets.

Chapter 8: Enhanced Alarm Methods

With a successful rationalization completed, it is quite natural to want to move directly to implement that wonderful new design. As tempting as that might be, this desire fails to recognize an essential aspect of the redesign work done thus far. The plant is comprised of a large number of interconnected and interacting elements: piping, vessels, utilities, and such. Up to this point, however, our rationalized alarm system consists only of a collection of single-point entities. Each alarm is separate and distinct from every other one. No alarm activation is designed to take into account the activation of any other alarm. No alarm is controlled or modified to take into account whether it is needed by the plant outside its normal operating situation (spare, operational, or shutdown).

Chapter 8 reviews the additional supportive technology that will be added to the alarm system in order to arrive at a complete alarm management solution. The goal of this added technology is to ensure that each alarm activation is worthy of operator action *in all plant operating states and for all operating conditions.* To do this, information and alarm control actions that incorporate plant operating states and conditions will be utilized. Enhanced alarming (sometimes referred to as logical processing or advanced alarm management) considers relevant plant variables and situations *before* permitting an alarm to be activated. Enhanced processing is implemented using logic and other advanced computations.

The literature on enhanced alarming approaches it with a large number of different methods, each with a unique name. This gives the impression of a "Chinese Menu" method of use. It suggests enhanced alarming is managed by picking a few that appear to be useful. However, the value of enhanced alarming is how it is deployed. Understanding must precede deployment. Thus this chapter lays out a clear and logical framework. It is organized into four basic areas:

1. Operator-enabled suppression
2. Preconfigured, simple suppression
3. Informative assistance
4. Knowledge based

This framework provides a more useful approach to understanding the purpose and uses of alarm control methods than conventional nomenclature-dominated work. The new approach is then related to the conventional nomenclature for completeness of coverage. The three conventional categories are standard, advanced, and other. Standard techniques include grouping, suppression (itself involving five categories), and equipment under test. Advanced techniques include intelligent fault detection, pattern recognition, neural networks, fuzzy logic, and other knowledge-based and model-based reasoning. The last category, other techniques, involves shelving, release, auto shelving, and handling of repeating alarms.

Part III: Implementing Alarm Management

Chapter 9: Implementation

Once rationalization is complete and all required advanced alarming designed, the job of alarm improvement is almost done. The next step is to ensure that the newly designed alarm system be able to be fit back into the existing facility infrastructure. Chapter 9 covers the additional parts that are necessary to turn a good alarm system design into a good alarm system in operation. Included are the management-of-change (MOC) issues and the importance of follow-up studies to ensure proper operation.

Up until now, the new process control alarm system is designed but not yet implemented. All design approvals are in place. The essential technical personnel are on board and standing by ready to implement the design. This chapter goes over the plan for the actual implementation. Implementation includes the following: (a) translating the design for the alarms into the specific process control system's configuration code ready for download; (b) additions and modifications to the process graphics; (c) preparation and review of new training materials; (d) training; (e) updating documentation; (f) updating and putting into practice all other complementary plant infrastructure changes such as modified maintenance practices, incident investigations, and so on; (g) physical download and activation of the new configuration and graphics changes; and (h) operability review to ensure things are working as designed and "as designed" fulfills the requirements of the enterprise for good and safe operation.

The chapter points out the importance of selecting an implementation sequence that is both compatible with the process control equipment architecture (data highway segmentation and such) as well as consistent with operator responsibilities. The chapter concludes with the design and execution of a plan to continually assess alarm system performance—and of course, that means plant performance—to assure continued, long-term benefits. It will include a program to periodically audit the alarm performance and make recommendations and modifications for improvement.

Chapter 10: Life Cycle Management

No design will remain fully functional over time, no matter how clever its creation or how dedicated its installation. This chapter focuses on a plan to continually assess alarm system and plant performance to assure continued long-term benefits. It will include a program to periodically audit the alarm performance and make recommendations and modifications for improvement.

Additionally, life cycle maintenance captures unique aspects of alarm management that provide extremely valuable insight into the effectiveness of the entire range of operations effectiveness. Alarm activation is a singular event. Each signals a situation that was planned to recognize abnormal situations. Beyond that, the pattern of activations and the pattern of operator responses to those activations point to systemic and infrastructure limitations, faults, and failures. It can be used as a key facilitator for long-term production improvement.

Chapter 11: Project Development

Alarm redesign involves straightforward, uncomplicated technology. It asks for judgments that are already being made as part of the safe operation of industrial production plants. On the other hand, it involves a technology and approach that is new, and in many cases, counter to current practices. Because the technology is so new, the discussion of specific project organization has waited until there is a good understanding of both what needs to be done and how to do it. With this understanding will come the insights necessary to recognize success factors and organize work to achieve the desired results. At this point, the working construction of the actual alarm management project will be examined. It was delayed until now for a very pragmatic reason: we needed to understand the full scope and breadth of what is needed for a good design before it would be possible to explore how such a large project might actually be worked.

By now, the scope and broad effects of alarm system design and performance will be apparent. For traditional manufacturing facilities utilizing 1000 to 1500 control loops (counting control valves usually provides a good loop count) an alarm improvement project might take from one to three years to complete. That work will likely cost (total cost, including outside tools and expertise and most internal expenses) somewhere in the neighborhood of $600,000 to $1,400,000. This represents a large undertaking by any measure. Depending on approval authority of local managers, other available resources, and the work practices of purchasing departments, it might be necessary to structure projects into several smaller components. There are a number of ways to efficiently do work of this size. A single, large project can be done, of course. Or the enterprise can be divided into individual operator areas and a separate project done for each area. On the other hand, the work of alarm redesign can be divided into its logical phased parts and each phase done across the entire enterprise for each operator area in turn.

Chapter 12: Situation Awareness

We ask operators to manage a production facility. Within their area of responsibility and authority, they can view and adjust every control loop, most equipment, and related support utilities as they see fit. For their primary information, they are provided only simple values of all variables, sometimes augmented by desired targets and more complex calculated parameters. Key to the understanding of the importance of this topic is the recognition that there is a dearth of useful tools and technology readily available to assist operators. Operators are provided trend plots for selected variables. They are provided with sketchy operational status for many pieces of equipment. And, of course, they are provided with an alarm system. From these simple tools, we ask operators to know what must be going on deep inside their operating unit, decide what is normal and what is not and if not, figure out what to do to return it to normal, and then take only the proper corrective actions to cause the return to normal. Few plants are so readily understood, and few operators are so favored by such an unusual talent.

Chapter 12 reviews the emerging industrial practices that enhance operator awareness of the true nature and condition of the plant in their charge. We call this situation awareness. This increased awareness, in part, can be enhanced through careful design of operator graphics and navigation. This chapter reviews the current best practices in operator interface design, organization, and navigation.

APPENDIX 6

Alarm Response Sheet

Alarm Summary Data

Tag ID: FI-2009	Alarm Priority: High
Point Descriptor: PB Feedwater Flow	Alarm Status: Enabled
Comments:	Operating Group: 12
Setpoint: 195 Units: KLbHr	Process Area: PB

HIGH ALARM	LOW ALARM	
Alarm Point: 200	Alarm Point: 100	Alarm Point:
Range Limit: 275	Range Limit: 0	Maximum:
Causes:	**Causes:**	**Causes:**
1. Failure of LVC/LIC 2011 2. Plugged flow element 3. Incorrect steam flow measurement 4. Instrument failure	1. Low boiler feedwater header pressure 2. Control valve failure 3. High steam drum level 4. Piping failure in exchanger 5. BFW pump failure 6. Mechanical failure	

(continued on following page)

Confirmatory Actions:	Confirmatory Actions:	Confirmatory Actions:
1. Check steam drum level LIC 2011 2. Check strip charts, check "periscope" 3. Check BFW header pressure FI 2078/2077, DEA Pressure PI 2098/2099 4. Check BFW pumps 5. Check flow control valve FCV 2009	1. Check steam drum level LIC 2011, strip charts, periscope and level control valve 2. Check BFW header pressure PI 2078 3. Check for piping failures 4. Check BFW pumps	
Consequences of Not Acting:	**Consequences of Not Acting:**	**Consequences of Not Acting:**
1. Boiler could overheat and seriously damage tubes 2. Insufficient steam flow, high press	1. Boiler could overpressure and rupture 2. Insufficient steam flow, low press	
Automatic Actions:	**Automatic Actions:**	**Automatic Actions:**
None	None	None
Manual Corrective Actions:	**Manual Corrective Actions:**	**Manual Corrective Actions:**
1. If steam drum level and steam pressure are high, put feedwater valve in manual and begin to close it 2. If steam drum level is low, manually increase opening of feedwater valve	1. Start another pump 2. Put feedwater valve in manual and open 3. If drum level is high, check level control valve, wait out upset; may have a "swell" if load INCREASED, be ready to INCREASE BFW flow 4. If problem with exchangers, check PDIC 2075, open bypass FDCV 2075	
Safety Related; Testing Requirements		
None	HazOp 12-4453; Yearly Testing	

APPENDIX 7

Metrics and Key Performance Indicators

The best is the enemy of good enough.

—Russian folk wisdom

PART I: RECOMMENDED REQUIREMENTS FOR ANALYSIS TOOLS

A7.1 Purpose

This document will lay out the design basis for structured queries to be used to retrospectively interrogate the real-time plant data stream for the purpose of producing evaluative conclusions of the performance of the enterprise. The evaluation will provide an understanding of the nature of the operational performance of the distributed control system (DCS). It is done through the collection and analysis of operational alarm data and other related operating data. The understanding is then used to provide an impetus for improvements in the alarm system and related operational tools.

It is expected that these requirements would be coded into the prestructured query database tools of the alarm analysis kit. These queries are either built by the tool providers or facilitated by guidance from the providers for construction by the end user.

A7.2 Background

The first important step in the alarm improvement process is to be able to evaluate the current state of design and performance of the existing alarm system. This evaluation centers primarily in the area of understanding the degree to which the alarm system is activating and the nature of the activations as it relates to plant condition and operator activity. The environment includes capture of the real-time activity log data, an understating of the structural nature of the production facility, and the operational backdrop present at the time the data are collected.

A7.3 Analysis Types

This section lays out the general design of the queries. For every query constructed and every query executed, the tool must automatically annotate with the unique and unambiguous identification of the tool and explicit data range and data modifiers used to select the data. Moreover, annotation must include ability of the user to add notes to further identify the purpose of the analysis or include any other contemporaneous label or intent information useful later on. The notes can be "picked" from both a pre-used list as well as a separate pick list.

Time-Domain

A first step in understanding the data is done via an overview of the entire real-time data from an historical time line perspective.

On a single timeline, present each of the essential capture items sorted by item class. The key capture items include but are not limited to the following:

- Alarm activation
- Alarm acknowledgment
- Alarm clearing
- Operator manual actions (mode change for controllers, controller setpoint change, controller output changes, etc.)
- Collective calculations (number of active alarms, number of unacknowledged alarms, number of inhibited alarms, time between activation and acknowledgment, and so on). There will likely be, of course, many collective calculations running concurrently.
- Permit display of this timeline data using the hierarchical feature extraction tool display engine discussed later on in this document.

Frequency Domain

This is the conventional way of extracting data from the database based on a defined data selection process. Typical selections include time ranges and/or data type classifications. Both time ranges and data type ranges may be compound. And the compounding ability must include the ability to construct unions (in the set theoretic sense) of data selections, individually as well as a whole. A simple example would be to just choose days at random (or via any other conscious or whimsical way). Included in this selection would be a sufficient number of "and," "or," and/or "but not" and so on.

Once the frequency data is calculated, it must be capable of display via the sorted attributes. Simple examples include, "in order of most to least," and "least to most." Moreover, the depth of display must include all the data. In most cases, it will not be possible to display the entire range of sorted/calculated data. Each screen image is limited. Therefore it must be possible to selectively work down the data range to expose whatever level and extent desired (subject to screen display area only). Again, refer to the discussion later in this document of the hierarchical feature extraction tool.

All displayed data must contain navigation links to enable the viewer to link back contemporaneously to both the master database (which might include background information, configuration information, and rationalization information) as well as the event database from which it was extracted.

A7.4 Queries

Here the extensive database is examined with the intent of extracting useful information sufficient to further understanding.

Ensure that comparisons are made against standardized alarm metrics as well as any site-defined good practices.

Alarm Activations

What alarms activate, when, and how often? Included are attributes of priority and activation point.

Item	Analysis	Information discovered
1	How do the activations relate to the general run of time?	Relative alarm processing load of the operator. Relative health of production process as it is currently being operated versus as designed or as intended to be operated.
2	Do activation rates correlate to process operations? If so, are the levels of critical nature of the operations directly related to the number of alarm activations? Are the more complex or difficult to control parts of the operation prone to more alarm activations? Higher alarm priorities?	Any particular production activities that are especially prone to upsets—hence large or significant of alarms—means a design or operational deficiency. The alarm system might be used to mask or otherwise inappropriately compensate for these inadequacies. If critical nature or complex nature is not related to alarm activations or/and not related to alarm activation priority, then this suggests a haphazard use of alarms and/or setting of priority and/or setting of limits.

3	What, if any, are the periods of high alarm activation rates (floods)?	If floods occur, do they relate to real events? If so, then the alarm system can be redesigned to aid this situation. If floods do not relate to real events in any substantial way, or occur often, then it suggests that there are substantial underlying plant design or operational deficiencies not amenable to alarm system improvement until fixed.
4	Are floods followed by short or long periods of process unrest/rest?	If followed by periods of unrest, then the process design or operational guidelines are inadequate or the process is inherently unreliable. If followed by periods of rest, then process somehow slipped into a difficulty; alarm system probably is basically sound, but improvement might be had by better alarming to either (a) uncover a problem situation before it becomes too serious, or (b) help in problem discovery and/or diagnosis.
5	Are floods preceded by short or long periods of process rest/unrest?	If preceded by long periods of rest, then alarm redesign should improve alarm flood management. If preceded by short periods of rest, then alarm system is probably a contributor to the problem and serious redesign is needed or process is significantly inadequately designed or operated.
6	Are there periods of "showers" where a high number of alarms are occurring?	Setting a threshold for "shower" to be the number of alarm activations occurring within a rolling time period based on EEMUA guidelines. Specifically, alarm activations occurring in excess of x per y minutes for at least z times y minutes constitutes the query. Use x = 5, 10, 20, 50; y = 1, 10, 30; z = 2, 10. Incidentally, this same template should be used to define flood. The parameters should be established to meet the desired thresholds. See Floods vs. Showers matrix later on.

7	Are there significant periods of little or no alarm activations?	This might indicate specific operating conditions or shifts where the activity is significantly different from the norm. This might also indicate situations where most if not all of the annoying alarms are either turned off (manually or automatically inhibited) or not challenged. If not challenged, something very unusual is going on.

Alarm Acknowledgments

What alarms are acknowledged and when?

Item	Analysis	Information discovered
1	Alarm acknowledges matches or parallels activations	If generally matches alarm activation, then operator understands and respects alarm system.
2	Alarm acknowledges doesn't match or parallel activations	If this varies noticeable from activations then it suggests that certain situations either distract operator from being able to service the alarm system or certain alarm events are of less consequence or interest that others. Suggests no consistent operational and alarm philosophy.

Alarm Clearings

What alarms cleared and when?

Item	Analysis	Information discovered
1	The patterns of alarm clearings generally match alarm activations.	Alarm system is generally in "balance." This does not mean that the design is okay, however.
2	Different pattern of alarm clearings than activations.	If consistent with difficulty of operations or significance of upset, then alarm system is generally in harmony, but still should be redesigned based on other alarm benchmark data. If inconsistent with difficulty of operations or significance of upset, it is strongly suggestive of no consistent alarm philosophy.

Time to Acknowledge

What alarms acknowledged and when?

Item	Analysis	Information discovered
1	Alarms acknowledged quickly	Operator is on the job. Alarm is important or alarm sound annoying. Operator is very lightly loaded and has little else to do. Operator has been overly sensitized to acknowledge alarms above other duties.
2	Alarms acknowledged slowly	Alarms generally of little importance. Operator is consistently overloaded. Plant has de-emphasized the importance of the alarms or the entire alarm system in general.
3	Alarms acknowledged haphazardly	Operator is sometimes very overloaded. Alarms are generally of little importance. Operator or plant culture is indifferent to alarm system and might obtain key information from other sources.

Time to Clear

What alarms cleared and how long did they take to clear? What is the relationship between the time to acknowledge and time to clear, if any?

Item	Analysis	Information discovered
1	Promptly acknowledged but takes a long time for alarms to clear	Suggestive of either a difficult to control process or inadequate operator diagnostic support. Operator too busy. Alarm acknowledgment done as a reflex, but operator cannot service the situation until later.
2	Alarms are not necessarily acknowledged promptly, but it takes a short time for the alarm to clear after being acknowledged, or after a return to normal before being acknowledged.	Alarm priority not representative of process needs. Alarm activation occurs too late for early correction. So the situations that lead to the problems that the activations announce are harder to manage. Process situation relatively minor and easy to fix. Probably could have been avoided in the first place by better operator management. Alarms are relatively unimportant. Operator too busy.
3	Time to clear is related to nature of upset	Normal operation of a well-designed alarm system.

4	Time to clear is not very much related to the nature of upset or particular alarm.	Alarm limits not properly set. Operator not sufficiently trained to handle problems. Alarm system not much help in managing upsets.

Time in Alarm

How long was it in alarm?

Item	Analysis	Information discovered
1	Some alarms around for a long time.	Unnecessary alarm. Most alarms are around for a long time. No alarm philosophy. Alarm important under some situations but not others, hence logic required to manage situation.
2	Few alarms around for a long time. Most alarms remain for a time that reflects their critical nature.	Normal operation of a well-designed alarm system.

A7.5 Alarm Remediation Analyses

The following alarm analysis categories represent both indications of insufficient alarm system design as well as more direct opportunities for early alarm improvement. A close examination of each alarm that stands out in the following examinations will usually lead to specific alarm configuration changes. Specifically, it will strongly suggest alarms that can be outright eliminated, alarms that might be combined with other alarms in a logic-related activation sequence, or alarms that point to specific underlying alarm system deficiencies that have specific remedy.

Standing Alarms

Item	Analysis	Information discovered
1	Alarm remains active long after activation and operator acknowledgement.	Alarm is unimportant. Alarm reflects a situation that the operator does not understand, but believes not to be urgent. Alarm reflects a situation that cannot be repaired or modified. Alarm reflects a situation that should not be alarmed; for example, used as a message.

Chattering Alarms

Item	Analysis	Information discovered
1	Alarm activates and then at times clears rapidly, other times not.	Normal operation of a well-designed alarm system.
2	Alarm activates and then often clears rapidly.	Improper alarm limit. Equipment or process problem that is not managed properly (faulty sensor, wiring problem, process design flaw, etc.)
3	Same alarm activates and then often takes a while to clear.	Production operating too near physical limitations. Control system not adequately configured or not sufficiently sophisticated for the situation.

Consequential Alarms

Item	Analysis	Information discovered
1	Alarm activation often followed by a predictable companion alarm activation. Alarm activation often followed by a predictable companion alarm, but the companion might differ from time to time or situation to situation.	Under specific situations, the companion alarm can be suppressed.
2	Alarm activation almost always followed by a predictable companion alarm activation. Alarm activation almost always followed by a predictable companion alarm, but the companion might differ from time to time or situation to situation.	The companion alarm is probably not necessary and should be eliminated.

Related Alarms

Item	Analysis	Information discovered
1	Alarm activation often followed by a predictable companion alarm activation. Alarm activation often followed by a predictable companion alarm, but the companion might differ from time to time or situation to situation.	Under specific situations, the companion alarm can be suppressed.

2	Alarm activation almost always accompanied by a predictable companion alarm activation. Alarm activation almost always accompanied by a predictable companion alarm, but the companion might differ from time to time or situation to situation.	One of the alarms is probably not necessary and should be eliminated. Retain the one with the most intuitive value.

Alarm Floods, Showers, and Drizzles

It is important to be able to identify time periods when significantly more alarms are occurring. This serves to both point to time periods when the operator might be very busy as well as to identify important performance problems with the alarm system itself. It is easy to think that if a lot of alarms are activating, then we have a shower. In a parallel manner, if a tremendous numbers of alarms are activating, then we have a flood. But these qualitative ideas must be reduced to tangible questions to ask the data. There are two "tuning" parameters associated with alarm activation database queries:

1. Number of alarm activations per minute (APM)
2. Duration, in minutes, of the base time period window (W)

A **drizzle** identifies those time periods during which the alarm system is indicating nothing much unusual about the production unit's health, but placing an annoying level of stress load on the operator. In general, a comprehensive "housekeeping" cleaning up of the controls and instrumentation and the elimination of the bad actors in the alarm system will be sufficient to dry up most of the drizzle periods. Those remaining will likely be associated with process movements (grade changes, equipment swings, etc.).

A **shower** identifies that time period during which the alarm system implies that the production facility is in a significant state of upset or abnormality. Things might be really wrong, but likely disaster is not imminent. Controlling of showers requires effort that goes quite deep into the alarm system's design. Generally, the efforts to control alarm flood will also significantly impact alarm showers. In fact, in a fully rationalized alarm system, a shower will usually occur only during severe production abnormality—that time period when the old alarm system design would have resulted in one or more floods!

A **flood** is used to identify that time during which the number of alarm activations is so large that the operator has no chance of understanding their meaning or properly responding to their importance. Basically, at this time, the alarm system becomes a problem in its own right, thus competing for the operator's attention and limited resources more needed to manage the production upset.

Designation	Number of alarms/ minute [APM]	Duration of calculation window [W]	Remarks
Drizzle	0.4 up to 1	10 minutes	
Shower	> 1 up to < 10	10 minutes	
Flood	≥ 10	10 minutes or more	Ends only when number of alarms drops below five in the time period

A7.6 Tools and Key Features

This section will describe special tools and/or features that need to be part of the tool kit design basis. Without these in place, the user either will have to work too hard to achieve reasonable benefits or will not be able to achieve certain analyses without extraordinary efforts.

Audit Log/Batch Construction

This is a usability feature that can also be used to produce a useful production tool. The archiving of both the audit log and the batches are kept in user-named files. Thus each user can arrange for the tool to perform and follow what he is interested in. He will not have to go back and undo others' work or undo his own if he goes down a blind alley.

Audit Log

Consider a user session as follows. A user opens up the diagnostic tool. Starting right then, the tool starts an audit of everything that the user does. It logs the database that it is connected to. It logs the diagnostic setup parameters, it logs the tests required, it logs (stores in a way reminiscent to the Windows XP "Restore Point") the results of the test. It goes on and on, following the user as he moves through the session. At the end of a session, or at the end of a certain length of inactivity, it will archive the session and give it a standard, well-understood name via a naming convention. The user, as part of a user-unique set of preferences, will structure this naming convention.

But this tool is far from described. For it is not just a text log, it is in fact a structured, object-oriented bidirectional pathway that can be examined and used. For example, after a session is complete, the user must be able to jump into the log and click on a specific step. This can also be done during a session, so don't just assume that the logs will not be iteratively used and entered during a session; except during session, the log part will suspend, of course. At that jump point, a separate frame will open and show the same results that were shown when the user was there the time the log was created. The user should be able to do all the things that he could have done had he arrived at this point during a primary session.

In addition, as the steps are logged, automatic annotation must be created that adequately describe what is being done and whether it is a primary step or just a secondary step. As an example, if the user produces a graph, then in addition to the text description, a "thumbnail" view will be produced to provide enough visual content to serve as a mini preview. Primary steps are the actual production of a diagnostic and/or the further refinement of the diagnostic as the user "homes in" on what he is looking for or trying to discover. Secondary steps are those where the user is setting up a diagnosis.

Batch Construction

Using the audit trail as an input, it must be possible for the user to "import" any audit, edit the audit, cut and paste edited parts of other audits, and produce a batch diagnostic command sequence. This sequence will be named and sequestered in a batch database. It must be possible for a batch to include other batches within itself, so long as the construction is not iterative. This batch will look exactly like an audit log but with a significant difference. The audit log actually points to a specific set of plant data; the batch will not necessarily. Therefore items like specific time ranges selected, alarm activation or configuration data selected, any shifts and such will appear either as hard coded or parametric. Refer to the cell construction capabilities of Excel for reference. There cells can be coded to be relative or explicit data, as with the batch.

The archiving, naming, and using of the batch commands will be done via conventional tool construction methods. Note that the invocation of a batch and/or a sequences of batches also interspersed with specific nonbatch tests will be captured via an audit log.

Parking Lot/Bookmark

Look at this as a (potentially set of) singular archiving of any specific query result. Again, it should auto annotate and auto name. Again, that function should follow the preferences set up by that user. This tool is especially useful during exploratory sessions where the user is trying to "get a handle on" the data. Note that bookmarks will not integrate into either audit logs or batches inasmuch as the trail leading up to the specific diagnosis will not have been captured. So while they will be "noted" in the audit log, it will be a notice-only entry and not linked to anything else.

Hierarchical Feature Extraction

This tool is designed to permit all important deviations in the display of parameters to be visible regardless of the extent of time compression of the display time axis. Think of it as an overview of the entire time series data. However, it is a special overview—one that exposes each and all anomalies that occur, regardless of the extent and magnitude of the event.

Construction

First, select the default time axis quantization level. Basically, this is the largest number of individual discrete time "buckets" on the abscissa that each can have a separate distinct ordinate value. Were the ordinate values to be displayed as bars, then the number of quantizations would be equal to the number of bars displayed, each at a separate time (though some of the bars might have a zero value, meaning zero height). This selection becomes the default **maximum** resolution of the time axis for all graphs and numeric data. But keep in mind that the actual time value assigned to each time "bucket" is an operational determination that is decided on contemporaneously. For example, if the total number of time buckets are 100, and the total time scale is 100 days, then each time bucket represents one day (full 24 hours, no matter how the day start and end is defined).

Second, decide the nature of the hierarchical levels. Simply stated, each level is determined in such a way to permit successively more narrow time windows to be viewed from the original data display, consistent with the overall data, yet provide significantly better discrimination of the data from a time perspective. In general, this decision will be natural and fixed at the ratio of 100::1 or 10::1 as the levels become finer. The first level is the entire timeline of the data. For example, the second level is 1/100th of the first level, and so there will be 100 of the second levels, ordered in the original time sequence of the first level. The third level is 1/100th of the second level, and so for each second level, there will be 100 third levels. And for completeness here, we state that there would be 10,000 third-level timeline graphs for each of the 100 first-level time buckets.

Third, determine the specific nature of the feature to be represented on the ordinate axis. Feature is defined as that single (ordinate) value that the tool will choose to represent all the (ordinate) data over the entire time period included in its "time bucket." If there are 627 data entries in this particular time bucket, then the feature will use 627 time value data points to calculate the single value representative of this specific individual time bucket. Remember that the number of time data points in each time bucket vary by time (along the time axis, since event messages are generated as needed, not by any sort of synchronous process) and by the hierarchical level of the particular view in which we are observing.

The features may be single or compound. Typical single-feature candidates include the following:

1. Average of all data points
2. Maximum or minimum of all data points
3. Median of all data points
4. Average of all data points, but replaced by the maximum (or minimum) whenever the maximum (or minimum) exceeds the average by a threshold percentage and so on.

Typical compound-feature candidates include the following:

1. Maximum and minimum of all data points

2. Average of all data points AND maximum and minimum of all data points
3. Maximum and minimum of all data points AND the average of all data points in the previous, for example, 5 time buckets and following 5 time buckets, and so on.

Fourth, establish the suite of display formats for the results. Since time is going to be fixed by the process laid out in the first step, this suite includes only ways to display the ordinate axes—that is, axes plural. Note that here one desires to construct a visual image sufficiently rich to encourage associative comparisons of the different data (e.g., alarm activation, alarm acknowledge, alarm clear, operator action; or perhaps alarm activation for low-priority alarms, alarm activation for high-priority alarms, and alarm activation for critical-priority alarms). Due to the two-dimensional constraints, this boils down into generally three categories:

1. Single time axis, single ordinate axis, and stacking the individual graphs one above the other, but with the scales on the ordinate axis repeating for each of the stacked graphs.
2. Single time axis, single ordinate axis and superimposing the (selected) individual graphs one atop the other. Clearly, this can be used only where there is sufficient ordinate separation of the graphs and each graph can be depicted by use of color and line representation sufficient to ensure understandability.
3. Single time axis, single ordinate axis for each individual graph and arranging all graphs in a "waterfall" configuration. Thus, each time and ordinate axis will be visible together with the data graphs, but the waterfall arrangement will facilitate comparisons of the various graphs.

PART II: METRICS

A7.7 Introduction

The following is a basic metrics table for identifying alarm assessment studies. The source is mostly the EEMUA 191 guide, although there are a number of others based on experience and an expansion of the guidelines.

These metrics apply to each operator console intended for a single operator. So long as the console is integrated and designed without completely isolated subdivisions, it is considered to be one console. An operator's area, or span of responsibility, constitutes the smallest plant unit for which alarm improvement projects should be considered and implemented. The user should observe that only the operating values of these metrics apply to subareas within an operator area as well. Configuration differences between subareas that don't naturally relate to differences in the nature of the individual operating areas can point to lack of consistency or lack of adequacy in the overall alarm philosophy.

The diagnostic tool should perform these analyses in a hierarchical manner—that is, the metrics should be able to be calculated from the results of a previous calculation in a sequentially "drilling down" into the data. The metrics are divided into the standard two categories:

1. Static or defined by the configuration and any configured logic or by explicit operator action that changes alarm-enabled status
2. Dynamic or computed from real-time operational databases that capture alarm activations, acknowledgments, and clearing as well as other operator actions

All metrics are to be calculated and interpreted as being per operator area (per console).

A7.8 Static (Configuration) Metrics

This section contains the definition and importance of metrics that are based on the control system configuration—that is, are based on the design of the controls as reflected by parameters within the control system infrastructure that determine the presence of an alarm. Whether that alarm ever activates is of no concern for these metrics.

Name	Definition	Authority	Significance to alarm management	Notes
Alarms (per control loop)	1 or fewer	EEMUA		
Alarms (per non-control measurement)	1 or fewer, for every 2 analog measurements	EEMUA		*Excludes* measurements directly feeding control loops (see alarms per control loop). *Excludes* highly repetitive measurements like arrays of tube-skin temperatures or multi-thermocouple arrays.
Alarms (per digital input)	1 or fewer, for every 5 digital inputs	EEMUA		

Alarms (by priority) Per Operator (about 300 loops)	EMERGENCY: fewer than 20 HIGH: 5% of total configured MEDIUM: 15% of total configured LOW: 80% of total configured	EEMUA		For three-priority levels, just ignore standard for "emergency." Alarm systems with more than four priorities are not recommended.
Disabled Alarms	1% or less but no more than 10			Alarms that have been disabled by explicit operator action.
Shelved Alarms	30 or fewer	EEMUA		Alarms that have been disabled by explicit operator action **Specifically related to a short time situation**.
Suppressed Alarms Inhibited Alarms	No limit			Includes only alarms that have been disabled by logic in order to reduce or eliminate unnecessary alarms and facilitate a "dark screen" (there is no operator involvement for this action)
Duplicate	none			Configured alarms that will always activate at the same time based on the configuration (assuming that they possess identical limits—different limits suggest very lax protocol)

A7.9 Dynamic (Activation) Metrics

This section contains the definition and importance of metrics that are based on the control system performance under actual operating conditions. Only those alarms and other related events that actually occur are considered and reflected in these metrics.

These metrics are used, both as individual events and as averages—that is, a site might wish to compute the number of consequential alarms within a selected data window as well as the average number of consequential alarms per any selected time period within that selected data window. For example, the selected data window might be a month. The desired key performance indicator (KPI) might be the average number of consequential alarms per day (or per shift). Most of the averages-type of metrics will not be specifically spelled out in the table below.

Name	Standard	Authority	Significance to alarm management	Definition and notes
Consequential	No standard; but working objective is 0.			One or more alarm activations that FOLLOW an (initiating) alarm activation more than 90% of the time. Percent is selectable, however 90% is recommended.
Repeating	No standard; working objective is 0.	D-RoTH		Ten or more activations of the same alarm within 15 minutes; regardless of whether any are acknowledged. This criteria is meant to capture annoying analog alarms; while chattering usually captures annoying digital alarms.
Chattering	No standard; working objective is 0.	TiPS PAS		Ten or more within one hundred seconds; regardless of whether any are acknowledged. Ten or more within one minute; regardless of whether any are acknowledged.

Consequential	No standard; working objective is 0.			PAS Five events of the same pair of alarms occurring in the same order, within 10 minutes of each other 50% of the time; regardless of whether any are acknowledged. Percentage is selectable, however 50% is recommended.
Related (duplicate)	No standard; working objective is 0.			Two or more alarms that co-occur within two seconds more than 90% of the time Sometimes confused with duplicate. Related is used for activity analysis, "duplicate" is used for configuration analysis.
Flood	No standard; working objective is 0.			EEMUA Ten or more alarms per ten minutes Note that if alarm rate exceeds 100 in ten minutes, operator will likely abandon use of alarm system for information.
Flood (serious)	No standard; working objective is 0.			Ten consecutive time periods of ten or more alarms in ten minutes 100 or more alarms in 10 minutes regardless of the number of periods rate is this high.
Activations per tag			Used to identify which tags are having trouble (either due to process problems or due to poor alarm practices)	How often each tag comes into alarm condition.

Normal operation	Manageable: One per 10 minutes Maximum target: < 6 per hour	EEMUA		Note, for a well-rationalized system, the target is no alarms!
Normal operation (by priority)	EMERGENCY: rare HIGH: One or fewer per hour MEDIUM: Two or fewer per hour LOW: Ten or fewer per hour	EEMUA		Ignore "emergency" distinction for alarm systems without that category.
Standing (long time period)	Ten or fewer that have been acknowledged but uncleared for more than 24 hours			Includes time periods of several shifts up and including one month EXCLUDES alarm flood conditions.
Standing (short time period)	Five or fewer within a 24 hour period that have been acknowledged but uncleared for 5 or more hours			Excludes long-term standing alarms, these are new ones for the shift Excludes alarm flood conditions.
Time to acknowledge	No standard (but expect one, and it will likely be different for different alarm priorities)			Time between an alarm activation and the first corresponding acknowledgment by the operator.
Time to clear (return to normal)	No standard.			Time between an alarm activation and the first corresponding alarm clear (return to normal).

APPENDIX 8

Alarm Management Pioneers

We make a living by what we get; we make a life by what we give.

—Winston Churchill

A8.1 OPENING NOTES

This appendix has been included as historical background into the people and organizations who had first hands in alarm management. It is a tribute to these many pioneers. It recognizes their innovative sprit and willingness to dig deeply into the soul of engineering, the power of technology, and the fragility and strength of human nature.

Father of Modern Alarm Management

David Strobhar, with one foot in human factors and the other foot in the aftermath of the Three Mile Island incident, pioneered what has now become alarm management. He recognized the power of process, the vital importance of information, and the intimate connection with the operator interface.

A8.2 ALARM MANAGEMENT TASKFORCE

On a warm afternoon in mid-October 1991, in Phoenix, Arizona, the first of many years of meetings of the Alarm Management Task Force was held. See Figure A8.2.1 for the original invitation to participate letter. Represented were oil-refining companies, chemical-manufacturing companies, a controls equipment manufacturing company, and power companies from almost all segments of their respective industries. From the very outset, there was the realization that prevention was the best alarm response possible.

Over the next few years they worked out the fundamentals of alarming and shaped the requirements for alarm analysis tools and controls infrastructure.

Pioneering Members

Here is the list of active players in the task force. As you will be able to see, the interest spans the industries from oil, chemicals, power, pulp and paper, and controls equipment manufacturing. Within these companies, the participants came from production operations, engineering, and research. The listings will use the company affiliations at the time of the task force though many have morphed significantly.

Participant	Company
A. Pasha Ahmad	BP Oil – RTD Development
Clarence Albus	Exxon USA
Jesse Chang	Mobil Oil Company
Mike Clark	Amoco Oil Company
Kenneth Emigholz	Exxon Research and Development
Jerry Gillman	Procter & Gamble
Roger A. Humphrey	Chevron Research and Technology Company
Ray Johnson	Pennsylvania Electric Company
Bob Ketcham	Union Carbide Chemicals & Plastics
Carol Lowrie	DuPont
Craig Moore	ARCO Cherry Point
Charles J. Ottino	Honeywell, Inc.
Lon Rollinson	West Vaco
Douglas Rothenberg	BP Research
Robert Stepp	Tennessee Eastman
Don Treat	Shell Oil Company
Jerry Trenta	Mobil Oil Company
Bob Van Atta	Chevron USA, Inc.
Carl Zimmerman	DuPont Engineering Department

Table A8.2.1. Members in the Alarm Management Task Force

Honeywell

Industrial Automation and Control
Honeywell Inc.

August 16, 1991

Mr. Doug Rothenberg
BP Research

Dear Doug,

During the 1991 User Group Symposium in June, the technical breakout sessions on alarm management resulted in a recognized need for a review of the alarming capabilities of the TDC 3000 system to identify areas of potential improvements. To optimize the alarm management assessment effort, a task force consisting of interested customers sponsored by appropriate Honeywell representatives would be formed to address the issue. You are hereby being invited to participate in that effort.

The scope of effort will encompass 2-1/2 days of meetings at a Honeywell facility and is scheduled for October 16,17, and 18th. The agenda for the meetings will address current Honeywell roadmap for alarm management developments and recommended directions in which Honeywell should consider as part of their future roadmap. Honeywell has prepared a list of key alarm management developments that are planned for the current and future product roadmap.

In conjunction with your interest in participating in the task force, we value your comments on the relative merit of the various ideas on the attached list. Please complete and forward the reply form by August 30, 1991 to the attention of Lori Hendon at the address above or FAX [redacted] You may reach her directly at [redacted] Once all responses have been received, we will forward all appropriate information regarding accommodations and specifics of this meeting.

If the above dates conflict with your schedule, your designation of a qualified delegate would be most appreciated. A list of your colleagues who have presently been selected to participate on the task force is included.

On behalf of Honeywell Industrial Automation and Control, we look forward to working with you on this important issue.

Regards,

Charles J. Ottino

Charles J. Ottino
Sr. Product Manager

33

Figure A8.2.1. Invitation to join the Alarm Management Task Force

Objectives for Work

It is very interesting to observe the thinking that was going on about alarms and alarm management at this early time, a time that was a full decade before alarm improvement became the front burner in the industrial manufacturing community. Below are the points of that early thinking:

Key items	• Mission: to identify alarm management needs • Alarm: notification to operator for which a response is required • Issues: make alarms meaningful, proper presentation; guide response, be robust and secure, what levels of response expected, definition of "operator"
Rules of alarms	• Objective is "dark screen" • Alarm must be a productive event and lead to action • Alarm must be significant (needed by the plant) • Alarm frequency of activation must match user abilities
Global look-ahead	• Be a "voice" as a major man-machine interface mode • May need modeling that is imbedded in system information base • Caliber of operator will be important • Operator will provide ability to understand problem needs and complement mechanical equipment • People will train the control system • Need to "spline" upset models into normal control models • Discrete alarms will go away • Better detection methods for slow-mode degradation • Better incipient failure detection • Self-correcting situations; operator will be involved only if system cannot correct itself • Advisory information needs to be integrated into coordinated "chucks" of knowledge • Control system needs to believe the operator

A8.3 ABNORMAL SITUATION MANAGEMENT CONSORTIUM

Using the collective knowledge from the Alarm Management Task Force, the Abnormal Situation Management (ASM) Consortium was formed in 1994. By 1998, they joined with leading technologists in Europe to support an EEMUA initiative. In 1999, EEMUA collated and published the first and most recognized guide to alarm systems assisted with the support of the Consortium.

Key Players

Table A8.3.1 provides a list of key players in the ASM Consortium that drove alarm management activities. Most had one or more alarm redesign projects done under their direction and were important drivers in the understanding and exploitation of alarm systems and their proper design and utilization. The participants are listed under their existing affiliation at the time.

Member	Company
Peter Bullemer	Honeywell
Jim Cawood	Celanese
Mike Clark	Amoco Oil
Kenneth Emigholz	Exxon Research and Development
Jamie Errington	NOVA Chemicals
Eddie Habibi	Plant Automation Services
Ed Huestus	Shell Canada
Tim Montgomery	Chevron
Ian Nimmo	Honeywell
George Pohle	Texaco
Douglas Rothenberg	BP Oil Research and Development
Charles J. Ottino	Honeywell, Inc.
Lon Rollinson	West Vaco
Douglas Rothenberg	BP Research
Robert Stepp	Tennessee Eastman
Don Treat	Shell Oil Company
Jerry Trenta	Mobil Oil Company
Bob Van Atta	Chevron USA, Inc.
Carl Zimmerman	DuPont Engineering Department

Table A8.3.1. Key alarm management drivers in the ASM consortium

Objectives for Work

The ASM Consortium is a very unusual consortium. Since their key reason for being is dramatically improved safe operation of industrial manufacturing, it was permitted for these competing companies to meet and share ideas and technology with each other. Out of this activity came the fully modern approach to alarm management. However, this was far from the fundamental objectives of the work. Actually, the foundations of their intent was to be able to understand the operations of their enterprise fully enough so that abnormal situations could be identified before they become severe enough to threaten good operation. Thus, alarms would only be needed as a last resort (but of course, this is figurative; the safety management hardware system would also be present and act as the final layer of protection).

This work was prescient in its scope and activities. The ASM Consortium has been active, approaching 15 years. Please see their Web site for more up-to-date information.

A8.4 ADDITIONAL CREDITS

Standards and Practice Organizations

For leading and showing us the way forward—**EEMUA**, **HSE**, **ISA**, **NAMUR**

Trainers and Consultants

For providing best-practice training and consulting—**4 Sight Consulting**, **D-RoTH**, **EEMUA**, **Invensys Process Systems**, **ISA**, **PAS**

Services Providers

For providing best-practice alarm improvement services to client industries—**Honeywell**, **Invensys Process Systems**, **Mustang Engineering**, **User-Centered Design Services**

Technology Providers

For designing and providing best-practice alarm monitoring and design software (in historical order)—**TiPS**, **PAS**, **Honeywell**, **Matrikon**

Industrial Controls Providers

For understanding the importance of transferring best practice alarm system designs back into their alarm system platform—**Honeywell**, **Rockwell Automation**

Personalities at Large

There are individuals, not previously mentioned, who stand out, not by virtue of their affiliation, but by their leadership and devotion to alarm improvement and safe plant operation—**David Beach, M. Bransby, Donald Campbell-Brown, Bridget Fitzpatrick, A. G. Foord, Bill Hollifield, Nick Sands**

A8.5 NOTE

1. ASM Consortium, http://www.asmconsortium.org.

APPENDIX 9

Qualitative Risk Method for Priority Assignment

Never worry about theory as long as the machinery does what it's supposed to do.

—Robert A. Heinlein

ACKNOWLEDGMENT

This material was provided through the courtesy of H. Porter of Chevron North America Exploration and Production in New Orleans, Louisiana.

A9.1 QUALITATIVE RISK

The qualitative risk method of assigning priority is really just another variation of maximum severity modified by urgency. What it brings to the table, if you will, is another method for assigning urgency. Not only is urgency determined by how fast the problem might go bad, but it also adds a component of whether there might be other forms of protection there to help the operator when intervention does not occur or occurs but not in time.

To use the qualitative risk matrix (Table A9), first pick the most severe consequence from the candidate alarm. That consequence identifies the column you will work in. Then, using the designated probability of an incident escalating (of your candidate alarm) in the first column, move over to the working column and read out the priority at the intersection. If we were to score this method on the summary table on page 580, it would score the same as maximum severity modified by urgency. Note that the cautions expressed in the Non-Weighted Maximum Severity with Urgency Direct to Priority method in chapter 7 apply to this method as well.

A9.2 PORTER'S DISCUSSION ON THE RATIONALES FOR THE QUALITATIVE RISK MATRIX FOR ALARM PRIORITIZATION

Goal

The goal of the risk matrix is to provide an easy-to-use tool for the assignment of priorities to specific alarms and to have some consistency in alarm priority assignments across different fields. This will allow control room operators to quickly identify what their focus should be. This is especially critical during upset conditions where traditionally there are many alarms that are displayed in a very short time period. Use of prioritized alarms reduces the chances that important alarms are missed due to an alarm overload and thus improving our OE performance.

Scope

This matrix applies only to alarms that are displayed in the Wonderware application. Prioritization of alarms available from Quick Panels or Panel Views is not addressed.

A9.3 DESCRIPTION OF MATRIX

The matrix format is a typical probability versus severity table used for general risk analysis that has been modified specifically for alarm prioritization.

Probability Axis

The probability axis is different from the typical risk matrix in that it is defined as the probability of an event occurring if an alarm were completely ignored as opposed to the probability of an event occurring in general. This is an important distinction because it represents what would happen if an alarm were missed during a barrage of alarms during an upset condition. Three considerations were used in determining the four different types of probabilities.

Likelihood

Likelihood of an event occurring is a subjective judgment. Instead of using a more complicated probability of failure on demand average (PFDavg), similar to that done in SIL evaluations, the matrix asks whether the event is likely. This requires a judgment call from the individual or team using the matrix to be based off sound reasoning, experience, and an understanding of the control system. The axis is designed such that likely incidents have a higher priority than unlikely ones.

Time Criticality

Time criticality is considered only for likely events. If an incident were likely to occur in a short time frame from the alarm generation, it would have a higher priority than if it were to occur after a longer time period. This will allow control room operators to focus their attention on the alarms that need immediate attention.

Last Line of Defense

Consideration is given to both likely and unlikely incidents regarding whether human intervention is the only thing that will prevent the incident from occurring once the alarm is generated. Alarms that are the "last line of defense" are given a higher priority because they may require human action and therefore need immediate attention.

Categories for the Probability Axis

Likely and Soon
For incidents that are likely to occur within 10 minutes of the generation of the alarm. The selection of the 10-minute time frame is arbitrary. Also for likely incidents that only human intervention can prevent.

Likely
For incidents that are likely but won't happen in next 10 minutes.

Unlikely
For incidents that are unlikely but are the "last line of defense" and may require human intervention. Time criticality is not considered.

Very Unlikely
For incidents that are unlikely and are not the "last line of defense."

Example of Probability Axis Category Selection

An LAH alarm (high level alarm, not shut down) on a separator is generated. Assuming that all conditions stay the same, the incident of an LSH occurring on the separator could be considered a likely event. The time criticality is subjective, but assuming the LAH setpoint is just below the LSH, it could be considered time critical. Again assuming that conditions remain the same, only human intervention (closing a well or inlet valve, adjusting a PID loop, setting the dump valve to manual full open, etc.) will prevent the LSH from occurring. Given these selections, the likely and soon category would be selected. The occurrence of the LSH is likely, time critical, and only human intervention will prevent it from occurring.

Severity Axis

The severity axis is very similar to the typical risk matrix with one exception. A production bullet has been added to help categorize incidents that are associated with downtime. Instead of focusing on downtime duration, total BOEG loss is considered. The reason for this is that the alarms are, in essence, in competition with each other for attention on the Wonderware alarm summary screen. Some Wonderware applications include alarms from many different locations, remote and central facilities. Comparing production incidents solely on downtime duration without consideration to production rates would probably produce prioritization assignments for alarms that do not reflect the true importance of the alarm. For example, a PSH on a remote gas lift well flowline may produce 24 hours of downtime, while a PSH on an incoming pipeline or bulk separator on a major facility may only produce a few hours of downtime. It's likely that the major facility being down for a shorter duration would produce a greater BOEG loss. The categories of the severity axis are determined by effects to personnel, the environment, facilities, and production loss. Major incidents are likely to affect all of these categories. The intent is that the user of the matrix select the column that contains the worst case of the four categories.

A9.4 DEFINITION OF PRIORITIES

Although EEMUA guidelines do not have fast and hard rules for priority selection or names, **Critical**, **High**, **Medium**, **Low**, **and Journal** are consistent with examples shown. Also, having five types of priorities allows a reasonable level of distinction between risk levels. Critical through Low prioritized alarms will appear on the Wonderware Prioritized Alarm Summary Screen (PASS). Journal prioritized alarms will not appear on the PASS but will appear on the Journal summary screen. The risk matrix has no Journal priority assignments. Some alarms have no risk associated with them and are, in essence, a message to to the operator. These types of alarms should be assigned to the Journal priority.

An important feature of the prioritization scheme is that higher-priority alarms should occur less frequently than lower-priority alarms so that they get the proper attention when they occur. Creating percentage of total alarm ratios for the higher priorities is a good metric that can be used across different fields. The following metrics are consistent with EEMUA recommendations:

- Critical Priority—less than .5% of all visible alarms
- High Priority—less than 5% of all visible alarms
- Medium Priority—less than 15% of all visible alarms

An important distinction between alarm rates that actually occur and alarm quantities that exist in the Wonderware database needs to be made. The metrics above represent the percentage of alarms that actually occur during operations, not the percentage of alarms that exist in the database. There may be some correlation between the two, but only empirical analysis will reveal it. The correlation may not be the same for different fields or Wonderware applications. Adjustments to the alarm priorities and possibly the matrix itself may be needed after a look back is completed post-alarm prioritization implementation.

SEVERITY OF INCIDENT **PROBABILITY OF AN INCIDENT* OCCURING IF AN ALARM WERE COMPLETELY IGNORED.**	**Major** • PERSONNEL—Fatality or permanently disabling injury. • ENVIRONMENTAL—Significant release with serious off-site impact, or spill greater than 100 bbl. • FACILITY—Major or total destruction to process area(s) estimated at a cost greater than $1,000,000. • PRODUCTION—Considering all effected production and duration, greater than 20 MBOEG loss total.
Likely and Soon Incident is likely to occur within the next 10 minutes or is likely and there are no other protections/alarms in place.	Critical
Likely Incident is likely to occur but not within the next 10 minutes.	Critical
Unlikely Incident is not likely to occur but there are no other protections/alarms in place.	High
Very Unlikely Incident is not likely to occur and there are other protections in place to prevent incident from occurring.	Medium

Table A9. Qualitative Risk Matrix

Serious	Minor	Incidental
• **PERSONNEL**—One or more severe injuries. • **ENVIRONMENTAL**—Significant release with serious off-site impact, or spill less than 100 bbl. • **FACILITY**—Major damage to process area(s) at an estimated cost greater than $100,000 but less than $1,000,000. • **PRODUCTION**—Considering all effected production and duration, between 2 and 20 MBOEG loss total.	• **PERSONNEL**—Single injury, not severe, possible lost time. • **ENVIRONMENTAL**—Release which results in Agency notification or Permit violation, or spill less than 1 bbl. • **FACILITY**—Some equipment damage at an estimated cost greater than $10,000 but less than $100,000. • **PRODUCTION**—Considering all effected production and duration, between 200 and 2,0000 BOEG loss total.	• **PERSONNEL**—Minor or no injury, no lost time. • **ENVIRONMENT**—Environmentally recordable event with no Agency notification or Permit violation, or spill less than 1/8 of a gallon. • **FACILITY**—Minimal equipment damage at an estimated cost less than $10,000. • **PRODUCTION** – Considering all effected production and duration less than 200 BOEG loss total.
Critical	High	Medium
High	Medium	Low
Medium	Low	Low
Low	Low	Low

*Incident for the purpose of this matrix is defined as any event that causes injury to personnel, the release of hydrocarbons/chemicals into the environment, damage to facility equipment, or loss of production due to an automated shut in.

Note: Alarms that provide only messaging value to the operator and have no risk associated with them should be assigned a Journal priority.

APPENDIX 10

Manufacturing Modalities and Alarm Management

Continuous versus Batch versus Discrete Manufacturing Alarm Management Issues

A10.1 INTRODUCTION

This is a discussion of the implications for alarm management that normally arise from differences in the manufacturing operations in vastly differing industrial production processes. In general, most industrial production operations are categorized as continuous, batch, or discrete. Refer to Table A10.1 below for brief descriptions. Process control platforms come generally in the formats of a distributed control system (DCS) and programming logic controllers (PLC). In general, most continuous manufacturing is controlled primarily by DCSs with some utilizing PLCs. Batch manufacturing is usually controlled by some DCSs but mostly PLCs. Discrete manufacturing is controlled almost exclusively by PLCs with some special-purpose modular controls (single case or custom) used.

Those readers who are conversant with batch and discrete manufacturing will undoubtedly recognize that this discussion is far from comprehensive. Rather than serving as a primer on the subject, the discussion is targeted to introduce readers to the usefulness of thinking about alarm management across the many modalities of manufacturing equipment. As applications of alarm redesign build up, it is hoped that the various practitioners will liberally share their experiences and insights.

A10.2 CHARACTERISTICS OF MANUFACTURING MODALITIES

Let's take a quick look at what the differing modalities of manufacturing are. Of course, there are lots of operations that may involve a combination of all of them. For each portion, the definitions here and the attributes in the next section apply.

Continuous	Raw materials are usually coming in a steady stream and product is usually produced in a steady flow. The flow path is generally the same. The scale of operations is usually quite large. Initial construction costs range from tens of millions of dollars to many hundreds of millions. Smaller works might come in at tens of thousands for specialized processes. Controlling operations are usually done from a centralized control center with dedicated inside and outside operators. Manufacturing usually takes place 24 hours a day, 7 days a week for years at a time; although, there are units that may operate only for hours at a time. Production interruptions vary from slight reduction in processing throughout to an entire shutdown lasting many months or even longer. Both the process of shutting down and restarting are major operational activities and may take many hours to days to effect. Generally, each manufacturing operation is distinct with few redundant facilities, save for spared critical equipment; although, it is not uncommon for plants to have a number of similar "lines" in order to make capacity or as the historical result of plant expansions.
Batch	Raw materials are used in batches and product produced in batches as well. The flow path can vary dramatically, depending on product. Scale of operations is usually modest. Initial construction costs usually range from tens of thousands of dollars to tens of millions. Controlling operations are usually done from one or more decentralized control centers with shared inside and outside operators. Each control center usually utilizes dedicated control equipment, with little or no cross-center communication. Manufacturing takes place anywhere from 8 hours a day, 5 days a week, to occasionally almost full time. However, each batch is distinct from the previous and the next. Any interruptions usually only affect the current batch. Most interruptions last from a few minutes to a few days, most lasting a few hours. Operations can be readily stopped, but restarting may be a complicated process of cleaning, testing, and other specialized operations. Manufacturing operations may be unique or done via a number of duplicate facilities.

Discrete	Raw materials are almost exclusively individual parts, components, or very small batches and product produced in parts or components, usually packaged. The flow path is generally fixed; however, it is common for some steps to be skipped as needed for the current production needs. Scale of operations is generally modest. Initial construction costs range from tens of thousands of dollars to a few million. Controlling operations are usually done from one or more decentralized control centers, small consoles, or custom controls with shared inside and outside operators. Manufacturing takes place anywhere from a few hours to almost full time. Each batch is distinct from the previous and the next. Any interruptions usually only affect a few components. Most interruptions last from a few minutes to a few hours and generally stop the complete production line. Operations can be readily started and stopped. Manufacturing operations are usually special purpose and therefore unique.

Table A10.1. Comparison of industrial manufacturing modalities

Because the specific manufacturing activities differ, the ability to start and stop operations differs, and the unit costs of both raw materials and finished products differ, these various manufacturing operations are constructed to different standards and manned and operated differently. Therefore it is quite understandable that the features and use of an alarm system might have differences as well. And they do. However, the basic principles and design parameters have more in common than they differ. The matrix below summarizes those differences and similarities. It is provided to both explain that there are some differences as well as to reinforce the power of their similarities.

A10.3 COMPARISON MATRIX

What follows is a list of alarm issues and attributes that are discussed for each of the manufacturing modalities individually.

Concept	Continuous industries	Batch industries	Discrete industries
Alarm activation situation	Alarms are set to identify individual process physical value going toward out of normal values Relatively few parametric changes: priority, enable, activation point, and so on	Alarms are set to identify individual process physical value going toward out of normal values Large number of parametric changes: priority, enable, activation point, and so on	Alarms are mostly set to identify one or more discrete features that have become abnormal Very few parametric changes: priority, enable, activation point, and so on
Alarm configuration parameters	Mostly constant Exceptions are mostly for out-of-service items	Highly variable depending on plant state	Mostly constant
Alarm responses	Relatively consistent over time	Highly changeable over a production run	Relatively consistent over time
Current alarm challenges	Better identification of operational state to facilitate alarm configuration changes to maintain alarm relevance and thus provide less distractive alarms	Alarm parametric changes and current plant state are easy to identify Alarms easy to customize for operational status	Development of reliable parametric predictors for abnormal situations that can be used to alarm, thus enabling the alarm system to maintain operation rather than restore operation after the problem
How alarm system notifies the operator	Relies on built-in capabilities of PCS At times utilize off-platform equipment to provide more diversity and/or better guidance information	Uses the built-in capabilities of the PCS (but they can be very limited) Dependent on staffing protocol may require some remote, in-plant horns and/or strobe lights and other equipment	Uses the built-in capabilities of the PCS (but they can be very limited) Generally requires remote, in-plant horns and/or strobe lights and other equipment

Automatic plant action after alarm activation	Automatic actions are rare; however, in some cases, alarms are misused; in this situation they are used exclusively as a convenient "switch" to initiate a process action (e.g., turn on a pump for 30 seconds).	Automatic actions are not unusual and are used to augment operator actions. An example would be to decrease pressure on a vessel until the operator could figure out what to do with another related part of the batch.	Almost always used to shut down the production operation completely. In rare cases, where there is sufficient surge capacity, connecting parts of the operation may continue to operate until the available storage is exceeded.
Identifying plant state	Difficult to identify but important	Clear to identify and important	Clear to identify but not much needed
Operating regime	Difficult to bring process to a safe stop Process is always at or near steady state Always operational, mostly the same product and production methodology	Easy to bring production to a safe stop, but product losses can be significant Process goes through a series of steady states as the run progresses Operating regime rather than equipment will change during a run	Easy to bring production to a safe stop without much product loss Process is almost always at a steady state; however, a "campaign" mode is the norm Operational at times, not at others Lots of "start" and "stop" operations Different behaviors for different plant areas since they use many different production methodologies
Operator location	Almost entirely at the central control board	Sometimes at the control board, sometimes in the unit	Always in the unit at different local control boards Lots of walking around tending to things

Table A10.2. Comparison of manufacturing modalities and alarm system issues (continued on the next page)

Operator relationship with alarm system	Alarms provide support to keep the process operational or return to operational from abnormal situations	Alarms provide support to keep the process operational or return to operational from abnormal situations Alarms also provide support during production mode changes	Alarms provide support to quickly modify the operating regime or restart the operation (as needed)
Operator responsibilities	Monitor during normal operations making little interventions Most problems involve operations that move out of a normal operating envelope, but plant keeps running	Monitor during normal operations making little interventions Most problems involve operations that move out of a normal operating envelope, but plant keeps running Adjustments to control settings and equipment often made during normal operation to accommodate production mode changes	Diagnose and remedy the problem that shut down the operation
Providing alarms that track operational state	Needed, but requires special effort to identify operational state	Required and easy to do since operational state is almost always clear	Not much needed
Rate of alarm activations	Modest, but during good operation, can be very few	Modest, but during good operation, can be very few	Mostly very few, but even a few can herald big problems with the process

Table A10.2. Comparison of manufacturing modalities and alarm system issues

APPENDIX 11

Notifications Management

AUTHOR'S NOTE: This is the original discussion document on this topic done with the Abnormal Situations Management (ASM) Consortium in February 1997. It was jointly authored by David Beach and Douglas Rothenberg. It is provided here for its historical importance. The original text follows:

What follows is a discussion of a new type of messaging. It was identified by the ASM Consortium in the course of trying to encapsulate an understanding of process alarms in the petrochemical industry. It is envisioned that as we significantly improve our ability to prevent upsets, notifications might surpass alarms as a primary communications vehicle to the operator from the control system.

The purpose of this document is to initiate discussion and serve as a starting point for any additional technical activities. It is a draft.

A11.1 INTRODUCTION

There are a considerable number of events that are not alarms that might require that people be notified by automatic equipment (control systems, computers, etc.). Notification can take any appropriate sensory form: text on a screen or hard copy, sound, visual, olfactory, and/or touch (vibration, etc.). These sensory forms can either occur singly or in combination. The notifications need to be in keeping with the nature of the situation as well as the location and status of the person (notifyee) to whom notification is intended.

It should be clear to the notifyee that the event, about which he is about to be notified, is separate and distinct from an alarm. A notification therefore is a message of any of the earlier mentioned forms—that is, not an alarm. By definition, an alarm is a situation that requires external (to the automatic equipment) action in order to have some process continue with proper operation. Notifications may or may not lead to actions; however, they do not by their nature require them to be timely.

A11.2 POINTS TO CONSIDER

1. Per se, notifications do not require direct action; however, this is not to say that a chain (or story; defined below) of notifications would not point to a situation that would eventually require action. Such action would usually be in advance of any alarm, and so would be a desirable proactive action.
2. Notifications can be
    a. Status events informing operators about where an automated process is in relation to conditions, production requirements, and so on.
    b. Early notification about what system, process events, or potentials are likely to take place in the near future.
    c. The inverse of alarm conditions; for example, the clear absence of any unusual situation(s) can itself be a condition that would be nice to know.
    d. Information that the operations team will likely need to know during the shift or other duty period.
3. Notifications can be single entities that are "mailed" out; or they can be continuous cumulative entities that depict a current situation that is dependent upon a history.
4. Notifications, while not requiring notifyee action, nonetheless, have a significant degree of temporal importance. Therefore their timely delivery is essential.
5. Notifications may or may not require acknowledgment. Though we think that few will require acknowledgment, acknowledgment should depend on the individual nature and needs of the notification.
6. Notifications have intrinsic value as single entities.
7. Notifications can have significant additional value when taken in combination. Such combinations are termed a "story." Therefore there is a need to have notifications contain within themselves a significant structure that would permit single ones to be put together into any number of collective stories.
    a. An example story might be a topic-based report such as "All notifications on the third shift on a certain date," or "All notifications relating to environmental emission weaknesses in our production of product 'X.'"
    b. Other examples might be to retrieve collective information about past activities that would be useful for the following:

   i. Trace performance of tools and/or usage of equipment and systems, and so on. Useful as a post-engineering aid.
   ii. Better plan for infrequent operations using experiences of earlier ones.
   iii. To identify production anomalies.
c. This means that if a notification is issued that suggests that something is amiss, it is required that another notification be issued when that situation has gone away.
d. Combinations might be a need-based grouping (structural tie that can make a type of story) in a free-format, dynamically sorted presentation.

8. Compared to alarms, the active lifetime of a notification would appear to be very long. The term *active lifetime* implies retrievability as well as continued linking to other notices and conditions (see #7). Thus storage, retrieval, and story generation capabilities must be in place for a significant time period.

A11.3 QUESTIONS AND ISSUES

1. We need to find a way to keep the perfect notification system from overwhelming the conscientious (or awfully poor) user.
2. What would a notification management tool kit look like?
3. What does the e-mail paradigm have to say about what we should and/or should not consider to be a part of a notifications system?
4. Would people be permitted to generate notifications? What about log messages? Shouldn't they be treated as notifications in order to permit them to be incorporated into a notifications system, or should they be separate?
5. If alarms are kept separate from notifications and both kept separate from human log messages, then what system will be responsible for putting together a total "story" from all this useful data?
6. What is the difference between a notification and any other one-way communication?
7. How will we know the difference between a notification and a quasi-alarm? A quasi-alarm is an event that is not related to any specific process tag but might indicate an undesirable operating condition that by itself might not be bad but if left to persist, might or might not be.
8. Should operational changes in the automated system be communicated as notifications? If so, why? If not, how do we communicate this information? Would it be dependent on the frequency of the events?

Examples could include the following:

a. New tuning parameters for loops that are automatically re-tuned.

b. Control restructuring as a result of predesigned changes that are dependent on how well the situation is being managed by the regular control system.

9. There might be a significant benefit to incorporation of notifications into the basic operator view. Hence notifications might be cast as pictorial icons that appropriately appear. (Draw an advanced control loop pentagon and populate it with a flow valve that is near saturation.)

10. How do we keep track of the condition of the notifyee (recipient of the notification) in order to tailor it to his needs?

11. What is the difference between importance and immediacy?

Index